ROBOT BUILDER'S
BONANZA

ROBOT BUILDER'S BONANZA

GORDON McCOMB

FIFTH EDITION

New York   Chicago   San Francisco   Athens   London
Madrid   Mexico City   Milan   New Delhi
Singapore   Sydney   Toronto

Library of Congress Control Number: 2018959521

McGraw-Hill Education books are available at special quantity discounts to use as premiums and sales promotions or for use in corporate training programs. To contact a representative, please visit the Contact Us page at www.mhprofessional.com.

Robot Builder's Bonanza, Fifth Edition

1 2 3 4 5 6 7 8 9   LCR   23 22 21 20 19 18

ISBN 978-1-260-13501-5
MHID 1-260-13501-2

The pages within this book were printed on acid-free paper.

Sponsoring Editor
Lara Zoble

Copy Editor
Kirti Dogra

Editorial Supervisor
Stephen M. Smith

Proofreader
Yashoda Rawat

Production Supervisor
Lynn M. Messina

Indexer
Michael Ferreira

Acquisitions Coordinator
Elizabeth M. Houde

Art Director, Cover
Jeff Weeks

Project Manager
Touseen Qadri, MPS Limited

Composition
MPS Limited

For Aria Lynn McComb—
your turn

About the Author

Gordon McComb has written about amateur and educational robot building for 30 years and is called "the godfather of hobby robotics" by *Make Magazine*. He is the author of all four previous editions of *Robot Builder's Bonanza*, which are among the most widely read books on hobby robotics.

CONTENTS

Part 2—Building Robots

Part 4—Make Your First Robot

Part 9—Online Robot Projects

FOREWORD

This is the fifth edition in the very popular *Robot Builder's Bonanza* series. It's all new: new projects, new ideas, new experiments. I'm happy to say I've been involved, in one way or another, with the writing and developing of all five editions in this series. My home is full of robots and robot parts!

Gordon McComb began tinkering at an early age. His sisters had to hide the screwdrivers to avoid having everything in the house disassembled. It's no surprise that for 40 years, Gordon has been writing about gadgets: taking apart gadgets, inventing new gadgets, building your own gadgets. In 1987, the first edition of *Robot Builder's Bonanza* launched a huge interest in the best gadgets of all: robots! The field of hobby robots, robots you could actually build and control in your own garage or on your own dining room table, was born.

I met Gordon in August of 1975, at a party neither of us planned to attend. We were both aspiring writers, and we talked for hours. It was so exciting to be with someone who was nearly exploding with ideas! My dad said, "That's a young man trying to drive a troika," a very difficult cart to manage, pulled by three horses! Gordon had so many interests, going in so many directions. I saw right away that he could do anything he wanted to do—with a little organization. Not long after, I became his cheerleader, his 100% supporter, his wife, his best friend.

Gordon died on September 10, 2018, shortly after writing this book. It was very sudden, very unexpected; my own state of shock has been mirrored in the robotics and electronics communities. Many people have written to me to say Gordon's books are the reason they chose a career in electronics, or robotics. He had, and will continue to have, an enormous effect on his readers.

He wrote over 70 books and hundreds of articles, mostly on robotics, electronics, and technology. What is so special about Gordon's writing is his style. He is comfortable to read. He is personal. He is conversational. He is very, very good at explaining technical information to those of us who are maybe just a bit techno-challenged. There is so much of Gordon's personality in his writing, we can know him. When we read him, we can hear his voice. We can continue to enjoy his wit, his creativity, his enthusiasm. E. B. White, one of our favorite writers, suggested that books are people, "… people who have managed to stay alive by hiding between the covers of a book." Oh, I sincerely hope so.

Because of Gordon's unexpected death, I have had to take on some of the jobs usually finished by the author between writing a book and its publication. For the fifth edition of *Robot Builder's Bonanza*, these jobs included checking and completing code, developing the Support Site, and creating and writing up the projects you'll find in Part 9 of the book. Gordon always left these jobs to the last. This has been difficult for me. My knowledge of electronics begins and ends with stepping on resistors and capacitors in my bare feet. If there are any omissions in this book, the fault is mine. I worked from his notes when I could and sought help when I couldn't. I would have been lost without my heroes on the Parallax Forum, especially Eric Ostendorff, Duane Degn, and Jon Hapeman, who rallied together and supplied the technical knowhow. I couldn't have done it without them. I am also very grateful to Touseen Qadri of MPS Limited for her patience as I struggled through the author review.

Personally, I miss Gordon terribly. He left behind a terrific daughter and son, a truly special grandson and granddaughter, and a brand-new great-granddaughter to whom this book is dedicated. We're all coping as best as we can. All of us, in the world of robots, will be all right in the long run. It's just the short run that is hard. To paraphrase E. B. White, it is not often that someone comes along who is a true friend and a good writer. Gordon was both.

Jennifer Meredith

ACKNOWLEDGMENTS

Since the first edition of *Robot Builder's Bonanza* in 1987, numerous friends and colleagues have contributed suggestions, wisdom, plans, ideas, and other nuggets of inspiration. As always I'm indebted to their generosity. For this edition, I thank my friends Eric Ostendorff, Jon Hapeman, Ken Gracey, and others for sharing ideas and making valuable observations.

I also thank Margot Maley Hutchison and my agents at Waterside Productions; Mike McCabe, Elizabeth Houde, Lara Zoble, and the others at McGraw-Hill Education; Touseen Qadri and her hardworking team at MPS Limited for whipping my manuscript into a published book; my wife Jennifer for her unrelenting support for my work; and of course my entire family for putting up with far too many late-night robot-building sessions.

Photo Credits

The author and publisher wish to thank the following companies and individuals for their photos and illustrations: Charmed Labs: Figure 33-13; SparkFun: Figure 33-1; VeeaR by RoboTech srl: Figure 33-4; Parallax, Inc.: Figures 24-3, 28-6, 28-8, 31-16, 36-7; Adafruit: Figures 25-2, 36-9, 36-11; iRobot: Figure 1-4; Pitsco: Figure 1-5; NASA: Figure 2-2; General Electric: Figure 2-3; Lynxmotion: Figures 2-6, 2-8, 13-7, 19-2; Efa: Figures 27-2, 27-3; OWI Inc.: Figure 6-11; Pololu: Figure 8-9; NextWave Automation: Figure 10-4; Glowforge: Figure 10-5; MakerBot: Figure 10-6. The author also wishes to thank OpenClipart, ShareCG, Daz, Turbosquid, FirstCAD, Corel, and other curators of royalty-free open-source and commercial art materials used in the preparation of this book.

INTRODUCTION

Consider robotics as your "homeroom for learning." Whether it's electronics, science, mechanics, mathematics, art, or technology, everything always comes back to a study of robots.

Bot makers are early adopters of neat new technologies used in everyday products like self-driving cars, smartphones, and the Internet of Things.

Robot makers are among the first to learn about breakthroughs in microcontrollers, miniature accelerometers, wireless data transmitters, GPS, artificial intelligence, and other cool stuff.

Robot makers get early exposure to the latest tech, a head start among peers in knowing how to leverage new scientific inventions beyond just robotics.

And to top it all off, robot makers enjoy a lifetime of continuous learning as innovations emerge.

Best Time to Get Started!

If you've ever wanted to build a robot, *now* is the perfect time to do it!

New technologies have dramatically driven down the cost of building a robot, not to mention the time it takes to construct one. It's never been easier and cheaper, and the results far exceed what was possible even five years ago.

For less than $50 you can construct a sophisticated, fully autonomous robot that can be programmed from your computer. You can easily change its behavior as you experiment with new designs.

Inside *Robot Builder's Bonanza*

Yet for all its advances, robotics is still a cottage industry. There's plenty of room for growth, with a lot of discoveries yet to be made. Maybe you'll be the one to make them?

If so, that's where this book comes in.

Robot Builder's Bonanza, Fifth Edition, is part tutorial and part reference. It tells you what you need to know to build a robot, plus a whole lot more about the art and science of robotics.

It's all about having fun while learning how to design, construct, and use small robots. Hands-on plans take you from building basic motorized platforms to giving the machine a brain—and teaching it to walk and talk and obey commands.

This book is about *inspiring* you—in the guise of a how-to book on constructing various robots and components. The modular projects in this book can be combined to create all kinds of highly intelligent and workable robots of all shapes and sizes. Mix and match your projects the way you like.

The projects in this book are a treasure chest of information and ideas on making thinking machines. You'll find what you need to know to construct the essential building blocks of a robot.

What You'll Learn

Robot Builder's Bonanza is divided into nine sections, each covering a major component of robot building.

Part 1: Art and Science of Robot Making. What you need to get started; setting up shop; how and where to get robot parts.

Part 2: Building Robots. Robots made of plastic, wood, and metal; working with common materials; converting toys into robots; mechanical construction techniques; how to build fast and cheap robots. Includes three full robot projects.

Part 3: Making Your Robot Move. Using batteries; powering the robot; working with different kinds of motors; powering motors from computerized electronics; mounting motors and wheels.

Part 4: Make Your First Robot. Fun projects and ideas for building robots with wheels, tracks, and legs.

Part 5: Robot Electronics. Circuitry for robots; common components and how they work; constructing circuits on solderless breadboards; making your own soldered circuit boards.

Part 6: Robot Brains. Smart electronics for your bot; introduction to microcontrollers; all about several popular microcontrollers for robotics, including Arduino, BBC micro:bit, Raspberry Pi, and Parallax Propeller.

Part 7: Robot Sensors. Adding senses to your robots, including collision detection and avoidance; sensing when objects are nearby; detecting magnetic fields; integrating sensors for navigation, light, sound, and distance.

Part 8: Interacting with Your Robot. Making your robot produce sound, music, and speech; operating your bot by remote control; ways to make your robot communicate with you with words, pictures, and light.

Part 9: Online Robot Projects. Fun community-driven projects that take a day or less to build, with full details on the RBB Support Site: constructing a light-seeking robot; hacking a remote control toy; constructing a line follower; and more.

Expertise You Need

Actually . . . you don't need any experience to use this book. It tells you what you need to know.

But if you happen to already have some experience—such as in construction, electronics, or programming—you're free to move from chapter to chapter at will. There are plenty of cross-references to help you expand your discovery zone.

This book doesn't contain hard-to-decipher formulas, unrealistic assumptions about your level of electronic or mechanical expertise, or complex designs that only a seasoned professional can tackle.

I wrote this book so that you can enjoy building a robot without lots of prerequisites. The projects in this book can be duplicated without expensive lab equipment, precision tools, or specialized materials—and at a cost that won't contribute to the national debt!

Free Bonus Content!

This book comes with free online content at the **RBB Support Site**. See the Appendix for more information!

What You'll Find in the Fifth Edition

This book is an updated revision of *Robot Builder's Bonanza*, first published in 1987, and then renovated again in 2001, 2006, and 2011. With this edition you'll find:

1. **Updated and expanded coverage.** Robotics is an evolving science, and here you'll find up-to-date coverage of the latest popular microcontrollers, including Arduino, BBC micro:bit, Raspberry Pi, and Parallax Propeller multi-core, plus hands-on projects using the latest motors, sensors, and modules.
2. **Kitchen table robotics.** Continuing with this edition is an emphasis on affordable and easily reproducible bots, made with simple hand tools and commonly available parts. More than ever, this book makes it easier for first-time builders—none of the projects in this book require expensive or elaborate tools!
3. **Enhanced online content.** You'll find even more how-to, plans, and projects on the new-and-improved RBB Support Site (see above), reorganized by topic and interest.
4. **1-day bonus projects.** You like step-by-step projects, so I've created a menagerie of online easy-to-build open-source plans, complete with parts lists and code. These are community-based robots where you're encouraged to participate in expanding and enhancing the designs!

. . . and lots more!

Better Than Cool

Think of *Robot Builder's Bonanza* as an adventure in technology—with a lot of fun thrown in for free.

As you construct your bot, you'll learn the latest technologies firsthand. It puts you way ahead of the curve, whether your interest in robotics is for school, for work, or as a hobby.

You can think of *Robot Builder's Bonanza* as a treasure map. No matter what path you follow, they all lead to the wonderful and rewarding art of robot building. Turn the page and start your adventure!

ROBOT BUILDER'S BONANZA

Art and Science of Robot Making

CHAPTER 1

Become a Robot Master

If you've just started building robots you're in for a wonderful ride! Watching your creation do something as simple as scoot around the floor or table can be exhilarating. You know that you've built something—however humble—with your own hands and ingenuity.

You know the time and effort that went into constructing your mechanical marvel, maybe something like the ones in the robo-zoo picture in Figure 1-1—all are homebrew hobby robots. Others may not always appreciate it, especially when it marks up the kitchen floor with its rubber tires. Still, you're satisfied with the accomplishment, and look forward to the next challenge!

While robot building brings much joy and satisfaction, it's not always without heartache and frustration. You know that not every design works, and that even a simple engineering flaw can cost weeks of effort, not to mention ruined parts. This book will help you—beginner and experienced robot creator alike—get the most out of your robotics hobby.

Why Build Robots?

In my youth the go-to robot movie was *The Day the Earth Stood Still* (the original, mind you, not the remake). I was both in awe and deathly afraid of Gort, the movie's super-powerful robot that could one day save the Earth, and the next, completely destroy it. Others have been similarly influenced from popular movies of the time, like *Star Wars, Short Circuit,* or *Terminator.*

No matter where the inspiration comes from, there are many reasons to build your own robot. Here are just some of them.

ROBOTICS: KEYSTONE OF MODERN TECHNOLOGY

Pull that smartphone out of your pocket. Did you know that many of its advanced features, like GPS, touch screens, and tilt sensors, were technologies robot builders have been toying with for *nearly two decades*?!

Figure 1-1 Some of the homebrew robots I've built over the years. Amateur robots can take many forms and sizes: mobile (moving) robots use wheels, tracks, or legs for propulsion; arms and grippers allow the robot to manipulate its world.

Figure 1-2 The modern microcontroller, like this Arduino Uno, provides an amazing amount of computational power for controlling a robot.

Robotics is a natural test bed for new ideas. Robot builders have been among the first amateurs to play with microcontrollers, accelerometers, digital compasses, voice control, electronic gyroscopes, global positioning satellite modules, speech synthesizers, solid-state imagers, vision recognition, tactile feedback, and many other cutting-edge technologies.

What's more, all of this is available at low cost. The pocket-size microcontroller circuit board in Figure 1-2 costs less than a dinner for two and rivals the thinking power of the computer that put Apollo astronauts on the moon. (You'll be learning lots more about this microcontroller in Chapter 25, "Using the Arduino.")

Whether you're a garage-shop tinkerer, a student, or an engineer working for a *Fortune* 500 company, experimenting with amateur robotics gives you ample opportunity to discover the technologies the world will be using tomorrow.

ROBOTICS: GATEWAY TO A CAREER

Robotics involves dozens of interconnected sciences and disciplines—mechanical design and construction, computer programming, psychology, behavioral studies, ecology and the environment, biology, space, micro-miniaturization, underwater research, electronics, and much more.

You don't need to be an expert in these fields just to build a robot. You can concentrate your studies on those things that most interest you, using your robot as a doorway to furthering your interests.

ROBOTICS: TO THE RESCUE

Science fiction has long painted robots as baddies—either on their own or as minions to mad scientist. Yet it turns out robots may be a way for people to live better, longer lives.

- Robots can venture where people can't go. Send a bot into a collapsed mine shaft, to the bottom of the ocean, or to the metal surface of an asteroid. It'll get the job done, and it doesn't need air or McDonald's breaks.
- Bomb-sniffing robots can save the lives of many people. They can locate the explosive and defuse it much more safely than humans can.
- Robots can act as nurses and doctors, even to those with highly contagious diseases.
- Kids respond to robots in ways that can help them to develop interpersonal skills. There are even some robots used as therapy for children with certain learning and social disorders.

ROBOTICS: MOST OF ALL, FUN!

It's okay to build robots just for kicks. Really!

Challenge yourself to a new project, and enjoy a hobby shared by others worldwide. Share your designs on a blog or forum. Enter a competition to see whose robot is fastest. Post a video of your robot on YouTube. Strut your stuff!

Less Expensive Than You Think

Homebrew robotics used to be expensive and time consuming. Weak electronic brains limited what the finished robot could do. No longer. Amateur robots are now:

- *Lower cost.* It's possible to build a fairly sophisticated autonomous robot for about $50, with only ordinary tools. Similar robots used to cost over $500 and required specialized tools to make.
- *Simpler.* Thanks to ready-made sensors, specialty electronics, and prefab parts, it's much easier to construct robots by putting together construction blocks.
- *More powerful.* Inexpensive microcontrollers add horsepower and functionality, with more memory, faster processing speeds, and easier interfacing to other components. If you have a PC or tablet with a USB port, you can start working with microcontrollers today—many cost just a few dollars and can control an entire robot.

Skills You Need

You don't have to be an expert in electronics and mechanical design to build robots. Far from it. Which of these best describes you?

- *I'm just starting out.* If you're an absolute raw beginner in all things robotics, start with the beginning chapters in this book and read all about the tools, materials, mechanics, and electronics of robot building.
- *I have some electronics or mechanical background.* Plow straight ahead to the construction guides and how-tos that follow. This book is organized into parts so that you can bone up on your skills and knowledge as you read.
- *I'm an experienced tinkerer.* If you are already versed in electronics and mechanics, you're well on your way to becoming a robot experimenter *extraordinaire*. You can read the chapters in the order you choose. There are plenty of cross-references among chapters to help you connect the dots.

ELECTRONICS BACKGROUND

Electronic circuits are what make your robots "thinking machines." You don't need extensive knowledge of electronics to enjoy creating robots. Start with simple circuits with a minimum of parts.

This book provides little in the way of electronics theory, just practical information as it relates to building bots. Many of the circuits in this book are in schematic diagram form, a kind of blueprint for how the parts of the circuit are connected. If you've never seen a schematic, there are a number of books and online guides that can get you started. See the *Selected Reading* list in the RBB Support Site, described in the Appendix.

The parts for the electronic projects in this book are selected to be widely available and reasonably affordable. I decided not to include vendor part numbers right in the book because these can change quickly.

Instead, you can visit the RBB Support Site for updated lists of parts used in this book and where to get them.

PROGRAMMING BACKGROUND

Modern robots use a computer or microcontroller to manage their actions. In this book you'll find plenty of projects, plans, and solutions for connecting the hardware of your robot to any of several kinds of ready-made robot brains.

Like all computers, the ones for robot control need to be programmed. Coding a robot may seem daunting at first, but in reality it's often easier because you only need to deal with a finite subset of programming concepts. If you're new to coding I provide some recommendations on the RBB Support Site.

MECHANICAL BACKGROUND

This book provides several start-to-finish robot designs using a variety of materials, from cardboard to space-age plastic to aluminum. If you're a workshop beginner, you'll find helpful tips on what tools to use and the best materials for constructing your robot bodies.

If you're one of those who just *hate* the idea of cutting a piece of wood, or drilling through plastic, there's good news: you'll find plenty of mail-order sources for purchasing bare-bones robot mechanics. You still need to assemble things, but you can get by with just a screwdriver. Much more about these options throughout the book.

Do It Yourself, Kits, or Ready-Made?

I've said it before but it bears repeating: There's never been a better time to be an amateur robot builder. Not only can you construct robots "from scratch," you can buy a robot kit—there are literally dozens to choose from—and assemble it using a screwdriver and other common tools.

MAKE YOUR OWN

This book is chock-full of robot projects made from wood, plastic, and metal. You can even hack a radio-controlled toy and turn it into an autonomous machine that operates on its own! One of the robots you can build is shown in Figure 1-3. This walking robot from Chapter 19, "Build Robots with Legs," uses only commonly available parts and materials.

READY-MADE BOTS

If you don't particularly like the construction aspects of robotics, you can purchase ready-made robot bodies—no assembly required. With a ready-made robot you can spend all your time connecting sensors to it and figuring out new and better ways to program it. An example is the iRobot Create 2, shown in Figure 1-4. It's like the iRobot's Roomba vacuum-cleaning robot, but without the vacuum cleaner. It's meant as a mobile robot platform for educators, students, and developers.

KIT O' PARTS

You know the Erector set—this venerable construction toy provides the hardware for building all kinds of buildings, bridges, cars, and other mechanical things. The traditional Erector set uses lightweight metal girders of different sizes, plus steel fasteners to put them all together.

Figure 1-3 Build your own bot with your own hands. They can roll, walk, or grind over the earth on tank tracks.

Figure 1-4 If the mechanics of robotics isn't your forte, you can learn by using a ready-made robotic platform. This one is designed to teach the principles of robot programming. (*Photo courtesy of iRobot.*)

Several robotics companies have taken the Erector set one step further, modifying the idea to handle the unique issues of robot building. Their construction kits can be used to create fully functional robots, or you can adapt the bits and pieces to your custom designs. Some are made of metal, like the set in Figure 1-5; others of plastic.

Figure 1-5 Some assembly required: precut and predrilled parts allow you to construct robots by piecing together manufactured components. (*Photo courtesy Pitsco Education.*)

CHAPTER 2

Anatomy of a Robot

When a science fiction movie depicts a robot, it often looks like a human. It has a head, body, arms, and legs. It walks like a human, and may even talk like a human. But this form is far from the most common type of real robot, where bots are far more likely to be bolted to the floor, or scoot around on a set of wheels.

This is a book about building real robots, after all, so the question is: What basic parts must a machine have before it's given the title "robot"? Let's take a close look at the anatomy of robots and the kinds of materials hobbyists use to construct them. For the sake of simplicity, not every type of robot is covered here, just the ones most often found in amateur robots.

Stationary or Mobile

When we think of robots most of us envision a machine that walks around on legs or rolls across the floor on wheels.

The most common robots work in factories. They stay put and manipulate some object placed in front of them. These *stationary* robots are often used to make things like cars, appliances, even *other* robots! Because these robots directly handle things around them, they are also called *manipulators*.

Conversely, *mobile* robots (see Figure 2-1) are designed to move from one place to another. Wheels, tracks, or legs allow the robot to traverse a terrain. Mobile robots may also feature an armlike appendage that allows them to manipulate objects around them. Of the two—stationary or mobile—the mobile robot is probably the more popular project for hobbyists to build. There's something endearing about a robot that scampers across the floor, either chasing or being chased by the cat.

Figure 2-1 Mobile robots move, typically using wheels or tracks but also legs and other forms of propulsion. Though this model also has a robotic arm to manipulate things, it's considered in a different class from a stationary robot.

As a serious robot experimenter, don't overlook the challenge of building both types of robots. Stationary bots typically require greater precision, power, and balance, since they are designed to grasp and lift things—hopefully not destroying whatever it is they're handling. Likewise, mobile robots present their own difficulties: maneuverability, adequate power supply, and avoiding collisions among them.

Autonomous versus Teleoperated

The first robots ever demonstrated for a live audience were fake; they were actually machines remotely controlled by a person discretely positioned offstage. It didn't matter. People thrilled at the concept of the robot, which many anticipated would be an integral part of their near futures. You know, like flying to work with your own jetpack, or colonies in outer space!

These days, the classic view of the robot is a fully autonomous machine, like Robby from *Forbidden Planet,* or that BB-8 thingie from *Star Wars*. With these robots (or at least the make-believe, fictional versions), there's no human operator, no remote control, no "man behind the curtain."

While many actual robots are indeed fully autonomous, many of the most important robots of the past few decades have been *teleoperated:* commanded by a human and operated by remote control. These are often used in police and combat situations. The typical *telerobot* uses a video camera that serves as the eyes for the human operator. From some distance—perhaps as near as a few feet to as distant as several million miles—the operator views the scene before the robot and commands it accordingly.

Figure 2-2 Human supervision of some telerobots may be superficial. The Mars rover *Sojourner* was commanded from over 140 million miles away. This hardy robot had a planned life of seven Earth days, but lasted 85 Earth days on the harsh Martian surface. (*Photo courtesy NASA.*)

The teleoperated robot of today is a far cry from the radio-controlled robots of the world's fairs of the 1930s and 1940s. Many telerobots, like the world-famous Mars rover *Sojourner*, shown in Figure 2-2, the first interplanetary dune buggy, are actually half remote-controlled and half autonomous.

In a semi-autonomous robot, the low-level functions of the robot are handled by a computer on the machine. The human intervenes to give general-purpose commands, such as "go forward 10 feet" or "hide, here comes a Martian!" The robot carries out basic instructions on its own, freeing the human operator from the need to control every small aspect of the machine's behavior.

On some teleoperated robots stereo video cameras give a human operator 3D depth perception. Sensors on motors and robotic arms provide feedback to the human operator, who can actually feel the motion of the machine or the strain caused by some obstacle. Virtual-reality helmets, gloves, and motion platforms literally put the operator "in the driver's seat."

Tethered versus Self-Contained

People like to debate what makes a machine a "real" robot. One side says that a robot is a completely *self-contained, autonomous* (self-governed) machine that needs only occasional instructions from its master to set it about its various tasks. It's complete in and of itself. It needs no *tether*—wires or strings—to humans in order to operate.

But this view disregards a wide swatch of very important robotics research. As noted, robots are fundamentally extensions of human endeavor. This concept allows bots to enjoy far wider appearance and operation.

Figure 2-3 This quadruped from General Electric was controlled by a human operator who sat inside it. The robot was developed in the late 1960s under a contract with the U.S. government. (*Photo courtesy General Electric.*)

Consider the experimental walking "lorry" from 1969 in Figure 2-3. A man sat inside the mechanism and operated it, almost as if driving a car. Cars aren't robots, yet this was a very important milestone in robot development.

The purpose of the four-legged machine was not to create a self-contained robot but to further research into *cybernetic anthropomorphous* machines, otherwise known as *cyborgs*. The idea was how to *amplify* human power through a machine. Never mind if a robot is tethered or self-contained, or "driven like a car." Today's human-controlled machine may be tomorrow's fully independent lifesaving robot.

So, What's a Robot, Anyway?

What makes a robot a robot and not just another machine? For the purposes of this book, let's consider a robot as *any device that—in one way or another—mimics a human or animal function.* The way the robot does this is of no concern; the fact that it does it at all is enough.

The functions that are of interest to us as robot builders run the gamut: from listening to sounds and acting on them, to talking and walking or moving across the floor, to picking up objects and sensing special conditions such as heat, flames, or light. So, when we talk about a robot, it could very well be—

- A self-contained automaton that takes care of itself, perhaps even programming its own brain and learning from its surroundings and environment.
- Or it could be a small motorized cart operated by a strict set of predetermined instructions that repeats the same task over and over again until its batteries wear out.
- Or it could be a radio-controlled arm that you operate manually from a nearby panel.

Each is no less a robot than the others, though some are more useful than others. As you'll discover throughout this book, you're in control in defining how complex your robot creations are, and where you intend to take your ideas once you experiment with them.

The Body of the Robot

For most robots, a robot's body holds all its vital parts. The body is the superstructure that prevents its electronic and electromechanical guts from spilling out. Robot bodies go by many names, including *platform, frame, base,* and *chassis,* but the idea is the same. The bodies of robots differ in size, shape, and style; their type of superstructure; and their overall construction.

ROBOT SIZE, SHAPE, AND STYLE

Robots come in all sizes. Some can fit in the palm of your hand, while others are so big it takes a crane to move them.

Homebrew robots are generally the size of a small cat or dog, although some are as compact as an aquarium turtle and a few as large as Arnold Schwarzenegger (hopefully your name isn't Sarah Conner!). The overall shape of the robot is generally dictated by the internal components that make up the machine, but most designs fall into one of the following "categories":

- Tabletop
- Rover
- Walking
- Stationary arms/grippers (also called manipulators)
- Android and humanoid

Smaller robots are both easier to build and more affordable. Their smaller size means smaller motors, smaller batteries, and smaller chassis—all of which tend to reduce price.

Tabletop

Tabletop robots are simple and compact, designed primarily for use on tables, desks, workbenches, small rooms, and similar areas. Another common name is *turtlebots*; their bodies somewhat resemble the shell of a turtle. Researcher F. Grey Walter used the term to describe a series of small robots he envisioned and built in about 1948. In popular usage, turtle robots also borrow their name from a once-popular programming language, Logo turtle graphics, adapted for robotics use in the 1970s.

Figure 2-4 Small tabletop or "turtle" robots represent a class that's easy and inexpensive to build. They are ideal for learning about robotic design, construction, and programming.

The desktop category represents the majority of amateur robots—Figure 2-4 shows one of them. It's popular among those involved in noncombat competitions—things like maze following or robotic soccer. Tabletop robots are most commonly powered by a rudimentary electronic circuit, or by a small single-chip computer or microcontroller.

You'll find a number of tabletop robot designs throughout this book, including Chapter 17, "Build Robots with Wheels."

Rover

The *rover* category is any of a group of rolling or tracked robots designed for applications that require some horsepower, such as vacuuming the floor, fetching a can of beer or soda, or mowing the lawn. These robots are too big to play with on a desk or workbench. Sizes range from that of a waffle iron and continue on up. The "death match" combat robots popular on TV are typical robots in the larger end of the rover spectrum, where weight is important to winning.

Because of their larger size, rover robots can be powered—brainwise—by everything from a simple transistor to a desktop computer. With the advent of small 32-bit computers no larger than a credit card, the processing power of these robots has skyrocketed hundreds of times beyond what homebrew robot makers built in the 1980s and 1990s. The rover in Figure 2-5 is powered by an Arduino Mega microcontroller, which costs less than a dinner for four at a modest restaurant.

Figure 2-5 Roverbots are built with both brains and brawn. Using heavier duty motors and frames than those of its smaller tabletop cousins, rovers are able to do heavier work in more diverse environments.

Figure 2-6 A rewarding project is a multilegged walking robot, such as this model that has six legs. It's operated using three individual motors per leg, for a total of 18 motors. (*Photo courtesy Lynxmotion.*)

Walking

A *walking* robot uses legs to move about. Most walker bots have six legs, like an insect, as the six legs provide excellent support and balance. However, gaining in popularity are two- and four-legged walkers, both for "scratch-build" hobby projects as well as for commercial kits.

An example of a six-legged walking robot kit is the T-Hex from Lynxmotion (Figure 2-6). It's made by combining radio control servo motors and prefabricated plastic and metal parts.

Figure 2-7 A robot *arm* mimics the basic architecture of the human arm. Motorized joints replicate the shoulder, elbow, and wrist. A *gripper* at the end of the arm serves as a limited "hand."

Walking robots require a greater precision in building. If you're just starting out you should opt for wheeled tabletop or even rover designs first to gain experience. And yes, I know: the walking robot looks cooler, but we all have to start somewhere.

That said, you'll discover several workable designs for walking robots in Chapter 19, "Build Robots with Legs."

Note that two-legged walking robots that resemble people are classified (see below) in their own category, considering the technological difficulties in designing and building them. Constructing a small two-legged robot that hobbles along the desk is one thing; creating a C-3PO-like robot is quite another, even if your name is Anakin Skywalker.

Arms and Grippers

The ability to manipulate objects is a trait that has enabled humans, as well as a few other creatures in the animal kingdom, to take command of their environment. Without our arms and hands we wouldn't be able to use tools, and without tools we wouldn't be able to build houses, cars, and—hmmm—robots.

Arms and *grippers* are used by themselves in stationary robots, or they can be attached to a mobile robot. An arm can be considered any appendage of the robot that can be individually and specifically manipulated; grippers (often called *end-effectors*) are the hands and fingers and can be attached either directly to the robot or to an arm. See Figure 2-7 for an example of a commercially available robot arm/gripper kit.

You can duplicate human arms in a robot with just a couple of motors, some metal rods, and a few other parts. Add a gripper to the end of the robot arm and you've created a complete arm-hand module. You can read more about robotic arms, and even make your own, in Chapter 20, "Build Robotic Arms and Grippers."

Android and Humanoid

Android and *humanoid* robots are specifically modeled after the human form. In current usage, *android* is a robot designed to look as much like a human being as possible, including ears, hair, and even an articulated mouth. A humanoid robot is one that shares the basic architecture of a human, but is not meant to be a physiological replica. Figure 2-8 shows an example humanoid bipedal robot that you can actually build.

Figure 2-8 Bipedal (two-legged) robots present special challenges, not only in construction, but in programming. Standardized metal brackets, like those used here, make building easier by not requiring you to have a complete metal-working shop in your garage—though if you do, by all means use it! (*Photo courtesy Lynxmotion.*)

SKELETAL STRUCTURE

In nature and in robotics, there are two general types of support frames: endoskeleton and exoskeleton.

- *Endoskeleton* support frames are the kind found in many critters—including humans, mammals, reptiles, and most fish. The skeletal structure is on the inside; the organs, muscles, body tissues, and skin are on the outside of the bones. The endoskeleton is a characteristic of vertebrates.
- *Exoskeleton* support frames have the "bones" on the outside of the organs and muscles. Common exoskeletal creatures are spiders, all shellfish such as lobsters and crabs, and an endless variety of insects.

The main structure of the robot is generally a wood, plastic, or metal frame, which is constructed a little like the frame of a house—with a bottom, top, and sides. Onto the frame of the robot are attached motors, batteries, electronic circuit boards, and other necessary components. In this design, the main support structure of the robot can be considered an exoskeleton because it is outside the major organs.

A shell or covering is sometimes placed over these robots, but the "skin" is for looks and maybe as protection of the internal components, not support. For the most part, the main bodies of your robots will have exoskeleton support structures because they are cheaper to build, stronger, and less prone to problems.

FLESH AND BONE—AND WOOD, PLASTIC, OR METAL

In the movies robots are made of some super-fantastic metal that even atom bombs can't destroy. Yes, metal is a common part of many kinds of robots, but the list of materials you can use is much larger and more diverse.

- *Foamboard.* Art supply stores stock what's known as *foamboard,* a special construction material typically used for building models. It's really a sandwich of paper or plastic glued to both sides of a layer of densely compressed foam. You can cut it with a knife or small hobby saw. Great stuff for quickie-made bots. I talk about foamboard and similar materials in Chapter 6, "Robots from Household Stuff."
- *Wood.* Wood is an excellent material for robot bodies, especially multi-ply hardwoods, like the kind used for model airplanes and model sailboats. Common thicknesses are 1/8″ to 1/2″—perfect for most robot projects. Read up on these creatures in Chapter 7, "Robots of Wood."
- *Plastic.* Plastic boasts high strength, but is easier to work with than metal. You can cut it, shape it, drill it, even glue it, with common, everyday tools. My favorite is PCV expanded plastic, but there are many other kinds of plastic. See Chapter 8, "Robots of Plastic," for more details.
- *Aluminum.* If you want to go metal, aluminum is the best all-around robot-building material, especially for medium and larger machines. It's exceptionally strong for its weight. Aluminum is fairly easy to cut and drill using ordinary shop tools. There's more on building robots with aluminum and other metals in Chapter 9, "Robots of Metal."
- *Tin, iron, and brass.* Tin and iron are common hardware metals that are used to make angle brackets and sheet metal (various thicknesses, from 1/32″ on up), and as nail plates for house framing. Cost: fairly cheap. See Chapter 11, "Putting Things Together," for ways to use common angle brackets in your robot construction plans.
- *Steel.* Although sometimes used in the structural frame of a robot because of its strength, steel (and its close cousin stainless steel) is more difficult to cut and shape without special tools. It's ideal for combat robots. As this book concentrates on small, lightweight, and easy-to-build tabletop bots, I'll discuss building bots using only small preshaped steel parts you can get at the local hardware store.

Locomotion Systems

Locomotion is how a mobile robot gets around. It performs this feat in a variety of ways, typically using wheels, tank tracks, and legs. In each case, the locomotion system is driven by a motor, which turns a shaft, cam, or lever.

1. *Wheels.* Wheels are the easiest and most popular method for providing robots with mobility. Robot wheels can be just about any size, limited only by the dimensions of the robot

and your outlandish imagination. You can read more about wheel designs in Chapter 13, "How to Move Your Robot," and the many hands-on projects in Chapter 17, "Build Robots with Wheels."

2. *Tracks.* The basic concept of *track-driven* robots is simple. Two tracks, one on each side of the robot, act as giant wheels. The tracks turn, and the robot moves forward or backward. Track drive is practical for many reasons, including the fact that it makes it possible to mow through all sorts of obstacles. Read more about these designs in Chapter 18, "Build Robots with Tracks."

3. *Legs.* Walking bots can have one or more legs. The fewer the legs, the more challenging the design. Two-legged (*bipedal*) robots use unique balancing methods to keep them from falling over. Four-legged robots (*quadrupeds*) are easier to balance, but good walking and steering often involve adding extra joints and some sophisticated math to make sure everything moves smoothly. You can read about building your own leg-based robot in Chapter 19, "Build Robots with Legs."

Power Systems

Most robots are powered using a set of batteries. Connect the batteries to the robot's motors, circuits, and other parts, and you're all set. There are many kinds of batteries, and Chapter 12, "Batteries and Power," goes into more detail about which ones are best for robots. Here are a few quick details to start you off.

Batteries generate voltage and come in two distinct categories: rechargeable and nonrechargeable. *Nonrechargeable* batteries include the standard zinc and alkaline cells you buy at the supermarket, but only the alkaline kind is truly useful in robotics.

Sensing Devices

Imagine a world without sight, sound, touch, smell, or taste. Our senses are an integral part of our lives. Same with robots. The more senses your bot has, the more it can interact with its environment, and go about on its own. Figure 2-9 shows how sensors form an integral part of the robot.

1. The sense of *touch* is provided by mounting switches and other sensors to the robot's frame, so that it can detect when it's run into something—or something has run into it. See Chapter 30, "Touch," for practical touch receptive designs.

2. Robots can use *proximity senses* to detect things around it. An example is bat-like *ultrasound*. The sensor emits a short *ping* of sound, then waits for its echo (see Figure 2-10). Another scheme uses invisible *infrared light* and a bit of mathematics to measure distances. Both ultrasonic and infrared detection are more fully explained in Chapter 31, "Proximity and Distance."

3. *Navigational sensors* use accelerometers for detecting tilt and motion, gyroscopes for determining speed, and global positioning satellite receivers for checking where on this Earth your robot currently is. Read more on these and other cool devices in Chapter 32, "Navigation."

4. It's easy to make a robot sensitive to *light* and *sound*; in fact, these are among the least-expensive sensors you can add to a robot. Chapter 33, "Environment," provides a number of plans for giving light sensing and hearing to a robot.

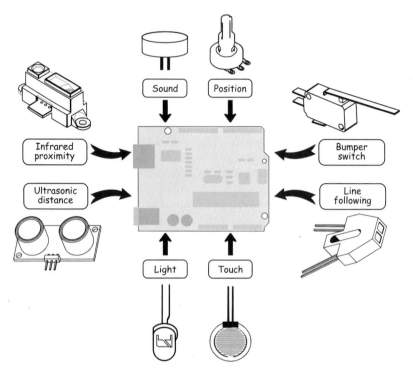

Figure 2-9   Sensors let your robot experience the world around it. Most robots use relatively simple technologies like ultrasonic sensors, light detectors, and bumper switches to monitor their environment. Sensors are integrated and processed by a control circuit, which then performs some action based on the sensory input.

Figure 2-10   Robots use ultrasonic sensors to measure distances of nearby objects.

Output Devices

Output devices are components that relay information from the robot to the outside world. Here are examples of output devices in robots:

- Light-emitting diodes (LEDs) that shine or blink, to indicate what the robot is doing—or trying to do.
- Liquid-crystal displays (LCDs), for showing complete messages.
- Sounds that allow your bot to communicate in nonvisual ways. You might add a speech synthesizer so your robot can talk to you. Or program it to wake you up in the morning by playing your favorite MP3s. You can record any sound and have your robot reproduce it instantly.

Read more on these and other ideas in Chapter 35, "Producing Sound," and Chapter 36, "Visual Feedback from Your Robot."

CHAPTER 3

The Safe and Sane Guide to Robot Building

Robotics is a safe and sane hobby, but only if you practice it with caution and respect. It can involve working with soldering tools that can burn, saws that can remove bits of flesh, and household current that can electrocute.

Any dangers involved in robot building are easily minimized by taking a few simple steps to practice safe working habits. Take your time and think things through; you may never have to burn yourself with a soldering iron, cut yourself with a knife or saw, or shock yourself with an exposed electrical wire.

Project Safety

If you plan on constructing the bodies and frames of your robots from scratch, you'll likely need to work with saws, drills, knives, and other building tools. Use all tools according to the manufacturer's instructions. Use tools only in well-lighted and well-ventilated areas. Wear proper clothing and shoes.

Power saws are especially dangerous, even those with safety guards and mechanisms. Never defeat them! Don't allow children in the work area. More:

- Wear eye protection at all times, especially when cutting and drilling material. The glasses or goggles should wrap around your temples to prevent stray debris from striking your eyes from the side.
- Make sure that saw blades and drill bits are sharp. If they're dull, sharpen or replace them.
- Some project plans require the use of sharp knives for cutting—making robot bases out of foamboard, for example. Use a sharp knife, and cut on an appropriate surface. Don't use your free hand to hold down the board; if the knife slips, you could get badly cut. Hold down the board with a straightedge.

Figure 3-1 Use proper protection when cutting and drilling any material. Safety glasses, respirator masks, and ear protection are cheap insurance against injury.

- Wear ear protection when using saws or any other power tools.
- Use a respirator when cutting plastic and wood to keep the sawdust out of your mouth and lungs (see Figure 3-1).

 I can't stress enough the importance of using adequate eye protection. I'm fortunate enough to have both of my eyes, but I came close to losing sight in one eye from flying debris when I was cutting some frame pieces with a miter saw. As a result, in my workshop eye protection is now mandatory for myself and all helpers.

Soldering Safety

Soldering electronic circuits requires that you use a *very* hot iron or pencil. Temperatures exceed 600°F, which is enough to give you third-degree burns after only momentary contact with the tool. This temperature is equivalent to an electric stove burner set at medium-high heat, so you can imagine the dangers involved.

If you plan on doing any soldering, keep the following safety tips in mind:

- Always place your soldering tool in a stand designed for the job. Never place a hot soldering tool directly on the table or workbench.
- Wear eye protection at all times when soldering; molten metal can spray in your face and seriously damage an eye.
- Mildly caustic and toxic fumes are produced during soldering. Maintain good ventilation to prevent a buildup of these fumes in your workshop. Avoid inhaling the fumes produced during soldering.
- Thoroughly wash your hands with soap and warm water after soldering to remove traces of lead, bismuth, and other metals.
- If your soldering tool has an adjustable temperature control, dial the recommended setting for the kind of solder you are using, usually about 650° to 700° for standard 60/40 rosin-core solder. Lead-free solder often requires a higher operating temperature.
- Always use rosin-core solder designed for use on electronics. Other kinds of solder could damage the circuit or your soldering tool.
- Do not attempt to solder on a "live" circuit—a circuit that has voltage applied to it. You run the risk of damaging the circuit, the soldering tool, and, most of all, you!

Fire Safety

Fire is a potential hazard during the construction and use of any electrical device. A hot soldering tool can ignite paper, wood, and cloth. A short circuit from a large high-current battery can literally melt wires. Although not common for the type of projects presented in this book, an electric circuit may develop too much heat, and could melt or burn its enclosure and surroundings. Proper construction techniques and careful review of your work will help prevent these kinds of mishaps.

If your project operates under house current, keep an eye on it for the first several hours of operation. Note any unusual behavior, including arcing, overheating, or circuit burnouts. If a circuit breaker trips while your project is on, you can bet that something is amiss with your wiring.

 None of the projects in this book involve constructing circuits that are directly powered by household AC current. When AC current is used it's always through a commercially manufactured power supply or certified wall transformer. Whenever possible, avoid AC-powered circuits and use low-cost commercially available power supplies instead.

For obvious reasons, you should always build and operate your projects away from flammable objects, including gasoline, lighter fluid, welding and brazing equipment, and cleaners. Always keep a fire extinguisher near you, and don't hesitate to use it if a fire breaks out.

Melting plastic can release highly toxic chemicals and gases. After you put out the fire be sure to ventilate the area thoroughly. Melted PVC plastic can release hydrogen chloride gas. Seek medical attention if you're not feeling well.

Battery Safety

Batteries used in robots may produce only a few volts, but they can generate lots of current—so much current that if the terminals of the battery are shorted, the battery could get *very* hot. If you're lucky, the battery terminals will only melt. But exploding batteries that cause fires are not unheard of.

Never short out the terminals of a battery, even to just see what'll happen. Store charged batteries so that the terminals will never come into contact with metal objects. Beware of loose wire ends that may short, as shown in Figure 3-2. The least that will happen is the battery will quickly discharge; the worst is it'll overheat and possibly catch fire. Always be sure to recharge batteries in a recharger meant for that type of battery.

Don't allow bare
leads to touch

Figure 3-2 Always store battery packs so
that bare leads will never touch, causing a short
circuit. Remove batteries from holders and wiring
snaps when not in use.

Avoiding Damage by Static Discharge

The ancient Egyptians discovered static electricity when they rubbed animal fur against the smooth surface of amber. When the materials were rubbed together they tended to cling to one another. While the Egyptians didn't comprehend this mysterious unseen force, they knew it existed. Today we fully understand static electricity and know it can cause damage to electronic components.

As a robo-builder, you must take specific precautions against *electrostatic discharge,* otherwise known as *ESD.* Damage from static discharge can be all but eliminated by taking just a few simple steps to protect you, your tools, and your projects from static buildup.

THE PROBLEM OF ELECTROSTATIC DISCHARGE

Electrostatic discharge involves very high voltages at extremely low currents. Combing your hair on a dry day can develop tens of thousands of volts of static electricity. But the current (akin to the *force* of the electricity) is almost negligible. The low current protects you from serious injury.

Many electronic components that are manufactured with semiconductor material are not so forgiving. These include transistors and integrated circuits, especially those that use what's known as a metal oxide substrate. These include:

- MOSFET transistors
- CMOS integrated circuits (ICs)
- Just about any microcontroller
- Any module (digital compass, sensor) that contains one or more of the above

You'll know a static-sensitive device when it comes to you in an antistatic plastic pouch or on antistatic foam. Keep these devices in their pouch or stuck into their foam until you use them.

USING AN ANTISTATIC WRIST STRAP

If you live in a dry climate or where static is an ongoing problem, consider using an antistatic wrist strap whenever you work with sensitive electronics. This strap grounds you at all times and prevents static buildup.

To use, put the strap around your wrist, then connect the clip to any grounded or large metallic object. A nearby computer (plugged in) or the frame of a metal desk or bookshelf are good choices. If you have an antistatic desk or floor pad it will likely have a metal stub on it that you can connect the wrist strap to. Even if the pad is itself not grounded, the idea here is that it dissipates the static because of its large surface area.

 Though you may read otherwise, it's not a good idea to go sticking your wrist strap into the ground hole of an electrical wall socket. It's too easy to accidentally plug yourself into one of the other holes and receive a shock.

STORING STATIC-SENSITIVE DEVICES

Static-sensitive electronics are best stored using one of the following methods. All work by connecting (grounding) the leads of the component together, thereby diminishing the effect of a strong jolt of static electricity. Note that none of the storage methods is 100 percent foolproof.

- *Antistatic mat.* This mat looks like a black sponge, but it's really conductive foam.
- *Antistatic pouch.* Antistatic pouches are made of a special plastic coated on the inside with a metallic layer. Many are resealable so you can use them again and again.
- *Antistatic tube.* Quantities of ICs are most often shipped and stored in convenient plastic tubes. They're treated with a conductive coating to help reduce static.

 Remove the chip or transistor from its antistatic storage protection only when you are installing it in your project. The less time the component is unprotected, the better.

USE ONLY GROUNDED SOLDERING TOOLS

A common source of ESD damage when building electronic circuits is using an ungrounded soldering iron or pencil. Ungrounded tools have only two prongs on their power cord, instead of three. The third (round) prong is the ground connection.

A grounded tool not only helps prevent damage from electrostatic discharge, but lessens the chance of a bad shock should you accidentally touch a live wire. Be sure to use only a grounded wall socket; if you use an extension cord, be sure it, too, is grounded.

Working with House Current

None of the projects in this book directly use AC house current, but let's cover the safety precautions just the same. A live AC wire can, and does, kill. Exercise caution whenever working with AC circuits. You can greatly minimize the hazards of working with AC circuits by following these basic guidelines:

- Always keep AC circuits fully covered. Always.
- Keep AC circuits physically separate from DC circuits. If necessary, construct a plastic guard within your project to keep the wiring separate.

- All AC power supplies should have fuse protection. The fuse should be adequately rated for the circuit but should allow a fail-safe in case of short circuit.
- Place your AC projects in a plastic box. Don't use a metal project box.
- Double- and triple-check your work before applying power. If you can, have someone else inspect your handiwork before you switch the circuit on for the first time.

First Aid

Despite your best efforts, accidents might happen. With luck, most will be minor, causing little or no injury. If injury does occur, be sure to treat it promptly. If necessary, see a doctor to prevent the condition from getting worse.

Consult a good book on medical first aid treatment for details on how to care for cuts, abrasions, and other minor injuries. You'll want to keep a first aid kit handy at all times, preferably right there in your shop, or conveniently located in a washroom or lavatory.

Purchase a first aid kit, and keep it in your workshop. If you use an item from the kit, be sure to replace it. Every year inspect your first aid kit and replace anything that is past its expiration or use-by date.

EYE INJURY FIRST AID

Perhaps most serious of all is injury to eyes. Be sure to wear adequate eye protection at all times—when soldering, when using shop tools, the whole bit. Chips of solder and metal leads can fly off when they are cut with nippers. If these get into your eye, not only is it excruciatingly painful, but it could damage your eye, temporarily or even permanently.

Should you get something in your eye, especially a piece of glass, metal, or plastic, see a physician *immediately*. Trying to remove the object yourself can cause further injury.

ELECTRIC SHOCK FIRST AID

Should you get a nasty shock from a circuit or battery, check the area of skin contact for signs of burns. Treat it as you would any other burn.

Immediately after an electrical shock, stop what you are doing and consciously calm yourself down. Monitor your pulse to make sure your heart isn't suffering from any aftereffects of the shock. If you feel anything is amiss, consult a doctor *immediately*.

USE COMMON SENSE—AND ENJOY YOUR ROBOT HOBBY

Common sense is the best shield against accidents and injury, but common sense can't be taught or written about in a book. It's up to you to develop common sense and use it at all times. Never let down your guard. Don't ruin the fun of a wonderful hobby or vocation because you neglected a few safety measures.

PART 2

Building Robots

CHAPTER 4

Getting Parts

Exactly where do you find robot parts? Your friendly neighborhood robot store would be a good place to start—but not many of us have such a store nearby! Fortunately, other online and local retail stores are there to fill in the gaps.

The fundamental stuff of bots includes:

- *Basic electronic parts* to control your robot. These include core electronic components such as resistors and capacitors, switches, wire, and relays.
- *Specialized electronic modules* to complete circuits that you connect with the rest of your robot. The options here are nearly limitless, but the main modules you'll be interested in include a microcontroller (holds programs for your robot, operates motors and other parts) and sensors.
- *Motors and wheels* for making your robot go.
- *Construction materials* for building the body, or base, of your robot. You can make the base from found material such as craft supplies and small plastic containers, or from wood, plastic, metal, foamboard, even heavy cardboard.
- *Fasteners, adhesives, and other building parts* to put everything together. These might include small screws and nuts, but also cable ties, Velcro, and hot melt glue.

In this chapter you'll learn where you can find these and other parts for your robot projects.

 For more information on finding and getting parts, refer to the RBB Support Site (see the Appendix), which contains parts lists for the projects in this book.

Local and Online Electronics Stores

Not long ago most towns had at least one local electronic parts store. Those days are gone; in the United States even the once-ubiquitous RadioShack, which at one time boasted over 7000 stores, is now represented in only a very small number of cities. Depending on where you live, your area may be supported by a bricks-and-mortar electronics retailer, such as Fry's, Microcenter, or Maplin. You can find these outlets by doing a local Web search.

Fortunately, there's no shortage of choices for online electronics stores; I provide a lengthy curated list on the RBB Support Site. While you have to wait for the parts to ship to you, the selection is better, and (often) so are the prices. Be sure to add Amazon and eBay as your regular haunts when finding bot building goodies. Many sellers offer free or low-cost shipping.

Specialty Online Robotics Retailers

With robotics now a hot hobby, a new type of online retailer has sprung up: the specialty electronics and robotics site. These sites offer one-stop-shopping for all—or nearly all—of the parts you need to build your robots. Examples of online retailers that specialize in both robotics and electronics include Parallax, Pololu, Lynxmotion, RobotShop, and SparkFun.

Inventory differs from one site to the next, and some products are exclusive to one company. No doubt you'll develop your favorite haunts. I regularly shop at several of these, looking for what's new and who has the most useful products at the best prices.

Once again, refer to the RBB Support Site for a regularly updated list of specialty online robotics outlets.

Craft Stores

Craft stores sell supplies for home crafts and arts. As a robot builder, you'll be interested in some of the aisles with:

- *Foam rubber sheets.* These come in various colors and thicknesses and can be used for pads, bumpers, nonslip surfaces, tank tracks, and lots more. The foam is very dense; use a sharp scissors or knife to cut it (I like to use a rotary paper cutter to get a nice, straight cut).
- *Foamboard.* Constructed of foam sandwiched between two heavy sheets of paper, foamboard can be used for small, lightweight robots. Foamboard can be cut with a hobby knife and glued with paper glue or hot glue. Look for it in different colors and thicknesses.
- *Parts from dolls and teddy bears.* These can often be used in robots. Fancier dolls use something called *armatures*—movable and adjustable joints—that can be applied to your robot creations. Look also for linkages, bendable posing wire, and eyes; they're great for building robots with personality!
- *Electronic light and sound buttons.* These are designed to make Christmas ornaments and custom greeting cards, but they work just as well in robots. Electric light kits come with low-voltage LEDs or incandescent lights, often in several bright colors. Sound buttons have a built-in song that plays when you depress a switch. Use these buttons as touch sensors, for example, or as a "tummy switch" in an animal-like robot.

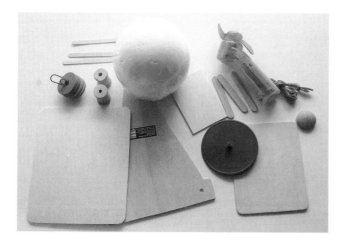

Figure 4-1 Craft stores are terrific repositories for wood and plastic bits for robot building. Be sure to browse every aisle.

- *Plastic and wood crafts construction material.* This can be used in lieu of more expensive building kits. For example, many stores carry the plastic equivalent of that old favorite, the ice cream stick. Or you can also get the original wooden ones if you prefer. The plastic sticks have notches in them so they can be assembled to create frames and structures. Look for other precut shapes of plastic and wood pieces, like in Figure 4-1, that you can use for your projects.
- *Model-building supplies.* Many craft stores have these, sometimes at lower prices than the average hobby-model store. Look for assortments of wood and metal pieces, adhesives, and construction tools.

Hobby Stores

Hobby stores are the ideal sources for small parts, including lightweight plastic, brass rod, servo motors for radio control (R/C) cars and airplanes, gears, and construction hardware. Most of the products available at hobby stores are designed for building specific kinds of models and toys. But that shouldn't stop you from raiding the place with an eye to converting the parts for robot use.

Browse the store and carefully look for parts that you can put together to build a robot. Some of the parts, particularly smaller components for R/C models, will be behind a counter, but they should still be visible enough for you to conceptualize how you might use them.

If you don't have a well-stocked hobby store in your area, there's always Internet mail order.

Hardware and Home Improvement Stores

Hardware stores and builder's supply outlets are a good source for hand tools for building your robots. They're also good for screws, nuts, and other fasteners, adhesives, and the materials for bot bases.

As you tour your neighborhood hardware store, keep a notebook handy and jot down what they offer. Then, when you find yourself needing a specific item, you have only to refer to

Figure 4-2   You can use small metal nailing plates as construction components for your robots. You can combine the pieces to make all sorts of robot bodies. Find the plates at any home improvement store.

your notes. Take an idle stroll through your regular hardware store haunts on a regular basis. You'll always find something new and laughably useful for robot design each time you visit. I know I do.

For example, in the house framing aisle you'll find steel nailing plates (see Figure 4-2). Although intended for building houses, sheds, and other constructions, many of these are small and lightweight enough for use in a robot. Look around for sizes and shapes that you can adapt for robot building. In fact, you'll build a robot from small and affordable steel plates in Chapter 9, "Robots of Metal"!

Shop Once, Shop Smart

Whether buying locally or through mail order you'll want to get as many of the parts as you need at once. This saves time, trouble, and expense.

When getting parts locally, there's the hassle of returning to the store for last-minute additions. That costs you time and gas. And when buying through the mail, repeat orders pile up the shipping costs. A transistor may cost only 25 cents, but add order minimums and shipping fees and you could easily be looking at $5 or $10 on your credit card.

Some ideas for savvy shopping:

- Keep an inventory of what you have, and think ahead. Plan your next several projects, and get as many of the parts for them at the same time as you can.
- Try to group purchases together when buying from the same store, even if the store doesn't have the lowest price. If the difference is minor, consider the additional costs of driving or shipping from another source.
- For very basic electronic parts try to get an assortment of standard values. Things like resistors and capacitors (discussed in Chapter 22, "Common Electronic Components for Robotics") are just pennies each. A $10 or $15 assortment of the most common values will save you time and money.
- Don't forget what you already have! See "Getting Organized," later in this chapter, on how to keep a good inventory of your stock.

Figure 4-3 Toys are among the best sources for robotic parts, especially motorized military tanks or construction vehicles. These can be turned into robots with rubber treads.

Other Useful Retailers

What specialty stores are useful to robot builders? Consider:

- *Toy stores.* Look for construction toys like Erector, Meccano, LEGO, and others. Check out their battery-operated toy vehicles, such as motorized tanks. Raid the motors and rubber treads, like the ones in Figure 4-3, for your robot. This is how I get parts for many of my tracked robots.
- *Sewing machine repair shops.* These are ideal for finding small gears, cams, levers, and other precision parts. Some shops will sell broken machines to you. Tear them to shreds and use the parts for your robot.
- *Auto parts stores.* The independent stores tend to stock more goodies than the national chains, but both kinds offer surprises in every aisle. Keep an eye out for small switches, wire, and tools.
- *Junkyards.* If you're into building bigger robots, old cars are good sources for powerful DC motors, which are used to drive windshield wipers, electric windows, and automatic adjustable seats (though take note: such motors tend to be terribly inefficient for battery-based bots).
- *Lawn mower sales-service shops.* Lawn mowers use all sorts of nifty control cables, wheel bearings, and assorted odds and ends. Pick up new or used parts for a current project.
- *Bicycle sales-service shops.* I don't mean the department store that sells bikes, but a *real* professional bicycle shop. Items of interest: control cables, chains, brake calipers, wheels, sprockets, brake linings, and more.
- *Industrial parts outlets.* Some places sell gears, bearings, shafts, motors, and other industrial hardware on a one-piece-at-a-time basis. The penalty: fairly high prices and often the requirement that you buy a higher quantity of an item than you really need.

Scavenging: Making Do with What You Already Have

You don't need to buy things to gather worthwhile robot parts. In fact, some of the best parts for hobby robots may already be in your garage or attic. Consider the typical used CD player or video cassette recorder from a secondhand shop. "These contain at least one motor, and possibly as many as five, numerous gears, and other electronic and mechanical odds and ends."?

Never throw away small appliances or mechanical devices without taking them apart and robbing the good stuff. If you don't have time to disassemble that CD player that's been skipping on all of your compact discs, throw it into a pile for a rainy day when you do have a free moment.

Likewise, make a point of visiting garage sales and thrift stores from time to time, and look for parts bonanzas in used—and perhaps nonfunctioning—goods. I regularly scout the local resale stores and for very little money come away with a trunk full of valuable items that I can salvage for parts.

Here is just a short list of the electronic and mechanical items to be on the lookout for and the primary robot-building components they have inside:

- Old *VCRs* are perhaps the best single source for parts. They may not be in plentiful supply as they once were, but you can still find the odd machine in thrift stores, flea markets, and garage sales. Figure 4-4 shows a vintage VCR with its top open, revealing the many goodies inside.
- *CD and DVD players* have optics you can gut out if your robot uses a specialty vision system. Apart from the laser diode, disc players have focusing lenses and other optics, as well as miniature motors.
- Old *fax machines* contain numerous motors, gears, small switches, and other mechanical parts.
- *Mice, printers, photo scanners, disk drives, and other discarded computer add-ons* contain valuable optical and mechanical parts. Ball-style mice contain optical encoders that you could use to count the rotations of your robot's wheels; printers contain motors and gears; disk drives contain motors; and so on.
- *Mechanical toys*, especially the motorized variety, can be used either for parts or as a robot base. When looking at motorized vehicles, favor those that use separate motors for each drive wheel.

Figure 4-4 Old VCRs (and fax machines, CD players, and other electronics of the previous century) are goldmines of parts for robot building.

Getting Organized

Make a conscious effort to maintain order in your shop, or else things rapidly get messy and increasingly expensive. Some ideas:

SMALL PARTS DRAWER CABINETS

Plastic drawer cabinets are really the mainstay for any activity that deals with small parts. You can get them in all styles and sizes, from little units with just a half dozen $1'' \times 2''$ drawers (these fit on bookshelves quite well) to much larger parts chests, with 20, 30, even 40 drawers of different shapes and sizes.

I like the cabinets with several drawer sizes, which accommodate parts of different bulks. For example, the cabinet might sport 24 or 30 small drawers of about $1.5'' \times 2''$ and 6, 9, or 12 large drawers of about $2'' \times 4''$. You might, for instance, place individual values of resistors in the smaller drawers and large capacitors in the large drawers.

HEAVY-DUTY STORAGE DRAWERS

A step up the size and cost ladder is the plastic heavy-duty storage bin. These are 1- to 3-foot-high units that sport between three and six large, heavy-duty plastic drawers. The drawer sizes may be different or all the same; find a model that suits your needs. The most robust models can readily hold heavy motors, metal gears, and other construction parts.

Larger still (and somewhat more flexible) are stacking drawers, with sizes big enough to hold bulky sweaters and pants. They're made to stack on top of one another, so you can get as few or as many as you want. (Avoid overdoing the height of your Tower of Babel, or else things could topple if you open a heavy drawer near the top of your stack.)

TOOLBOXES AND TOTES

Sometimes you have to build while on the go, and you'll need to bring your tools and supplies with you. For the big stuff, a standard metal or heavy-duty plastic toolbox is the best choice.

For lighter jobs, a plastic fishing tackle box or tote makes it easy to lug your supplies. The typical tackle box has a storage drawer on the top for small parts. When you open the top, the drawer slides up and over, and you can reach into the bottom of the box for larger tools and supplies.

KEEPING TRACK OF YOUR INVENTORY

Unless you have perfect memory you'll need some system to keep track of what has gone where. On the low end of the scale is the old "Magic-Marker-on-the-side-of-the-parts-bin trick." (You can use any type of felt-tip marker; a Sharpie is my favorite.)

Instead of writing directly on the plastic, you can instead tape an index card to the container and write on the card. If you change the contents of the container, just peel the card off and start over. Wide clear packing tape works well, too.

For smaller parts drawers, an electronic labeler is a great way to keep track of parts. The better machines can accommodate labels of many different widths.

 Need lots of storage bins for your garage? Try corrugated cardboard shelf bins. They lie flat until you need them. Fold a flap here, tuck a flap there, and your bin is ready for use. If you don't need the bin anymore, untuck the flaps and it'll store flat again.

AD-HOC STORAGE SOLUTIONS

Not everything needs a fancy plastic storage bin. Sometimes simple—and free—is better. Here are some storage ideas for when you don't need a fancy solution.

- *Cardboard shoe boxes* still make great storage containers. Keep the lid so you can stack up the boxes, or just use the box open if you need quick access.
- *Zipper locking food storage bags,* particularly the heavy-duty ones for freezer use, make ideal containers for odd-sized items. Mark the contents with a Sharpie. Tip: The wide-bottom bags stand up on their own.
- *Baby food jars,* plastic or glass, are still an excellent storage solution for very small parts, such as 2-56 size hardware. (If you can get 'em, use the glass jars for electronics parts, as they don't generate static.)
- *Empty egg crates* and egg boxes are also useful for holding small parts, but take care not to overturn the crate or box, as the lid doesn't close over the hollow for the egg. If you're not careful, your parts may spill out or get mixed up.

CHAPTER 5

Robot Building 101

So we know robots are a clever combination of mechanics and electronics. The mechanical part constitutes the "body" of the robot—call it a frame, a base, a platform, it's all the same. Constructing a robot body comprises three straightforward tasks:

- Choose the material (you can mix and match).
- As needed, cut it to shape, and drill holes for mounting parts.
- Put it all together.

In this chapter you'll learn the basics about the materials used in constructing amateur robots, the household tools you need to build them, and important mechanical construction techniques for putting everything together.

Then, the remaining chapters in Part 2 give you explicit and detailed instructions for using these various materials. Chapters include hands-on plans for several robust robot platforms, which you can use as starting points for your robot creations.

 Expensive, fancy tools are not required to complete any of the designs in this book. All you need are basic shop tools—you probably already have them in your garage, or you can borrow them from a friend or relative.

Picking the Right Construction Material

Your robots can be made from wood, plastic, metal, even heavy-duty cardboard or foamboard. But which to choose?

That all depends on the design of the robot: how big and heavy it is and what you intend to use it for. Your budget, construction skills, and tools also influence your choice of materials. It takes a lot less sweat to use a 1/8″ sheet of plywood or plastic than it does to cut or drill metal.

- *Wood*. Wood is easy to work with, can be sanded and sawed to any shape, doesn't conduct electricity (avoids short circuits) unless wet, and is available everywhere.
- *Plastic*. Depending on the type of plastic, it's more durable than wood, but a little harder to cut and drill. Not as universally available.
- *Metal*. By metal I'm talking mostly aluminum, though copper and brass sheets and tubes are commonly available in hobby stores. Steel is handy if you need something heavy-duty, like a large combat robot.
- *Lightweight composite materials*. Heavy-duty cardboard, paper laminated foamboard, and lightweight plastic board used to make signs are good contenders for fast-and-easy bots.

Let's take a closer look at these. Even more details on these construction materials are found in later chapters.

WOOD

If Noah can build an ark out of wood, it's probably good enough for robots—and for bots you don't have to know what a cubit is (if you *really* have to know, it's roughly 18″, or about the length of an adult forearm). Wood is reasonably inexpensive and can be worked using ordinary tools.

Avoid soft "plank" woods, like pine and fir; and instead select a hardwood plywood designed for model building. They're available at hobby and craft stores.

See Chapter 7, "Robots of Wood," for more details on using wood in your robots. You'll find wood is among the most easiest to use and least expensive robot building material.

PLASTIC

There are literally thousands of plastics, but don't let that alarm you. For robotics, you're likely to use just a few.

- *Acrylic* is used primarily for decorative or functional applications, such as picture frames or salad bowls. It's usually clear but also comes in solid and translucent colors. You have to be careful of cracking when drilling and cutting.
- *Polycarbonate* is similar in looks to acrylic but is considerably stronger. Because of its increased density it's much harder to work with and is more expensive.
- A special type of PVC that comes in sheets, called *expanded PVC*, is ideal for making small and medium-size robots. An example robot made with PVC is shown in Figure 5-1. It's easy to work with.

See Chapter 8, "Robots of Plastic," for the lowdown on making plastic robots. Using plastic is a great way to help your robots stand out.

Figure 5-1 Expanded PVC is an ideal construction material for robotics. It's inexpensive and lightweight, and it cuts and drills like wood. It comes in a variety of colors and thicknesses to match the needs of your project.

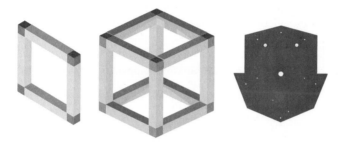

Figure 5-2 The three basic types of robot bases: *flat square frame* (it can actually be made of wood, plastic, or metal), *box frame*, and *shaped base*, which provides the structure for the robot's components.

METAL

Metal is the archetypal material for what we image as a "robot." Major downside: it's not as easy to work with as wood or plastic. But if you need sheer strength metal is your best bet.

For robots, aluminum and steel are the most common metals. Aluminum is a softer metal, so drilling and cutting is easier. There are two general approaches to metal construction in robots, shown in Figure 5-2:

- A *flat frame* provides the base of the robot and lends it support. A *box-shaped frame* is just what its name implies: a 3D box with six faces. It's well suited for larger robots or those that require extra support for heavy components.
- A *shaped base* is a piece of metal cut in the shape of the robot. The metal must be rigid enough to support the weight of the motors, and other parts without bending or flexing.

Robots made of wood, plastic, or other materials may nevertheless use metal (typically aluminum) in their construction. Common metal parts include brackets, to hold pieces together, and nuts, screws, and other fasteners.

See Chapter 9, "Robots of Metal," for more details on using aluminum, steel, and other metal in your robots.

LIGHTWEIGHT MATERIALS

Not every robot needs to withstand a howling sandstorm on *Rigel VII*. A technique known as *rapid prototyping* uses lightweight materials that are cut with basic hand tools. With rapid proto-typing you can make a robot in less time, for less money.

Several projects in the book use rapid prototyping materials; for now let's just concentrate on some of the typical materials used:

- *Cardboard*, either *regular* or *heavy-duty*. Heavy-duty cardboard, which is thicker and heavier than everyday cardboard used for shipping, is surprisingly strong, yet easy to cut and drill. It makes for decent robot bodies.
- *Laminated composite* materials include foamboard, which is a piece of plastic foam inside two sheets of heavy paper. Other kinds of laminated composite sheets may use a combination of wood, paper, plastic, even thin metal.
- *Corrugated plastic* is a favorite among sign makers. They use sheets of it to make lightweight (and very affordable) indoor and outdoor signs. These look like cardboard but they're entirely made of plastic.

IN REVIEW: SELECTING THE RIGHT MATERIAL

Let's review the main construction materials for building robots and compare their good and bad sides.

Material	Pros	Cons
Wood	Widely available; reasonably low cost; easy to work with using ordinary shop tools; hardwood plywoods (recommended wood for most robot bases) very sturdy and strong	Not as strong as plastic or metal; can warp with moisture (should be painted or sealed); cracks and splinters under stress
Plastic	Strong and durable; comes in many forms, including sheets and extruded shapes; several common types of sheet plastic (acrylic, polycarbonate) readily available at hardware and home improvement stores; other types can be purchased via mail order	Melts or sags at higher temperatures; some types of plastic (e.g., acrylic) can crack or splinter with impact; PVC and many other plastics are not dimensionally stable under stress so they can bend out of shape; can be more expensive than wood
Metal	Very strong; aluminum available in a variety of convenient shapes (sheet, extruded shapes); dimensionally stable even at higher loads and heats	Heaviest of all materials; requires power tools and sharp saws/bits for proper construction; harder to work with (requires more skill); can be expensive
Lightweight	Lower weight and very easy to cut and drill using ordinary tools; great for small robots and testing new ideas; inexpensive; come in many thicknesses	Not as strong as other materials; composites made with paper or wood can be damaged by moisture; some types may not be harder to find except at specialty stores or online

Basic Tools for Constructing Robots

Construction tools are the things you use to fashion the frame and other mechanical parts of your robo-buddy. These include such mundane things as a screwdriver, a saw, and a drill.

Take a long look at the tools in your garage or workshop. You probably already have everything you need to build your robot. Here's a list of the most important tools used for building your bots. There are other tools for constructing robot electronics. These tools are detailed in their own chapter; see Chapter 21, "Robot Electronics—the Basics."

TAPE MEASURE

You need a way to measure things as you build your robot. A retractable 6- or 10-foot tape measure is most convenient. Nothing fancy; you can get one for a few dollars at a discount store. Graduations in both inches and metric is helpful, but it's not critical.

A paper or fabric tape measure—one yard long, available at yardage stores, often for free—can substitute in a pinch but may not be as accurate. They're handy for measuring in tight places.

SCREWDRIVERS

You need a decent set of screwdrivers, with both flat and Phillips (cross) heads, as shown in Figure 5-3. These come in sizes; get #0 (small) and #1 (medium) Phillips, and small- and medium-tip flat-blade drivers. Magnetic tips are handy, but not necessary. Be sure to purchase a good set. Test the grips for comfort. The plastic of the grip should not dig into your palm. Try soft (rubber) coated grips for extra comfort.

HAMMER

None of the designs in this book call for pounding nails into wood, but you might still use a hammer for tapping parts into alignment or for using a center punch to mark a spot for drilling a hole. A standard-size 16-oz claw hammer is perfect for the job, but a ball-peen hammer also works, as long as it's not too large. You don't want a sledgehammer when a gentle knock is all you need.

PLIERS

Pliers hold parts while you work with them. A pair each of standard and needle-nose pliers is enough for 94.5 percent of all jobs. Don't use either as a wrench for tightening nuts; they'll slip and round off the corner of the nut, making it harder to remove later on. Instead, use a nut driver, detailed below.

A set of "lineman's" pliers can be used for the big jobs, and they provide a sharp cutter for clipping thicker wire.

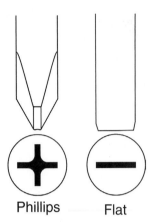

Phillips   Flat

Figure 5-3 The basic assembly tool for making a robot is the screwdriver. There are two common types, Phillips and flat.

HACKSAW

The hacksaw is the mainstay of robot building. Look for a model that allows quick blade changes. Common blade sizes are 10″ and 12″ in length. The smaller blade length is recommended when working with metal. Purchase an assortment of carbide-tipped blades in 18 and 24 teeth per inch (referred to as *tpi*, or *pitch*).

ELECTRIC DRILL AND BITS

You use a drill to make holes; an electric drill makes the whole hole process easier. Pick an electric motorized drill with a 1/4″ or 3/8″ chuck—the chuck is the part where the drill bit is inserted. Chuck size determines maximum diameter for the shank of the bit. The vast majority of work on small robots will require bits of 1/4″ or smaller.

 Spring for an adjustable-speed, reversible drill. The slight added price is well worth it. Adjusting the speed is important when working with different kinds of materials, as some (like metal) need a slower tool.

A drill is what turns a bit; the bit is what actually makes the holes. Drill bits come in different sizes, measured either in fractional or metric. In the United States and other locations where people still use inches, drill bits are measured in fractional sizes. The typical fractional drill bit set contains 29 bits (give or take), in sizes from 1/16″ to 1/2″, in 64ths-of-an-inch steps. For most robotic creations, you'll use only a third of these, but it's nice to have the full set in case you ever need the others.

Drill bits under 1/4″ are also identified by their numerical size. For example, a #36 bit is the same as 7/64″. See the RBB Support Site (refer to the Appendix) for a handy chart comparing number drill bits with their fractional counterparts.

The least expensive drill bits for robot building are made of *high-carbon steel*. Better bits are made from *high-speed steel, tungsten carbide,* or *cobalt*. These are more expensive but they stay sharp longer.

Drill bits can have different kinds of coatings, which extend their life. *Black oxide coating* is the least expensive, and is useful for wood, soft plastic, and thin aluminum. Various *titanium* coatings greatly extend the life of your bits; the coating is useful as long as you don't resharpen the bit.

Save money! Get a standard fractional drill set in standard high-speed steel, then augment that set with specific sizes of more-expensive longer-lasting bits. The most commonly used bit in my shop is 1/8″, so I get extras of those made of cobalt or coated with titanium.

HOBBY KNIFE

Hobby knives include the X-Acto brand, sporting interchangeable blades. They're ideal for cutting cardboard, foamboard, and thin plastics. A word of caution: The blades in these knives are *extremely* sharp. Use with care.

NUT DRIVER

Nut drivers look like screwdrivers, but they're made to tighten hex-head (six-sided) nuts. Drivers are easier to use than wrenches. They come in metric and imperial (inch) sizes. On the imperial front, common driver

sizes are:

Driver	For Hex Nut
1/4″	#4
5/16″	#6
11/32″	#8
3/8″	#10
7/16″	1/4″
1/2″	5/16″

OPTIONAL BUILDING TOOLS

There are a couple more tools that are nice to have but aren't absolutely critical to build small robots. Add them as your budget allows.

- *Miter boxes* help you make straight and angled cuts into tubing, bars, and other "lengthwise" material. Attach the miter box to your worktable.
- A *vise* holds parts while you drill, cut, or otherwise torment them. What size vise to get? One that's large enough for a 2″ block of wood, metal, or plastic is about right.
- A *drill press* sits on your workbench or table. It helps you make smoother, more accurate holes. Lower the bit and drill the hole by turning a crank.

Hardware Supplies

A robot is about 70 percent hardware and 30 percent electronic and electromechanical components. Most of your trips for robot parts will be to the local hardware store. The following sections describe some common items you'll want to have around your shop.

SCREWS AND NUTS

Screws and nuts are common fasteners used to keep things together. Screws (called bolts when they're bigger) and nuts come in various sizes, either in metric or in imperial (inch) units. For this book I stick with imperial sizes, because that's still what we use in the United States, and it's what I'm used to.

Anyway, here are the very basics of what you should have to build your robots. You can read more about fasteners in Chapter 11, "Putting Things Together."

- Use 4-40 size screws and nuts for the typical tabletop robot. The "4" means it's a #4 fastener; the "40" means there are 40 threads per inch. Screws come in various lengths, with 3/8″, 1/2″, and 3/4″ being the most useful for small robotics. I use 4-40 × 3/8″ and 4-40 × 1/2″ screws the most.
- For bigger parts and bigger robots, use 6-32, 8-32, and 10-24 screws and nuts. The most commonly used screw lengths are 1/2″, 3/4″, 1″, and 1-1/2″. These and other sizes are available at any hardware or home improvement store, and you can pick up what you need when you need it.
- For very heavy duty work, you want 1/4″-20 or 5/16″ hardware (1/4″-20 means the screw is 1/4″ in diameter and has 20 threads per inch, the standard for this size). Get these only when plans call for them.

Locking nuts are a special kind of nut you'll use a lot in robotics. They have a piece of nylon plastic built into them that provides a locking bite when they are threaded onto a screw.

WASHERS

Washers are used with screws and nuts and help to spread out the pressure of the fastener. They're also used when you want to make sure the nut and screw don't come apart. The most common is the *flat washer,*

which is typically used to keep the head of the screw (or the nut) from pulling through the material. Tooth and split *lock washers* apply pressure against a nut to keep it from coming loose.

ANGLE BRACKETS

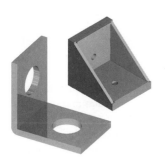

Also ideal for robot construction is an assortment of small steel or plastic brackets. These are used to join two parts together. They come in different sizes, the smaller sizes being perfect for use on desktop robots—smaller brackets weigh less. I often use 1-1/2" × 3/8" flat corner steel brackets when joining two pieces cut at 45° angles to make a frame.

Mechanical Construction Techniques

More than likely, at one time or another, you've used tools to build or fix something. But even with that experience, you may not know the ins and outs of using tools to work with robot-building materials. If you're a die-hard workshop expert, that's fine; you can skip this chapter and move on to the next. But keep reading if you lack basic construction and tool use know-how, or if you need a refresher course.

REMEMBER SAFETY!

Cutting and drilling tools can be dangerous. Exercise caution.

- Use eye and ear protection when using any hand or power tool.
- Don't wear loose clothing. This goes for long-sleeve shirts, ties, baggy pants, everything. The loose material could get caught in the rotating parts of the saw.
- Keep your tools sharp. Dull drill bits and saw blades can cause you to apply too much force against the tool, which can result in slipping and serious personal injury.

FIRST THINGS FIRST: EYE AND EAR PROTECTION

Always wear eye protection when using any tool, power, or hand. This helps prevent harmful debris from flying into your eyes, possibly causing injury.

Don't forget your ears when using power tools. High-speed tools, especially power saws and air tools, create a high volume of sound. Wear sound-suppressing ear protection—wraparound earmuffs designed for shop use or even basic earplugs that you can get at the local drugstore.

Read more about how to safely enjoy the robot-building hobby in Chapter 3, "The Safe and Sane Guide to Robot Building."

PLAN, SKETCH, MEASURE, MARK

Start with a plan, and take a moment to visualize how your robot will look and the parts that will go into it. Then,

- *Work up a quick budget.* If you're just starting out, you won't have many parts and supplies. Research what you need—you can start at the RBB Support Site—and make a list of their names, part numbers, and prices.
- *Sketch out your plan* on paper or by using a computer vector graphics program. I prefer Inkscape (it's free).

If you need to cut and drill, mark directly on the materials. For wood, cardboard, and foamboard use a #2 soft lead pencil. For everything else use a black Sharpie or similar marker; a fine tip works best.

Use a measuring tape to ensure accuracy. If you're using a tape measure marked in inches (see Figure 5-4), the graduations are every 1/2", 1/4", 1/8", and 1/16". It's easy to miscount the subdivisions and get the measurement wrong, so double-check every measurement. If you're using a metric tape, it'll be marked in simple-to-follow meters, centimeters, and millimeters.

DRILLING HOLES IN THINGS

Except for rapid prototypes (like those in Chapter 6, "Robots from Household Stuff"), many of your robots will need some holes drilled into them so you can mount things like battery holders, motors, and electronics.

Regardless of the material (wood, plastic, metal), the basic concepts of drilling are the same: you put a bit into the drill (hand or power), mark where you want the hole to be, and drill there. Good drilling involves following some simple procedures, covered here.

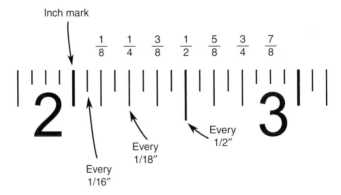

Figure 5-4 Learning to read the graduations of a tape measure is the first step to mastering any construction project. Most tape measures marked in inches have graduations down to 1/16 or 1/32 of an inch.

Selecting the Proper Drill Bit

That 29-piece drill bit set you've been wanting for your birthday offers a fine selection of the most common drill sizes you'll need. But in actual practice you'll probably end up using only a small handful of them regularly and the rest only very occasionally. That's how it is with me.

I use these five drill bits for the vast majority of the holes I drill for my desktop robots. I keep them in a large block of wood for quick access.

Bit Size	For Drilling
@tb:5/64″	Starter or *pilot* holes
1/8″	Holes for 4-40 size screws
9/64″	Holes for 6-32 size screws
3/16″	Holes for 8-32 size screws
1/4″	Odds and ends (e.g., holes for feeding wires through)

- When drilling into metal or hard plastic (acrylic, polycarbonate), first use the 5/64″ bit to make a pilot hole. Then mount the bit for the hole size you want.
- Unless you're making really large holes, when drilling into wood and soft plastic (PVC or ABS) you can go directly to the bit for the hole size you need.
- When drilling holes larger than 1/4″—and especially when working with metal—start with a smaller bit and work your way up. For example, if drilling a 3/8″ hole, begin with a 1/8″ pilot, then switch to 1/4″, and finish with the 3/8″ bit.

Selecting the Proper Drilling Speed

Different materials require different speeds for drilling. High-speed drilling is fine for wood but leads to dull bits when used with metal and cracks when used with plastic.

Material	Drilling Speed
Softwood (pine)	High
Hardwood (birch)	Medium to high
Soft plastic	Medium
Hard plastic	Slow to medium
Metal (aluminum)	Slow
Metal (steel)	Very slow

Most drill motors lack a means to directly measure the speed of the tool, so you just have to guess at what's high, medium, or slow. Go by the sound of the tool. If you're using a portable variable-speed drill, pull the trigger all the way in for full speed. Then let it out and estimate half (medium) and quarter (slow) speeds by listening to the sound of the motor.

If you're using a drill press where you adjust the speed by changing the position of a rubber belt, place the belt into the *High, Medium,* and *Low* positions accordingly. As changing the belt is a hassle, you may want to just set it to *Low,* and drill everything that way. It's better to drill wood using a slow bit than metal using a fast bit.

Proper Use of the Drill Chuck

The *chuck* is the mechanical "jaws" that hold the bit in the drill motor. While some electric drills use an automatic chuck system, most chucks are operated using a chuck key: insert the key into one of the holes in the side of the chuck, and loosen or tighten the jaws. Keep the following in mind when using the drill chuck:

Insert only the smooth shank of the bit into the chuck and none of the flutes. Otherwise, the bit might be damaged.

Be sure the bit is centered in the jaws of the chuck before tightening. If the bit is even slightly off-center, the hole will come out too large and distorted.

Don't overtighten the chuck. Too tight makes it harder to loosen when you want to remove the bit.

Tighten the chuck using at least two key holes. This evens out the torque applied to the chuck and makes it easier to loosen the chuck when you're done.

Controlling Drill Depth

Most of the holes you drill will be completely through the material, but sometimes you need to drill only partway in, then stop. For example, you might want a limited-depth hole in the jaw of an android robot for setting a pin that's used as a hinge.

There are various methods to control the depth of the hole in thicker materials. My favorite method is to wrap masking tape around the shaft of the drill bit (see Figure 5-5). Place the edge of the tape at the depth you want for the hole.

Another type of depth-limited hole is the counterbore, where the drill hole is in different diameters. Counterbore holes can be created by first drilling with a small bit, then using a larger

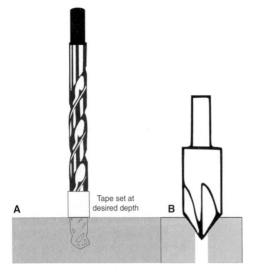

Figure 5-5 (A) Prepare holes of a specific depth by wrapping a piece of masking tape around the drill bit. Position the bottom of the tape at the depth of the hole you want to make. (B) For counterbores drill a "pilot" hole then enlarge with a counterbore bit or reamer.

bit or reamer only partway through the material. The technique is often used with tapered screws, when you want the bottom of the hole to be smaller than the top.

Aligning Holes

Drilling with any handheld tool will naturally produce a certain amount of error. Even the most skilled worker cannot always drill a hole that is at exact right angles to the surface of the material.

When precisely aligned holes are a must, use a drill press or drill-alignment jig. The latter can be found at better hardware and tool stores that stock specialized carpentry accessories, but, frankly, I've never found them all that convenient. The drill press is the much better tool for ensuring alignment. A basic no-frills bench model is under $100. I'm still using the one I bought over 30 years ago!

Using Clamps and Vises

When drilling with power tools, exercise care to hold the part in a clamp or vise. Avoid holding the part just with your hands, especially if it's small. Why? As you drill, the bit may "dog" into the material, causing it to get caught. If you're holding the part, it could get violently yanked out of your hands, causing injury.

Hold-down clamps and vises come in various shapes, sizes, and styles. There is no one type that works for every occasion. Spring-loaded clamps (they look like giant tweezers) are useful for very small parts, while C-clamps are handy for larger chunks of material. A vise is required when drilling small parts of any type with a drill press. For very small and lightweight pieces you can use a pair of large lineman's pliers or a pair of Vise-Grips.

Tips for Drilling

Here are some handy tips for drilling wood, metal, and plastic. All speeds are in RPM.

	Wood	Plastic	Metal
General	Wood is readily drilled using a motorized drill, either handheld or drill press. Speed depends on the size of the bit and the density of the wood. Following are general speed recommendations: • Larger than 1/8": 2000 • 1/8" to 1/16": 4500 • Smaller than 1/16": 6000	For soft plastics (like PVC), speed settings same or slightly slower, as for wood. For harder plastics (acrylic, polycarbonate), reduce drill speed by 50 percent.	Metals should be drilled using a motorized drill. Small parts are more readily drilled using a drill press. Following are general speed recommendations for aluminum and other soft metals. For harder metals, reduce speed by 50 to 70 percent. • Larger than 1/8": 500 • 1/8" to 1/16": 1000 • Smaller than 1/16": 1500

	Wood	Plastic	Metal
Bits	Wood bits should be ground to 118° (pretty much the standard). For cutting all but very dense hardwoods (e.g., oak), regular carbon twist drills are adequate.	Use wood bits for soft plastics. For hard plastics, use a pointed bit designed for acrylic and polycarbonate.	For aluminum and other soft metals, bits should be ground to 118, the standard. For harder metals like steel (and even hard plastic), use 135° bits. For longer life, consider titanium- and cobalt-coated bits.
Cooling	Air cooling is sufficient. If the wood is very hard and thick, pause every 30 seconds to allow the bit to cool down.	Air cooling is sufficient, but if plastic remelts into the hole, slow down the bit, drill smaller pilot holes first, or splash on some drops of water.	Use cutting oil for metal (thin aluminum can usually be drilled without oil). The idea is to avoid excessive heat, which dulls the bit.

CUTTING THINGS TO SIZE

Wood, metal, and plastic can be cut using hand-operated or power tools. For all but the lightest materials, however, you will find that power tools make short work of the job.

In the realm of hand tools, practical choices are:

For wood, a *backsaw*; for metal and plastic, a standard *hacksaw*. The hacksaw uses a replaceable blade, which is required when working with harder materials.

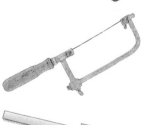

A *coping saw* allows you to cut tight-radius corners in wood, plastic, and softer metals. A coping saw is similar to the hacksaw, except the blades are smaller. Replace the blade when it's dull.

A *razor saw* is used with thin woods and plastics. Its shape is like that of a backsaw, but much smaller. You can find razor saws at the hobby store.

And for power tools, practical choices are:

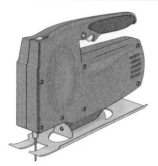

Jigsaw for wood. You need one with adjustable speed unless you work only with soft wood (not really recommended for robotics anyway).

Circular saws and table saws are useful for cutting long, straight cuts in wood and metal. Be sure to use the proper blade, or else damage to the material and/or blade could result.

A *circular miter saw* is useful when cutting aluminum channel and bar stock. The alternative is a hacksaw, a miter block, and sweat. When cutting metal be sure to use clamps to hold the material in the miter!

There are plenty of other saw types not mentioned here, like the scroll saw, which can cut very intricate shapes in wood and metal and some plastics. If you have a preference for one type of tool over another, by all means use it.

Tips for Sawing

Here are some handy tips for sawing wood, metal, and plastic.

	Wood	Plastic	Metal
General	Use medium pitch (teeth per inch) blade; for motorized tools set at high speed.	For motorized tools set speed to no more than 50 percent.	For motorized tools reduce speed to 25 percent.
Blade	Match the blade with the thickness and grain of the material. Circular saw blades are often classified by their application (such as "crosscut"). Use this as a guide.	For circular saw: If possible, use a "nonmelt" or high-quality plywood blade. The wider the kerf the better. For hacksaw or scroll saw: Use an intermediate pitch blade (18 to 24 teeth per inch). A wide-kerf is best.	As a general rule, 3 to 5 teeth should engage the metal. Select the blade according to the thickness of the material you're cutting. Use an abrasive cutoff tool for heavy-gauge ferrous metals.
Cooling	Air cooling is sufficient, though beeswax can be used if the wood is very dense.	Air cooling is usually sufficient. If remelting occurs, direct 50 to 75 psi air from a compressor over cutting area.	Use cutting oil or wax for heavy-gauge metals.

In the preceding table, the term kerf means the width of the cut made by the blade. Many blades use cutting teeth that protrude to either side (often referred to as set). This makes for a wider cut (kerf), but helps keep the blade from binding into the material. When cutting plastic, a wide kerf helps prevent melting.

Limiting Cutting Depth

When using power saws, you can readily limit the depth of the cut by changing the height of the blade within the tool. This is useful when you want to add channels or grooves in the material but not cut all the way through. The following apply to sawing wood or plastic, and when the thickness of the material is at least 1/4"—any thinner and it doesn't much matter how deep the cutting is.

Set the blade to route by just grazing into the material. Use this technique to score very hard material, like polycarbonate plastic; once scored, you can snap the pieces apart using a wood dowel placed under the score mark.

Cut grooves into thicker materials by setting the blade depth to about 1/3 to 1/2 thickness. The width of the groove is the kerf (or set) of the blade— the width of the blade itself, plus the right and left offset of the cutting teeth.

With a full cut, the blade penetrates completely through the material.

Figure 5-6 A tubing cutter is easier to use than a saw when sawing a tube in half. It's for so-called thin-wall tube, not thick pipe. Use a small tool (also shown) for tubing under 1/2"; it's available at many hobby stores.

OTHER WAYS TO CUT MATERIAL

While a saw is the most common means of cutting materials, there are other methods as well. Select the method based on the material you are cutting and the demands of the job.

- Very thin (less than 1/8") hard plastics can be cut by scoring with a sharp utility knife. Place the score over a 1/4" dowel, and apply even pressure on both sides to snap apart the material.
- Thin (to about 20 gauge) metal can be cut with hand or air snips. Pneumatic air snips make the work go much faster. Manual snips are available for straight cuts, left-turning cuts, and right-turning cuts.
- For higher gauges of metal, and for long, straight cuts in thinner gauges, cut using a bench shear (also called a metal brake).
- Thin-wall tubing (aluminum, brass, copper) can be cut using a tubing cutter, like those in Figure 5-6. Tubing cutters are easier to use, and they do a better job than using a saw.
- Foam-based materials that are not laminated (e.g., Styrofoam) can be cut using a hot wire. Hot wire kits are available at most craft stores.

CHAPTER 6

Robots from Household Stuff

Not every robot needs heavy-duty construction. Sometimes, all you're after is a general idea that your design is workable. Rather than use traditional construction with wood, plastic, or metal, you build a fast prototype using common household materials.

This concept is technically known as *rapid prototyping,* and despite its fancy-sounding name, to us it means building robots fast and on the cheap. Construction takes less time, and it's less expensive. While rapid prototyping can also be used to test the merits of a design, it can also be used to build finished robots that don't require sturdy long-life construction.

In this chapter you'll learn about making robots using lightweight materials, existing products such as toys, and "found parts" like plastic storage containers, small trash cans, computer mice, and various other bits and pieces you can find around the house.

Fast Robots from Lightweight Materials

The ideal "fast proto" material is lightweight yet reasonably strong. But you're also looking for stuff that can be cut with a knife, a razor saw, or even a pair of scissors. Candidate materials include cardboard, corrugated plastic, laminated paper, and foamboard.

From a technical perspective, these kinds of materials are often referred to as "substrates," because they're used as an underlayment for things like indoor and outdoor signs, walls for temporary booths at trade shows, and posters for hanging stuff on your wall.

Substrates often (but not always) have multiple layers of complementary materials—each layer contributes to adding strength to the substrate. An example is shown in Figure 6-1. The "sandwich" construction places a top and bottom layer (say, of thin paper or plastic) with a very lightweight core, such as Styrofoam. Individually each layer is quite flimsy, but when combined the material is surprisingly strong and sturdy.

Core
Foam, plastic, paper

Thin overlay material
Plastic, paper, metal, etc.

Figure 6-1 Substrates are constructed of sandwiching layers of materials together, each layer reinforcing the other. For very lightweight substrates, a foam core is layered top and bottom with a thick paper or plastic sheet.

CARDBOARD—PLAIN OR HEAVY-DUTY

Cardboard is the most basic of all rapid prototyping substrates. You can build a robot base out of the cardboard of an ordinary shipping box. Next time you order from Amazon save the box and use it for your next robot. Cardboard gets its strength by laminating two thinner sheets of paper to a corrugated core that has alternating ripples and grooves.

For a more sturdy bot opt for heavy-duty cardboard. It's available in thicknesses from 1/8″ to over 1/2″. You can find it in larger sheets or simply cut up a used heavy-duty shipping box, like the kind used to package heavier objects. To make an even stronger material, laminate several pieces of cardboard to make it thicker and stiffer. "Criss-cross" the corrugation of the inner layers of the cardboard for greater strength. Use a good paper glue or contact cement for a solid bond.

An even heavier-duty cardboard uses a stiff honeycomb-like inner layer. It's more expensive than ordinary cardboard, but when used properly it can hold over 50 pounds. You can often find this type in packing materials for shipping very heavy objects, such as automobile engines. For this, you want a hand or power saw; it's too thick to be safely cut with a knife.

CORRUGATED PLASTIC

Corrugated plastic is a common staple in the sign-making biz. It's used for temporary outdoor signage, restaurant menu boards, that sort of thing. The plastic is composed of several layers, all bonded together during manufacture. To give the material its strength, the inner core is corrugated, like cardboard. Corrugated plastic comes in a variety of thicknesses, with 1/4″ being common. You can cut it with a knife or even a heavy-duty scissors. A quick mock-up or prototype can be roughed out in minutes, and with simple tools.

Corrugated plastic gets its rigidity from its "fanfold" design. It's meant to be used as a backing for temporary outdoor signs, so it's not particularly hardy. If you need a stiffer substrate, you can use several layers of the plastic, sandwiching the layers at 90°. This orientation increases the rigidity of the material.

FOAMBOARD

Foamboard (aka Foam Core, a brand name) is likewise a good candidate for quick prototypes. This material is available at most craft and art supply stores and is constructed out of a foam laminated on both sides with stiff paper. Most foamboard sheets are about 1/4″ thick, but other thicknesses are available, too, from 2mm to half an inch. You can find foamboard in colors at any art or craft supply store. Colored boards are more expensive, but you really only need white.

Cut with a knife or small hobby saw. Make holes with a hand drill. Because the board is laminated with paper, you can use any of a number of paper glues to try out different designs. For safety, cut the board on a piece of cardboard as a backing; don't cut over carpet or a hard floor. Use a metal straightedge as a cutting guide.

Cutting and Drilling Substrate Sheets

The idea with all of these materials is that they are easy to cut and drill. In most cases, ordinary hand tools are all you need. Holes can be drilled with a hand drill, making these materials better suited for robot projects involving young learners. (Give younger children pieces already cut to size, to avoid having them handle a sharp knife.)

When cutting cardboard, foamboard, or corrugated plastic, bear in mind that small pieces are inherently stiffer than larger ones. A 2- × 4-foot sheet of corrugated plastic looks awfully flimsy when you hold it, but cut down to the sizes you'll most often use—4" to 8" round or square—and suddenly the stuff is remarkably more rigid than you thought.

If, after cutting to size, you think the material is too thin to do its job, consider doubling up the thickness or gluing on reinforcement strips out of thin wood or metal from the hobby store.

CUTTING WITH A KNIFE

When cutting substrate with a knife, always use a sharp blade. Dull tools make you press down too hard, which ruins the cut and increases the possibility that the knife will slip and cause you injury.

Use a metal straightedge that you can hold down with one hand. I made mine out of 2"-wide aluminum I purchased at a hardware store for $5. Put a plastic handle in the middle. (Be sure to countersink the holes on the bottom, and use flat-headed screws.) You can also glue on a block of wood or plastic as the handle. You don't need anything fancy. Put masking or drafting tape on the underside of the straightedge to keep it from slipping.

To cut:

1. Use a #2 pencil to draw the line you wish to cut, even if you're using a straightedge. The line helps you know you're cutting in the right spot.
2. Place the substrate material on a flat surface, covered with some kind of "sacrificial" backing board—a piece of discarded paperboard or cardboard will do. You'll cut into this back instead of into the table.
3. Position the straightedge just to the side of the line, and hold down firmly.
4. Make an initial score (shallow) cut with the knife. Press the knife down just enough to break through the surface of the substrate. See Figure 6-2.
5. Repeat again for a thorough cut with the hobby knife. For deep cuts (1/8" or more), draw the knife over the substrate several times, being sure to retrace the same route each time.

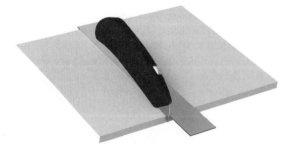

Figure 6-2 Use a metal straightedge as a cutting guide when using a utility knife.

USING A MAT CUTTER

A mat cutter is designed for cutting out picture frame mats. It also works well with softer substrates like foamboard. Most mat cutters are designed with a built-in straightedge. The cutting tool is enclosed in a heavy-duty handle, which slides along the length of the straightedge.

The advantages of mat cutters are that they're generally safer to use than hobby knives (though they can still cause injury if misused) and they're more accurate. Select a cutter that allows for depth adjustment.

Not all mat cutters cut straight lines. Some are designed for curves and even circles. A circle cutter lets you make small, round bases—up to about 20″ diameter in some. Adjust the cutter to a smaller size and you can even pop out your own wheels!

Putting Things Together with Hot Melt Glue

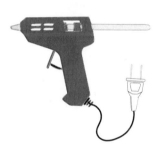

Everyone knows the benefit of glue. Glop on some sticky stuff, squeeze the pieces together, and sooner or later they're stuck like, well, glue! While liquid glue has its place in rapid prototyping, an even better alternative is hot melt glue. Rather than wait minutes or even hours for liquid glue to set, hot glue forms a strong bond in seconds as the material cools to room temperature.

You'll read more about hot melt and other glues in Chapter 11, "Putting Things Together," but it's worth a special note here. Simply apply a bead of glue to one part like that in Figure 6-3, and before the glue has a chance to cool, press the part into place. While hot melt glue offers many advantages, be aware of some of its shortcomings, too:

Figure 6-3 Hot melt glue allows you to assemble your robot quickly and easily.

- Hot glue forms a near-permanent bond when used with paper substrates, such as foamboard. If you need to later remove a part to reposition it, pulling it off the substrate will likely destroy the substrate.
- The strongest bonds are made using high-temperature glue sticks and glue guns. These guns melt the glue plastic to about 400°F, enough to cause serious burns. Don't touch the glue until it has cooled!
- For small robots and robots made by kids, opt for the low-temperature glue sticks and gun. These heat the glue to about 250°F, still hot but not quite as injurious should the glue come in contact with tender skin.
- Avoid applying too much glue to the bonded surfaces. That can actually make the glued joint less strong. If glue oozes out the sides when you press the pieces together, you've used too much.

Fast Construction with Semipermanent Fasteners

Permanence is the degree that something stays the same as when it was first created. Robots that use rapid building techniques are generally considered to be semipermanent; it's not meant to last years and years. Tape, ties, hold-downs, cable clamps, and hook-and-loop are used to produce semipermanent bots.

What follows applies to consumer-grade tape, Velcro, and other products. If you can get your hands on the industrial-grade stuff, it'll work the same but will give you greater holding power. Naturally, it's harder to find, more expensive, and often available only in bulk. Try specialty industrial suppliers, some of which are listed in the resource sections of the RBB Support Site.

Don't forget you can use regular machine screw and nut fasteners with rapid prototypes. Assembly can still be fast, especially when you use a motorized screwdriver. A word of advice: When using cardboard, foamboard, and other soft substrates, add flat washers on both sides of the fasteners. This helps prevent the screw and nut from popping through the material.

HOOK-AND-LOOP FASTENERS

Velcro was discovered when its inventor noticed how burrs from weeds stuck to the fur of his dog. The construction of Velcro is a two-part fabric: one part is stiff (the burrs) and the other soft (the dog). Attach them together and they stick. The term "Velcro" is a combination of the French words "velour" and "crochet."

Velcro is a trade name for a kind of *hook-and-loop fastener*. It or a generic equivalent is available in a variety of sizes and types, from ordinary household Velcro you already know about to heavy-duty industrial strips that can support over 100 pounds. Figure 6-4 shows some Velcro in action on a robot. It's being used to mount a wheel caster.

Figure 6-4 Temporarily attach parts to your robots using hook-and-loop (Velcro) fabric. Get the heavy-duty kind for holding larger components.

Among the most useful hook-and-loop products is the continuous strip, which you can cut to the desired length. The strip comes in packages of 1 foot to several yards, in several common widths, with 1/2″ and 1″ wide being popular. The strips come with a peel-off adhesive backing.

While Velcro may be the best-known hook-and-loop material, it's not the only kind. A great alternative—sold in the tool department of many department stores—is 3M Dual Lock, a unique all-plastic strip that is composed of tiny plastic tendrils. Dual Lock has no separate "hook" and "loop" components. It sticks to itself.

STICKY TAPE

Sticky tape is a broad family of products that have an adhesive on one or both sides of the tape. Sticky tape is cheap, easy to use, and bonds nearly instantly to the surfaces you apply it to. While sticky tape makes for handy construction material, remember that the tape adhesive is gummy and can leave residue on the parts. Use denatured alcohol to remove the residue. (But test first to ensure that the alcohol doesn't dissolve the parts of your robot you want to keep!)

Most sticky tapes are not dimensionally stable; that is, under stress and load—like a drive motor—parts may shift under the adhesive. This can cause a "creep" that will result in parts becoming misaligned over time. Use sticky tape for noncritical applications only, or when you provide another way (such as mechanical stops) to keep things aligned.

DOUBLE-SIDED FOAM TAPE

A common staple in any robot builder's workshop is a roll of *double-sided foam tape*. This is like the aforementioned sticky tape, but thicker. This tape is composed of a layer of springy foam, usually either 1/32″ or 1/6″ thick, and from 1/4″ to over 1″ wide. The tape is coated with an aggressive adhesive on both sides. To use, peel off the protective paper and apply the tape between the parts to be joined. The adhesive is pressure-sensitive and cures to a strong bond within 24 hours.

Many consumer-grade foam tapes are engineered with an adhesive that never fully cures. It stays gummy so that the tape can be more readily removed from walls. Maybe this is what you want, or maybe not. For a more permanent bond, look for industrial-grade double-sided foam tape; it's available at better hardware stores, as well as industrial supply mail-order outlets. One such product is 3M VHB self-adhesive tape.

PLASTIC TIES

Intended to hold bundles of wire and other loose items, *plastic ties* can also be used to hold things to your robot. The tie is composed of a ratcheted strip and a locking mechanism. Loop the strip into the mechanism, and pull the strip through. The locking mechanism is one-way: you can tighten the strip, but you can't loosen it (this applies to most plastic ties; some have a releasable lock).

Plastic ties are made of nylon and are very strong and durable. They're available in a variety of lengths, starting at 100mm (a little under 4″) to well over 12″. Anchor the tie into a hole you've drilled in your robot, or use one of several mounts specifically designed for use with plastic ties (see Figure 6-5). I prefer mounts designed for use with hardware fasteners; they provide a stronger hold.

Figure 6-5 Plastic cable ties and mounts can be used to secure pieces to your robot. Use several ties for larger components. The mounts may be stuck onto the robot with self-adhesive tape or mounted using small fasteners.

CABLE CLAMPS

Motors need to be fastened to the base of the robot in such a way that they won't easily come off or go crooked. Motor shafts akimbo result in misaligned wheels, which make your bot harder to control and steer.

When using round motors (the most common kind), look for suitably sized plastic cable clamps, available at hardware stores and online at computer accessory outlets. These clamps can accommodate cable thicknesses from 1/4″ to over 1″, and are secured to a surface using screws. Use one or two clamps as a motor mount; if the motor is a bit too small for the clamp, wrap electrical tape around it to thicken things up a bit.

When using just a single clamp per motor, you'll need a way to keep the clamp from pivoting at its fastener hole. You can try tightening the fastener as far as it'll go, but a better method is to put "stops" in front of and behind the clamp. The stops—which can be something as simple as a screw head sticking out of the robot's base—prevent the clamp from moving. For larger motors you can use two clamps, with the mounting holes on either side of the motor.

Constructing High-Tech Robots from Toys

Ready-made toys can be used as the basis for surprisingly complex homebrew hobby robots. Snap or screw-together kits, such as the venerable Erector or Meccano set, let you use premachined parts for your own creations.

Some kits, like LEGO and K'NEX, are even designed to create futuristic motorized robots and vehicles. You can use the parts in the kits as is or cannibalize them. Because the parts already come in the shape you need, the construction of your own robots is greatly simplified.

ERECTOR SETS

(Note: Everything in this section also applies to the Meccano brand and similar metal beams-and-girders construction sets sold by others.)

Many Erector sets come with wheels, construction beams, and other assorted parts that you can use to construct a robot base. Motors are typically not included in these kits, but you can readily supply your own.

I've found the general-purpose sets to be the best bets. Among the useful components of the kits are prepunched metal girders, plastic and metal plates, tires, wheels, shafts, and plastic mounting panels. You can use any as you see fit, assembling your robots with the hardware supplied with the kit or with your own 4-40 or 6-32 machine screws and nuts.

You can use the Erector pieces directly by fastening them together with the included screws and nuts, or as malleable metal cut, drilled, and bent into custom shapes. In Figure 6-6 you can see the skeleton of an animatronic robot bird—affectionately called the Ardweenie Bird—constructed using standard Erector parts, small motors to give moment to the head and wings, and covered with a white cockatoo hand puppet.

Find more pictures and discussion about the *Ardweenie Bird* on the RBB Support Site (see Appendix). The robot works with the popular Arduino microcontroller to process movement and sounds.

The prepunched metal girders also make excellent construction components, such as motor mounts. They are lightweight enough that they can be bent, using heavier pliers or a vise, into a U-shaped motor holder. Bend the girder at the ends to create tabs for the machine screws, or use the angle stock provided in an Erector construction set.

Many Erector sets of cars, trucks, tractors, and other vehicles have wheels but no motors. Don't be afraid to cannibalize these and combine with motors from other toys.

CANNIBALIZED TOY VEHICLES

Many cheap car and truck toys share one motor with two wheels—the contraption goes forward but doesn't turn. By using two toys, you can create a robot base that steers. To convert these motor drives for use on a robot platform, remove one wheel from each side of the motor, as shown in Figure 6-7.

- For the right motor, pull off the left wheel.
- For the left motor, pull off the right wheel.

You now have two independent motor drives. Use girders or other parts from the Erector set to mount the motors to the base. Figure 6-8 shows a "4WD" car used as a motor base. Start by removing the car body, then pull off the wheels. Add standoffs (or hot glue) to mount the car chassis to a platform.

Figure 6-6   The elusive *Ardweenie bird,* constructed using parts from a $20 Erector set.

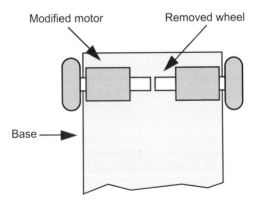

Figure 6-7   Robot bases powered by two motors and wheels on either side may be constructed using a pair of inexpensive motorized toys. For each toy, remove the opposite wheel(s) from the motor mechanism.

Figure 6-8 Simple steps in "borrowing" motors from a small toy to make an inexpensive robot base.

FISCHERTECHNIK

The Fischertechnik kits, made in Germany and imported into North America by Amazon and a few educational companies, are the Rolls-Royces of construction toys. Actually, "toy" isn't the proper term because the Fischertechnik kits are not just designed for use by small children. In fact, many of the kits are meant for high school and college industrial engineering students, and they offer a snap-together approach to making working electromagnetic, hydraulic, pneumatic, static, and robotic mechanisms.

All the Fischertechnik parts are interchangeable and attach to a common plastic baseplate. You can extend the lengths of the baseplate to just about any size you want, and the baseplate can serve as the foundation for your robot. You can use the motors supplied with the kits or use your own motors with the parts provided.

K'NEX

K'NEX uses unusual half-round plastic spokes and connector rods to build everything from bridges to Ferris wheels to robots. You can build a robot with just K'NEX parts or use the parts in a larger, mixed-component robot. For example, the base of a walking robot may be made from a thin sheet of aluminum, but the legs might be constructed from various K'NEX pieces.

A number of K'NEX kits are available, from simple starter sets to rather massive special-purpose collections (many of which are designed to build robots, dinosaurs, or robot-dinosaurs). Several of the kits come with small gear motors so you can motorize your creation. The motors are also available separately.

CONSTRUCTION WITH SNAP-TOGETHER COMPONENTS

I'm far from the purist. I don't mind reusing things like LEGO bricks, MEGA Bloks, and K'NEX (see above) in my robots. "Parts is parts," as they say. Unless you're going after an unusual design, there is no cutting or drilling involved—just pick the piece you want to use, and snap it into place.

Making Joints (More or Less) Permanent

Snap-together components are by their nature temporary. They are made to be taken apart and reused. This may be your aim with your latest robot creation. Also bear in mind that temporary constructions can come apart when you don't want them to, especially if the robot is mishandled, takes a fall from the workbench, or bangs into objects or other robots.

Though snap-together parts are most often used in robotic constructions with or without adhesives, it is also perfectly acceptable to use other binding techniques with them, including double-sided foam tape and nylon tie-wraps. By no means are you limited in any way in how you lash the goodies together. As the variations are endless, I'll just leave the discussion at that, and let your creativity come up with interesting alternatives.

If you decide gluing the parts together is the method, pick the glue to match the kind of plastic used for molding the pieces.

Construction Parts	Temporary	Permanent
LEGO, K'NEX	White glue	ABS plastic solvent cement; 2-part epoxy
MEGA Bloks, plastic models (e.g., model cars or airplanes)	White glue, hot melt glue	Plastic model-building (polystyrene) solvent cement; 2-part epoxy
Fischertechnik, most other plastic construction toys	Hot melt glue	ABS-PVC solvent cement; 2-part epoxy

- For a strong but less permanent bond, use only very small amounts of solvent cement or epoxy.
- You can also make nonpermanent bonds by using very small amounts of cyanoacrylate (CA) adhesive. Super Glue is a common brand of CA adhesive. Apply to one side only. Remove the parts by giving them a good twist.
- Low-temperature (as opposed to high-temperature) hot melt glue provides a good middle ground between temporary and permanent constructions. Use sparingly if you wish to disassemble the parts later. The glue can usually be peeled off.
- Flexible adhesives, such as Shoe Goo or any silicone-based RTV adhesive, also make for strong yet temporary bonds.

Depending on the design of the construction toy, you can also use mechanical fasteners to hold things together. Drill holes in your LEGO, MEGA Bloks, or K'NEX parts, for instance, and secure them with miniature 2-56 or 4-40 machine screws and nuts. The plastic is easy to drill through, and the fasteners can be readily removed if you need to take things apart.

Using Snap-Together Parts to Make Modules
You're not limited to using snap-together construction parts just for the body or structure of your robot. You can use extra or discarded LEGOS and similar construction pieces for making customized accessories. The snap-on nature of these parts allows you to easily reuse these accessories for different prototype projects.

For example, you might glue the flat bottom of a standard R/C servo motor to some LEGO blocks. (It's okay to use glue here because you're able to reuse the motor as much as you'd like. Simply pull it off the LEGO plate when you're done.) Use the servo motor to quickly and effortlessly attach a sensor turret to your base. Just add a bracket and ultrasonic or infrared sensor on top of the servo, snap the servo into place on the base, and you're done.

Figure 6-9 You can combine LEGO pieces with homemade plastic parts for such things as making mounts for radio control servo motors.

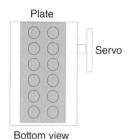

The concept doesn't stop with LEGO beams, plates, and other parts. You can also use plastic or metal construction set pieces from an Erector set. One benefit here is that if you make accessories using these parts, they are easily transferred to full robots once you have successfully prototyped the design. For instance, you can permanently mount a short LEGO Technic beam to a bracket for a servo motor (see Figure 6-9), and use it as a mounting bracket for a cardboard-based prototype.

SPECIALTY TOYS FOR ROBOT HACKING

Some toys and kits are just made for *hardware hacking* (retrofitting, remodeling) into robots. Some are already robots, but you may design them to be controlled manually instead of interfacing to control electronics. Here are just a couple of ways you can use inexpensive toys to make homebrew robots.

Tamiya

Tamiya is a manufacturer of a wide range of radio-controlled models. They also sell a good selection of educational construction parts in kit form. Chief among the offerings are motor gearboxes that you can use for your robot creations. One of the most useful for starter bots is the Twin Motor Gearbox (item number 70097), which consists of two small motors and independent drive trains. You can connect the long output shafts to wheels, legs, or tracks.

One sample robot using this motor is shown in Figure 6-10. I made it in just 15 minutes by taking a 5″ diameter plastic base, and cutting opening for the wheels. The Twin Motor Gearbox secures with just two screws and nuts; the wheels are also sold by Tamiya. The balancing caster on the opposite end of the motor was yanked off a toy car bought at the nearby dollar store.

The Twin Motor Gearbox is also used in several projects in this book. Other Tamiya kits are covered in various chapters, including single motors with selectable gear ratios, wheels, casters, and more.

OWIKIT and MOVITS

The OWIKIT and MOVITS robots are precision-made miniature robots in kit form. A variety of models are available, including manual (switch) and microprocessor-based versions. The robots can be used as is, or they can be modified for use with your own electronic systems.

For example, the OWIKIT Robot Edge Kit (Figure 6-11) is normally operated by pressing switches on a wired control pad. With just a bit of work, you can connect the wires from the arm to a computer interface (with relays or transistors, for example) and operate the arm with your

Figure 6-10 Tamiya's educational construction kits make is easy and affordable to build robots of all kinds of shapes and sizes. This robot cost under $15.

Figure 6-11 The OWIKIT Robot Edge is a robotic arm kit with five degrees of freedom (joints) plus a gripper. (*Photo courtesy of OWI Inc.*)

own software control. Or if you don't feel up to that challenge, you can also connect the arm to a PC using a separately available USB interface and program it with provided software.

Most of the OWIKIT/MOVITS robots come with preassembled circuit boards; you are expected to assemble the mechanical parts. Some of the robots use extremely small parts and require a keen eye and steady hand. The kits are available in three skill levels: beginner, intermediate, and advanced. If you're just starting out, try a beginner-level kit.

MAKING ROBOTS FROM CONVERTED VEHICLES

You've already seen how to cobble together a nice robot base by combining the motor mechanisms from two dollar toys. With many higher-end motorized vehicles you can directly convert them to robot service by simply hacking into their motor connections. Quick and simple!

And let's not forget that you can rob parts from nonmotorized toy vehicles. I've gotten some of the best stuff off of cheapo "dollar store" toys! Push-around toys with rubber tank tracks are an especially nice find. Rob the tracks and put 'em on your own robot base.

Converting Steerable Vehicles

Let's start with inexpensive radio-controlled cars. These have a single drive motor and a separate steering servo or mechanism; the setup doesn't lend itself well to robot conversion. In many cases, the steering mechanism is not separately controlled; you "steer" the car by making it go in reverse. The car drives forward in a straight line but turns in long arcs when reversed. These are impractical for use as a robot base and you should spend your attention elsewhere (they're okay for stripping off parts).

On the other hand, most radio- and wire-controlled tractor (farm, military tank, construction) vehicles are perfectly suited for conversion into a robot. Remove the extra tractor stuff to leave the basic chassis, drive motors, and tracks.

You can keep the remote control system as is or remove the remote receiver (or wires, if it's a wired remote) and replace it with new control circuitry. In the case of a wired remote, you can substitute relays or an electronic circuit for the switches in the remote. Of course, each toy is a little different, so you'll need to adapt this wiring diagram to suit the construction of the vehicle you are using.

The most common electronic circuit used as a substitute for switches is the transistorized H-bridge, discussed more fully in Chapter 14, "Using DC Motors." When replacing manual switches with an H-bridge, be sure the motors don't draw more current than the H-bridge can handle, or damage to the electronics could occur.

Using Just the Parts from Toys

It's called *repurposing*. Toys can be a terrific source of parts that would otherwise cost a lot more if purchased as honest-to-goodness "robot accessories." This is especially true of wheels and tank tracks.

Because of the economies of volume production, a $10 toy may contain four wheels that would otherwise sell for $5 each—a savings of 50 percent. The same is true of rubber, plastic, and even metal tank tracks, which are hard to find in any case. A pair of rugged rubber tracks specifically for robotics could retail for between $30 and $50, yet a toy tank with the same tracks might sell for $20.

Figure 6-12 shows a motorized remote control tank outfitted with rubber tracks, drive sprocket, and "idler wheels" that keep the track in place. It's operated by two motors, one for each track—there's even a third motor on this toy, used to swivel the cannon turret back and forth.

The majority of these toys are imported from China, where stock comes and goes, so you never know what will be available, or for how long. That can be frustrating, but if you're on your toes, you can snatch a bargain when you least expect it.

I bought four of these tanks for experimenting, and within six months the source for them was dry; the item replaced with some other toy vehicle (that unfortunately wasn't as good— that's the way it goes). The moral: Always be on the lookout for motorized toys that could make for good robot platforms. Toy vehicles such as this one are ideal starter bases for your robots. Remove the remote control electronics and attach your own to the motors already in the tank.

Figure 6-12   A store-bought tank toy contains motors, rubber tracks, drive sprockets, and idler wheels—all can be repurposed for a desktop tracked robot.

Building Bots from Found Parts

"Found parts" are things you find around the house—or garage or hardware store or anywhere else—that are just *begging* to be used in your next robot. Or used *as* your next robot! Found parts can help reduce the costs of building a robot. And if the found part can be used as is, without any special cutting, it makes building the robot easier because you don't need as many tools for construction.

There's practically no limit to the number and type of found items you can press into using with your robotics projects—either as the body of the robot itself or as part of a subsystem. Rather than even attempt to cover them all, this chapter explores the concepts of using found parts to stimulate your creativity.

10 IDEAS TO GET YOU STARTED

There are plenty of everyday objects you can use for robot building—all it takes is looking at them a bit differently than the objects' manufacturers intended. Some examples to whet your appetite (all of these have been turned into robots, either by me or by someone I know):

Plastic storage containers: Available in square, round, and other shapes, these durable plastic boxes—available in the housewares section of any department store—can be used with or without their press-on lids. Plastic boxes are available from small snack size to big shoe boxes.

Small "dorm-size" trash cans: Just large enough to hold a Big Gulp, these trash cans have a convenient cylindrical shape and removable top. Great for building miniature R2-D2 bots. The plastic trash cans are easy to drill through and cut, for mounting motors and other parts.

Computer mice: A discarded computer mouse makes a great body for a micro-miniature robot. Almost all mice can be disassembled by removing one or two screws on the bottom. After removing the circuit board, mouse ball, cable, and switches, you can install small motors, a small battery, and a one-chip brain.

Compact discs and DVDs: Save the world's landfill and use these 4.7"-diameter discs for robot bases and other parts. Use care when drilling holes in the plastic: the material can shatter into sharp pieces.

Solderless breadboards: Solderless breadboards are used to experiment with circuits before using more permanent solder and wire-wrap construction. Mount motors and wheels on the underside of your solderless breadboard, and you create a versatile and ever-changeable mobile robot.

Plastic project boxes: These boxes, sold by RadioShack and other electronics stores, are made to hold custom electronics projects. The boxes come with removable metal or plastic lids to allow access to the inside. The plastic is easily drilled for mounting motors and other parts.

Clear or colored display domes: Also called hemisphere or half-round domes, display domes can be purchased in sizes from about 2" to over 12" in diameter. The dome can be used as the body of the robot or as a cover to protect its electronics. A "robotic ball" can be made by gluing two domes together. The wheels of the robot spin the ball, which in turn rolls on the floor.

Metal hardware parts: These include T-braces used for lumber framing in houses. Sizes and shapes vary greatly; take a stroll down the aisles at the hardware store and you're sure to find plenty of candidates. There are lots of sizes to choose from, for making palm-sized robots to large 50- to 75-pound rovers. More about this idea later in the chapter.

Wide-mouth beverage bottle caps: Looking for cheap and easy wheels for your robot? Try the plastic cap of that beverage drink you just finished. Aim for the wide-mouth bottles, the ones with caps measuring 1-1/2 in diameter. These wheels are just about the perfect size for use with modified radio control servos. Mount a round servo horn to the inside of the cap. Hint: Steal the fat rubber band off a broccoli stalk for the tire.

PVC irrigation pipe: All forms of polygonal frames can be constructed using PVC irrigation pipe. Most hardware and plumber supply stores carry PVC pipe in various sizes and wall thicknesses. Select the pipe based on the size and weight of the robot. Obviously, you'll need larger and thicker pipe for the big and heavy robots.

USING WOOD AND PLASTIC SAMPLES

Walk through a well-stocked home improvement store and you're bound to find free or low-cost samples of wood and plastic products that you can reuse in your robots. For example, hardwood flooring samples are about the right size for a small robot. If not free then the cost for samples is quite low, maybe a dollar or so for a piece of wood that measures about 4″ × 8″ (dimensions vary depending on the manufacturer).

Most hardwood flooring is a laminate of a thin veneer over a sheet of high-density board. Thickness: 1/4″. The sample usually includes the tongue-and-groove edges used to assemble the wood to make flooring; you'll want to cut or sand this part off. You'll also want to round off the corners to keep them from chipping.

If the board samples are too small, you can lash several together using Erector set parts or metal framing plates available at your nearby home improvement store. Use short wood screws to hold things together or, for a less permanent construction, heavy-duty double-sided foam tape.

Other small-piece samples (usually available free or for a very small charge) can often be found in the kitchen cabinet department of the home improvement store. Look for 2″ × 3″ or larger samples of countertops made with Formica, resin, or other materials. These may be too small for building a robot base, but they're useful as housings for small sensors, backing material for touch switches, and other routine requirements.

CHAPTER 7

Robots of Wood

If billionaire Howard Hughes could build the world's largest powered airplane out of spruce wood, how hard could it be to construct a small robot out of the stuff? Wood may not be high-tech, but it turns out it's an ideal building material for hobby robots. Wood is available just about everywhere, it's relatively cheap, and it's easy to work with.

In this chapter, you'll look at using wood in robots and how you can apply simple woodworking skills to construct wooden robot bodies and platforms, and then take what you learned and build a simple wood-based robot!

Using Hardwood or Softwood

While there are thousands of types of trees (and, therefore, wood) in the world, only a relatively small handful share the traits that make them ideal for building robots. Wood can be broadly categorized as hardwood or softwood. The difference is what kind of tree the wood is from.

Hardwood is produced from trees that bear and lose leaves (deciduous), and *softwood* trees bear needles (coniferous) or do not undergo seasonal change (nondeciduous). In general, deciduous trees produce harder and denser woods, but this is not always the case. A common hardwood that's very light and soft is balsa, often used in craft projects.

Planks or Ply

Unless you're Abraham Lincoln, building a robot from lumber hewn from the forests of Kentucky, you'll most likely purchase milled wood in either plank or laminate (ply) form. (Milling means it's cut and formed to size.)

Planks are lengths of wood milled from the raw lumber stock. They are available in standard widths and thicknesses and come in either precut lengths (usually 4, 6, and 8 feet) or are sold by the linear foot.

Plywood is made by sandwiching one or more types of wood together. The grain—the direction of growth of the original tree—is alternated at each ply for added strength.

Both plank and ply are made from softwoods and hardwoods. Depending on where you live, your local home improvement store is likely to have only softwood ply and just a few types of hardwood plank (mostly oak). For a wider variety you need to shop at a hardwood specialty store or mail order.

USING PLYWOOD

The best overall wood for robotics use, especially for foundation platforms, is *plywood*. Plywood gets its strength by sandwiching two or more pieces of wood together, where the grain (growth pattern) of the wood alternates direction. Each thin slice strengthens and reinforces the other.

Don't use general-purpose plywood you find at the local lumber yard. A better choice is specialty hardwood plywoods, available at craft and hobby stories. They're more desirable because they are denser for their thickness and less likely to chip.

This type of hardwood plywood comes in two "grades": aircraft and craft. A typical *aircraft plywood* uses birch and consists of from 3 to 24 plies—the more plies, the thicker the wood. It's rated for use in model airplanes, where structural strength is critical. A less expensive variation is *craft plywood,* and isn't quite as strong. For most robotics use, the craft plywood is fine. Hobby stores commonly carry plywoods in 12″ × 12″ squares.

Plywood comes in various thicknesses starting at about 1/8″ and going up to over 1″. Thinner sheets are good for making a small robotics platform, but only if the plywood is made from hardwood. Typical thicknesses are:

	Thickness	
Metric	**Inch**	**Plies**
2.0mm	0.7874	3 or 5
2.5mm	0.9843	5
3.0mm	1.1811	7
4.0mm	1.5748	8
6.0mm	2.3622	12
8.0mm	3.1496	16
12.0mm	4.7244	24

3.0mm = approx. 1/8″

6.0mm = approx. 1/4″

12.0mm = approx. 1/2″

The Woodcutter's Art

You don't need special tools or techniques to cut wood for a robot platform. The basic shop cutting tools will suffice: a handsaw, a backsaw, a coping saw—you name it.

- When cutting plywood use a wood handsaw or (if you have one) table saw fitted with a fine plane.
- Some parts of your robot may use narrow planks (about 1/4″ to 1″ of hardwood. For these use a backsaw (my favorite), a handsaw, or a jigsaw.

Make sure the blade is made for cutting wood. When using a power saw opt for plywood-paneling blade. These have more teeth per inch and produce a smoother cut.

CUTTING A BASE

The easiest shape of all is square, and that's what you'll begin with, even if you plan on something more elaborate. Start with a wooden square or rectangle that's about the size of the finished base. If you have them, you can use power tools to make short work out of cutting the wood to the basic square/rectangle shape.

You can readily turn the square into octagons, hexagons, and pentagons simply by lopping off the corners. Unless you're an expert at the table saw, do these cuts with hand tools; it'll provide extra accuracy. Figures 7-1 and 7-2 show some variations on a theme. These more elaborate shapes don't take that much longer to produce—just a few minutes per cut, and you'll make a better robot.

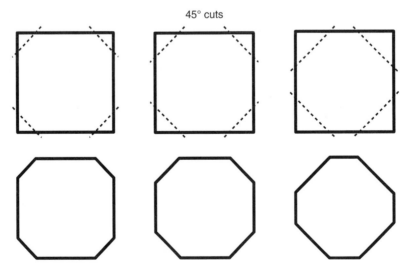

Figure 7-1 Lop off the corners of square wood pieces to streamline the shape of your robot. With all four corners cut, you end up with an octagon. The shape of the octagon depends on how much of each corner you remove.

- To make an octagon (eight-sided) base, cut the corners off at 45°.
- To make a hexagon (six-sided) base, cut the corners off at 60°.
- To make a pentagon (five-sided) base, cut the corners off at 72°.

 If you have a heavy-duty motorized sander, you can lop off the corners by sanding rather than cutting. Start with a coarse sandpaper (see the section later on in this chapter about sanding). As a final step, use a fine sandpaper to make the edges smooth.

Lopping Off Even More Corners

You can approximate near circles by cutting off more corners. Eight-sided octagons with their corners lopped off (chamfered) make 16-sided "circlettes," as shown in Figure 7-3. Chamfered pentagons produce 10-sided shapes; chamfered hexagons make 12 sides.

To make these cuts, mark directly on the wood (you don't have to be precise—to the nearest 1/4″ is usually fine), or use a piece of graph paper lightly glued to the wood. You can use paper paste—the kind you ate in grade school—or a nonpermanent glue stick.

Making Cutout Wells for Wheels

The robot bases you've cut so far don't have special cutouts ("wells") for wheels. Wheel wells are nice to have, because they allow you to place the wheels flush with the contours of the body or without having to raise the level of the base to clear the wheels. Figure 7-4 demonstrates the basic idea.

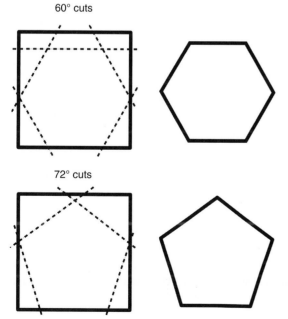

60° cuts

72° cuts

Figure 7-2   By making alternative cuts at the corners you can produce hexagon and pentagon bases. You'll probably want to use a simple protractor (available at any school or office supply store) to measure the angles.

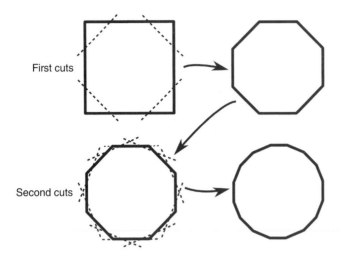

First cuts

Second cuts

Figure 7-3 For near-circular bases you can cut off the four corners of a square base, then trim off the eight corners you just made. It's a little extra effort, but worth it.

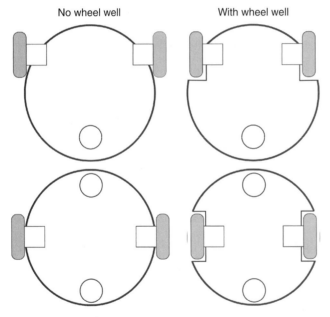

No wheel well

With wheel well

Figure 7-4 Wheel well cutouts allow the robot to have a slimmer profile for any given size of base. The wheels fit into the cutouts, rather than jut out to the sides.

Cutting wheel wells is easier when the wheels are placed at one end of the robot. When in the middle of the robot you need to make multiple cuts to literally "carve out" the well. Well cutting works best when you use either a power jigsaw or a coping saw. The coping saw has a small blade for tight corners.

Figure 7-5 shows a simple three-cut approach. Start with a jigsaw or backsaw and make two cuts perpendicular to the side of the base.

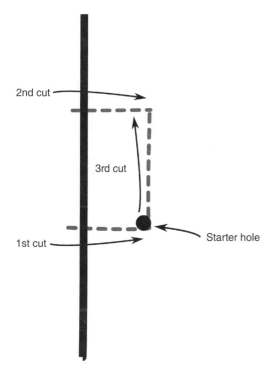

Figure 7-5 The basic wheel well is created by marking a rectangle into the base and cutting as shown. Drill a hole to insert the jigsaw blade.

1. Cut in only as deep as you want to make the wheel—an inch or two is usually enough.
2. Use a 1/4″ bit and drill a hole at one of the inside corners of the well. Position the hole so that it's on the inside of the well; that way you won't have the remnants of the hole after the well has been cut out.
3. If using a jigsaw, insert the blade into the hole and cut toward the opposite end of the well. If using a coping saw, remove one end of the blade from the saw frame and pass the blade through the wood. Reattach the blade to the frame and begin cutting.

Figure 7-6 shows another technique that's ideal when using only hand tools.

1. Use a backsaw to make cuts 1, 2, and 3. A backsaw is preferred because it will make straighter cuts.
2. Use a coping saw for the fourth and final cut.

CUTTING A FRAME

Frame construction allows you to make larger but lighter robots. The frame provides the overall skeletal structure of the bot, and over the frame you can place some light material to support the components

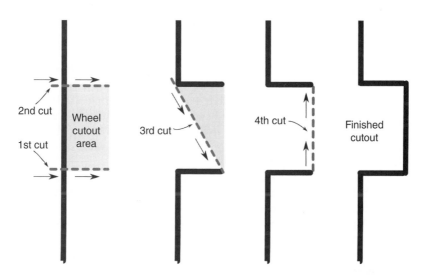

Figure 7-6 A wheel well can be cut using only hand tools, such as a coping saw. Begin with two parallel cuts for the side of the well. The coping saw blade is fine enough to make tight turns.

of your machine. Frames also allow you to build tall robots in the same way they construct multistory buildings. You can stack multiple frames on top of one another, with the equivalent of pillars between them.

Wood frames can be constructed using strips of hardwood plank. Most hobby stores stock birch and mahogany, both excellent choices.

Frames are best made using a small miter box and a backsaw. You can purchase both of these for under $20 total. It's a small but wise investment. The miter box helps you make the straight and angled cuts necessary for frame construction.

Accurate Measuring a Must

The miter box will help you to make proper 45° angle cuts, but you still have to be careful that each of the four pieces of the frame is the exact proper length. Otherwise, the frame will not be a perfect square—or a perfect rectangle, whatever the case may be.

Follow these steps to ensure proper measuring and cutting of your frame.

1. Determine the outside dimension of the frame. For example purposes, we'll assume you're making a 9″ square robot.
2. Let's start with the *top* frame piece. Using the miter box, cut a "left-hand" miter at the tail end of a wood strip. I refer to it as a left-hand miter because the joint will be on the left side of the frame.
3. With the tape measure placed at the top corner that you just cut, measure exactly 9″. Using a pencil, make a mark to the immediate right of the 9″ mark on the tape.
4. Again using the miter box, cut a "right-hand" miter, lining up the top of the cut with the mark. When cutting, be sure the blade is just to the right of the mark. This is called "keeping the line," and it ensures the length of the piece will be exactly 9″, even considering the width of the saw blade (as you will recall from Chapter 5 this is called the *kerf*).
5. Now for the *bottom* frame piece. Pick up the strip that was left over from step 4. It'll already have the proper (right-hand) miter cut. Measure exactly 9″ and mark.
6. Turn the piece over and place in the miter box. Cut a left-hand miter, being sure to leave the line as you did in step 4. I'm having you turn the wood strip over so that the mark at 9″ is along the top. It's easier to see that way.
7. Now, compare the two frame pieces you've cut. If they are slightly different lengths, carefully sand the longer piece down. When sanding be careful not to round off the cut, or else the mitered corners won't meet well.
8. Repeat steps 2 through 7, this time for the left and right pieces. These are also 9″, so the steps are duplicated exactly.

Assembling the Frame

Once the pieces are cut out, they are then assembled to make a frame. The frame pieces are held together using L-shaped angle brackets and steel fasteners.

Figure 7-7 An L angle bracket connects the frame pieces together. Use machine screws, nuts, and washers for a solid construction.

Start by assembling the upper-left corner, using one side piece and one top piece. With a bracket as a guide, mark a hole for drilling at the corner of one of the frame pieces. Do just one "leg" of the bracket at a time; don't try to mark multiple holes. With the wood marked, move the bracket out of the way and drill the hole (see the next section for details about drilling). Assemble the bracket on the frame piece following the example in Figure 7-7. Use a flat washer next to the nut.

Now for the adjoining frame piece. Align it to the corner you just assembled. Using the other "leg" of the bracket, mark a hole in the second piece. As you did before, drill the hole and finish the corner by adding the fasteners.

Put this corner aside, and repeat the steps for the lower-right corner. When done you will have the top left side of the frame as one piece and the bottom right side as the other piece.

Complete the frame the rest of the way by carefully lining up the corners, marking, drilling, and assembling with fasteners.

I don't recommend assembly techniques using nails or staples. The finished frame isn't strong enough for robot use, and the brackets and fasteners let you disassemble the frame should you decide to rebuild it.

DRILLING WOOD

Holes are for mounting things to your robot, and holes are made with the drill and bit. You can use a hand or electric power drill to make holes in wood. Electric drills are great and do the job fast, but you can also use a hand drill if you feel uncomfortable with power tools. Either way, it's important that you use only sharp drill bits. If your bits are dull, replace them or have them sharpened.

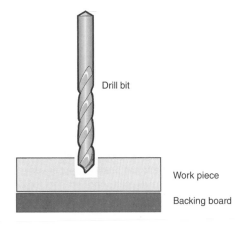

Drill bit

Work piece

Backing board

It's important that you drill straight holes, or your robot may not go together properly. If you have a drill press, and it's large enough, you can use it to drill perfectly straight holes in plywood and other large wood stock.

To prevent splintering, place a piece of scrap wood behind the piece you're drilling. Use a soft wood, like pine, so you can "feel" when the bit has penetrated the wood you're drilling. If you're still getting splintering on the underside of the wood as you drill, try pushing the bit through more slowly. And be sure you're using a sharp bit. Replace if necessary.

FINISHING WOOD

Make better-looking bots with simple wood finishing. Wood finishing involves sanding, which smoothes down the exposed grain, then painting or sealing. Small pieces can be sanded by hand, but larger bases benefit from a power sander.

Shaping, and Sanding

You can shape wood using rasps. A rasp is the same as a file—both are covered with a surface of sharp teeth to grind down the wood—but the teeth of a rasp are much more coarse. You can also shape wood using a heavy-duty drum or disc sander, outfitted with very coarse sandpaper (see the section immediately following, on sandpapers).

Sandpapers are used to smooth wood, removing saw marks, chips, and other imperfections. Sandpapers are available in a variety of *grits*—the lower the grit number, the coarser the paper. With a higher coarseness you remove more wood at a time.

The recommended approach is to start with a coarse grit to remove splinters and other rough spots, then finish off with a moderate- or fine-grit paper. For wood, you can select between aluminum oxide and garnet grits. Aluminum oxide lasts a bit longer. Sandpapers for wood are used dry. For hand sanding, wrap the paper around a wood or plastic block to provide even pressure.

	Grit			
Use	**F**	**M**	**C**	**EC**
Heavy sanding			•	•
Moderate sanding		•		
Finish sanding	•			

Grit Key	**Name**	**Grit**
F	Fine	120 to 150
M	Medium	80 to 100
C	Coarse	50 to 60
EC	Extra coarse	30 to 40

Painting

Wood can be painted with a brush or spray. Brush painting with acrylic paints (available at craft stores) takes longer but produces excellent results with little or no waste. One coat may be sufficient, but two may be necessary. Woods with an open grain may need to be sealed first using a varnish, primer, or sealer, or else the paint will "soak" into the wood. You may also opt to skip the painting step altogether and apply only the sealant.

Spray paints are an alternative to brush painting. Be sure to use the spray can according to the directions on the label. Use only outdoors or in a well-ventilated area.

Hands-On: Build a Motorized Wooden Platform

Now that you've learned how to work with wood you can put that knowledge to good use to build the PlyBot, a simple, affordable, and expandable robot base using just a small sheet of 1/8″ plywood.

The construction of the PlyBot requires no special tools. You need a saw for cutting the wood and a drill for making holes. Plus: a screwdriver and small pliers for assembling the pieces. Parts for the PlyBot are available online and at better-stocked hobby stores.

In the text that follows I provide the exact model numbers for the pieces you need. But remember that you're free to substitute with something else if you already have it or if you've found substitutes that are cheaper or easier to get. See also the RBB Support Site (check out the Appendix) for additional parts sources and alternatives.

MAKING THE BASE

Refer to Table 7-1 for a list of parts.

Figure 7-8 shows the completed PlyBot, with wooden base, twin gear motors, wheels, and support caster. The robot measures 5″ × 7″ overall. The base is constructed from 1/4″-thick

Table 7-1 PlyBot Parts List	
Robot Base:	
1	5″ × 7″ 5-ply 1/4″-thick birch plywood*
2	Tamiya Worm Gear kit, model 72004†
1	Tamiya Narrow Tires (pair), 58mm diameter, model 70145†
1	11/16″ diameter ball transfer (i.e., ball caster), McMaster-Carr, item #5674K57‡
6	Assembly hardware: 4-40 × 1/2″ machine screws, 4-40 nuts

* Birch plywood is available at hobby and craft stores. Look for 1/4″-thick aircraft-grade plywood.

† Tamiya motors and wheels are available at most online hobby stores, such as Tower Hobbies, as well as Amazon and many robotics specialty sites.

‡ You may substitute most any ball caster with a 1″ to 1.5″ flange-to-wheel depth. There is nothing special about this particular caster; feel free to substitute.

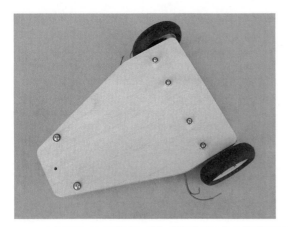

Figure 7-8   The PlyBot uses two Tamiya gear motors and wheels, plus a ball caster (or skid) for balance. It's made using a single piece of quarter-inch-thick 7″ × 5″ aircraft plywood.

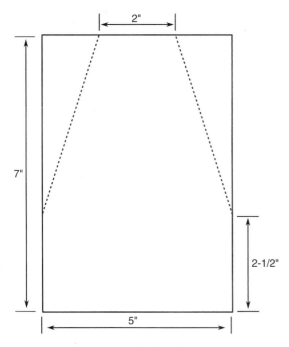

Figure 7-9   Cutting guide for the PlyBot. Dimensions are flexible, and you can scale the robot up in size if you need something a bit bigger.

5-ply birch plywood and is driven by two Tamiya worm gear motors; the motors come in kit form, and assembly takes 10 to 15 minutes for each motor. The wheels, also made by Tamiya, securely lock onto the axles of the motors using hardware that's included.

The "third wheel" of the PlyBot is a ball caster—more technically called a ball transfer, as they are used on conveyor belts to literally transfer goods (boxes, palettes, machinery) throughout a warehouse or factory. I used a ball caster/transfer from McMaster-Carr (*www.mcmaster.com*), a large and well-known online industrial supply retailer. The caster costs under $5, and you can substitute another for it if you like, as long as it's about the same size. Your caster should have a flange-to-ball depth of about 1″ to 1.5″.

Cutting and Drilling

Begin construction by cutting the wood to 5″ × 7″. Then, using Figure 7-9 as a guide, cut the sides of the wood to create a narrowing shape.

Once the base is cut, use sandpaper to round off the four corners and to smooth the surface of the wood.

Consult the drilling template in Figure 7-10. Use a 1/8″ drill bit for all holes. The location of the holes are not supercritical *except* for the distance between the two holes for each of the motors. These need to be fairly accurate.

BUILDING AND ATTACHING THE MOTORS

The PlyBot uses two Tamiya #72004 worm gear motors. They're called worm gear because of the type of gearing mechanism that's used.

While assembling the motor you may select either of two ratios: 216:1 or 336:1. I built the prototype PlyBot using the lower, 216:1, ratio. Whatever ratio you choose, be sure both motors are assembled the same way, or your PlyBot will run around in circles! Assembly instructions are included with the motor. You'll need a small Philips-head screwdriver and pair of needle-nose pliers to build it.

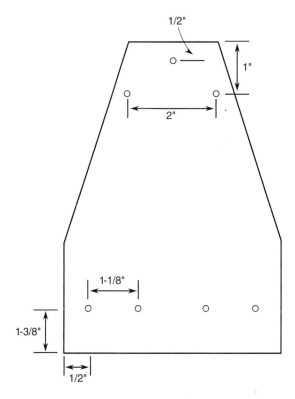

Figure 7-10 Drilling template for the PlyBot. The extra hole at the top is for the optional skid, should you wish to not use a ball caster. Turn the robot over so the clearance between the ground and the base is reduced.

Before assembly of the motors, use the motor flange itself to mark the spacing on a piece of paper. You can then transfer the marks to the wooden base using a nail or small punch.

The motor comes with two long wheel axles. You want the axle with the holes drilled into each end. You can save the other axle for some other project.

The axle is secured to the gearbox gears using a set screw. Initially position the axle so that about 1″ sticks out of the side of the motor (see Figure 7-11). You'll be adjusting the position of the axle after you've mounted the motors, so for now just lightly tighten the set screw.

You'll be building one left motor and one right motor. That means the 1″ of axle should point to the left on one of the motors and to the right on the other.

Important! Prior to inserting the motor into the gearbox, manually rotate the gears so that the set screw on the wheel axle points upward (away from the mounting flanges). This allows you to fine-tune the position of the axle once the motors have been mounted. If you don't do this now, you might not be able to access the set screw.

Attach each motor to the underside of the plywood base, as shown in Figure 7-12. Use two 4-40 × 1/2″ machine screws and nuts. Feed the screws from the motor side, and tighten the nuts on the top of the base. You may use #4 washers (if you choose) on the nut side.

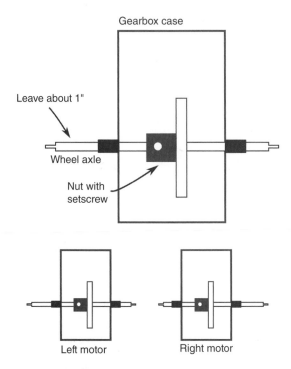

Figure 7-11 Tamiya worm gear motors, showing the alignment of the wheel axle through the output gear. Leave about an inch of the axle showing for mounting the wheel. Make a "left" and a "right" motor, as shown.

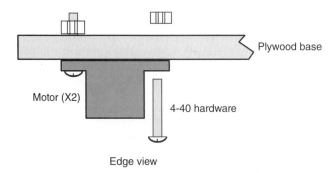

Figure 7-12 Mounting the motors using 4-40 hardware fasteners.

BUILDING AND MOUNTING THE WHEELS

The Tamiya wheels specified in the parts list are designed to directly attach to the axle of the worm gear motor. The wheels are in kit form and come with two different hubs. You want the hubs for the 4mm round axles. These hubs are visually identified with a thin slot that runs through the center. Assemble the wheels according to the instructions that come with the set.

Before mounting the wheels, use a pair of pliers to insert the small spring pin (included with the motor) into the hole at the outside (wheel) end of the axle. The pin engages into the slot in the wheel hub. Without the pin, the wheel will just freely rotate over the axle.

The wheel securely mounts to the axle using the supplied nut. The Narrow Tire wheel set comes with a small plastic wrench for use in tightening this nut over the axle. The wheel set only comes with two nuts, so don't lose these! They can be hard to replace because of their small size.

ATTACHING THE BALL CASTER

The ball caster attaches to the front of the robot using two 4-40 × 1/2″ screws and nuts. The screws should be inserted from the top side of the base; tighten the nuts against the mounting flange of the ball caster. Figure 7-13 shows the 4-40 hardware used with the ball caster, and Figure 7-14 shows the caster attached to the bottom of the PlyBot base.

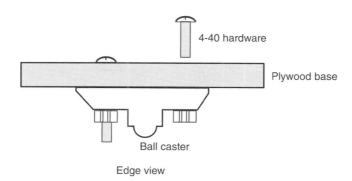

Figure 7-13 Mounting the ball caster using 4-40 hardware fasteners.

Figure 7-14   Ball caster (aka ball transfer) as specified in Table 7-1. You can substitute any other caster mechanism that is of the same approximate size.

USING THE PLYBOT

The Tamiya motors used in the PlyBot are rated for 3 to 6 volts. You can rig them up to switches to manually control the motor (on/off and direction) or use electronic control. These topics are covered in Chapter 14, "Using DC Motors."

The PlyBot has plenty of room on top (and underneath, too) for mounting electronics, batteries, sensors, and other paraphernalia. Because the base is made of wood, it's easy to drill additional holes for mounting components.

CHAPTER 8

Robots of Plastic

In the old days, billiard balls were made from elephant tusks. By the 1850s, the supply of tusk ivory was drying up and its cost had skyrocketed. So in 1863, Phelan & Collender, a major manufacturer of billiard balls, offered a $10,000 prize for anyone who could come up with a suitable substitute for ivory. A New York printer named John Wesley Hyatt was among several folks who took up the challenge.

Hyatt didn't get the $10,000. The material he promoted, a hardened form of nitrocellulose, carried with it too many problems—like occasionally exploding during its manufacture. Hyatt's name won't go down in the billiard parlor hall of fame, but he'll be remembered as the man who helped start the plastics revolution. Since the introduction of celluloid, plastics have taken over our lives and is a great material for building robots.

Read this chapter to learn more about plastic and how to work with it. After you learn about which plastics is best for bots, I'll show you how to construct an easy-to-build "turtle robot"—the PlastoBot—from inexpensive plastic parts.

Main Kinds of Plastics for Bots

Plastics represent a large family of products. They often carry a fancy trade name, like Plexiglas, Lexan, Acrylite, or Sintra. Some plastics are better suited for certain jobs, and only a relatively small number of them are appropriate for robotics.

You can use plastic for the entire robot or just parts of it—mixing and matching materials is not only allowed in robot construction, it's encouraged. Here's a short rundown of the plastics that I find most useful in robotics:

ABS is short for "acrylonitrile butadiene styrene," should you care about this sort of thing. ABS is most often used in sewer and wastewater plumbing systems; it's the large black pipes and fittings you see in the home improvement store. But it can be in any color, and it can come in any shape; LEGO building blocks are made of ABS plastic. You can get ABS plastic in sheets.

Acetal resin is called an "engineering plastic," because it is designed to be used for various engineering applications, such as making prototypes. It's often referred to by one of its brand names, Delrin. Acetal resin comes in sheet and block form.

Acrylic is the mainstay of the decorative plastics industry. It can be easily scratched, but if the scratches aren't too deep they can be rubbed out. Acrylic is somewhat tough to cut because it tends to crack, and it must be drilled carefully. In the United States, acrylic is often referred to by its most popular trade names, Plexiglas and Lucite. In the United Kingdom, it's Perspex.

Polycarbonate plastic looks similar to acrylic but it's more durable and resistant to breakage. It comes in rods, sheets, and tubes. Often sold as a replacement for glass windows, polycarbonate is fairly hard to cut and drill. Common brand names include Lexan, Hyzod, and Tuffak.

Polyethylene ("polythene" in the United Kingdom and elsewhere) is lightweight and translucent and is often used to make flexible tubing. A variation of this plastic, called high-density polyethylene, or HDPE, is used to make very durable kitchen cutting boards. I like this stuff, but it's very difficult (darn near impossible) to glue anything to it.

Polystyrene is a mainstay of the toy industry. Although often labeled "high-impact" plastic, polystyrene is brittle and can be damaged by low heat and sunlight. It's only modestly useful in robotics and is mentioned here only because hobby model stores carry the stuff in sheet form.

PVC is short for polyvinyl chloride, an extremely versatile plastic best known as the material used in freshwater plumbing. Usually processed with white pigment, PVC can be in any color. Great stuff for robots, especially PVC sheets; much more about this stuff later in this chapter.

Best Plastics for Robotics

Let's summarize the best of these plastics and compare how they are typically used in amateur robots. And while we're at it, let's note how workable—ease of cutting, drilling, and gluing—each plastic is.

Plastic	Typical Robotics Uses	Workability	Gluability
ABS	Base material; small parts cut to size and shape	Easy to drill and cut	Excellent
Acrylic	Base material; small parts cut to size and shape—however, avoid any application where repeated impact or stress might cause cracks or breaks	Medium workability; difficult to cut and drill without cracking	Excellent
Polycarbonate	Base material	Difficult to cut and drill, but not as prone to cracking as acrylic	Good
HDPE—high-density polyethylene	Base material; structure for framing	Somewhat easy to drill; moderately easy to cut, but requires power tools	Poor
PVC	Base material; small parts cut to size and shape	Very easy to drill and cut	Excellent

Where to Buy Plastic

Some hardware stores carry plastic, but you'll be sorely frustrated at their selection. The best place to look for plastic—in all its styles, shapes, and chemical compositions—is a plastics specialty store or plastics sign-making shop. Most larger cities have at least one plastics supply store or sign-making shop that's open to the public. Look in the Yellow Pages under *Plastics—Retail*.

Another useful source is a plastics fabricator. There are actually more of these than retail plastic stores. They are in business to build merchandise, display racks, and other plastic items. Although they don't usually advertise to the general public, most will sell to you, either full sheets or remnants (ask nicely and they may give you some of their discards). If the fabricator doesn't sell new material, ask to buy the leftover scrap.

The Benefits of Rigid Expanded PVC

Rigid expanded PVC is the robot maker's dream material, and I use it extensively, more than any other type of plastic. Figure 8-1 shows a prototype robot I made using 1/4"-thick expanded PVC plastic.

Rigid expanded PVC is commonly used for sign making, so it's relatively cheap, lightweight, and available in a rainbow of colors. It's manufactured by mixing a gas with molten plastic, which reduces its density and weight, while also making it easier to work with. Expanded PVC is sometimes referred to as "foam PVC" or "foamed PVC," but it is far more rigid than ordinary foam.

PVC sheet goes by many trade names, such as Sintra, Celtec, Komatex, Trovicel, and Versacel, but it's probably easiest if you just ask for it by its generic *expanded PVC* or *foamed PVC* moniker.

Figure 8-1 With its ease of drilling and cutting, expanded PVC plastic lets you create all sorts of designs for your robots. This one uses two "decks" or levels that are merely rectangles with the corners cut off. The robot is completed using brackets for the four servo motors.

Choices in Sheet Thickness

Plastic sheets are commonly available in any of several fractional inch or metric sizes. Here are some of the more common thicknesses you'll find:

- 3mm, or roughly 1/8"
- 6mm, or roughly 1/4"
- 10mm, or roughly 13/32"

Weight is important in building robots. The following table details the weight of a $12'' \times 12''$ rigid expanded PVC sheet, at various thicknesses, compared to the same size of acrylic plastic. Weights are representative, as some brands are lighter or heavier than others.

Thickness	Weight (lb/sq ft)	
	Expanded PVC	**Acrylic**
.080 (5/64")	.287	.547
.118 (1/8", or 3mm)	.429	.729
.197 (3/16")	.722	1.09
.236 (1/4", or 6mm)	.858	1.46
.393 (3/8")	1.03	2.19
.500 (1/2")	1.30	2.91

How to Cut Plastic

For more about saws and materials cutting in general, be sure to read Chapter 5, "Robot Building 101."

Soft and thin plastics (1/16" or less) may be cut with a sharp utility knife. When cutting, place a sheet of cardboard or art board on the table. This helps keep the knife from cutting into the table, which could ruin the tabletop and dull the knife. Use a carpenter's square or metal rule when you need to cut a straight line. Prolong the blade's life by using the rule against the knife holder, and not by the blade.

CUTTING BY SCORING

Harder plastics like acrylic can be cut in a variety of ways. When cutting acrylic plastic less than 3/16" thick, one way is to use the score method (see Figure 8-2):

1. Use a sharp utility knife and metal carpenter's square to "score" a cutting line. By scoring you're cutting in only part of the way—making a deep scratch. If necessary, use clamps to hold down the square. Most sheet plastic comes with a protective peel-off plastic on both sides. Keep it on when scoring.
2. Carefully repeat the scoring from 5 to 10 times, depending on the thickness of the plastic. The thicker the plastic, the more scoring lines you should make.

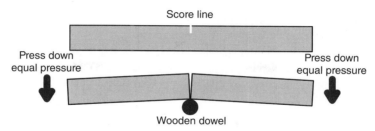

Figure 8-2 Cut acrylic, polycarbonate, or other hard plastic by scoring a line with a sharp knife. Then break at the line over a round wooden dowel.

3. Place a 1/2"- or 1"-diameter dowel under the plastic so the score line is on the top of the dowel. With your fingers or the palms of your hands, carefully push down on both sides of the score line. If the sheet is wide, use a piece of 1-by-2 or 2-by-4 lumber to exert even pressure.

 Cracks are most likely to occur on the edges, so press on the edges first, then work your way toward the center. Don't force the break. If you can't get the plastic to break off cleanly, deepen the score line with the utility knife.

CUTTING BY SAWING

Thicker sheet plastic, as well as extruded tubes, pipes, and bars, must be cut with a saw. If you have a table saw, outfit it with a fine-tooth blade designed for nonferrous metals when cutting acrylic or other hard plastic, a plywood-paneling blade when cutting PVC. When cutting acrylic and other hard plastic, slow down the feed rate—the speed at which the material is sawed in two. Forcing the plastic or using a dull blade heats the plastic, causing it to deform and melt.

When working with a power saw, use fences or pieces of wood held in place by C-clamps to ensure a straight cut.

You can use a handsaw to cut smaller pieces of plastic. A hacksaw with a medium- or fine-tooth blade (24 or 32 teeth per inch) is a good choice. You can also use a coping saw (with a fine-tooth blade) or a razor saw. These are good choices when cutting angles and corners as well as when doing detail work.

You can use a motorized scroll saw to cut plastic, but you must take care to ensure a straight cut. If possible, use a piece of plywood held in place by C-clamps as a guide fence. Be sure the material is held down firmly. Otherwise, the plastic will vibrate against the cutting tool, making for a very rough edge and an uneven cut.

 Slow down the speed of the scroll saw to prevent the plastic from melting at the cut as the blade gets hot. If the plastic melts back into the cut even with the saw at its slowest setting, choose a coarser blade.

How to Drill Plastic

Wood drill bits can be used with expanded PVC and ABS, but for acrylic and polycarbonate plastic the best bet is to use a drill bit that's specially made for the job. What happens with regular bits is that the flutes may suddenly get caught in the plastic, causing the tool to grab the plastic and crack or break it.

Needless to say, not only does this wreck the piece, it's dangerous, as tiny bits of plastic fragments may fly through the air. That's why you should always wear eye protection, even when doing something as simple as drilling a hole.

START WITH A PILOT HOLE

If you can't afford a specialty bit made for drilling in plastic, try starting the hole with a small bit, say 5/64", then gradually increasing the bit size until you get the hole you want. This minimizes the "dogging" that can occur when the flutes of the bit get caught in the plastic. You can use an ordinary high-speed steel bit when drilling into PVC.

RIGHT SPEED WITH POWER TOOLS

When making holes in plastic with a power tool, the drill should have a variable-speed control. Reduce the speed of the drill to about 500 to 1000 RPM. When drilling acrylic and polycarbonate, always back the plastic with a wooden block. Without the block, the plastic is almost guaranteed to crack. As with cutting, don't force the hole and always use sharp bits. Too much friction causes the plastic to melt.

Expanded PVC doesn't usually require backing with wood, but be sure not to force the drill bit through.

 For more about drill tools and drilling in general be sure to read Chapter 5, "Robot Building 101."

Making Plastic Bases

I don't hide that my unabashed favoritism in building robots is for plastic. For its size, plastic is stronger than most woods, and it's easier to work and cheaper than metal. My preferred types of plastics are expanded PVC, ABS, HDPE, and polycarbonate, in that order. Expanded PVC is used in industry as an alternative to wood (example: wood molding), and it cuts and drills much like wood.

FOR STARTERS, SEE WOOD BASE DESIGNS

Many of the techniques that you can use with wood to make handsome robot bases also apply to plastics, especially sheet PVC. Refer to the "Cutting a Base" section in Chapter 7, "Robots of Wood," for details on how to cut out useful shapes for robot bases.

One exception is the use of a motorized jigsaw. Unless you're experienced, the jigsaw may produce too much vibration as it cuts, especially when working with the thinner 1/8" plastic sheets. At best, the vibration will cause rough edges in the cut pieces; at worst, the plastic may crack or even break.

Here are some additional ideas for making bases using plastic sheet.

BASES FROM STRAIGHT CUT PIECES

Chapter 7 detailed the benefit of cutouts for the wheels of your robot—so-called wheel wells. Because plastic is a bit stronger by thickness than wood, it's possible to construct bases with wheel wells without fancy cutting. It can be done just with squares and rectangles. Figure 8-3 shows the idea of making ersatz wheel wells with three separate rectangles of 3mm expanded PVC plastic.

1. Begin by cutting the center piece. This piece is the full length of the robot, but only as wide as the inside edges of the wheel wells.
2. Cut two end pieces. These measure the total width of the base, but their length is from the wheel well to end of the base.
3. Since avoiding square edges is a good idea (when possible), chamfer two corners of each end piece as shown. You can cut the chamfers with a saw or use a rasp or coarse sandpaper. If you use a motorized sander, such as a drum or disc sander, push the plastic in slowly; otherwise, it'll melt.
4. Assemble the three pieces using glue or fasteners. Ordinary household glue will work, or you can use contact cement or solvent cement (as discussed later in this chapter) designed for the plastic you're using.

Using two 3mm sheets of plastic, the main center portion of the assembled base is 6mm (about 1/4") thick. That's usually sufficient for bases under about 10" in size. Note that the corners of the base are only 1/8"; that's okay, as the corners aren't structurally relevant, and you are unlikely to mount heavy components there.

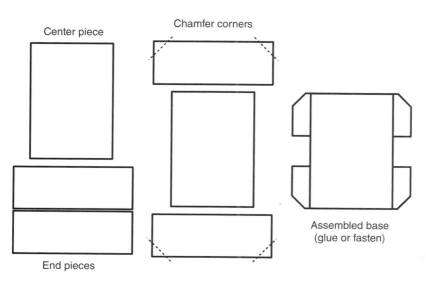

Center piece

Chamfer corners

End pieces

Assembled base
(glue or fasten)

Figure 8-3 If cutting out wheel wells proves difficult with the material or tools you have, you can achieve the same result by assembling individual rectangular pieces together.

Making Plastic Frames

You read about making full-bodied frames out of wood in Chapter 7, "Working with Wood." The majority of plastics available to consumers are rather flexible—this is especially true of expanded PVC. This makes them less than adequate for the job of creating robot frames.

However, given a thick enough material (I'd say 3/8" or larger), the flexing is minimized. You can combine strips of 6mm PVC to make 1/2" framing material. See Chapter 7 on how to use a miter box and backsaw to cut the ends of the frame pieces to 45°. Assemble using the same kind of flat angle brackets and steel fasteners.

How to Bend and Form Plastic

Many kinds of plastics can be formed by applying low localized heat. A sure way to bend sheet plastic like PVC and acrylic is to use a strip heater, available at ready-made plastics supply outlets. Look through your Yellow Pages for local stores that carry plastics and plastic tools.

When you plug in the tool, a narrow electric element applies a regulated amount of heat to the plastic. When the plastic is soft enough, you can bend it into just about any angle you want.

Plastic heaters are easy to use, but getting good results takes lots of practice. Try with some scrap pieces until you get the hang of it. None of the projects in this book require plastic bending and forming, but feel free to try it on your own if you want to experiment. A couple of tips to get you started:

- Be sure that the plastic is pliable before you try to bend it. Otherwise, you might break it or cause excessive stress at the joint (a stressed joint will looked cracked or crazed).
- Bend the plastic *past* the angle that you want. The plastic will "relax" a bit when it cools off, so you must anticipate this. Knowing how much to overbend will come with experience, and the amount will vary depending on the type of plastic and the size of the piece you're working with.
- Use the right heat setting for the plastic. Don't apply too much heat, or you'll be sorry. Expanded PVC has a very low melting point—about 165 to 175°F. This stuff emits a very noxious and corrosive gas (hydrogen chloride) when it burns, so treat it with care!

How to Smooth the Edges of Plastic

After cutting, your plastic parts may need a bit of smoothing to remove any rough edges. As with wood, you can apply a light sanding with a fine-grit aluminum oxide (not garnet) sandpaper. For soft plastic like PVC, use the sandpaper dry. For hard plastic, you can use the paper dry or get it wet with water.

You can also shape the plastic—to remove sharp corners in square bases, for example—by using a very coarse sandpaper.

Recommended Grits: Aluminum Oxide

Use	Very Fine (160 to 200 grit)	Fine (120 to 150 grit)	Coarse (50 to 60 grit)	Extra Coarse (30 to 40 grit)
Shaping			•	•
Smoothing	•	•		

Grit represents the coarseness of the surface of the sandpaper. The higher the number, the finer the grit.

How to Glue Plastic

When building bots, my preference is always to use mechanical fasteners … nuts, screws, that type of thing. The reason: It's not uncommon to want to disassemble a robot, either to reuse parts or to allow someone else to rebuild it, and learn from the experience. When parts are glued, disassembly—and, of course, reassembly—is much harder.

Still, there are plenty of reasons to cement pieces together, and most plastics (especially PVC, ABS, and acrylic) make gluing pretty easy. The kind of glue you use depends on the kind of plastic you're working with and the type of bond you want.

- PVC, ABS, acrylic, and polycarbonate plastics can be bonded using a solvent-based cement. The cement contains chemicals that actually melt the plastic at the joint. The pieces become fused together—that is, they shall be whole, the two made one, like the *Dark Crystal*.
- Household adhesives can be used for gluing plastic together, with varying degrees of success, depending on the glue and the plastic. Experiment.

Table 8-1 lists adhesives I recommend for bonding popular plastics with themselves and with other materials. When all else fails, try epoxy cement.

APPLYING SOLVENT CEMENT

There are different solvent mixtures for the different plastics. The best is to use a solvent-based cement specifically made for the kind of plastic you're gluing—PVC solvent for PVC plastic, and so on.

Table 8-1 Plastic Bonding Guide			
Plastic	Cemented to: ...itself, use	...other plastic, use	...metal or wood, use
ABS	ABS-ABS solvent	Rubber adhesive	Epoxy cement
Acrylic	Acrylic solvent	Epoxy cement	Contact cement
Polystyrene	Model glue	Epoxy	CA glue*
Polystyrene foam	White glue	Contact cement	Contact cement
Polyurethane	Rubber adhesive	Epoxy, contact cement	Contact cement
PVC	PVC-PVC solvent	PVC-ABS (to ABS)	Contact cement

* CA stands for cyanoacrylate ester, sometimes known as "Super Glue," after a popular brand name.

Otherwise you can try an "all-purpose" or "universal" solvent cement, though these may not provide as strong of a joint. Most home improvement stores stock at least one type.

You don't need some fancy-schmancy solvent cement when working with expanded PVC sheets. The same cement made for PVC irrigation pipes can be used with expanded PVC. It's clear and has a good consistency for brush-on application. You can find it at any home improvement store.

When using a solvent-based cement to PVC or ABS plastic, either brush it on the surfaces to the bonded or squirt it into the joint by using a bottle applicator.

You must be sure that the surfaces at the joint of the two pieces are perfectly flat and that there are no voids where the cement may not make ample contact. After applying the cement, wait several minutes for the plastic to re-fuse and the joint to harden. Disturbing the joint before it has time to set will permanently weaken it.

APPLYING HOUSEHOLD ADHESIVE

Solvent cements work well when bonding together similar types of plastics, but they do little or nothing when trying to glue plastic to wood, metal, and other materials.

For these, you can try a household adhesive. As noted in Table 8-1, contact cement works well when bonding plastic pieces to metal or wood. You can get contact cement at the home improvement store. Two-part epoxy (you mix liquid from two tubes) is used for the same things, and is good when you need a very strong bond.

If you are joining pieces whose edges you cannot make flush, apply a thicker type of glue— contact cement, epoxy, and household glue are good contenders. You may find that you can achieve a better bond by first roughing up the joints to be mated. You can use coarse sandpaper or a file for this purpose.

 Some plastics don't like to be glued. You can just about forget bonding nylon and HDPE, unless you use an industrial cement that you can't get anyway because they're expensive and require special applicators. Instead, use mechanical fasteners to hold these pieces together.

USING HOT GLUE WITH PLASTICS

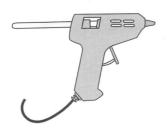 Perhaps the fastest way to glue plastic pieces together is with hot glue. You heat up the glue in a glue gun, and when the glue is all melty, squeeze the trigger to spread it out over the area to be bonded.

Hot melt glue and glue guns are available at most hardware, craft, and hobby stores in several different sizes. The glue is available in a "normal" and a low-temperature form. Low-temperature glue is generally better with most plastics because it avoids the "sagging" or softening of the plastic sometimes caused by the heat of the glue.

How to Paint Plastics

Sheet plastic is available in transparent or opaque colors, and this is the best way to add color to your robot projects. The colors are impregnated in the plastic and can't be scraped or sanded off. But you can also add a coat of paint to the plastic to add color or to make it opaque. Most all plastics accept brush or spray painting.

Spray painting is the preferred method for all jobs that don't require extra-fine detail. Carefully select the paint before you use it, and always apply a small amount to a scrap piece of plastic before painting the entire project. Some paints contain solvents that may soften and ruin the plastic.

Among the best all-around paints for plastics are the model and hobby spray cans made by Tamiya. These are specially formulated for styrene model plastic, but work with many other plastics, too. You can purchase this paint in a variety of colors.

Build a Motorized Plastic Platform

This chapter details construction of the PlastoBot, the base for a small but peppy robot that's constructed using 1/8"-thick plastic—most any plastic will do, so if you already have a piece in your garage feel free to use it. I built the prototype PlastoBot using 3mm (about 1/8") expanded PVC.

PlastoBot is basically a square, and making it involves only straight cuts. The corners of the plastic are chamfered—lopped off at a 45° angle—to enhance the looks and to prevent the base from snagging on things. You can build the robot with or without cutouts for the wheels. The cutouts can be tricky to make, but they allow the robot to retain its sleek design by keeping the wheels inside the profile of the base.

The PlastoBot as described here is 4" square, but the design is scalable. As desired, you can make it larger ... or smaller, if you have teeny-tiny motors and wheels. The practical maximum is about 10" square. For any base larger than 5" or 6" you should double the thickness of the plastic—from 1/8" to 1/4".

 I provide the exact model number for each of the pieces you need, but remember that you're free to substitute parts for others if you already have them or if you've found substitutes that are cheaper or easier to get.

MAKING THE BASE

Refer to Table 8-2 for a list of parts.

Figure 8-4 shows the completed PlastoBot. It measures 4" square and is balanced on one end with a small plastic ball caster. (If you don't want to use a ball caster, you can provide the bot with a static skid, using an 8-32" × 1" machine screw and acorn nut.)

While the plans call for 1/8" plastic, you may use another thickness if that's what you have. For reasons of weight, you'll want to avoid using a dense plastic (acrylic or polycarbonate) if it's over 3/16" or so. The thicker plastic is also harder to cut and drill.

Begin with a 4" by 4" square, and lop off the corners. How much material you remove is up to you, but a 1/2" chamfer is sufficient—to measure this amount, just make a set of marks 1/2" from each corner, and cut a 45° diagonal across the marks.

The PlastoBot design calls for rectangular cutouts for the wheels—what I refer to as *wheel wells*. For the wheels specified (see Table 8-2), the cutouts measure 1-3/8" by 1/2"; both are placed centerline in the base. Refer to Figure 8-5 for a guide. You can see the benefit of the cutouts in Figure 8-6, which shows the differences in the profile of the robot, given wheels on the outside of the base versus wheels within the area of the base.

Table 8-2	PlastoBot Parts List
1	4"- × 4"- × 1/8"-thick plastic[*]
2	Pololu micro gear motors, 100:1 gear ratio, item 992[†]
1	Pololu mini motor brackets (pair), item 989[†]
1	Pololu wheels with rubber tires (pair), 32mm diameter, item 1087[†]
1	Pololu ball caster, 1/2" ball, item 952[†]
Misc.	Assembly hardware for these parts comes with the parts; if you substitute, you may need an assortment of 2-56 or 4-40 × 1/2" machine screws and nuts.

[*] Use expanded PVC, ABS, polycarbonate, acrylic, or most any other plastic. Acrylic and polycarbonate may be found at better-stocked home improvement stores; PVC and ABS plastic is available at specialty plastic outlets and online.

[†] Motors, motor mounts, wheels, and ball caster are available at Pololu.com, and many of its distributors. You may also use most any other miniature motors and wheels if you have another source for them. The motors are approximately 5/8" square by 1" long. The wheels are 32mm (about 1-1/4") in diameter.

Figure 8-4 The finished PlastoBot, using two micro motors, wheels, and a small plastic ball caster for balance. The robot measures 4" square.

ATTACHING THE MOTORS

Pololu

The PlastoBot uses a pair of micro-miniature motors. These are highly precise motors that use metal gears, yet their price isn't much more than most other small motors for robotics.

The motors I selected are sold online through Pololu.com, as are the plastic mounts for attaching the motors to the base. Also provided by them are the wheels and the plastic ball caster described below. The same or similar motors and wheels are available at other robotics specialty stores; see the RBB Support Site for alternatives. (In many cases these other stores resell the Pololu products, and in other cases, they offer competing merchandise.)

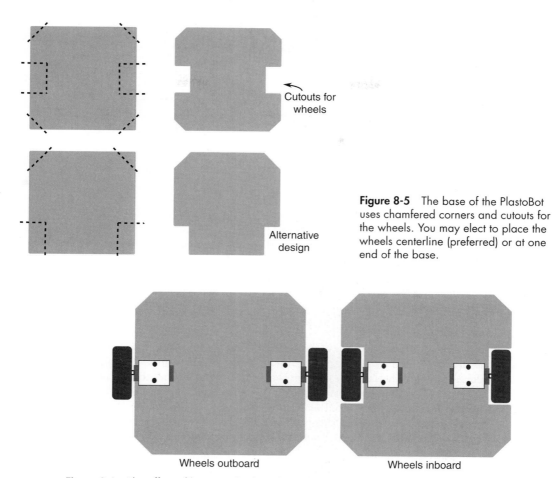

Cutouts for wheels

Alternative design

Figure 8-5 The base of the PlastoBot uses chamfered corners and cutouts for the wheels. You may elect to place the wheels centerline (preferred) or at one end of the base.

Wheels outboard   Wheels inboard

Figure 8-6 The effect of having wheels outboard of the base and inboard. When inboard, the shape of the robot is more streamlined, and it's less likely to snare on objects it encounters.

The micro motors measure 0.94" by 0.39" by 0.47", and have a 3/8"-long 3mm D-shaped shaft that directly couples to the specified wheels. The motors are too small to have a mounting flange to directly attach them to the base, but brackets are available that make the job a cinch.

Follow the drilling guide in Figure 8-7A. The spacing of the holes requires a modest amount of accuracy, so measure twice. The brackets have two holes spaced 18mm (about 11/16") apart, and they come with their own miniature 2-56" steel machine screws and nuts. The nut fits into a shaped recess inside the bracket; the screw comes up through the base to secure the motor in place (see Figure 8-7B). For now just finger-tighten the screws.

Both the motor brackets and the ball caster use millimeter sizes. I've included the spacings in both millimeters and the closest fractional inch, accurate to 1/16".

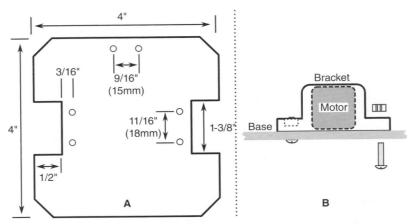

Figure 8-7   Drilling layout and motor mount detail for the PlastoBot. Miniature fasteners come with the motor mounts and are inserted as shown.

FITTING THE WHEELS

PlastoBot's wheels are molded plastic with a removable rubber tire. The wheels are available in a variety of sizes, and I've picked one of the smallest, because of the diminutive stature of the robot. If you scale up the PlastoBot to bigger dimensions you'll probably want to opt for a larger wheel—the 80mm or 90mm wheels can be used for bases three and four times the size of PlastoBot.

To fit the wheels, merely press them into place over the motor shaft. The wheels have a D-shaped hub, matching the flatted shaft of the motors.

Once the wheels are in place, slide the motor into the bracket, adjust the position of each motor so the wheels are the same distance from the base, and tighten the two screws until the motor is snug.

ATTACHING THE BALL CASTER

The support caster uses a 1/2″ plastic ball (see Figure 8-8), which smoothly rotates inside a cavity. Like the motor brackets, the ball caster comes with its own 2-56 fasteners. It also comes with several plastic spacers; use the thicker of the two spacers to increase the height of the caster to better match the diameter of the PlastoBot's wheels. The screws must be inserted so that the head fits into the body of the case. Position the nuts on the top side of the base.

Figure 8-9 shows the underside of the PlastoBot, with caster, motors, and wheels attached.

USING THE PLASTOBOT

The PlastoBot uses high-efficiency micro motors that operate at between 3 and 9 volts, with 6 volts being nominal (normal). You can rig them up to switches to manually control the motor (on/off and direction) or use electronic control. These topics are covered in Chapter 14, "Using DC Motors."

These motors draw low current for the torque (turning power) they provide, allowing you to use most any kind of electronic control. When free-running (no load), the motors consume only

Figure 8-8 Miniature ball caster, with 1/2" plastic ball. The ball rotates against small rollers inside the caster body. (*Photo courtesy Pololu.*)

Figure 8-9 Underside of the PlastoBot, showing motors in their mounts and the ball caster.

40 milliamps; at stall (the motors are physically prevented from turning), current rises to a still-respectable 360 milliamps. The low current demand of the motors lets you pick from a wide variety of motor drive circuits.

See Chapter 13, "How to Move Your Robot," for more on motor torque, milliamps, current measurements, voltage, and other motorific specifications.

If you've built the centerline wheels version of the PlastoBot, be mindful of maintaining a weight balance that slightly favors the end with the ball caster. This prevents the robot from tipping over on the end without the caster.

CHAPTER 9

Robots of Metal

Before the modern dependency on plastic, metal was the mainstay of the construction material world. Toys were not made of plastic, as they are today, but of tin.

Metal is a good material for building robots because it offers extra strength that other materials cannot. In this chapter you'll learn how to construct robots out of readily available metal stock, without resorting to welding or custom machining.

All About Metal for Robots

Metal is routinely broken into two broad categories: ferrous and nonferrous.

- *Ferrous* metals are made from iron (*Fe*, from which "ferrous" is derived, is the symbol for iron in the Periodic Table of the Elements).
- Metals other than iron are *nonferrous*. This includes copper, tin, and aluminum.

When you buy a piece of metal, you're seldom buying the stuff in its pure form. Instead, metal is almost always processed with other metals. The resulting material is called an *alloy*. Different alloys provide different properties for the metal. For example, there are aluminum alloys specifically designed for casting, and others intended for machining parts.

ALUMINUM

Aluminum is the most common metal used in robot construction projects, partly because of cost and partly because it is strong yet lightweight. It's also one of the easier metals to cut and drill, and it requires only a modest assortment of tools. The aluminum you buy at the hardware store is actually an alloy; raw aluminum (which is manufactured from bauxite ore) has little commercial value as a finished metal. Rather, the alumina metal is alloyed with other metals.

Aluminum alloys are identified by number. A common aluminum alloy is 6061, which boasts good machinability (it's not difficult to drill or saw), yet is still lightweight and strong. Common forms of aluminum useful for robotics include thin sheets and extruded shapes, such as squares, tubes, angle Ls, and U-channel. More about shaped aluminum later in this chapter.

STEEL

Iron is used to make *steel*, which is further classified by the amount of carbon added to it during processing. The kind of steel you can routinely find at the local welder's shop or home improvement store has low amounts of carbon and is called *mild steel*.

Stainless steel is a special formulation of steel with small amounts of chromium added in. The chromium develops a microscopic film on the metal that helps it to resist rust and corrosion. Because stainless steel is very difficult to work with, is hard to weld, and is more expensive than most other metals, it's seldom used in amateur robots (exception: combat robots).

COPPER

Copper is available as just copper metal, but it's also often alloyed with other metals to make something else. When combined with zinc, copper makes *brass*; when combined with tin, copper makes *bronze*. All three are soft metals and are relatively easy to cut and drill.

Copper and its alloys can be readily *annealed*, which is the process of heating the metal to high (but not melting point) temperatures, then allowing it to cool very slowly. Annealing is used to change such properties as the softness of the metal. Annealing is commonly used to make metal springs and metallic spring strips.

ZINC AND TIN

Zinc and tin are often employed as an alloy ingredient or as a coating. *Tin*, which resists corrosion and inhibits rust, is also a common material in some crafts and home decorating. Tin is a soft metal, and if the sheet is thin enough, it can be cut with a pair of large scissors. It's easy to make hold-down straps and other parts using tin sheets designed for "punch" crafts.

Measuring the Thickness of Metal

There are many ways to denote the thickness of metal. The most common are:

- *Fractional inch/millimeters*. Metal primarily sold to consumers may be specified using inches or metric units. Thickness is expressed in fractions, such as 3/64″ or 1/16″. When in metric, the thickness is in millimeters.
- *Decimal inch*. There's more precision in measuring thickness as a decimal value, down to the thousandths of an inch. A thickness of .032 means the metal is 0.032″ thick—or "thirty-two thousandths."

- *Mil.* A "mil" is a unit of measure equal to one thousandth of an inch. A decimal thickness of 0.032″ is equivalent to 32 mils. By comparison, plastic trash can liner is usually 1 to 4 mils.
- *Gauge.* This is a pseudo-standard used to specify the thickness of various materials, especially sheet metal and wire. Gauges of different types of materials aren't always the same (e.g., between plastic and metal), so you can't readily compare one to another. Purely as an example, 20-gauge aluminum is 0.032″ thick.
- Complicating matters: When used with metal sheet, gauge varies depending on the metal. That hunk of 20-gauge aluminum is equivalent to something between 21 and 22 gauge for steel, and 13 gauge for zinc.

You need a machinist's micrometer to accurately measure the thickness of metal. Digital micrometers can be switched between decimal inch and other units of measure (typically millimeters), but less expensive models just have a mechanical scale, which can take a bit of getting used to.

What's This About Heat Treatments?

When shopping for metal bits and pieces, you may come across references to how the material was heat treated. Heat treating is used to enhance certain physical properties of the metal. As heat treatments can affect the price of the metal, there's no reason to pay for something you don't need. So you can be an informed shopper, here are the most common treatments in a nutshell:

Hardening strengthens the metal and literally makes it harder. The process also makes the metal more brittle. Tools are commonly made of hardened steel, and the hardening process makes it very difficult to cut or drill the metal.

Annealing softens the metal and makes it more workable—a metal that is *ductile* or *malleable* is easier to work with, usually because it's been annealed. Copper is routinely annealed, but many other metals, like aluminum, can be annealed, too.

Tempering removes some of the hardness and brittleness of steel and in doing so makes it even tougher. An example of a tempered aluminum alloy that you may encounter is 6061-T6, which has several times the strength of the same alloy untempered.

Case hardening is a coating process for soft steels and allows relatively low-carbon steels, such as wrought iron, to be hardened. It's frequently used with steel to make tools.

Hardening and tempering can be accomplished in the home shop, useful for the die-hard combat robot enthusiast. The subject is beyond the scope of this book, but if you're interested in the concept, check your local library for a good tutorial on home metalworking.

Where to Get Metal for Robots

You'll find most metals for robot building at these local sources. If they don't provide the materials you need, try a Web search to locate mail-order suppliers of the metals you want:

Girder-based construction toys like Meccano/Erector are a terrific source for small and lightweight metal parts. Assemble the pieces using the included steel fasteners or hot melt glue. Tip: save money with a multi-model kit that comes with a variety of girders, beams, and other pieces.

Specialty machined and "stamped" parts for robots are metal pieces specifically designed for robotics applications. Machines pieces are precision made on mills and lathes while stamped parts are constructed by huge hydraulic metal presses and punches. A bit more expensive than other metal pieces but time-savers. Examples of useful stamped metal is the servo brackets shown in Figure 9-1.

Hardware and home improvement stores carry some aluminum and steel sheets, but look for angle brackets, rods, and other shapes. See the section "Metal from Your Home Improvement Store," below, for more details.

Hobby and **craft** stores sell aluminum, brass, and copper, in small sheets, rods, tubes, and strips. A common brand sold by stores in North America is K&S Engineering. The metal is sold in small quantity, which makes it more expensive, but more convenient. Read more in the section "Metal from Hobby and Craft Stores."

Metal supply shops that cater to welders are usually open to the public and offer all kinds of useful metal. Many sell stock in large pieces, which you can have cut so you can get it home in your car. Tip: Check out the "remnant" bin for odds-and-ends sizes.

Restaurant supply stores, most of which are open to the public, sell many aluminum and steel materials. Look for spun bowls, cookie and baking sheets, unusually shaped utensils, strainers, and other items that you can adapt to your robot creations. Metal is metal.

Figure 9-1   Stamped metal brackets make it a snap to mount for model radio control servo motors to any robot design.

METAL FROM YOUR HOME IMPROVEMENT STORE

Your local home improvement or hardware store is the best place to begin your search for metal. Depending on the store, here's what you may find:

Extruded Aluminum

Aluminum comes in all kinds of forms, including sheets, bars, rods, and something called *extrusions*. It's the last form that's really—and I mean *really*—useful in robot making. I started using it for my bots back in the 1980s, when most folks were still cutting out sheet metal to form robot bodies. Ugh! Extruded aluminum has now become the number one metal for making homebrew robots.

Extruded aluminum is made by pushing molten metal out of a shaped orifice, as those Play-Doh Fun Factory play sets do. As the metal cools it retains the exact shape of the orifice. Extruded aluminum generally comes in various lengths, from 12 inches to 12 feet. My local stores routinely sell it in 4-foot lengths. It's often used in home improvement jobs to trim the edge of wood or tile and to make channels for shower doors.

There's lots of variety to choose from, allowing you to pick the extrusion to match your project. Common shapes include equal and unequal L-angles, U-channels, and flat bars; see Figure 9-2.

I've found the following extrusions among the most useful in my robot building. Try these for starters, but don't be afraid to select other shapes and sizes for your projects.

- *Channel—equal and unequal sizes.* A popular size is 1/2″ × 1/2″ × 1/16″ channel, which is sold as trim edge for 3/8″ plywood.
- *Angle—equal and unequal.* The 1/2″ × 1/2″ × 1/16″ (equal) angle stock is especially useful for use in robots up to about 15″ in size.
- *Bar and rod.* Example: 3/4″-wide × 1/16″-thick flat bar.

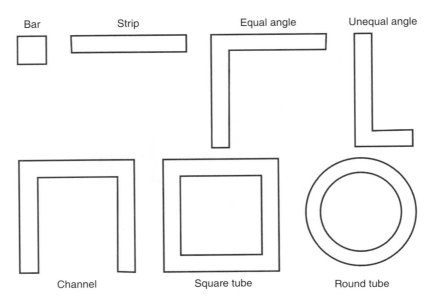

Figure 9-2 Extruded aluminum is available in a variety of handy shapes for building robots. Strips, angles, and channels are among the most useful.

Nail or Tie Plates

Nail plates (also called tie plates) are designed to strengthen the joint of two or more pieces of lumber that are nailed together. The plates are made of steel but are *galvanized* to resist rust. The galvanizing gives the metal a noticeable mottled look.

Check for this stuff in the lumber area of your home improvement store. Much of what they'll offer is preformed shaped for specific jobs, such as attaching 2-by-4 lumber to a ceiling beam. What you want is the flat plates (or flat plates with a flanged edge; those are useful, too), available in several widths and lengths. Common widths are 3″ to 5″ and lengths from 3″ to 7″. See Figure 9-3 for a size comparison of popular 3″ × 5″, 3″ × 7″, and 3″ × 9″ plates.

You can use the plates as is or cut them to size. The plates have numerous predrilled holes to help you hammer in nails, but the metal is thin enough (usually 20 gauge or so) that you can drill new holes. Just be sure to use sharp drill bits.

In North America the most common brand of nail plate is the Simpson Strong-Tie. Visit the company's Web site (*www.strongtie.com*) for sizes and specifications of their various products.

Steel Tubes and Angles

Most hardware stores carry a limited quantity of extruded round and square tubes, as well as angles and other shapes. These are solid and fairly heavy items made with 14-gauge steel. Use them when you need lots and lots of support. Many are predrilled with holes along the sides, saving you the effort. You still have to cut them, though. For cutting, use a new hacksaw with a fine-toothed blade, as detailed later in this chapter.

METAL FROM CRAFT AND HOBBY STORES

Hobby and craft stores are another handy outpost for your metal needs. Sheets of stainless steel, copper, and even bronze are common. Some of the metal products are meant to be used for decorative purposes, so they're very thin and not as useful as structural parts in a robot. These are more like thick aluminum foil than metal sheets. But other, heavier pieces (22 gauge and above) are suitable for making bases and frames.

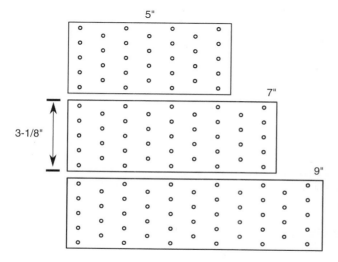

Figure 9-3 Nail plates are made of galvanized steel and are useful as base or construction metal for your robots. They come in different sizes; avoid unnecessary cutting by matching the size to best fit your intended use.

Table 9-1 Summary of Metals for Robot Building			
Metal	**Common Robotics Applications**	**Main Benefits**	**Main Drawbacks**
Aluminum	Bases; arms; all structural parts	Reasonably priced; lightweight but strong; easy to cut and drill using proper tools	Plethora of alloys makes picking the right one difficult; can be hard to weld
Brass	Brackets and straps; structural parts for small robots; nonstructural parts for larger bots	Commonly available at hobby and craft retail stores	Relatively soft; low tensile strength
Copper	Brackets and straps; nonstructural parts	Easy to cut, bend, and shape; readily machined	Somewhat expensive; tarnishes easily
Steel	Heavy-duty frames	Very strong; inexpensive	Rusts if not protected; can be hard to drill and cut
Stainless steel	Heavy-duty frames	Resists rust and corrosion	Can be hard to cut and drill; more expensive than other metal choices
Tin*	Sheet metal bodies; easy-to-make parts like motor mounting straps	Soft and malleable; thin sheets ideal for robot bodies; relatively low melting point (450°F to 725°F)	Can be hard to find locally; look online for "punch tin" sheets used for craft projects

* Some sources sell thin sheets of tin-plated steel and call it "tin." Though not the same, the materials may be interchangeable depending on the application.

Besides sheet metal there are metal strips, bars, and tubes, both round and square. The tubes are sized so that they "telescope"—they fit within one another. You might use these to make ersatz shaft couplers, for example, to match a small shaft to a larger wheel hole. In fact, Chapter 16, "Mounting Motors and Wheels," discusses this very topic.

Recap of Metals for Robotics

See Table 9-1 for a review of metals that are particularly well suited for the construction of robotics. Each metal is noted with its common use, main benefits, and principle drawbacks.

The Metalsmith's Art

One reason to use metal for your robot is that it's stronger. It also looks pretty cool—though to be fair, you can get plastic that looks like metal, but that's beside the point. Another key benefit is the longevity of a metal or frame; it simply lasts longer. Even if your robot falls off the table or gets mauled by the dog, at least its body will remain intact.

But while metal provides a resiliency that wood and plastic cannot match, it's harder to work, costs more, and weighs more. Choices, choices.

As first discussed in Chapter 5, "Robot Building 101," to cut metal you should use a hacksaw outfitted with a fine-tooth blade—24 or 32 teeth per inch. Coping saws, keyhole saws, and other handsaws are generally engineered for cutting wood, and their blades aren't fine enough for metal work.

You'll probably do most of your cutting by hand. Use a miter box when cutting pieces for a frame. Be sure to get a miter box that lets you cut at 45° both vertically and horizontally. Firmly attach the miter box to a workbench using hardware or a large clamp.

You'll always have better-than-average results if you use sharpened, well-made tools. Dull, bargain-basement tools can't effectively cut through aluminum or steel stock. Instead of the tool doing most of the work, *you* do. That's no fun, and robotics is supposed to be fun.

> ⚠ *Cutting, drilling, and finishing metal are hazardous!* Be sure to always wear eye and ear protection. Don't disengage the safety device on any tool. Follow all manufacturer instructions. When using power tools be sure to hold smaller pieces in a vise or clamp while you work with them. DO NOT *ever* hold the work with your bare hands!

CUTTING A BASE

Use sheet metal to make robot bases. Mild steel and aluminum are the most common for use in bots, and they are relatively inexpensive. For both, select a thickness that will support the size and weight of the robot, but without adding undue weight of its own. Consider the following starting points in your design:

For small robots, under 8":

- Aluminum: 1/32" to 1/16" (0.03125" to 0.0625")
- Steel: 22–20 gauge

For medium robots, 9" to 14":

- Aluminum: 1/16" to 1/8" (0.0625" to 0.125")
- Steel: 20–18 gauge

For large robots, 14" and over:

- Aluminum: 1/8" to 1/4" (0.125" to 0.250")
- Steel: 18–16 gauge

Cutting sheet metal by hand is a chore, but it can be done with a bit of work. These are some specialty hand tools for cutting sheet metal:

Aviation snips are suitable for aluminum under 1/8" (0.125") thick, or 18-gauge thickness for steel. When cutting out circular bases, consider using a right- or left-hand tool; it'll make better cuts. "Right-hand" and "left-hand" in this case refer to the direction of the cut, not whether you are right- or left-handed.

A nibbler tool takes out little bites to cut and shape the metal.

Pneumatic shears make quick work of cutting sheet metal. I don't know what I'd do without mine! The shears themselves are relatively inexpensive (most under $40), but you need a suitable air compressor to power them. An alternative is all-electric sheet metal shears, but these cost more. With both, the maximum practical thickness is 18-gauge steel.

The basic steps are the same whether you use snips, nibbler, or shears:

1. Use a scribe or construction pencil (anything that writes on metal) to mark the cut you want to make.
2. Hold the metal sheet in one hand—use a work glove if you need to—and the cutting tool in the other, or clamp the metal in a vise.
3. Cut along the marked path. If you need to make sharp turns, you may have to cut (or nibble) away from the scrap portion of the work, then approach the path from another angle.
4. Smooth the edges with a metal file or fine-grit sandpaper, as detailed later in this chapter.

Cutting Thick Sheet Metal

For thicker stocks, specialized metalworking tools are needed. On large pieces, a metal brake is used to make straight cuts. You probably don't own one of these, but if you attend school, ask the shop teacher if you can use the school's for a few minutes.

Very thick steel (1/4″ or more) can be cut with a torch, then ground down as needed with an electric grinder. Obviously, this requires a cutting torch and the experience that goes with using it.

Cutting Thin Sheet Metal

On the other side of the spectrum, the thinner aluminum stocks (1/32″ and under) can be cut to shape using aviation snips (see above) or even a pair of heavy-duty scissors. You can also use a scroll saw, as long as it is a variable-speed model (use the slower speeds) and you've equipped it with a metal cutting blade.

CUTTING A FRAME

Robot frames can be constructed using aluminum extrusions, described above in "Metal from Your Home Improvement Store." These pieces are meant to be used for such things as trim for a shower door in the bathroom, but it won't mind if it's used for making a robot.

*   *Channel (equal and unequal).* Channel is available with equal and unequal dimensions. Pick the style best suited for your project. It's available in thicknesses starting at 1/32″.
*   *Angle (equal and unequal).* Available with equal and unequal sides, frames can be constructed using standard bracket hardware, pop rivets, and machine screw fasteners. Thicknesses start at 1/32″.
*   *Flat bar.* Available in different widths and thicknesses, use this stock to make your own brackets and support columns. It's also useful as a reinforcement strap.
*   *Tubing.* Available in square, rectangle, or round shapes. These tend to be the most expensive, and are harder to build frames from.

Sources for these structural shapes include:

Larger dimensions: Your nearby hardware and home improvement store is likely to have a good selection. Online is another choice if your local store is out of stock.

Smaller dimensions: Hobby and craft stores are your best bet. Some hardware and home improvement stores also carry a limited assortment of the smaller pieces. As noted earlier in the chapter, a popular maker of metal structural components is K&S Engineering. Check their Web site at *www.ksmetals.com* for a description of their products.

Using a Backsaw and Miter Box

The process of cutting a frame from aluminum (or other) stock is the same as it is when making a frame out of wood. So instead of repeating those steps here, please refer to Chapter 7, "Robots of Wood," for the lowdown details.

A small, lightweight but sturdy frame can be constructed using 1/2″ × 1/2″ × 1/16″ U-channel stock, cut to length with miters. The 1/2″ dimensions are for the outside of the stock; inside it's 3/8″, so you can connect the pieces using 3/8″-wide L angle brackets, also available at your hardware store.

To assemble, use 4-40 machine screws and nuts. The screws can be 3/8″ or 1/2″ in length. You need only one screw per "leg" of the angle bracket—this makes it easier for you to assemble the frame and saves on weight. In a pinch, you can use 6-32 machine screws and nuts, but these are bigger and heavier.

Creating a Box Frame

Box frames are three-dimensional bodies for your robots. They can be constructed using two (or more) square frames, anchored together with metal or plastic "pillars," as shown in Figure 9-4.

For good strength but less weight, I like to use 6mm PVC for the pillars. These are cut to about 3″ wide. Then,

1. Drill a pair of holes near both the top and the bottom of the pillar. Space the holes no closer than 1/4″ from the top (and bottom), and no closer than 1/2″ from the sides.
2. Drill corresponding holes at the approximate center of each side of the frame. (See "Drilling Metal," next section.)
3. Assemble, using 4-40 × 1/2″ machine screws and nuts. Use flat washers on the screw-head side.

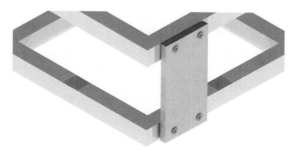

Figure 9-4   By using pillars (also called risers or columns), you can literally stack frames together to make boxlike body shapes for your robots.

Figure 9-5 Machine framing provides very durable frame structures that slip or bolt together.

When building large frames—say, over 14″ square—use two pillars per side; that means eight total. Position each pillar about 1″ to 2″ from the corners of each frame. This gives you room in the middle of the box to work.

Using Machine Framing

All those difference shapes of aluminum shown in Figure 9-2 are made by extruding molten metal through a special die. The "profile" of the aluminum takes on the size and shape of the die. *Machine framing* is a special kind of extruded aluminum with a far more complicated profile— cuts, bumps, ridges, and other shapes that help to add extra strength.

Figure 9-5 shows a boxlike body made out of machine framing and specialty angle connectors. On less expensive framing the connectors just slip on, but with the higher prices stuff the connectors secure tightly with small socket screws or other metal hardware.

Machine framing is sometimes also called *T-slot,* because the profile of the extruded aluminum often takes the shape of the letter T. A premier online seller of machine framing is 80/20 Inc. Most providers sell the aluminum framing pieces by the inch, custom cut to length, but you can often save money by getting a kit of premade stock and angle connectors.

DRILLING METAL

Metal requires a slow drilling speed. That means, when using a drill, you should set it to no more than about 25 percent of full speed. Variable-speed power drills are available for under $30 these days, and they're a good investment. Use only sharp drill bits. If your bits are dull, replace them or have them sharpened. Quite often, buying a new set is cheaper than professional resharpening. It's up to you.

Punching a Starter Hole

You'll find that when you cut metal, the bit will skate over the surface until the hole is started. You can eliminate this skating by using a center punch tool prior to drilling. (In a pinch, a nail works, too.) Use a hammer to gently tap a small indentation into the metal with the punch.

Dabbing on Some Oil

When drilling aluminum thicker than 1/16", or most any thickness of steel, first add a drop of oil over the spot for the hole. If you're drilling a very thick piece, you may need to stop periodically and add more oil. There's no rule of thumb, but one drop per 1/16" or 3/32" of thickness seems about right. The purpose of the oil is to keep the bit cool, which makes it last longer.

Using a Drill Press

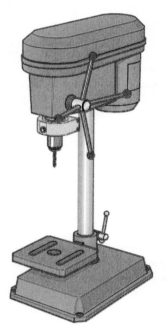

When it comes to working with metal, particularly channel and pipe stock, a drill press is a godsend. It improves accuracy, and you'll find the work goes much faster. Always use a proper vise when working with a drill press. *Never* hold the work with your hands. The bit can snag as it's drilling and yank the piece out of your hands. If you can't place the work in the vise, use a pair of Vise-Grips or other suitable locking pliers.

If using a drill press, your model may require you to change the position of belts in order to alter the speed. Set the speed to the lowest available.

BENDING METAL

Most metals can be bent to another shape. Sheet metal is typically transformed into curves and angles using a tool called a metal brake. Hollow pipe, such as metal conduit for electrical wiring, is bent using (get this!) a pipe bender.

Not all kinds of metals are as easily bent, at least not without specialty tools. The thickness of the metal and its overall size, shape, and tempering qualities greatly affect how easy (or hard) it is to bend. Strips of softer metal like copper and brass bend easily. The thinnest pieces can be bent by hand; otherwise, you can use a bench vise and rubber mallet to bang it into shape.

Mild steel up to about 1/16" thick that has not been heat treated (tempered) to make it harder can be bent using a shop vise and brute force. But you get better results with a metal brake.

Tube bending, especially thin-wall aluminum or brass, requires either skill or a special tool to keep the tube from collapsing. The idea is to prevent a sharp kink at the bend. Benders are available for different sizes of tube. Smaller tubing available at hobby stores can be worked using a tightly wound spring. With practice, you can also bend lightweight tubing by first filling it with sand.

Bending causes stress in the metal, which can weaken it. Some aluminum alloys are engineered to be bent, but most of the stuff you find at hardware and home improvement stores has been hardened. Bending it to more than 20° or 30° can seriously degrade the structure of the metal.

FINISHING METAL

Cutting and drilling often leave rough edges, called *flashing* and *burrs,* in the metal. These edges should be finished using a metal file or very fine sandpaper. Otherwise, the pieces may not fit together properly, and the rough edges can scratch skin and snag on carpet. Aluminum flash comes off quickly and easily; you need to work a little harder when removing the flash from steel or zinc stock.

If there is a lot of material to remove, use a small grinding wheel attached to a drill motor or hobby tool, such as the Dremel. Use the standard Dremel mandrel. It's composed of two pieces: a shaft and a screw. Then purchase an aluminum oxide grinding wheel in the shape you need. Example: Dremel item 541 is a set of two 7/8"-diameter flat wheels made to be mounted with the mandrel.

Using Metal Files

A metal file is the same as any other file, excepts its teeth—the part that removes the material—are finer. You should always use a metal file for metal; never use a wood file, because the teeth on a wood file are far too coarse. Unless you plan on doing lots and lots of metalworking, you can purchase an inexpensive set with a variety of files in it or get just what you need. I'd start with just one or both of the following:

- For aluminum, get a single-cut mill, half-flat, half-round file; that is, on one side the file is flat, and on the other it's rounded. You don't need a handle on the file unless you want one.
- For steel, get the same, except make it double-cut.

The number of teeth per inch defines the smoothness of the finished work. For general-purpose deflashing and deburring, opt for files with 30 to 40 teeth per inch. This equates roughly to "bastard" and "second cut" files, if the tools you're looking at are marked that way instead.

For small pieces, you might want to invest in a set of needle files, so-called because they're small like knitting needles. The set has different sizes and shapes.

Most files have teeth that face forward, away from the handle. So they do their work as you push it into the work, not as you draw it back. Use this fact to make your work easier. Don't try to use the file like you're cutting a loaf of bread with a knife. Make even strokes, and bear down only on the forward stroke.

Using Sandpaper

Use sandpaper if you want the smoothest edge possible. This is not just for looks, but for function, too: the surfaces need to be like glass on pieces that slide against one another.

You need aluminum oxide or silicon carbide sandpapers for working with metal. For general deburring and cleaning, use a fine or medium aluminum oxide paper; for finishing/polishing, use a fine emery cloth (doesn't use a paper backing), and dip it in water as you work. The higher the grit, the finer the finish.

Grit Key	Name	Grit
M	Medium	80 to 100
F	Fine	120 to 150

For a final smooth finish, buff the metal using 00 or 000 steel wool. You can get this at a hardware or home improvement store. Look in the paint section. Which brings us to ...

Painting Metal

Bases and frames of mild steel should be painted, to prevent rust. No painting or other treatment is needed if the aluminum is already anodized—it'll have a silvered, black, or colored satin appearance. For bare aluminum, the metal can be left as is, but you may prefer to paint it. You can paint using a brush or spray can.

Before painting clean the metal with denatured alcohol to remove grease, oils (including skin oil), and dirt.

Build the CrossBot: A "No-Cut" Metal Platform

Of all the aspects of robot building, cutting stuff up is my least favorite, especially when it involves metal. Most designs use stock metal of some kind: U-channel, tubing, strips, or large plates that must be cut down to size.

But what if you could find metal already in the size and shape you need for building robots? You can, but this stuff is found in a different part of the hardware store than the stock metal bins. And with it, you can construct "no-cut" metal platforms that require no (or very little) cutting to form into usable sizes and shapes.

The basic idea behind the no-cut is to use base materials that are already the proper size and shape. The parts of the robot—the motors, sensors, batteries, and so forth—can then be attached using fasteners, glue, hook-and-loop, double-sided foam tape, tie-wraps, or other techniques.

A prime source for materials for no-cut bases is any hardware store that sells galvanized steel nail straps for framing lumber pieces in a house or barn. Keep your eyes open, and you'll note many ready-made nail straps that can be used without any additional sawing or sanding. Following is an example of a no-cut mobile robot design using commonly available (and inexpensively priced) metal nail straps.

SHEET METAL FOR LUMBER STRAPPING

Let's first talk about these nail straps. Find them in the lumber or building section of your local hardware or home improvement store. Some of what you'll find is not terribly useful for robot bases—specially formed and bent pieces for things like hanging 2-by-4 joists in an attic or garage. Look instead for the simpler shapes and the flat metal straps uses to nail pieces of wood together.

Figure 9-6 shows four commonly available nail plates. The model numbers are specific to Simpson, a popular brand in the United States, but similar products should be available in other areas. Most of these are made of 14-gauge steel, so they are quite sturdy. (They're also a wee heavy, so be careful about adding too much extra weight.)

66T T strap: T-shaped cross plate measures 5" × 6". We'll use just this one piece for the CrossBot. If you need bigger, there's the 1212T which measures 12" on each side.

66L L strap: L-shaped plate measures 6" on each side. If you need bigger, there's the 88L, which is 8" on each side. Example uses: mounting brackets for larger robots; outriggers for motors.

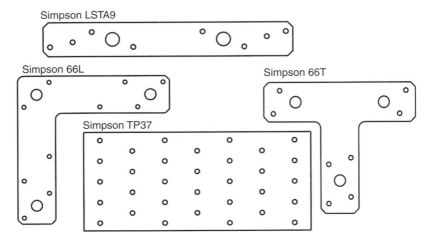

Figure 9-6 Outline shapes of four common nail plates. Like T straps, these are found in the lumber section of local hardware or home improvement stores.

LSTA9 strap tie: Measures 9″ × 1-1/4″. Example uses: center rail in a walking robot; connecting strap for wood, metal, or plastic bases; side angle bracket for tracked bases.

TP37 and similar tie plate: Flat plate with different lengths to suit various applications. Width for all is 3-1/8″: TP35 length: 5″; TP37 length: 7″; TP39 length: 9″. Example uses: robot base; mounting plate for heavy parts (large motors, batteries) on framed robots; side panels.

 You may elect to cut or trim some of the pieces, but since they're already in the basic shape you need, there is less work required overall. Sheet metal for lumber strapping is typically 18 or 20 gauge and can be cut with a hacksaw, metal snips, or motorized (electrical or air-powered) shears.

MAKING THE MINI CROSSBOT

As noted the Mini CrossBot is made from a 6″ strapping T (or *tee*), commonly used in lashing together pieces of lumber in a home. Strapping Ts are available in numerous sizes; the 6″ size is the smallest that I've been able to locate, but they are also available up to 16″. The size measures the top of the T; the vertical portion of the T is in various lengths, depending on the design.

While I've specified the Simpson 66T the brand and even the size doesn't matter. Like most strapping Ts the 66T has holes in it for nailing; these holes are offset and most will not line up with hardware you want to hang on your robot. You'll need to drill new holes. A power drill or, better yet, a drill press is recommended for drilling the holes.

 As an alternative to drilling is hot melt glue. Use a high temperature gun and glue sticks for the best bond. Be sure all the parts are scrupulously clean before gluing. You may want to rough up the metal surfaces with a file to provide for grip for the glue.

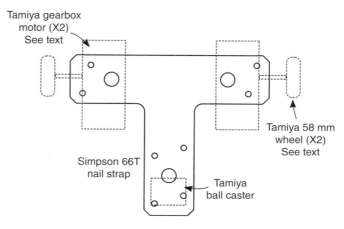

Figure 9-7 Design layout for the Mini CrossBot. It uses a 6" T-shaped galvanized steel strap, found in the lumber section of your nearby hardware or home improvement store.

MAKING THE MINI CROSSBOT

The basic layout template for the Mini CrossBot is shown in Figure 9-7. The robot uses the following parts, in addition to the strapping T and assorted fastening hardware. You are, of course, free to substitute others you may have on hand or like better.

- Tamiya worm gear motors, #72004 (two)
- Tamiya narrow tires, 58mm diameter, #70145 (one set of two tires)
- Tamiya ball caster, #70144 (comes in sets of two, only one used)

The Tamiya parts can be purchased from many online hobby retailers. The motors are mounted on the ends of the cross, and the caster is mounted at the base. The Tamiya caster offers the option of two heights; I selected the taller height to better match the wheelbase afforded by the motors and tires. The two drive motors are at the front of the robot; the caster acts as a kind of tailwheel.

Only a few holes need be drilled in the strapping T. I used a 5/16" drill to make holes for 4-40 × 1/4" machine screws (see Figure 9-8). The small fasteners and the somewhat larger holes provide some "slop" in mounting. With some wiggle room, you can better align the caster (not critical) and the two motors (critical).

Figure 9-9 shows the assembled Mini CrossBot with motors, wheels, and ball caster. With the caster made to its taller height the CrossBot has a kind of jacked-up stance, like a 60s dune buggy. To complete the CrossBot you'll need batteries (Chapter 12), a dual motor bridge (Chapter 14), and a microcontroller or other electronic brain (Chapters 24 to 29).

Total weight of the Mini CrossBot prototype, with 66T strapping T, motors, wheels, caster, battery holder, battery, breadboard, and assorted small switches, is 17.5 ounces (that's 496 grams for you metric folks). Note that the four AA batteries alone contribute 3.5 ounces (about 100 grams) to the weight of the robot.

Figure 9-8   The motors are attached to the CrossBot using 4-40 × 1/4" machine screws.

Figure 9-9   The finished Mini CrossBot, ready for electronics, and a floor to run on.

For your reference, here are the specifications of the most commonly available sizes of Simpson Strong-Tie strapping Ts, and their weight in ounces and grams. Larger robots can be built using bigger strapping Ts. The 1212T strap weighs almost a pound, so you need bigger motors (and batteries) to haul around that kind of weight.

Model	Material	L	H	W*	Weight
66T	14-gauge galvanized	6"	5"	1-1/2"	5 oz; 142 grams
128T	14-gauge galvanized	12"	8"	2"	11 oz; 312 grams
1212T	14-gauge galvanized	12"	12"	2"	14 oz; 397 grams

* Width is the width of the strapping metal.

GOING FURTHER

Of course, the concept of the no-cut extends beyond the Mini CrossBot or the other strapping products detailed here. You can use the same idea for other robot designs made out of different metal materials, no matter where you find them. The key points to keep in mind are:

- The material should already be in the size you need, so no cutting is required.
- Drilling may be needed. Avoid materials that already have lots of holes. The holes may not line up with the motors and other components you wish to add, and the existing holes can cause trouble when drilling new ones so close by.
- Avoid very thick materials for small robots, as they add unnecessary weight.
- Consider sheet materials that can be bent to create unusual robot base shapes.

CHAPTER 10

Building Bots with Digital Fabrication

Your robot may operate under its own power, but you probably designed and built it yourself by hand. That can take lots of time, and the results are very dependent on your shop skills. An alternative: Streamline the construction process by using digital fabrication, which include CNC routers, laser cutters, and 3D printers.

In this chapter you'll learn the basics of creating construction layouts using computer-aided design, then how to use the digital artwork to produce your own robot creations by machine.

Making Drilling and Cutting Layouts

Everything goes better when you have a plan.

Even without construction machines like a laser cutter you can *use computer-aided design* (CAD) software to produce printed templates for drilling and cutting by hand. Draw your idea in the program, print it out, and use the paper copy as a template.

Using a CAD program greatly aids in accuracy. It also makes changes a snap. For example, with a computer it's easy to make the layout slightly smaller or larger, in case you want to adjust the size.

AN INTRODUCTION TO CAD

There are two general types of graphic images, and therefore two types of programs for making this images: bitmap and vector. The difference is how the program stores the shapes you draw.

- *Bitmap graphics* is composed of a series of dots, like the dots in a newspaper picture. Windows Paint is a good example of a bitmap graphics program. Pick a drawing tool, and it creates a swath of dots in some distinct shape, size, and color.
- *Vector graphics* is composed of lines and other shapes. You make drawings by combining different shapes—squares, rectangles, lines, and so on—together, like that in Figure 10-1.

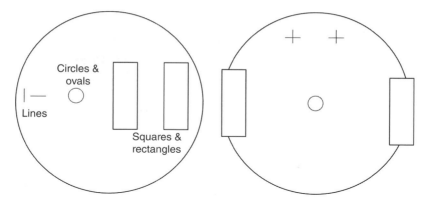

Figure 10-1   Prepare a construction template using simple shapes by combining circles, squares, rectangles, and lines.

Changes are harder with bitmap graphics, because once the bits for a shape are laid down on the digital canvas, the shape itself can no longer be edited.

Vector Graphics Best Choice

Of the two, vector graphics programs are by far the most useful in drafting your robots. You can use the program to create the overall design—basically a drawn picture of how your robot will look when finished. Or you can use the program to create drill and cutout templates—the same idea but better execution to layouts drawn by hand.

You might already have a vector graphics program handy—even Microsoft Word comes with vector drawing tools. But there are plenty of free and open-source vector graphics applications you can try, and some are better suited for robot building. These include Inkscape, which has become something of an industry standard.

Using Inkscape for Robot Design

Inkscape serves as a good reference for using a vector graphics program to design robots. Figure 10-2 shows the Inkscape program window. Here's a rundown of the important parts of the program interface.

> **Menu bar and tool bars:** These control the program using commands in menus and on various toolbars.
>
> **Canvas:** Your drawing goes in the middle portion, the canvas. Zoom controls let you see your drawing from a distance or up close.
>
> **Drawing tools:** You create or edit the drawing using this small selection of tools on the left side of the screen. Vector graphics programs like Inkscape are based on what's known as *Bezier curves*, whereby any shape is composed of one or more lines. Each line can be bent into sharp or smooth curves—a circle is really a line that bends back to itself.
>
> **Color palette:** Solid shapes and their lines can be filled with color using a handy palette of preset shades, located at the bottom of the screen. Millions of other colors can be set using specialized tools.

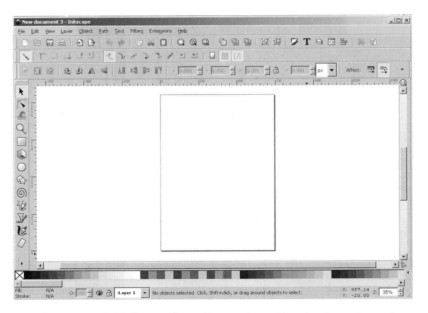

Figure 10-2   Inkscape (available for Windows, Macintosh, and Linux) is free software for creating and printing vector graphics. Its interface is simple and easy to learn.

   Inkscape also supports writing text on your drawings. Use the text tool to label parts or provide dimensions. Inkscape lacks features for automatically inserting dimensions (this is more the domain of a CAD program; see the next section), but you can readily add the text yourself.

Figure 10-3 shows a drill and cutting layout created in Inkscape using just simple shapes. A circle is used to define the shape of the robot's base, and rectangles indicate cutouts for motors. Small circles with thin crosshair marks show where holes go, and the size of the circles indicates the approximate diameter of the holes. The design is printed at 100 percent so that it can be directly used as a drilling and cutting template.

To use the template, just print it out, then tape or transfer it to the material used for your base. Use the same techniques detailed previously under "Creating Layouts by Hand."

The idea behind CAD is that not only can you draw a square, you can draw a square that is precisely 1″ by 2″. Absolute measurements are stored with the CAD file and, when used with the appropriate printer, produce highly accurate renditions of your drawings.

Basic CAD Functionality

Most CAD programs (including Inkscape) share the same basic functionality, things you need to get the job done.

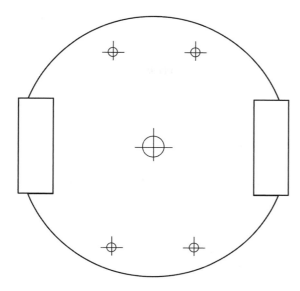

Figure 10-3 The cutting and drill template created by Inkscape and then printed to scale.

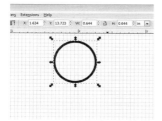

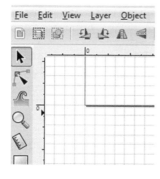

- *Drawing setup.* Here you define the drawing size and drawing scales (e.g., 1:1, 1:12, etc.), unit of measurement, as well as grid size and drawing resolution. For most robotics projects, you'll want a 1:1 scale, a grid of 1/4" or 1/8", and a resolution of 2- or 3-decimal places—that is, down to the hundredths or the thousandths of an inch.

- *Drawing tools.* Only a few shapes are used for typical robot layout drawings: line (and/or polyline), circle, and rectangle. Lines are used to mark cutting layouts. A polyline is a set of lines that share at least two vertices (corners), and it is used whenever you want to cut out more complex shapes. Circles are typically used to denote holes for drilling. A rectangle or square is a closed polyline shape and can be produced using the line, polyline, or rectangle tool.

- *Editing/sizing tools.* You can adjust the size and look of the shapes by using the mouse or by entering values at a command-line prompt. The mouse is good for "eyeballing" the design, but the command-line entry is handy when you need accurate placement.

- *File saving/printing.* Once done with the drawing, you can save it for future use or print it out. Any supported printer will do, such as a laser or ink-jet printer. CAD programs don't have to be used only with pen plotters anymore.

- Drawings are placed on a *workplane*. With 2D CAD, a simple *X* and *Y* coordinate system is used to denote the origin (the start point in virtual space) of the drawing. With most CAD programs the origin is the lower-left corner of the drawing and is denoted as *0,0*. The first digit is the *X* axis; the second digit is the *Y* axis.

- Each circle, rectangle, line, or other shape is an *object*. Objects are placed on the drawing area of the program in an overlapping back-to-front fashion, like playing cards stacked on a black-jack table. This order can be important when using CNC routers and many laser cutters; the machine will cut the shapes in the order they appear in the document.

FILE FORMATS FOR 2D CAD

You're probably familiar with GIF, JPG, and PNG graphics files. These are all bitmap file formats and typical of images you see on Web sites.

Likewise, vector graphics have their own file formats. As many were created for a special purpose or graphics program, they tend to be unique to themselves and often are not compatible with one another (you can sometimes use a converter program to exchange one format for another, but the results can leave much to be desired).

Of the varied vector graphics formats, these that follow are the most common, and the ones you'll likely work with and share with others.

- *SVG*—Scalable Vector Graphics, now the uber-standard for vector-based images, and supported by Wikipedia. This is Inkscape's default file format.
- *EPS*—Encapsulated PostScript is an interchange format promoted by Adobe. It contains definitions of the graphic elements as PostScript command codes, plus (usually) a medium-resolution bitmap for use by those programs that can only import a bitmap. Programs have varying support for EPS, and some—like Inkscape—need additional plugins to understand the format.
- *DXF*—The Drawing Exchange Format is most commonly found on CAD programs and was originally developed by AutoDesk to allow AutoCad to share its 2D files with other applications. It's now become a de facto standard for all CAD apps.
- *AI*—The native file format for Adobe Illustrator, AI is one of the premier commercial vector graphics applications. Saving and sharing this format is acceptable as long as everyone has Illustrator and, better yet, similar versions. Otherwise, use SVG.

Using a CNC Router

See that cutting and drilling template in Figure 10-3? Because it's made to exact scale, you could use the drawing to have your robot base professionally produced with a computer-driven cutting tool.

The *CNC router* was one of the first tools for automating the drilling and cutting of materials. *CNC* stands for Computer Numerical Control, a process that uses a computer to operate motors; these motors in turn move a cutting tool—the router—that makes holes and cuts.

With a CNC router you can transform a 2D drawing created with Inkscape (or other CAD program) into a real-world object. You have your choice of materials: wood, plastic, paperboard, and more. The router drills and cuts with a precision that far exceeds what most people can do using manual tools, and do it faster.

While CNC routers can be big, loud, and frightfully expensive, there are many models designed for garage-shop consumers;

Figure 10-4 shows the Shark II, a CNC router designed for the serious hobbyist or semi-pro. The machine cuts pieces up to 13" × 25" × 7. You supply the router depending on the type and density of material you're cutting.

Figure 10-4 With a CNC router you can cut 2D shapes out of wood, plastic, paperboard, and even metal. (*Photo courtesy NextWave Automation.*)

The process of using a CNC router is fairly straightforward:

1. **Prepare and as necessary import the CAD drawing.** Most CNC software can directly open DXF and SVG files.
2. **Adjust toolpath widths according to the cutting bit you are using.** CNC routers use bits similar to drill. The diameter of the bit dictates the minimum hole size that can be drilled. When cutting, the bit removes material on either side of the cut. This is called the *kerf*, and is the same as when using a saw to cut a piece of wood. To compensate for the loss of material you must set a tool "offset" so that the dimensions of the cut pieces are the size you expect them to be.
3. **Secure material on the work table.** The routing process creates a lot of vibration and movement, so the material being cut must be firmly secured.
4. **Adjust the cutting parameters.** A CNC router can cut materials of varying thicknesses. Prior to cutting you must specify the cutting depth and other settings.
5. **Cut material.** During the cutting process sawdust is created that must be vacuumed away from the cutting table.

Using a Laser Cutter

Gaining popularity as an automated fabrication tool for small robotics is the laser cutter. These are now regularly found in schools, maker spacers, even in the dens of lucky robot builders. Rather than cut with a high-speed rotary tool, these machines use a high-intensity pencil-thin light beam to make precision holes and cuts in wood, plastics, and even some metals.

The process of preparing and importing a CAD file for a laser cutter is similar to using a CNC router. Because the laser beam has virtually no width (about 0.01″ on many models), it's not as

Figure 10-5 The Glowforge Pro has an effective cutting area of 11″ × 19.5″ and supports a passthrough slot for cutting from larger stock. (*Photo courtesy Glowforge Inc.*)

critical to adjust for kerf—exception: you're creating some very high-precision parts. Figure 10-5 shows a Glowforge laser cutter with a passthrough slot front and back. With the passthrough you can cut your pieces from larger plastic stock. This is a nice feature to have, but costs extra.

Some tips:

- Be sure to read and understand the instructions for submitting files. Many laser cutters prefer files using the SVG format (these have a .svg extension). Any self-respecting CAD or vector graphics program can save files in this format.
- Some laser cutters have software that will scan and vectorize bitmap art, including hand drawings. These are great if you're working from stylized designs.
- Use just one line thickness. Don't use a "fill" on any of the shapes. Holes should be circles; the hole will be the diameter of the circle.
- The lower left of your drawing should start at 0,0. No part of your drawing should go below the 0 marks, or it may not be cut.
- There are technical limitations to the thickness and type of materials you can cut with a laser cutter. Unless it's a higher-end commercial machine, forget cutting aluminum or other metal. Cutters for home use are typically limited to etching into metal, but not cutting it.
- The best plastics to use are acrylic and polycarbonate; PVC of any type is a serious no-no. The fumes emitted during the cutting process can make you very sick and permanently damage the laser cutter. For plastics stick with acrylic and polycarbonate.

Using a 3D Printer

Ever thought about making stuff out of the ooze from a hot melt glue gun? That's the idea behind a 3D printer like the MakerBot Replicator Mini+ shown in Figure 10-6. The machine heats up a thin filament of plastic to get it into a near-liquid form, ready to come out of a nozzle. A series of motors then moves the nozzle in X/Y/Z space, depositing the plastic gloop in successive

Figure 10-6   The MakerBot Replicator Mini+ can make parts measuring up to 4″ × 5″ × 5″, using black or colored plastic filament. (*Photo courtesy MakerBot.*)

layers. The technique is often referred to as *additive manufacturing,* successively adding a small amount of material to build up the desired finished part.

The result is a 3D shape been slowly sculpted one little bead at a time. It's time consuming, but since a machine does it, you can start the process and go have dinner. When you return the part is done for you, and the machine is ready for its next job.

Of all the features of a 3D printer the size of its *print volume* or *work area* is perhaps the most important. The print volume represents the largest single physical object you can create. An area of 3″ cubed is considered sufficient for making parts for small robots, though the larger the print volume the better. Bear in mind that bigger projects are possible by joining together smaller chunks.

MODELING OBJECTS FOR 3D PRINTING

3D printing requires a 3D graphics program. Shapes are defined by creating a *mesh,* an editable outline of the object you want to produce with the machine. This process is called modeling. As with a CAD program, you create meshes using tools that define basic shapes, but these are 3D, likes cubes and spheres. From these shapes you can mold them as if they're made from Plasticine.

Where 2D CAD is fairly easy to pick up in an afternoon, the learning curve for a 3D design tool can be fairly steep. You'll likely want to start by using a free 3D program such as TinkerCad, FreeCAD, or Blender. Start with a simple project to gain experience.

If you're starting out make it easy on yourself by trying out some ready-made object files from an open source repository, such as Thingiverse. Robotics is a favorite subject so you'll find plenty to keep you busy. Don't be afraid to experiment.

3D FILE FORMATS

As with 2D CAD, 3D graphics entails its own specialized file formats. The most common open standards are STL and OBJ. Most 3D modeling programs and printing software understand at least one of these formats.

There are plenty of proprietary 3D file formats, too, like 3DS Max, Blend (for the Blender open source modeling software), Inventor, and Solidworks. These are very powerful 3D CAD programs. If you use these formats you may wish to restrict the program features to just those supported by open standards or else you could wind up with a file that cannot be imported by your 3D printer software.

To make sure your project can be used by your printer export your project to STL, then import it back again to make sure the model is still properly formed. If the model is corrupted it may be an issue with a non-supported feature you used.

CHAPTER 11

Putting Things Together

A robot is an exotic mixture of parts both great and small, important and seemingly inconsequential. We don't forget the big stuff: motors, wheels, batteries, bases. But it's easy to overlook how the robot is put together—the way it's assembled.

How your robot is put together is no less important than its motors, wheels, batteries, and bases. So in this chapter, I'll show you how to assemble the raw materials of your robots to make full-fledged creations. After all, even the Frankenstein monster used bolts to keep his head from falling off.

All About Screws, Nuts, and Other Fasteners

Mechanical fasteners are the most elementary of all assembly hardware. They are the nuts, screws, and washers used to hold pieces together. Fasteners are favored because they are cheap, easy to get, and simple to use. Most fasteners can also be undone, so you can disassemble the robot if need be.

There are dozens of fastener types but just a few that are practical for amateur robotics. These are nuts, screws, and washers, shown in Figure 11-1.

- *Screws* are designed for fastening together the parts of machinery, hence the name. *Wood screws* and *sheet metal screws* have a pointed end and drive right into two (or more) pieces of material to cinch and hold them together. *Machine screws* do not have a pointed end; they're designed to be secured by a nut or other threaded retainer on the other end.
- *Nuts* are used with machine screws. The most common is the hex nut. The nut is fastened using a wrench, pliers, or hex nut driver. Also handy in many robotics applications is the *locking nut*, which is like a standard hex nut but with a nylon plastic insert. The nylon helps prevent the nut from working itself loose.

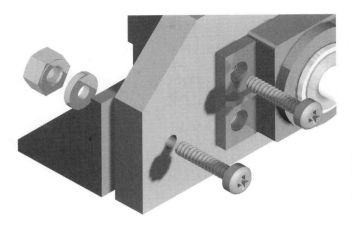

Figure 11-1 Nuts, screws, and washers are the fundamental pieces used in mechanical fastening. Nuts and screws are threaded to match; washer size is based on the diameter of the screw.

● *Washers* act to spread out the compression force of a screw head or nut. The washer doubles or even triples the surface area of the fastening pieces. Washers are available in sizes to complement the screw. Variations on the washer theme include *tooth* and *split lock washers*; these provide a locking action to help prevent the fastener from coming loose.

FASTENER SIZES

Fasteners are available in common sizes, either in metric or imperial.

Imperial

Imperial (also referred to as American, standard, or customary) fasteners are denoted by the diameter, either as a reference number or in fractions of an inch. For machine screws and nuts, the number of threads per inch is also given. For example, a machine screw with a size of

$$6\text{-}32 \times 1/2''$$

has a diameter referred to as #6, with 32 threads per inch and a length of 1/2″. Diameters under 1/4″ are indicated as a # (number) size; diameters 1/4″ and larger can be denoted by number but are more commonly indicated as a fractional inch measurement—3/8″, 7/16″, and so on. See Figure 11-2 for the sizing parameters of the typical machine screw fastener.

The number of threads per inch can be either *coarse* or *fine*. With few exceptions, your local hardware store carries just coarse thread fasteners. The one major exception is the #10 machine screw, which is routinely available in either coarse (24 threads per inch) or fine (32 threads per inch). Be careful which machine screws you buy, because nuts for one won't fit the other.

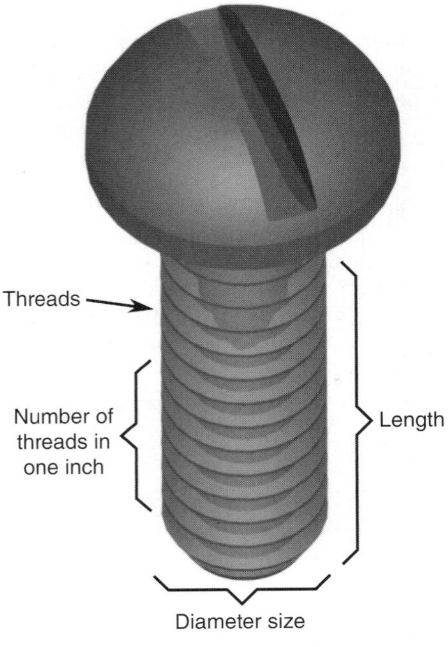

Threads ⟶

Number of threads in one inch

Length

Diameter size

Figure 11-2 The size of a machine screw is specified by its diameter (in inches, millimeters, or numbered scale), threads per inch (or millimeter), and length.

Metric

Metric fasteners don't use the same sizing nomenclature as their imperial cousins. Screw sizes are defined by diameter; the *thread pitch* is the number of threads per millimeter. This is followed by the length of the fastener. All in millimeters. For example:

M2-0.40 × 5 mm

means the screw is 2mm in diameter, has a pitch of 0.40 threads per millimeter, and has a length of 5mm. Check out the RBB Support Site (see Appendix) for handy charts that compare the sizes of both imperial and metric fasteners.

SCREW HEAD STYLES

When selecting screws you have a choice of a variety of *heads*. The head greatly contributes to the amount of torque that can be applied to the screw when tightening it. These are the most common you'll encounter:

	Pan	Good general-purpose fastener. However, the head is fairly shallow, so it provides less grip for the screwdriver.
	Round	Taller head protrudes more than pan head, so it provides greater depth for the screwdriver. Good for higher-torque applications when it doesn't matter if the head sticks out.
	Flat (or countersunk)	Used when the head must be flush with the surface of the material. Requires that you drill a countersunk hole (or if the material is soft, like PVC plastic, drive the head down into the material).
	Fillister	Extra-deep head for very high torque. The top of the head is rounded. Often used in model airplanes and cars for miniature mechanical parts.
	Hex head	Doesn't have a slot for a screwdriver, and requires a wrench to tighten. The best for high-torque applications.

SCREW DRIVE STYLES

Most screws available at the hardware store come made for different types of drivers (i.e., the tool you used to tighten or untighten the screw). For all types, different sizes of drivers are used to accommodate small and large fasteners. In general, the larger the fastener, the larger the driver.

	Slotted	Made for general fastening and low-torque drive; screwdriver may slip from the slot.
	Phillips	Cross-point drive resists drive slippage, but the head is easily stripped out when using an improperly sized driver. My personal choice; they're particularly well behaved when using motorized screwdrivers.
	Socket	Hexagonal-shaped wrench resists slippage. Can be made very small for ultra-miniature screws.
	Hex	Uses a wrench or nut driver to tighten or untighten. Nut drivers look like screwdrivers, but with a hex socket at the end. You can use these tools to tighten nuts, too.

GOING NUTS OVER NUTS

When using machine screws you need something to tighten the screw against. That's the job of the machine screw nut. (You can also use tapped threads in the material itself; check out a good book on shop practice for the details.) Nuts must be of the same size and thread pitch as the machine screw they are used with. That is, if you're using a 4-40 screw, you need a 4-40 nut.

There's more than one kind of nut. Some of the more common are:

- *Standard hex* nuts are the most common, so named because they have six sides (hexagonal = six sides).
- *Acorn* or *cap* nuts are like hex nuts, but with no hole on the other side. You can use them for decorative purposes, but a more common application is as a balancing "skid" for a two-wheeled robot.

- *Nylon insert* nuts have a nylon plastic core. They're used to create self-locking mechanisms. To use, tighten the nut into place. The plastic inside keeps the nut from working loose.
- *Threaded couplers* are like very long nuts. Use them as spacers or, if very long, as "risers" for the separate levels of your robot base.

For all, you may use a wrench or a nut driver to tighten.

WASHERS AND WHEN TO USE THEM

Washers are disc-shaped metal or plastic, used with fasteners. They aren't fasteners themselves, but they augment the job of screws and nuts.

Washers come in two general forms: flat and locking. *Flat washers* are used as spacers and to spread out surface area. Each size screw has a corresponding "standard-size" washer. Specialty washers are available with larger diameters and thicknesses. For example, a *fender* (or *mudguard*) washer has a very large diameter in comparison to the screw it's used with.

Locking washers, or simply lock washers, come in two basic styles: tooth and split. Tooth-style washers dig into the material and/or fastener to keep things in place; split lock washers use compression to keep pressure against the fastener.

Which type of lock washer should you use? Split washers provide the highest locking power. But to do their job, you have to tighten the fastener enough that the split in the washer is compressed. That means they're best suited for metal.

Toothed lock washers are best used with softer materials like wood, plastic, or even aluminum. The teeth dig into the material to hold it.

SHOPPING FOR FASTENERS

You can save considerable money by purchasing fasteners in quantity. If you think you'll make heavy use of a certain size of fastener in your robots, invest in the bigger box and pocket the savings. Of course, buy in bulk only when it's warranted. As you build your robots, you'll discover which sizes you use the most. These should be the ones you purchase in bulk.

Robot builders gravitate toward favorite materials, and fasteners are no exception. I can't tell you which sizes of fasteners to buy, because your design choices may be different from mine. But I can tell you what is used the most in my robot workshop. Perhaps that'll give you a starting point if you're just now stocking up.

For small tabletop robots I try to use 4-40 screws and nuts whenever possible, because they're about half the weight of 6-32 screws and, of course, they're smaller. I use 4-40 × 1/2" screws and nuts to mount servos on brackets, and 4-40 × 3/4" screws for mounting small motors. Larger motors (up to about a few pounds) may be fastened using 6-32 or 8-32 hardware.

I try to keep a few dozen of the following fasteners in stock at all times:

- 4-40 steel machine screws in lengths 1/2", 3/4", and 1"
- 6-32 steel machine screws in lengths of 1/2" and 3/4"
- 8-32 steel machine screws in lengths 1/2" and 1"
- #6 × 1/2" and #6 × 3/4" wood screws, for fastening together panels of rigid expanded PVC

And, of course, I keep around corresponding stock of nuts, plus some flat washers and lock washers in #4, #6, and #8 sizes. For all other sizes I buy them when needed.

FASTENER MATERIALS

The most common metal fastener is steel plated with zinc but other fastener materials may include brass (typically used for decorative purposes), stainless steel, and nylon. Nylon fasteners are lighter than steel fasteners, but can't be used where high strength is required.

All About Brackets

Brackets are used to hold two or more pieces together, usually—but not always—at right angles. You might use a bracket to mount a servo motor to a robot base, for example. Brackets come in a variety of shapes and materials. Here's what you need to know.

ZINC-PLATED STEEL BRACKETS

Though intended to lash two pieces of wood together, steel zinc-plated hardware brackets are ideal for general robotics construction. These brackets are available in a variety of sizes and styles at any hardware store. You can use the brackets to build the frame of a robot constructed with various stock. See the "Cutting a Frame" section in Chapter 7, "Robots of Wood," for ways brackets are used in robot construction.

The most common brackets are made of 14- to 18-gauge steel; the *lower* the gauge number, the thicker the metal. In order to resist corrosion and rust, the steel is zinc plated, giving the brackets their common "metallic" look. (Some brackets are plated with brass and are intended for decorative uses.)

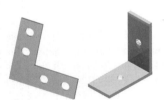

Common sizes and types of steel brackets include:

- 1-1/2" × 3/8" flat corner (or L) brackets. Use these with wood, plastic, or metal pieces cut at 45° miters to make a frame.
- 1" × 3/8" or 1" × 1/2" corner angle brackets. Typical uses are for attaching pieces at right angles to base plates and for securing various components (such as motors) to the robot.

PLASTIC BRACKETS

Metal brackets can add a lot of extra weight to a robot. Plastic brackets add only a little weight and for small robots are just as good. The bracket is made of a durable plastic, such as high-density polyethylene (also known as HDPE).

To add strength, the bracket uses molded-in gussets that reinforce the plastic at its critical stress points. The result is a bracket that is about as strong as a steel bracket but lighter. Figure 11-3 shows a plastic bracket used to secure a servo motor to its mounts. Brackets make it easy to assemble and disassemble the parts of the robot.

Selecting and Using Adhesives

Glues have been around for thousands of years—stuff like tree sap, food gluten, and insect secretions. Modern glues—more accurately called *adhesives*—are chemical concoctions designed to bond two surfaces together.

While there are bazillions of adhesives, only the ones available to consumers are covered here. Intentionally left out are the glues that need expensive tools to apply, are very dangerous to use, or are only available in 55-gallon drums.

SETTING AND CURING

All glues bond by going through a number of phases. The main phases are setting (also called fixturing), then curing. During the first phase, *setting*, the adhesive transforms from a liquid or paste to a gel or solid. Though the adhesive may look to be "hard" when set, it is not yet very strong. This requires *curing*. Setting times for most adhesives are measured in minutes or even seconds. But curing takes a lot longer—typically 12 to 24 hours (see Figure 11-4), and often more.

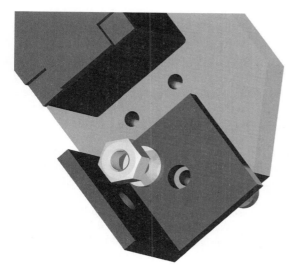

Figure 11-3   Plastic brackets may be used to secure parts to robot bases, like this pair of brackets used to secure radio control servo motors to the underside of a robot.

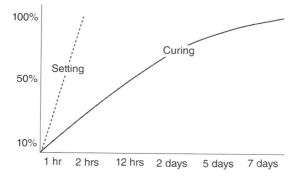

Figure 11-4   Adhesives have both a setting time and a curing time. So-called instant glues have a short setting time, but their curing time can be minutes, hours, even days.

For *most* adhesives, curing time is greatly dependent on several factors:

- *Surface temperature.* Warm surfaces tend to promote faster curing. This is most notable when gluing metal.
- *Adhesive volume.* The more adhesive that is applied to a joint, the longer it takes to cure. That's why you shouldn't apply too much glue.
- *Air temperature.* The warmer the air, the faster the curing (and setting time, for that matter).
- *Air humidity.* Adhesives differ in their affection for moisture in the air. Some, like Super Glue, cure faster when the air is moderately humid. Others, like epoxy, cure faster when it's dry.

LEARNING ABOUT "HOUSEHOLD" ADHESIVES

The term "household glue" is a large, diverse, and not very accurate way to describe glues that you'd use for normal household chores, like fixing broken plates or mending a busted chair. They're also good for most robotics chores because they're easy to get, inexpensive, and most won't kill you the moment you uncap the bottle.

1. **PVAc**-based adhesives are among the most popular general-purpose glues now available, and they are often sold as white and yellow "woodworking" glues. They are water-based, easy to clean up, and inexpensive. They're best with porous material, like wood.
2. **Silicone**-based adhesives are used for both gluing and sealing. They can bond most any nonporous surface to another, such as metal to hard plastic. A common trait of silicone adhesives is that they remain elastic. Use only in well-ventilated areas. After use, be sure the bottle or tube is recapped tightly so that no moisture can enter.
3. **Contact cement** is based on various volatile organic compounds (gives me headaches!). As its name implies, this adhesive is designed to bond quickly on contact. This is accomplished by applying a thin layer of the cement on one or both surfaces to be joined, then briefly waiting for the cement to partially set up. Applying pressure to the joint aids in creating a strong bond.
4. **Solvent cement** uses a chemical that dissolves the material it is bonding. It can be tricky to use because if the solvent isn't precisely matched with the material, nothing happens! The most common solvent-based adhesive is for bonding PVC irrigation pipe, which can also be used with expanded PVC sheets, detailed in Chapter 8, "Robots of Plastic." Other solvent-based cements are available for ABS plastic, polycarbonate, acrylic, and styrene.

BEST PRACTICES: APPLYING HOUSEHOLD ADHESIVES

With very few exceptions, household adhesives (those grouped above, at any rate) use single-part chemistries, so there is nothing to mix. Just open the container and apply the adhesive to the surfaces to be joined.

Figure 11-5   Two-part epoxy can be quickly and accurately applied when using a two-tube plunger.

- Use all adhesives sparingly. A common mistake is to think that if a little bit of adhesive will do the job, a lot must be better. In fact, the reverse is true.
- For adhesives with a watery consistency, apply with a small brush or cotton-tipped swab. For thicker consistencies, apply directly from the tube or with a wooden toothpick or a manicure (orange) stick.
- Very few adhesives will stick to grease and dirt, so be sure to always clean the surfaces to be joined. Household-grade rubbing (isopropyl) alcohol does the trick.
- Avoid moving the glued joint until the adhesive has had a chance to set. As needed, clamp them or tape the bonded parts together.

THE MIRACLE OF TWO-PART EPOXY ADHESIVES

When normal household adhesives aren't enough, you need to turn to the "big guns": two-part epoxy adhesives. They're called two-part because at the time of application you combine a separate *resin* and a *hardener* (also called a catalyst). Separately, these materials remain in a liquid form. But when mixed the hardener reacts with the resin and the mixture sets quickly, usually within 5 to 30 minutes.

The typical package of epoxy adhesive consists of two tubes or bottles: one is the resin, and the other is the hardener. The tubes are separate in some products, and in others they are joined as one unit, with a single "plunger" (see Figure 11-5) in the center for accurately metering the resin and hardener.

So-called 5- or 30-minute epoxies represent the setting time, not the curing time. It takes 6 to 24 hours for most epoxies to cure to 60 to 80 percent, then the remainder over a period of *several days*. Once cured, the bond achieves its maximum strength.

Check the directions that came with the epoxy for mixing instructions. Most call for a 1:1 mix for both resin and hardener. Apply equal length of beads of both liquids to a 3-by-5 index card. Use a wood coffee stirrer to thoroughly mix the resin and hardener. Mix only as much as you need, as unused mixed epoxy *must* be discarded.

SUPER GLUE AND YOU

Super Glue is a trade name, but it's often used as a generic term for a family of adhesives known as ethyl cyanoacrylates, or *CAs*. It's well known as being able to bond to most anything within seconds. Cyanoacrylate is great stuff as long as you know its limitations:

No gaps, please. The most common cyanoacrylate is water-thin and unable to fill in any gaps between the materials to be joined (if you need that, get the thicker "gap-filling" kind).

Keep it clean. CA glues are very susceptible to ruined bonds from even minor dirt and oil. Prior to gluing, clean surfaces with isopropyl alcohol.

Don't use too much. Applying too much CA glue is far worse than not using enough. Try a few drops on the bonded surface only.

Keep bonded joints away from heat and sunlight. Otherwise, the bond will become weak, and it may even spontaneously come apart.

Don't use with natural fibers. CA glues can produce a reaction when in contact with wood or cotton. Do not wear cotton gloves when using CA glue.

Check the date. Cyanoacrylate glue has a relatively short shelf life. Toss product that's more than a year old. Keep the unused portion in a cool, dry place.

Don't get any glue on your skin, or you could wind up cementing your fingers together. If you have an accident, acetone is a good solvent for CA glues. Seek medical help *immediately* if any CA glue gets in your eyes.

USING HOT MELT GLUE

I've saved the best for last: hot melt glue.

Hot melt glue comes in stick form (at least the kind we're interested in) and is heated by a special gun-shaped applicator. Melting temperatures range from 250°F to 400°F. As it turns out, hot melt glue isn't glue at all, but plastic. Adhesion occurs when the molecules of the plastic contract and harden.

The main benefit of hot melt glue is that it sets quickly—in less than a minute—yet yields a strong bond. To apply the glue, allow the gun to reach full operating temperature; it usually takes just a few minutes. Aim the nozzle of the gun over one of the parts to be jointed, squeeze the trigger to apply the gloop, like in Figure 11-6, and then immediate set the parts together to make the bond.

Hot melt glue sticks and guns are available in two general types:

- Low-temperature, suitable for general crafts where the strength of the bond is not as critical. Low-temperature guns and glue are ideal for younger robot builders, as the glue doesn't get as hot.
- High-temperature, best for structural elements and other joints that may be stressed during use. An example is motors, casters, and other moving parts. They operate at much higher temperatures, so exercise care to avoid getting burned.

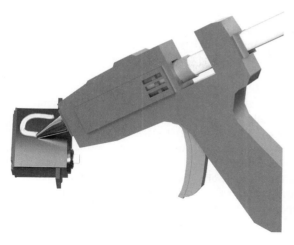

Figure 11-6 Hot melt glue guns are easy to use, and the bonded parts set in just a few seconds. This makes for very quick construction of your robot.

Some glue sticks are "all temperature," meaning they will work in both low- and high-temperature guns. They do this by having a wide melting point. Other sticks are formulated for working with wood, and there are even craft sticks with glitter and coloring already embedded inside. You're free to try these alternatives, though avoid using a glue stick containing metal particles near or on electronic components.

While hot melt glue is a boon to most bot building, it's not without its limitations. First and foremost is that the glue may not bond with everything. Certain kinds of plastics, such as nylon, may be too slippery for the glue. You'll need to use hardware fasteners or a solvent-based adhesive for these.

Hot melt glue may also not provide a high bond when used on very smooth nonporous surfaces. If you find the parts come apart too easily, try adding a reinforcing bead of glue around the edges of the joined pieces. You can also try using sandpaper to rough up the surfaces of the bonded parts prior to gluing.

CLAMPING AND TAPING GLUED JOINTS

It takes time (minutes or even hours) before a liquid glue will hold the pieces together on its own. For very quick bonds—on the order of a minute or two—it's acceptable to manually hold the pieces until they are set. Longer setting times may require clamping or taping.

Woodworking clamps are adequate for larger parts. But for smaller pieces, taping the joint is the most effective. After applying adhesive and mating the joint, tape is applied to keep the joint together. Masking tape works well in most situations, but if you need something stronger, white bandage tape can also be used. It's available in widths of 1/2" and wider.

USING JOINT REINFORCEMENTS

Critical to the strength of any bond is the way the pieces are aligned and positioned. The weakest are "butted" joints (no jokes please!), where two materials are bonded end to end. The reason: There is little surface area for the joint. As a rule, the larger the surface area, the more material the adhesive can join to, and therefore the stronger the bond.

For a stronger joint, you will want to apply any of a variety of reinforcements that increase the surface area of the bond. Joint types and reinforcement techniques are explained below (see Figure 11-7).

Butted joint. The typical butted joint provides minimal surface area for a strong bond and is the weakest of all. Avoid it when you can.

Overlapped. Overlap the pieces themselves, instead of butting them end to end. This is not always feasible, but it is a quick-and-easy method when the option is available. You can readily adjust the amount of overlap as needed.

Reinforced butted joint. Overlap an extra piece of material along the seam of the joint. Use the widest overlap piece you can, in order to increase the surface area. Apply adhesive to this extra piece, and clamp or tape until set. You can also reinforce with small fasteners.

Mitered. Increase the surface area of the joint by mitering the ends. This is most practical with materials that are 1/4″ or thicker. The technique is particularly helpful when joining wood.

Gusset reinforced. Use gusset pieces on the top and/or bottom.

IN SUMMARY: SELECTING THE BEST GLUE

With so many types of adhesives to choose from, it can be hard to select the right one. Table 11-1 summarizes the most common adhesive families, along with their pros and cons, and the bonds they are best used for. Table 11-2 provides various bonding recommendations for each major adhesive family.

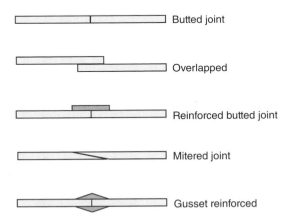

Figure 11-7 Use a joint reinforcement to add strength to bonded pieces. The weakest is the simple butted joint. Overlap or add reinforcing pieces to make the joint stronger.

Table 11-1   Selecting an Adhesive

Adhesive Type	Pros	Cons	Best Used For
Contact cement	Very fast adhesion	Careful assembly of parts required; toxic fumes	Laminating flat substrates
Cynaoacrylate (e.g., Super Glue)	Good adhesion to rubber or plastics	Poor gap filling; poor impact resistance; hard to accurately dispense with consumer tubes and bottles	Bonding porous and nonporous materials that are not subject to impact
Epoxy	Strong bond when prepared and applied correctly	Toxic fumes; curing sensitive to temperature and moisture; must be mixed correctly	Bonding all materials except silicone, Teflon, and other "slippery" materials
Hot melt	Readily available, fast setting, no harmful fumes	Can weaken under heat; poor impact resistance; accurate metering of adhesive difficult with consumer guns	Bonding wood, plastics, and light metals; use a high-temperature glue stick for a better bond with heavy materials
PVAc	Commonly available household adhesive	Not for use when both materials to join are nonporous	Bonding porous to porous and nonporous bonds (e.g., wood to metal, foam to plastic, etc.)
Solvent cement	Extremely strong bond in plastics and rubber	Requires matching the solvent to the material	Bonding plastic and rubber
Silicone	Remains flexible after curing	Toxic fumes; low strength	Bonding rubber and plastics; creating semiflexible seals

Table 11-2   Recommended Adhesives, by Bonding Material

	Bonding to:			
Adhesive	**Metal**	**Plastic/Foam**	**Rubber**	**Wood**
Contact cement	OK	Recommended	OK	Recommended
Cynaoacrylate	OK	Recommended	OK	OK
Epoxy	Recommended	Recommended	OK	OK
Hot melt	OK	Recommended	—	OK
PVAc	—	OK	—	Recommended
Solvent cement	—	Recommended	OK	—
Silicone	OK	OK	OK	OK

— Means not applicable or noneffective.

PART 3

Making Your Robot Move

CHAPTER 12

Batteries and Power

Forget miniature atomic piles or motivator power cells. The robots in science fiction are seldom like the robots in real life. With few exceptions, today's robots run on batteries—the same batteries that power a flashlight, portable CD player, or cell phone. To robots, batteries are the elixir of life, and without them, robots cease to function.

To be sure, batteries may not represent the most exciting technology you'll incorporate into your robot. But selecting the right battery for your bot will go a long way toward enhancing the other parts that *are* more interesting. Here's what you need to know.

This chapter makes reference to common electronic components, such as *resistors* and *capacitors*. If you're new to these concepts, be sure to check out Chapter 21, "Robot Electronics—the Basics," and Chapter 22, "Common Electronic Components for Robotics."

Overview of Practical Power Sources

Before getting waist-deep in the big muddy of battery selection, let's first review the practical power sources available for use with mobile robots. Note the word *practical*: there are plenty of potential power sources available in the world, but not all are suitable because of their size, safety, or cost.

- *Windup mechanisms* provide power using tension that is slowly released. A common type is based on the idea of a clock mainspring. These use a metal coil as a tension spring. The coil powers a shaft or other movement as its tension is relieved. For robotics, the typical windup mechanism is confined to small toys, particularly older collectable toys.

Figure 12-1   Desktop robot (this one uses tank tracks) hides its battery supply under an expansion panel. There's room for additional batteries (as needed) or a larger battery holder.

- *Solar cells* get their power from the sun and other light sources. A disadvantage of solar cells is that power is directly related to the intensity of the light. Robots that use solar power are often equipped with a rechargeable battery or a large capacitor; both store the energy collected by the solar cell for later use.
- *Batteries* are by far the most common and among the least expensive methods of powering any mobile device. Batteries can be grouped into two broad categories: nonrechargeable and rechargeable. Both have their place in robotics, and cost and convenience are the primary factors dictating which to use. These issues are discussed throughout the chapter. Figure 12-1 shows a robot and its power source—ordinary household batteries in a convenient holder.

Batteries for Your Robots

While there are hundreds of battery compositions, only a few are ideal for amateur robots.

CARBON-ZINC

Carbon-zinc batteries are also known as garden-variety "flashlight" cells: because that's the best application for them—operating a flashlight. They're an old technology and not up to the task of running a robot. We'll disregard them from here on.

ALKALINE

Alkaline batteries offer several times the operating capacity of carbon-zinc and are the most popular nonrechargeable battery used today. Robotics applications tend to discharge even alkaline batteries rather quickly, so a bot that gets played with a lot will run through its fair share of cells. Good performance, but at a price.

Alkalines are also available in a super-duper form; these go by a variety of self-descriptive names, like *Monster* and *Ultra*. High-capacity alkalines are made for loads demanding higher power. They're pretty expensive, though, making them best suited as emergency backup power, in case your regular robo batteries get unexpectedly worn out.

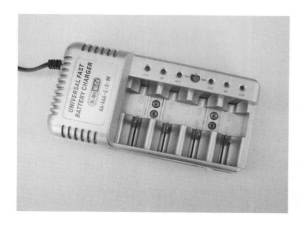

Figure 12-2 This universal charger works with both NiCd and NiMH batteries. You flip a switch depending on which kind you are recharging.

Also available from some retailers are rechargeable alkaline batteries, a special alkaline formulation that allows the cell to be recharged. Rechargeable alkalines are not commonly available, and they require a recharger designed for them. They can be revived dozens or hundreds of times before discarding.

Rechargeable alkalines are probably the best choice as direct replacements for regular alkaline cells. The reason: Most rechargeable batteries put out about 1.2 volts per cell; both rechargeable and nonrechargeable alkalines are rated at 1.5 volts per cell. See the section "Understanding Battery Ratings," later in this chapter, for more details about cell voltage.

NICKEL-CADMIUM

Being able to recharge your batteries rather than throw them out saves money and the planet. An older such technology is nickel-cadmium, or NiCd. These batteries are available in all standard sizes, plus special-purpose "sub" sizes for use in sealed battery packs for consumer products—things like rechargeable handheld vacuum cleaners, cordless phones, and so forth. Most battery manufacturers claim their NiCd cells last for a thousand or more recharges, though in actual use the reliable service life is less than this.

NICKEL METAL HYDRIDE

Nickel metal hydride is the replacement formulation for consumer rechargeable cells. Not only do they offer better performance than NiCds, they don't make fish, animals, and people (as) sick when they are discarded in landfills. They are the premier choice in rechargeable batteries today, including robotics, but they're not cheap. They require a recharger made for them (Figure 12-2).

NiMH batteries can be recharged 400 to 600 times and have what's known as a low internal resistance. That means they can deliver high amounts of juice in a short period of time. Unlike NiCds, NiMH batteries of any type don't exhibit a memory effect. NiMH cells can't be recharged as many times as NiCd batteries: about 400 full charge cycles for NiMH, as opposed to 2000 cycles for NiCd.

LITHIUM-ION

Lithium-ion (Li-ion) cells are frequently used in the rechargeable battery packs for laptop computers and radio control (R/C) toys, but they are also available as replaceable cells that fit into battery holders. Li-ion cells require specialized rechargers. Li-ion batteries actually use a wide variety of chemistries. One of the more common is lithium polymer, nicknamed LiPo or LiPoly.

There are three important aspect of Li-ion you should know about:

- If not properly recharged, Li-ion batteries can become damaged by *overheating*. Always use the proper type of recharger.
- Li-ion battery cells are available with and without *protection circuitry*. This circuitry, internal to the cell, is designed to prevent damage in case the battery is subjected to rapid charging or discharging. Under these circumstances the battery can get very hot, possibly catching fire and even exploding! Protected Li-ion batteries are more expensive, but recommended.
- Li-ion batteries have a much *higher per-cell voltage* than their alkaline, NiMH, or NiCd counterparts. Li-ion cells are available in the popular AA battery size. While you can use them in cells holders meant for conventional AA batteries, you must exercise care not to simply exchange the same number of Li-ion cells for other types of batteries. Otherwise the electronics or motors you connect to the Li-ion cells may be permanently damaged.

Another popular Li-ion cell size is the 18650, which looks like an AA battery that should go on a diet. The 18650 will not fit an AA holder, and ideally should be used with a battery holder specially made for it. (I prefer the self-contained kind with its own charge and recharge protection circuitry.)

There are also lithium-based batteries that are not rechargeable. These are commonly used in so-called long life applications, such as smoke detectors and "keyfobs" for keyless auto door locks. These lithiums don't have much use in robotics.

SEALED LEAD-ACID (SLA)

Sealed lead-acid batteries are similar to the battery in your car, and they work in much the same way. SLAs are "sealed" to prevent most leaks, though in reality the battery contains tiny pores to allow oxygen into the cells. SLA batteries are rechargeable using simple circuits, and are the ideal choice for very high current demands, such as battle bots or very large robots.

SLA batteries are most often sold in 6- and 12-volt packs. Inside the battery are multiple 2-volt cells—the cells are combined to create the desired voltage. Three cells are used to make a 6-volt pack, for example.

SO WHICH ONE SHOULD I PICK?

Most experienced builders select from a small palette of battery types based on the size and application of their robot.

- Alkaline, NiMH, and Li-ion (specifically LiPo) batteries are the most common among small tabletop robots. When using alkalines, you may choose between rechargeable and non-rechargeable types. Consider that alkaline and NiMH/LiPo cells produce slightly different voltages, and this can affect the operation of your robot. See the section "Understanding Battery Ratings" for more details.

- Midsize "rover" robots use larger NiMH or Li-ion cells; bigger robots still are ideally suited for SLA batteries. SLA batteries are available in a wide variety of capacities, and the capacity largely determines the size and weight of the battery. Lead weighs a lot, so reserve SLA batteries only for the bigger brutes.

Understanding Battery Ratings

Batteries carry all sorts of ratings and specifications. The two most critical are voltage and capacity.

VOLTAGE

The importance of *voltage* (or *V* for short) is obvious: the battery must deliver enough volts to operate whatever circuit it's connected to. A 12-volt system is best powered by a 12-volt battery. Lower voltages won't adequately power the circuit, and higher voltages may require voltage reduction or regulation, both of which entail some loss of efficiency.

Nominal Voltage Level

Battery voltage is not absolute. The voltage of a battery may—and usually does—diminish as it is used.

Alkaline (chargeable and nonrechargeable) cells are rated at 1.5 volts, though when new the actual voltage produced may be on the order of 10 to 30 percent higher. When fully charged, the typical 1.5-volt cell may deliver 1.65 volts. When fully discharged, the voltage may drop to 1.2 volts.

Batteries are rated at a *nominal* voltage (see Figure 12-3). Nominal means "in name only" and when used with batteries simply represents the predicted value. The battery delivers this specific voltage for only a portion of its discharge.

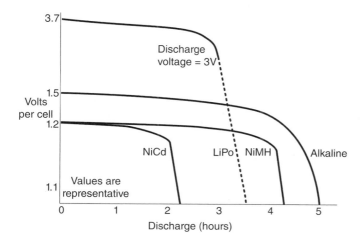

Figure 12-3 Simplified but representative discharge curves for several popular types of batteries used in robotics: alkaline, NiMH, lead-acid, and Li-ion.

Battery Type*	Nominal Voltage
Alkaline	1.5 volts
NiMH (and NiCd)	1.2 volts
Sealed lead-acid	2 volts
Li-ion	3.7 volts (typical)

* Individual cells.

To achieve higher voltages, you can link cells together, like lights on a Christmas tree. See the section "Increasing Battery Ratings," later in this chapter, for more details. By linking together six 1.5-volt cells, for example, your battery "pack" will provide 9 volts.

When Decreasing Voltage Becomes a Problem

The varying voltage of a battery as it discharges doesn't usually present a problem. That is, unless the voltage falls below a certain critical threshold. That depends on the design of your robot, but it usually affects the electronic subsystems the most.

Batteries are considered dead when their power level reaches about 70 to 80 percent of their rated voltage. That is, if the cell is rated at 6 volts, it's considered "dead" when it puts out only 4.8 volts. Some equipment may still function below this level, but the efficiency of the battery is greatly diminished.

Most electronics systems in robots use a voltage regulator of some type, and this regulator requires some overhead . . . usually several volts. As the battery voltage drops below that needed for the regulator, the electronics go into a "brownout" mode, where it still receives power, but not enough for reliable operation.

UNDERSTANDING BATTERY CAPACITY RATINGS

Current in a battery determines the ability of the circuit it's connected with to do heavy work. Higher currents can illuminate brighter lamps, move bigger motors, propel larger robots across the floor, and at higher speeds. Because batteries cannot hold an infinite amount of energy, the current of a battery is most often referred to as an *energy store,* and is also referred to as *capacity,* abbreviated C.

Capacity Expressed in Amp-Hours

Battery capacity is rated in *amp-hours,* or roughly the amount of amperage (a measure of current) that can be delivered by the battery in a hypothetical 1-hour period. In actuality, the amp-hour rating is an idealized specification: it's really determined by discharging the battery over a longer period.

What exactly does the term *amp-hour* mean? Basically, the battery will be able to (again, theoretically) provide the rated current for 1 hour. This current is expressed in *amps*—short for *amperes*—the common unit of expressing current flow from one part of an electrical circuit to another. If a battery has a rating of 5 amp-hours (expressed as "*Ah*"), the battery can theoretically provide up to 5 amps continuously for 1 hour, 1 amp for 5 hours, and so forth.

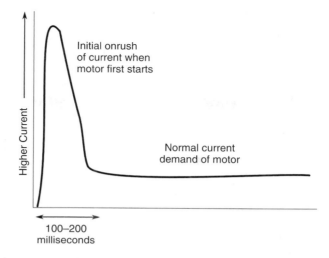

Figure 12-4   Current demand is the highest when electronic components, especially motors, are first switched on. The high onrush of current lasts a fraction of a second but can cause problems for the robot's electronics.

Plan for Extra Capacity

When choosing a battery, select one that has a capacity of at least 30 percent (preferably more) than the highest current demand of your robot. Design the robot with the largest battery you think practical.

Some components in your robot may draw excessive current when they are first switched on, then settle to a more reasonable level. Motors are a good example of this. A motor that draws 1 amp under load may actually require several amps at start-up, as shown in Figure 12-4. The period is very brief, on the order of 100 to 200 milliseconds.

The Dangers of Overdischarging

A battery produces heat as it discharges. Not only is the heat destructive to the battery, but it alters its electrical characteristics. The faster the discharge (such as a direct short of the battery terminals), the higher the heat that is generated. Very high heat can destroy the battery, and poses a fire hazard.

Capacity Ratings for Smaller Batteries

Smaller batteries are not capable of producing high currents, and their specifications are listed in *milliamp-hours*. There are 1000 milliamps in 1 amp. So a battery that delivers half an amp is listed with a capacity of 500 milliamp-hours, abbreviated *mAh* (or less accurately as *mA*).

Even larger batteries might be rated in milliamp-hours, as is the case with rechargeable NiMH cells, where it's not uncommon to see them listed as 2000 mAh—that's the equivalent of 2 amp-hours.

Recharging Batteries

Nickel metal hydride, rechargeable alkalines, and rechargeable Li-ion batteries all require special rechargers. Avoid substituting the wrong charger for the battery type you are using, or you run the risk of damaging the charger and/or the battery—and perhaps causing a fire.

Batteries are recharged by applying voltage and current to their power terminals. Exactly how much voltage, and how much current, depends on the type of battery. Some general tips and observations:

- Rechargeable alkaline, NiMH, and Li-ion must use a battery charger designed for them.
- Always observe polarity when recharging batteries. Inserting the cells backward in the recharger will destroy the batteries and possibly damage the recharger.
- Most lead-acid batteries can be recharged using a simple 200- to 500-mA battery charger. The charger can even be a DC adapter for a video game, as long as the output voltage of the DC adapter is slightly higher than the voltage of the battery. Remove the battery from the recharger after 18 to 24 hours.
- Li-ion batteries can *catch fire* if they are incorrectly recharged. *Only* use a recharger specifically designed for the cell or battery pack you are using.

Robot Batteries at a Glance

As you've seen, batteries can be rechargeable or nonrechargeable. And different battery types also vary by the volts per cell. Table 12-1 shows, in a nutshell, the common battery types most often used in robotics, the nominal voltage they deliver per cell (when fully charged), and other selection criteria.

Common Battery Sizes

Battery sizes have been standardized for decades (see Figure 12-5), though most consumers are familiar with just a few of the more common types: N, AAA, AA, C, A, and 9-volt. There are many other "in-between" sizes as well.

For the most part, the size of the battery directly affects its capacity—assuming the same types of batteries are compared. For example, because a C battery provides roughly double the internal area of an AA battery, it stands to reason that the capacity of a C battery is about twice that of the AA cell. (In actual practice, size versus capacity is more complicated than this, but it'll do for a basic comparison.)

 There's more about batteries, battery sizes, and selection on the RBB Support Site!

Table 12-1	Batteries and Their Ratings				
Battery	Volts per Cell*	Application	Recharge†	Internal Resistance	Notes
Carbon-zinc	1.5	Low demand, flash-lights—not robots	No	High	Cheap, but not suitable for robot-ics or other high-current applications
Alkaline	1.5	Small appliance motors and electric circuits	No	High	Available every-where; can get expensive (replacement costs) when used in a high-current application like robotics
Rechargeable alkaline	1.5	Substitute for non-rechargeable variety	Yes	High	Good alternative to nonrecharge-able alkalines
High-capacity alkaline	1.5	Same as other alka-line cells, but can han-dle larger current demands	No	Low	More expensive than standard alkaline; keep for emergencies
NiCd	1.2	Medium and high cur-rent demand, includ-ing motors	Yes	Low	Slowly being phased out because of their toxicity
NiMH	1.2	High current demand, including motors	Yes	Low‡	High capacity; still a bit pricey
Li-ion	3.7§	High current demand, including motors	Yes	High	Expensive, but lightweight for their current capacity
Lead-acid	2.0	Very high current demand	Yes	Low	Heavy for their size, but very high capacities available

* Nominal volts per cell for typical batteries of that group. Higher voltages can be obtained by combining cells.

† Some nonrechargeable batteries can be "revitalized" by zapping them with a few volts over several hours. However, such batteries are not fully recharged with this method, and are redischarged very quickly.

‡ Internal resistance of NiMH batteries starts out low when the cell is new, but increases significantly as it is recharged over many times.

§ Li-ion cells have different voltage characteristics, depending on manufacturer; 3.7 volts per cell is common but is not considered a standard. Li-ion batteries are frequently used in packs containing multiple cells.

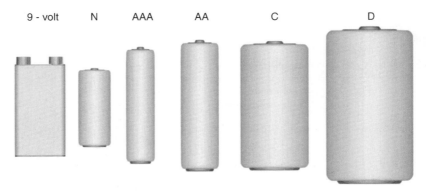

Figure 12-5   Comparison of consumer battery sizes, from N to D cell, plus 9-volt.

Increasing Battery Ratings

Through the magic of electricity, you can get higher voltages and current by connecting two or more cells together. Increasing volts or current depends on how you wire the batteries together:

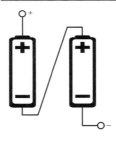
To increase voltage, connect the batteries in series. The resulting voltage is the sum of the voltage outputs of all the cells combined. This is the most common way to connect batteries.

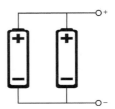
To increase current, connect the batteries in parallel. The current is the sum of the current capacities of all the cells combined.

Take note: When you connect cells together, not all cells may be discharged or recharged at the same rate. This is particularly true if you combine two half-used batteries with two new ones. The new ones will do the lion's share of the work and won't last as long.

Therefore, you should always replace or recharge all the cells at once. Similarly, if one or more of the cells in a battery pack are permanently damaged and can't deliver or take on a charge like the others, you should replace it.

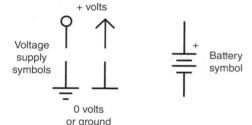

+ volts

Voltage
supply
symbols

0 volts
or ground

Battery
symbol

Figure 12-6 Various common battery symbols used in electronic diagrams.

Power and Battery Circuit Symbols

So far you've learned about using batteries in your robot; now it's time to discover how to actually use them in your robot designs. There's a little bit more to it than simply strapping a couple of cells to your robot and sticking the wires into some motors.

Let's start with the symbols used in electronic circuit designs to denote power and battery sources. Schematics are the road maps of electronic circuits. And in most circuits, it all starts with the power connection. For robotics, that power typically comes from batteries. Figure 12-6 shows the most commonly used symbols for power and batteries.

- When power comes from an arbitrary source (batteries, connection from wall transformer, whatever), it's often shown as a small circle, or sometimes an upward-pointing arrow. To complement the power connection, the circuit will also show another connection for *ground*. The exact form of the ground symbol is varied, but the set of three lines tapering to a point is among the most common.
- When power comes from a battery or battery pack, the symbol indicates the positive connection with a + (plus) sign. The – (negative) connection is inferred.

There are lots of different names for the ground connection. You may also encounter *common, negative, 0 volts, 0V, return, chassis,* and *earth*. Technically these terms don't always mean the same thing, but unless otherwise noted in the description of the circuit you can assume the terms are interchangeable.

Using Battery Cells in a Battery Holder

Perhaps the most convenient method of using multiple batteries with your robot is with a battery holder. Electrical contacts in the holder form the proper connections from the cell-to-cell; the voltage at the holder terminals is the sum of all the cells.

Holders are available for all the common battery sizes; 1, 2, 4, 6, and 8 cells are the most common. Battery holders often must conform to the shape of the object they are used in, so there are plenty of variations in how the cells are laid out—for example a four-cell holder may orient the cells all in a single row, or it might pack them side by side. See Figure 12-7 for some examples of single-side battery holder layouts. There are even more variations for battery holders with cells on both sides.

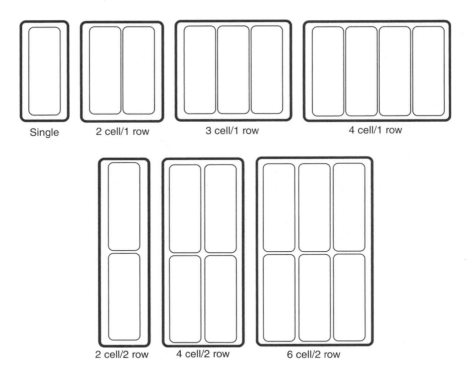

Figure 12-7 Sampling of battery holder layouts. Only a single side of the holder is shown here; the same layouts are available in double-sided holders, which carry twice the number of cells.

Ways to mount the holder: Fasteners provide a solid mounting, but you can also use Velcro (or similar hook-and-loop material), double-sided tape, or hot melt glue.

Try to mount the holder in a location that allows for ready access to the battery cells, so you can get to them. As necessary, consider the underbelly of the robot. Most holders secure the cells with a tight fit, so you can mount the holder upside down and the batteries should not (normally) fall out.

SNAPS AND CLIPS FOR 9-VOLT BATTERIES

Use a polarized battery snap for 9-volt batteries. Secure the battery to your robot with a 9-volt battery clip. These are available for either front or side mount. I like the metal holders, as you can manually squeeze the tangs together to make for a tighter fit against the battery.

Mount the clip to the robot using double-sided foam tape, hook-and-loop, or fasteners. I prefer the fastener approach when a permanent installation is required and I want to make sure the battery stays put.

Alternative: If you prefer to skip the holder, and the battery is not replaced regularly, stick it on with a small piece of Velcro or double-sided tape, as shown in Figure 12-8.

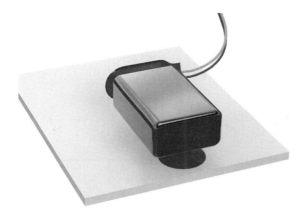

Figure 12-8 Secure 9-volt batteries by sticking them into place with Velcro or double-sided tape. If the battery is replaced often place it into a metal or plastic clip, and attach the clip to the robot.

BATTERY HOLDERS FOR "IN-BETWEEN" VOLTAGES

Sometimes you may want an in-between voltage. Just use an odd number of cells. For instance, five 1.2-volt rechargeable cells provide 6 volts; five nonrechargeable cells, 7.5 volts. While they do make battery holders for odd numbers of cells, they're not always easy to find, and they tend to be more expensive.

One alternative is to use a standard even-cell holder and "bridge" one of its cell "pockets" to produce an odd-cell voltage. By removing one cell from the holder, and replacing it with a jumper wire, you can make any odd-cell holder you wish.

Figure 12-9 shows the concept of bridging a cell pocket in a six-cell holder. For a temporary bridge use an ordinary jumper clip. For a permanent job you'll want to solder a length of wire between the battery posts in the last cell pocket.

There's more about batteries, battery holders, and battery packs on the RBB Support Site.

Using a Rechargeable Battery Pack

Many of today's consumer products use rechargeable battery packs. So can your robot. The battery pack can use cells that fit into all shapes and sizes of compartments. Wires and a quick-disconnect plug allow easy removal of the batteries, for things like recharging or replacement.

You can rob the battery pack out of a discarded consumer electronics device you no longer need. You'll then have batteries for your robot, as well as the charger that came with the phone.

Or you can purchase new packs. You're best off using battery packs designed for R/C applications, as detailed next, as these typically use standardized voltages compatible with consumer rechargers.

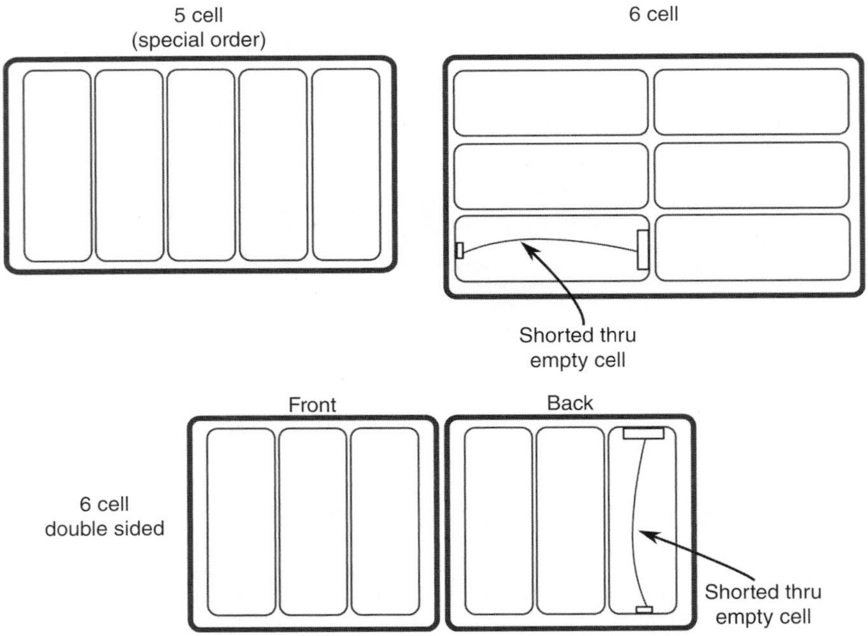
5 cell
(special order)
6 cell
Shorted thru
empty cell
Front
Back
6 cell
double sided
Shorted thru
empty cell

Figure 12-9 Create your own "odd-number cell" battery holders by bridging across the terminals of one of the cell pockets. This maintains electrical continuity for the holder but lets you use fewer batteries.

USING PREMADE R/C BATTERY PACKS

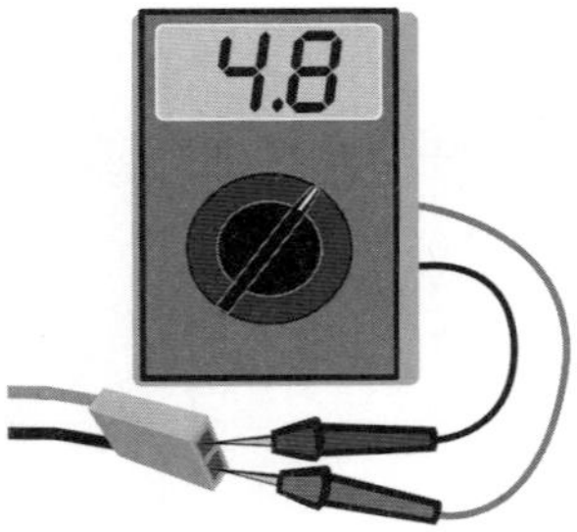
4.8

R/C applications, like model airplanes and model cars, are power hungry. High-capacity rechargeable battery packs are the norm. The packs are available in a variety of voltages and capacities from any local or online store specializing in R/C cars and planes.

Common packs voltages for NiMH and NiCd cells are:

4.8 volt
7.2 volt
9.6 volt

The 7.2-volt packs are perhaps the most useful for robotics. Note the fractional voltages in this list. These are the result of the 1.2-volts-per-cell batteries used inside the packs. Current capacities range from about 350 milliamp-hours to over 1500 milliamp-hours. The higher the current capacity, the longer the battery can provide juice to your robot. Unfortunately, higher-capacity batteries also tend to be larger and heavier.

You should always pick the capacity based on the estimated needs of your robot, rather than just selecting the biggest brute of a battery that you can find. Bigger batteries weigh more and they cost more. As you work more with robotics you'll get a sixth sense for the sizes of the batteries you need to power your creations.

Many newer model electronics (R/C toys included) come with rechargeable Li-ion batteries, which may have different voltage ratings than NiMH and NiCd. Common packs voltages for Li-ion cells are:

3.7 volt (one cell)
7.4 volt (two cells)
11.1 volt (three cells)

Note: the exact voltage depends on the type of Li-ion batteries inside the pack. Actual pack voltage may be a few tenths of a volt higher or lower.

MAKING YOUR OWN RECHARGEABLE BATTERY PACK

Suppose you don't want to use an already-made battery pack. You can assemble your own packs using affordable NiMH or NiCd cells. I recommend NiMH batteries as they provide for high capacities, with ratings of 900 to 3000 mAh, and over.

Both NiCd and NiMH battery packs require rechargers designed for them. The better battery rechargers work with a variety of pack voltages.

You can arrange the cells in whatever layout best suits the space constraints of your robot. Remember that when batteries are connected one after the other in a string (series), the volts of each cell (1.2 volts for both NiMH and NiCd) are added together. Most battery packs use cells in series. Start with one cell, and wire its positive terminal to another cell's negative terminal. Continue until all the cells are connected.

If you need to attach the cells using wire, match the gauge of the wire with the current demand from the battery. Thicker wire can handle more current. A good starting size is 14- or 16-gauge stranded (not solid) wire. See the RBB Support Site for a handy reference chart of wire gauges.

After you've soldered the batteries together, use "battery shrink-wrap," a PVC plastic tube to enclose everything into one nice package. Use a hair dryer on high to shrink up the plastic so it makes a snug fit around the batteries. Battery shrink-wrap is available at many hobby stores catering to builders of R/C airplanes and cars, as well as online battery outlets.

To complement the pack, you'll need a compatible battery charger that is designed not only for the total voltage of all the cells, but for the type of cells and their amp-hour capacity. Rechargers are available for either NiMH or NiCd (some are switchable), with common voltage ratings as follows:

Voltage	Number of Cells	Voltage	Number of Cells
2.4	2	9.6	8
4.8	4	12	10
7.2	6	14.4	12

In-between voltages are also possible, using an odd number of cells. Seven cells of either NiCd or NiMH batteries is 8.4 volts, for example. You just need to make sure your battery charger is adjustable for that voltage.

Use a polarized connector so you can easily hook up your battery to the rest of your robot. *Never* leave the wires bare, as this increases the chance of a short circuit. Shorting freshly charged batteries can cause fire or burns and can permanently damage the battery.

Best Battery Placement Practices

For obvious reasons, the battery in your car is intentionally mounted for easy access. The same obvious reasons apply to robots. When possible, the battery holder or pack for your bot should be located where it affords quick and reliable access.

One ideal location for the battery pack or holder is on the underside of the robot. This assumes there is enough ground clearance between the robot and the floor.

Mounting on the bottom allows for quick access to the cells, either for replacement or recharging: just turn the robot over. And it saves space for electronics and other parts. Avoid any location that requires you to dismantle lots of parts of the robot just to access the batteries.

Wiring Batteries to Your Robot

There are lots and lots of ways to wire batteries to the rest of your robot. Picking the right method depends on the design and complexity of your bot.

In all cases, it's important to use a wiring system that eliminates the chance of short-circuiting the terminals of the batteries. It's also a good idea to use a connection scheme to prevent damage if the battery pack is connected to the robot circuitry in reverse, as detailed in the following section.

Simple bots, replaceable cells. For the most basic robots, using battery holders with individual cells, you can merely solder the battery pack to the electronics and motors. To replace or recharge the batteries, just remove them from the holder.

Solderless breadboards. Rather than solder to the electronics, insert the wires from the pack into the + and − rails of the breadboard. You'll need to solder a short (half-inch) length of 22-gauge solid conductor wire to the end of the battery pack leads, so you can plug into the breadboard.

Rechargeable packs with a standard 0.100″ two-prong female plug. Use corresponding male header pins on your circuit or breadboard so you can readily connect and disconnect the batteries. Danger! This poses a chance of reverse polarity, so consider one of the solutions under the section "Preventing Reverse Battery Polarity" (later in this chapter).

Battery packs with a polarized plug. Use the corresponding socket on your circuit board or breadboard. A popular polarized connector is the coaxial barrel plug, discussed in the next section. It can also be used on battery holders; use a six-cell (7.2- or 9-volt) holder and a 2.1mm barrel plug to power an Arduino board with batteries.

Battery packs with a customized polarized plug. You can make your own polarized connector systems or adapt one from the battery packs used in R/C model airplanes and cars. These are also discussed in the next section.

Preventing Reverse Battery Polarity

At all costs, you want to avoid reversing the connection of your batteries when you plug them into your robot. At best, your bot won't function; at worst (and more likely), you'll instantaneously blow out some or all of the electronics.

There are two general choices: mechanical interlock and electronic polarity protection.

MECHANICAL POLARIZED CONNECTIONS

You can reduce and even eliminate the possibility of hooking up your batteries the wrong way simply by using a "polarized" connector. A coaxial barrel connector used with many low-voltage DC wall transformers is by far the simplest and most common. You need a mating socket on your circuit or breadboard.

Barrel plugs and sockets come in different diameters, sized in millimeters. You need to make sure the plug on your battery pack is the right size. The Arduino microcontroller boards, for example, use a 2.1mm barrel plug; see Figure 12-10. Many sellers of the Arduino offer battery packs and 9-volt battery connectors with the 2.1mm barrel plug already on them.

The center connector on the plug can be + or –. When soldering a barrel plug to your battery pack (or using a wall transformer with a barrel plug already on it), *be absolutely sure to observe the correct polarity*! Center positive is the most common, but never assume that's the case. Always double-check.

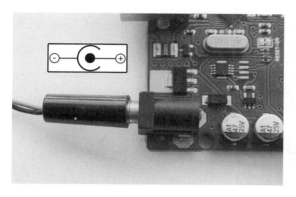

Figure 12-10 A polarized barrel plug assures that the power to your microcontroller or other electrics is never reversed—positive for negative.

Most wall transformers with barrel plugs indicate the polarity of the plug, using a pictogram like that in Figure 12-10. Double-check the polarity with your digital multimeter. If you're not sure how to check voltages with your meter, see Chapter 21, "Robot Electronics—the Basics" for details.

You can make your own bare-bones polarized connections using ordinary 0.100" header pins and sockets. Some soldering is required. These also work with solderless breadboard circuits. The idea is to use three or four pins—rather than just two—and wire up the pins in a way that it's impossible to connect the battery incorrectly.

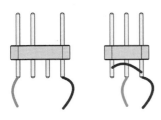

Three-wire polarized plug. Use a block of three 0.100" header pins. Solder the + lead from the battery pack to the center pin. Solder the − lead to the other pins.

Four-wire polarized plug. Use a block of four 0.100" header pins. Solder the + and − leads from the battery pack to the two outside pins. Cut off one of the inside pins, and insert the cut pin into the corresponding female connector on your circuit or breadboard. The cut pin will prevent you from reversing the plug.

See Chapters 21 and 22 for more ideas on working with 0.100" header pin connectors.

You can also buy polarized connectors that restrict or disallow you mating them backward. Polarized connectors and wiring harnesses with polarized connections are available at local and online R/C model stores.

ELECTRONIC POLARITY PROTECTION

When you can't use mechanical means to avoid physically connecting the batteries backward, you can turn to some simple electronics that will protect the rest of your circuitry.

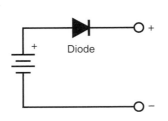

The simplest method is to put a diode inline with the incoming + connection from the battery. Diodes restrict current flow to one direction only. If you reverse the power leads from the battery, the diode will stop current from entering the circuit. Nothing will work, but your circuit won't be damaged.

Choose a diode with the current-carrying capacity for the circuit. The 1N4001 silicon diode handles up to 1 amp; the 1N5401 diode, up to 3 amps.

The downside to this method is that there is a voltage drop across the diode. For silicon diodes the drop is about 0.7 volts. For Schottky diodes the drop is only about 0.3 volts. See Chapter 22, "Common Electronic Components for Robotics," for information about different kinds of diodes.

Adding Fuse Protection

Big lead-acid, NiMH, and Li-Po batteries can deliver a most shocking amount of current. If the leads of the battery accidentally touch each other, or there is a short in the circuit, things could melt and a fire could erupt.

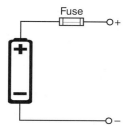

Fuse

That's why fuses were invented. Fuse protection helps eliminate the calamity of a short circuit or power overload in your robot. Connect the fuse in line with the positive terminal of the battery, as near to the battery as possible. You can purchase fuse holders that connect directly to the battery wire.

Choosing the right value of fuse can be a little tricky. It requires that you know *approximately* how much current your robot draws from the battery during normal and stalled motor operation. You can determine the value of the fuse by adding up the current draw of the various parts of your robot, then tacking on 20 to 25 percent overhead.

Assuming you're not building a big-brute combat bot, you can use your digital multimeter to determine the power draw of your robot. It's easiest if your meter has a 10-A (10-amp) current setting. See Chapter 13, "How to Move Your Robot," for more details on how to connect your multimeter to read current draw.

Let's say that the two drive motors in the robot draw 2 amps each, the main circuit board draws 500 milliamps (0.5 amp), and other parts draw less than 1 amp. Add all these up and you get 5.5 amps. Installing a fuse with a rating of at least 6 amps will help ensure that the fuse won't burn out prematurely during normal operation. Adding that 20 to 25 percent margin calls for a 7.5- or 8-amp fuse. You may have to get the next-highest standard value.

Recall that motors draw lots of current when they are started. To compensate for the sudden inrush of current, use a "slow-blow" glass-type fuse. Otherwise, the fuse may burn out prematurely.

Fuses don't come in every conceivable size. For the sake of standardization, choose the 3AG fuse size—these fuses measure 1-1/4″ × 1/4″. Holders for them are easy to find at any electronic parts seller.

An alternative to glass fuses is the *resettable PPTC fuse*. PPTC stands for "polymer positive temperature coefficient," and are miniature electronic components that react to the heat caused by high currents. If too much current flows through the fuse, it "trips" momentarily, causing a break in the circuit. When the circuit fault (like a short circuit) is removed, the device cools back down, and it reconnects the circuit. PPTC (also referred to as PTC) fuses are smaller than standard glass fuses and are used in the same way.

Like standard glass fuses, you need to match the current rating of the resettable fuse to the highest acceptable current draw for your circuit. Select the fuse based on its *trip* current, the maximum current you want to allow. Devices are available with trip currents as low as 100 milliamps (0.1 amp) to over 50 amps.

Regulating Voltage

The parts of your robot may need specific voltages to operate properly. This is most often the case with electronics, which typically require 5 or 3.3 volts. The exact voltage can vary a bit, but not much. For example, on a 5-volt system, the acceptable voltage levels may be from 4.5 to 5.5 volts. No higher, no lower.

DC motors typically don't require voltage regulation. Most run fine over a wide range of voltages. Increasing the voltage has the effect of making the motor run faster, and vice versa. R/C servo motors are an exception. Unless otherwise specified, they need to be operated at between around 4.5 and 7.2 volts.

Voltage regulation is accomplished in a number of ways. Here are the five most common; the first four are explained in more detail in this chapter.

- *Zener diodes* clamp the voltage to a specific level and won't let it get any higher.
- *Linear voltage regulators,* the most common variety, are cheap but relatively inefficient. In effect, they "step down" voltage from one level to another; the difference in voltage is dissipated as heat.
- *Switching voltage regulators* are more efficient—some boast up to 95 percent. They're recommended over linear voltage regulators, but they may require more external components to implement in your designs. Many switching regulators can increase voltage—boost 3 volts to 5 volts, for example—as well as produce negative voltages from a positive voltage source.
- *Modular DC-DC converters* are self-contained voltage changers. Internally they use one or more switching regulators, and they also include all the additional components required. They're more expensive, but to use one you just add it between your robot's battery and its electronics.

ZENER DIODE VOLTAGE REGULATION

A quick and inexpensive method for providing a semiregulated voltage is to use zener diodes. A typical hookup diagram is shown in Figure 12-11. You can use zener regulation for circuits that don't consume a lot of power—say, under an amp or two.

With a zener diode, current does not begin to flow through the device until the voltage exceeds a certain level. This level is called the *breakdown voltage.* Any voltage over this level is then "shunted" through the zener diode, effectively limiting the voltage to the rest of the circuit. Zener diodes are available in a variety of voltages, such as 3.3 volts, 5.1 volts, 6.2 volts, and others. A 5.1 zener is well suited for use on circuits needing a +5 volt supply.

Zener diodes are rated by their tolerance—1 percent and 5 percent are common. If you need tighter regulation, get the 1 percent kind.

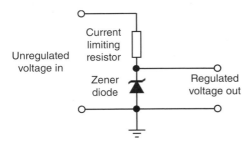

Figure 12-11   Zener diodes provide a simple form of voltage regulation. The value of the resistor depends on the incoming and regulated voltage levels and the amount of current the circuit is expected to draw. See the text on how to select the proper resistor value.

They're also specified by their power rating, expressed in watts. For low-current applications, such as operating a very small circuit, a 0.25- or 0.5-watt zener should be sufficient; higher currents require larger 1-, 5-, and even 10-watt zeners. Note the resistor shown in Figure 12-11. It limits the current through the zener.

To calculate the value of this resistor, you need to know the maximum current draw of your circuit. You have to look up the specifications of the components you are using. Add them together and that's the expected current draw of your circuit.

You then do a bit of math:

1. Calculate the difference between the input voltage and the voltage rating of the zener diode. For example, suppose the input voltage is 7.2 volts, and you want to use a 5.1-volt zener:

    7.2 − 5.1 = 2.1 volts

2. Determine the current draw of your circuit, as noted above. You want to add an overhead margin of about 150 to 200 percent. If, for example, the circuit draws 100 mA (milliamps), then

    $0.1 \times 2 = 0.2$

    In the preceding equation, 0.1 is 100 milliamps.

3. Determine the *value* of the resistor by dividing the current draw by the dropped-down voltage:

    2.1/0.2 = 10.5 ohms

    The nearest standard resistor value is 10 ohms, which is close enough.

4. Next, determine the *wattage* of the resistor. You do this by multiplying the difference in voltage from step 1, by the current draw of the circuit.

    $2.1 \times 0.2 = 0.42$ watts

    Resistors are rated in watts and fractional watts: 1/8, 1/4, 1/2, 1, 2, and so on. Pick a resistor wattage at or above the calculated value. For this example, a 1/2-watt resistor will work.

5. Finally (whew!), determine the power dissipation for the zener diode. This is done by multiplying the current draw of the circuit by the voltage rating of the zener:

    $0.2 \times 5.1 = 1.02$ watts

You should use a zener rated at no less than 1 watt.

There is some simplification used in these calculation formulas—the reverse current of the zener is ignored, for example. But there are many approximations anyway, such as the overhead margin and use of standard component values. If you find your components get too hot, use a higher-wattage zener and resistor. If the voltage is too low, slightly decrease the value of the resistor.

LINEAR VOLTAGE REGULATION

Solid-state linear voltage regulators provide much more flexibility than zener regulators, and they're relatively inexpensive—a few dollars at the most. They are easy to get at any electronics parts outlet, and you can choose from among several styles and output capacities.

Two of the most popular voltage regulators, the 7805 and 7812, provide +5 volts and +12 volts, respectively (other voltages are available—just change the last two digits). You connect them to the + (positive) and – (negative or ground) rails of your robot, as shown in Figure 12-12. In normal practice, you also place some capacitors across the input and output of the regulator. These act to smooth out any instantaneous fluctuations in the voltage.

The smallest linear regulators are provided in the compact TO-92 transistor package (see Chapter 21 for more on common electronic component sizes). In fact, they look just like small transistors. For the 7805 and 7812 regulators, and depending on the manufacturer, these are often identified with an "L" within the part number, for example, 78L05. The TO-92 regulators are limited to use in circuits drawing 100 mA or less of current.

More current is provided by regulators in the TO-220 style "power transistor" package. Use these when powering circuits that consume less than 500 milliamps. Other variations of the TO-220 style regulator can handle 1 amp or more. Check the datasheet that comes with the part you order.

 Pinouts of voltage regulators can vary. Negative voltage regulators often have different pin assignments than positive voltage regulators. Reversing the connections to a voltage regulator will burn it out, so *always* double-check with the datasheet for the device you are using. Always.

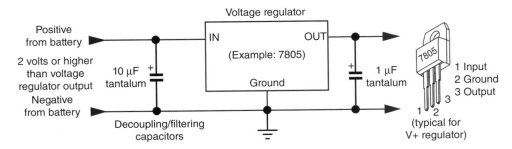

Figure 12-12 A linear voltage regular requires no external parts, though it is customary (as good design) to include capacitors to help stabilize the voltage provided by the regulator.

If you need more current, linear regulators are provided in larger TO-3 style transistor packages. They provide current outputs to several amps, depending on the exact device.

You can also get adjustable voltage regulators. These are just a few of the more common ones:

- The LM317T is an adjustable regulator in a TO-220 package. With some external parts (see its datasheet), you can adjust it to deliver from 1.2 to 37 volts, at a maximum of 1.5 amps.
- The LM317L offers the same voltage span, but is designed for use on circuits demanding 100 milliamps or less.

 In order to enjoy the full current capacity of a voltage regulator you need to mount it on, or attach it to, a heat sink. Expect somewhat less current-carrying ability when a heat sink is not used.

Linear regulators require that the input be about 2 volts over the expected output voltage. For example, for a 5-volt regulator, the unregulated supply should be about 2 volts higher, or 7 volts. By the same token, avoid applying too much voltage in relation to the output. The regulator throws off the extra voltage as heat, which is not only wasteful of energy but potentially harmful to the regulator. For that 5-volt regulator, the input voltage should be between 7 and 12 volts.

SWITCHING VOLTAGE REGULATION

Linear regulators, just described, basically take an incoming voltage and clamp it to some specific value. Linear regulation isn't very efficient; as the voltage is stepped down to its desired level, excess energy (the difference between the input voltage and output voltage) is wasted in the form of heat.

An alternative to linear regulators is the *switching* (or switching-mode) voltage regulator, which is more expensive but more efficient. Most high-tech electronics equipment uses switching power supplies. They're common and not frightfully expensive.

SWITCHING VOLTAGE REGULATION BOARD

Making a switching voltage regular can be a *fiddly* affair; most regular ICs require an inductor and several well-chosen electrolytic capacitors. While building a switching regulator is certainly not beyond the home builder, to be honest most of us take the easy road and buy a finished switching regulator.

You can find miniature regulator boards with one or more voltage outputs—3.3, 5, and 9 volts are not uncommon—at 2 to 3 amps. A step-down switching regulator is also called a *bucking* or *buck* converter. It efficiently steps down the voltage while stepping up the current from the supply. Figure 12-13 shows one such bucking regulator that's smaller than a postage stamp. It supports selectable output voltages.

A special class of switching voltage regulators can increase a lower voltage to a higher one. These are called *step-up* or *boost* regulators. One typical application is turning 3 volts from a pair of 1.5-volt alkaline cells into 5 volts, to run some microcontroller or other circuit. The downside of stepping up the voltage is a comparative reduction in current from the supply.

Figure 12-13 This small and affordable step-down switching regulator has multiple outputs, including a variable output where you can finetune the voltage.

 As you might surmise, some switching regulators combine the step-down and step-out functionality in one package. These are called *buck-boost* regulators: they provide a specific voltage regardless of whether you provide a higher or lower input.

Dealing with Power Brownouts

Robots don't use power in an even, predictable manner. One moment the robot's sitting still, using very little power, waiting for the right time to pounce on your poor unsuspecting cat. The next moment it's screaming down the hall after the furry feline, burning amps like they're going out of style.

Every time the motors kick in, a high amount of current is drawn from the batteries. This increase in current consumption can make the voltage delivered by the batteries momentarily dip. If these same batteries also supply the electronics in your robot, a condition called *brownout*, or *sag,* could occur.

As the battery voltage drops below that needed for the regulator, the electronics go into a brownout mode. They get enough power to stay on, but not enough for reliable operation. To avoid possible brownouts:

- Use separate batteries for the electronics and the motors. This is the single best way to avoid brownout problems. Use a small-capacity battery or pack for the electronics, and a pack with larger AA, C, or D cells for the motors. Note: To make this work, *the ground leads of both battery supplies must be connected*!
- Use batteries with a higher per-cell voltage to ensure enough overhead for proper voltage regulation. If you're using NiCd or NiMH cells, for instance, which put out 1.2 volts per cell, switch to rechargeable alkaline. These produce 1.5 volts per cell.
- Use one or more additional batteries to increase the voltage provided by the pack. Though nonstandard—and sometimes a bit hard to find—use a five-cell battery pack with your 1.2-volt NiCd or NiMH batteries. That increases the pack voltage from 4.8 to 6 volts.

- Power the electronics from a single 9-volt battery. Use appropriate voltage regulation, of course. The Arduino microcontroller boards have onboard regulation that will take the 9 volts input and provide the necessary 5 volts.
- Don't let your batteries get so discharged that they can't provide even the minimal operating current for your bot.
- Design for lower-voltage electronics. Many newer microcontrollers operate at 3.3 volts.

On the Web: Bonus Content

Check out the RBB Support Site for:

- Step-by-step soldering guide on how to solder a barrel plug onto a battery holder or DC wall transformer. Make your own polarized power connectors.
- Using single batteries versus multiple batteries
- Building voltage monitor circuits
- Tips on best battery care and use

CHAPTER 13

How to Move Your Robot

By definition, all mobile robots *move*. To propel themselves across the floor, a robot might use wheels, or perhaps tank tracks, maybe even legs. Moving your robot is called *locomotion,* and how these wheels, tracks, and legs are arranged is called the *drive geometry.* There are many variations of drive geometries, some relatively easy to achieve, and others not.

Selecting the right locomotion system and drive geometry involves figuring out what you want your robot to do. But it also takes assessing the mechanical requirements of constructing the drive mechanism. It's easy to "bite off more than you can chew," and design a robot propulsion system that is not practical for you to build. A good example is robots with legs. These are much harder to build than wheeled bots, and they often require regular maintenance to keep everything tight and aligned.

In this chapter you'll learn about locomotion systems and drive geometries, and how to select the best one for the robot you're building.

 Chapters 17 through 19 in Part 4 contain additional specific details about using locomotion principles to power three common types of robots: wheeled, tracked, and legged. This chapter introduces you to the basic concepts, plus issues common to all three varieties of locomotion systems.

Choosing a Locomotion System

Let us take a quick look at the three main locomotion systems used with mobile robots. Don't worry if some of the terminology in the comparison table is new to you; I'll explain it to you as we go along.

Locomotion	Drive Considerations	Mechanical Considerations
Wheels	• Most common arrangement is 2 wheels on opposite sides of the base, with 1 or 2 casters or skids for balance. Typical variations include 4- and 6-wheeled bases. These do not require balancing casters/skids. • Size of wheel greatly influences traveling speed of robot. Larger wheels (for a given motor speed) make the robot go faster. • On 2-wheeled robots with support caster/skid, the wheels can be mounted centerline in the base or offset to the front or back. • Distance measuring (*odometry*) more reliable with wheeled bases. Accurate travel distance calculations are difficult with tracked and legged robots.	• Mounting wheels to motors or wheels to shaft is the hardest part of building a wheeled base. R/C servo motors provide a consistent means for mounting small wheels to them, so these types of motors are quite common in small mobile robots. • Modest degree of accuracy needed in mounting the wheels to avoid *run-out*, side-to-side wobble as the wheel rotates.
Tracks	• The tracks form a wide base that enhances stability of the vehicle. The mechanics of the tracks creates a "virtual" wheel with a very large surface area that contacts the ground. • No need for a support caster or skid. • Though not as common, the tracks may be augmented by wheels—similar to the half-track military vehicle. The tracks are shorter and support only one end of the base.	• Suitable track material can be hard to find; most common approach is to hack a toy tank. • The large surface area of tracks greatly increases friction; tracked vehicles can have trouble making turns, and the tracks can pop off if they are made of flexible rubber. • Rubber track tracks (the most common on hacked toys) can stretch over time. A track tensioner mechanism is recommended.
Legs	• Variations include 2, 4, 6, and even 8 legs; 6 legs (*hexapod*) is the most common. • Most legged robots use *static* balance, meaning the arrangement of the legs on either side of the robot base prevents it from toppling over. More rare is *dynamic* balance, where weight on the base is shifted to compensate for stepping. • Joints of each leg are defined as *degrees of freedom* (*DOF*): the more DOF, the more agile the platform, but the more difficult to build.	• Of all locomotion types, legs require the greatest degree of machining and assembly. • Flexing of legs can cause stress in material; acrylic plastics can break over time. • Legs with independent articulation (each leg can move separately and independently) are the most difficult to construct. An easier alternative is the "linked gait" articulation, where the movements of legs are linked together. Fewer moving parts and motors required.

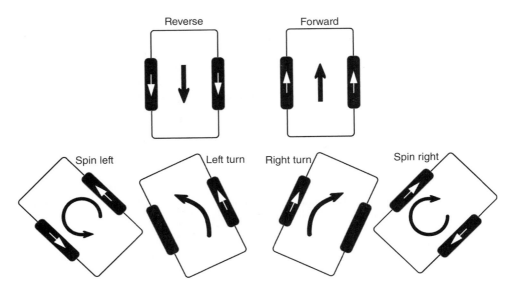

Figure 13-1 Differential steering involves using two motors on either side of the robot. The robot steers by changing the speed and direction of each motor.

Locomotion Using Wheels

The drive geometry for robots that use wheels is defined by how each one is steered. There are a lot of choices.

DIFFERENTIAL STEERING

The most common way to move a robot is with *differential steering*. The basic form consists of two wheels mounted on either side of the robot, as shown in Figure 13-1. It's called differential steering because the robot is steered by changing the speed and direction ("difference") between these two wheels.

A feature of most differentially steered robots is that they use one or two casters or skids, placed centerline over the robot in the front and/or back, to provide support for the base. See Chapter 17, "Build Robots with Wheels," for more information on selecting and using casters and skids with your robot designs.

Variations of the two-wheeled base include four or six wheels (4WD, 6WD), but the idea is the same. On bases that use more than two wheels, support casters or skids aren't generally needed.

One of the key benefits of differential steering is that the robot can spin in place by reversing one wheel relative to the other, as shown in Figure 13-2.

CAR-TYPE STEERING

Pivoting the wheels in the front is yet another method of steering a robot. Robots with *car-type* steering are not as maneuverable as differentially steered bots, but they are better suited for outdoor use, especially over rough terrain.

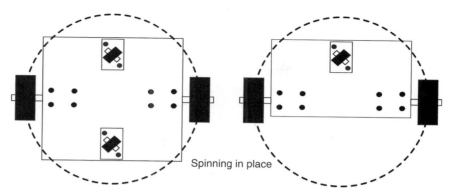

Figure 13-2 Differentially steered robots can turn in a circle within itself. This is called the steering circle, and the size of the circle depends on the dimensions of the robot and the placement of the wheels.

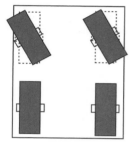

Why even bother with twentieth century car-type steering? One of the greatest drawbacks of the differentially steered robot is that it will veer off course if one motor is even a wee bit slow. You can compensate for this by monitoring the speed of both motors and ensuring that they operate at the same RPM. This, of course, adds to the complexity of the robot.

THREE-WHEELED TRICYCLE STEERING

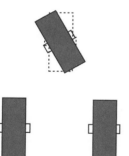

Car-type steering makes for fairly cumbersome indoor mobile robots; a better approach is to use a single drive motor powering two rear wheels, and a single steering wheel in the front; the arrangement is just like a child's tricycle.

The robot can be steered in a circle just slightly larger than the width of the machine. Be careful of the wheel base of the robot (distance from the back wheels to the front steering wheel). A short base results in instability in turns, causing the robot to tip over in the direction of the turn.

Tricycle-steered robots require a very accurate steering motor in the front. The motor must be able to position the front wheel with subdegree accuracy. Otherwise, there is no guarantee the robot will be able to travel a straight line. Most often, the steering wheel is controlled by a servo motor; servo motors used a "closed-loop feedback" system that provides a high degree of positional accuracy.

There are two basic variations of tricycle drives:

- *Unpowered steered wheel.* The steering wheel pivots but is not powered. Drive for the robot is provided by one or two other wheels.
- *Powered steered wheel.* The steering wheel is also powered. The two other wheels freely rotate.

A subvariant of the tricycle base design reverses the functionality of the wheels: two wheels in the front of the robot steer, and a third wheel in the back provides support. The third wheel can even be a simple caster or omnidirectional ball (see the section on caster types, below).

OMNIDIRECTIONAL (HOLONOMIC) STEERING

All of the steering methods described so far are known as *nonholonomic*. This basically (and simplistically) means that in order for the robot to turn, it has to change the orientation of its body. A good example of nonholonomic steering is a car. It can turn, but only by following a circle described by the axis of its four wheels. The car cannot instantaneously move in any direction of the compass.

Holonomic drives are distinctive in that they allow motion in any direction, at any time. They can go straight ahead, then suddenly move 90° sideways—all without changing the orientation of the vehicle. A ball demonstrates holonomic movement: it can instantaneously travel straight, then move in any direction of the compass.

The common trait of holonomic steering systems is that the robot is omnidirectional, able to move in both the x and y directions with complete freedom. The most common form of holonomic robot base uses three motors and wheels, arranged in a triangle.

How Omnidirectional Steering Is Achieved

How exactly does this work? In the majority of robots that use holonomic steering, the secret is in the wheels. There are three wheels, each driven by a separate motor. Each wheel has rollers around its circumference—wheels within the wheels, as shown in Figure 13-3. The rollers are at some angle to the main wheel. The rollers provide traction to the wheel when the wheel is turning, but the rollers also let the wheel "slip" sideways to make turns.

The robot moves "forward" by activating any two motors; it turns by adjusting the speed or direction of any and all three of the motors (see Figure 13-4). These types of wheels were originally designed for materials handling, as a substitute for conveyor belts, but recently they've found new use in robot propulsion.

Figure 13-3 An example wheel used in a holonomic robot. Instead of a solid rubber tire, the rim of the wheel is composed of several rollers that are set at opposite angles to the rotation of the wheel.

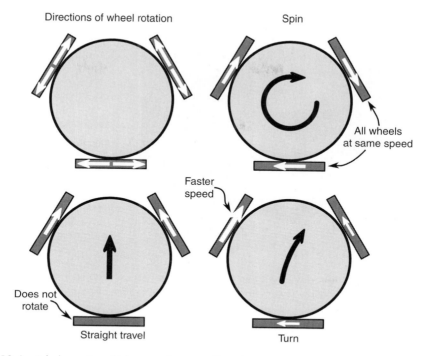

Figure 13-4   A holonomic vehicle steers by controlling the speed and direction of all of its wheels. For example, it goes forward by equally powering two wheels and letting the third "drag." It spins by equally powering all three wheels. Turns are achieved by altering the speed and/or direction of each wheel.

Locomotion Using Tracks

Since World War I the tank has become the symbol of military battles. The hulking mass of iron plowing over the earth, the shrieking sound of metal crushing against the ground, the blasts of fire from its cannon, all combine a sense of awe and respect. Little wonder that the tank design is popular among robot builders. The same principles that make a military tank superior for uneven terrain apply to robots or, in fact, any other type of tracked vehicle.

Too, a number of robots from science fiction films run on tracks. There's Number Five from the movie *Short Circuit,* Robot B-9 from *Lost in Space,* and many others. Unlike their walking cousins, these robots actually look feasible. Their wide tracks provide a solid footing over the ground. Figure 13-5 shows plastic tracks on a small plastic robot.

Like the common two-wheeled bot, tracked robots are also differentially steered. Two long, chainlike tracks are mounted parallel to each side of the vehicle. A separate motor powers each track in either direction via a sprocket; the toothed design of the sprocket ensures that the drive mechanism doesn't just spin if the track gets jammed. The tracks are kept inline by the use of idler rollers, placed at intervals along the sides of the vehicle.

Figure 13-5 Individual links can be added or removed to lengthen or shorten these solid plastic tracks. Since the tracks don't stretch, they won't easily come off the drive sprocket of the robot. That can be a problem with rubber tracks.

Tracks turn by skidding or slipping, and they are best used on surfaces such as carpet or dirt that readily allow low-friction steering. Very soft rubber tracks will not steer well on smooth, hard surfaces. To reduce friction, one approach is to always steer by reversing the tracks directions.

Because of the long length of the track, tank bots don't need a support caster or skid. The track acts as one giant wheel, one on each side.

- If both tracks move in the same direction, the robot is propelled in a straight line forward or backward.
- If one track is reversed, the robot turns (Figure 13-6).

This method is often referred to as *tank steering* or *skid steering,* but at the end of the day it's the same as differential steering.

The main benefit of a tracked vehicle is its ability to navigate over rough terrain. The tracks enhance the "grip of the road," allowing the robot to travel over loose dirt, sand, grass, and other surfaces that a wheeled robot can only dream about.

See Chapter 18, "Build Robots with Tracks," for additional information on track selection and use. You'll also find several hands-on projects for building robots with tracks.

Locomotion Using Legs

Thanks to the ready availability of smart microcontrollers, along with the low cost of R/C (radio-controlled) servos, legged automatons like the one in Figure 13-7 are becoming a popular alternative for robot builders. Robots with legs require more precise construction than the average wheeled robot. They also tend to be more expensive.

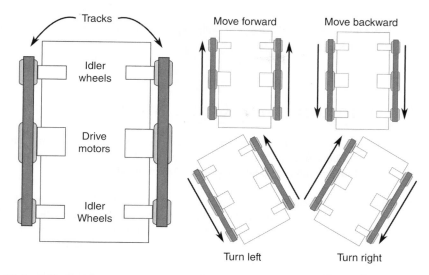

Figure 13-6 Differential steering using tracks is the same as with wheels, except that, to make a turn, one track needs to move in the opposite direction. If one track is stopped, the robot may "skitter" in the turn or the tracks may pop off.

Figure 13-7 This six-legged hexapod robot uses 3 motors per leg, for a total of 18 motors. Because of the high number of motors, these kinds of robots are more difficult and more expensive to build. (*Photo courtesy Lynxmotion.*)

Even a "basic" six-legged walking robot requires a minimum of 2 or 3 servos, with some six- and eight-leg designs requiring 12 or more motors. At about $12 per servo (more for powerful higher-quality ones), the cost can add up quickly!

Obviously, the first design decision is the number of legs:

- Robots with one leg ("hoppers") or two legs are the most difficult to build because of balance issues. In most two-legged bots the "feet" are oversized to offer the largest balancing area possible.
- Robots with four and six legs are more common. Six legs offer a static balance that ensures that the robot won't easily fall over. At any one time, a minimum of three legs touch the ground, forming a stable tripod.
- Walking robots with eight (or more) legs are possible, but their construction cost and problems with higher weight make them largely impractical as a springboard for amateur robotics.

Check out Chapter 19, "Build Robots with Legs," for more information on constructing multilegged robots.

Locomotion Using Other Methods

While robots with wheels, tracks, and legs are the most popular among robot builders, that doesn't mean there aren't other ways of moving a robot. Here are just a few alternatives that you might develop as you build up your robot construction skills:

- *Whegs* combine the action of wheels and legs into one unit. They're a favorite at the Biologically Inspired Robotics Laboratory at Case Western Reserve University, where they've adapted the idea from several robots designed for space and military use. An attribute of most (but not all) whegs is that they are compliant, meaning there is built-in flexibility to conform to the terrain.
- *Flippers* are similar to whegs but are intended primarily for locomotion across very sandy terrain or water. Whegs and flippers share common traits, allowing for amphibious robots that can go from land to water.
- *Multisegment* robots mimic the locomotion of caterpillars, snakes, and other crawling creatures. The robot crawls by systematically moving each segment a little bit at a time.
- *Hydra* robots combine two or more locomotion styles into one. For example, the robot may use articulated legs with wheels or small tracks as feet. Other types of hydra robots may combine the functionality of two front legs to serve as a gripper. The name hydra comes from Greek mythology; the Hydra was a beast with seven heads, who guarded the entrance to the underworld.
- *Unpiloted underwater* robots use various propulsion means, such as propellers and thrusters, to move around in fresh and saltwater.
- Similarly, *unpiloted flying* bots are scale versions of planes and helicopters.

On the Web: Managing the Weight of Your Robot

Check out the RBB Support Site for free articles on how to manage the weight of your robot and best distribute that weight to avoid tip-overs and other embarrassing design problems.

Choosing the Right Motor

Motors are the muscles of robots. Attach a motor to a set of wheels, and your robot can scoot around the floor. Attach a motor to a lever, and the shoulder joint for your robot can move up and down. Attach a motor to a roller, and the head of your robot can turn back and forth, scanning its environment.

There are many kinds of motors; however, only a select few are truly suitable for homebrew robotics. Let's learn about them now.

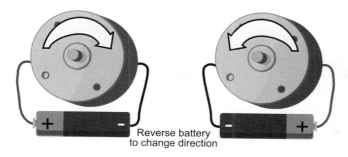

Figure 13-8 The direction of rotation of most DC motors may be altered simply by reversing the polarity of its power connections. With a battery or other DC power source connected one way, the motor spins clockwise. Reverse the polarity, and the motor spins counterclockwise.

Reverse battery to change direction

AC OR DC MOTOR?

Direct current—DC—dominates robotics; it's used as the main power source for operating the onboard electronics, for opening and closing relays, and, yes, for running motors that propel a robot across the floor. The alternative motor type, alternating current (AC), is seldom used in robotics. AC motors are best suited for things like household fans and other applications where power comes from a wall socket.

When looking for DC-suitable motors, be sure the ones you buy are reversible. Few robotic applications call for motors that run in one direction only. DC motors are inherently bidirectional, but some design limitations may prevent reversibility, so this is something you have to be on the lookout for.

The best and easiest test for reversibility is to try the motor with a suitable battery, as shown in Figure 13-8. Apply the power leads from the motor to the terminals of the battery or supply. Note the direction of rotation of the motor shaft. Now reverse the power leads from the motor. The motor shaft should rotate in reverse.

CONTINUOUS OR STEPPING MOTOR?

DC motors can be either continuous or stepping. Here is the difference: with a *continuous motor*, the application of power causes the shaft to rotate continuously (hence the name). The shaft stops only when the power is removed or if the motor is stalled because it can no longer drive the load attached to it.

With *stepping motors*, the application of power causes the shaft to rotate a few degrees, then stop. To keep rotating the shaft, power must be pulsed to the motor. Stepping motors are used when you want to control how far a motor turns, in either direction. They don't have the mechanical torque that a continuous DC motor has, but they're useful for certain tasks that don't need brute force.

With increased use of servo motors (see following section), stepper motors seem less used in amateur robotics. If you're interested in stepper motors be sure to read the bonus material on the RBB Support Site.

SERVO MOTORS

A special subset of continuous motors is the servo motor, which in typical cases combines a continuous DC motor with *electronic feedback* to ensure the accurate positioning of the motor. A common form of servo motor is the kind used in model and hobby radio control (R/C) cars and planes. We use these a lot in robotics, so this is a vital motor to get to know.

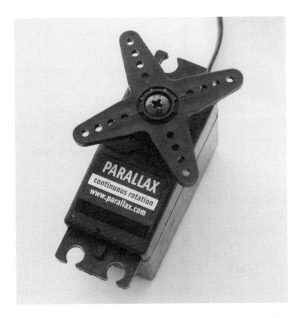

Figure 13-9 Servo motors for radio control (R/C) model airplanes and cars are DC motors, but with added control electronics and built-in gear reduction.

Compare the R/C servo motors in Figure 13-9 with the DC motors in Figure 12-12. The first thing you'll notice is that R/C servo motors come in a neat little rectangular box. This is one of their most alluring traits. Sizes of R/C servos are standardized (more or less), and they even have standardized mounting flanges, allowing you to easily add them to your robot creations.

R/C servos can be—and often are—used for the same jobs as stepper motors. Because they're cheaper, easier to find, and simpler to use than steppers, R/C motors have almost completely replaced the stepper motor in amateur robot designs.

For this reason, we devote a separate chapter just to them. See Chapter 15, "Using Servo Motors," for more information on using R/C servo motors—not only to drive your robot creations across the floor but to operate robot legs, arms, hands, heads, and just about any other appendage.

Motor Specs

Motors come with extensive specifications. But of all the specs for motors, only a small handful are truly meaningful to the amateur robot builder, so I'll just concentrate on those. These specifications are operating voltage, current draw, speed, and torque.

OPERATING VOLTAGE

All motors are rated by their *operating voltage*. With some small DC "hobby" motors, the rating is typically a range, something like 4.5 to 6 volts. For others, a specific voltage is specified. Either way, most DC motors will run at voltages higher or lower than those specified for it. A 6-volt motor is likely to run at 3 volts, but it won't be as powerful, and it will run slow.

Many motors will refuse to run, or will not run well, at voltages under about 40 or 50 percent of the specified rating. Similarly, a 6-volt motor is likely to run at 12 volts. As you may expect, as the speed of the motor increases, the motor will exhibit greater power.

 I don't recommend that you run a motor continuously at more than 200 percent its rated voltage, at least not for long. The electrical windings inside the motor may overheat, which can cause permanent damage to the motor. Motors not designed for high-speed operation may turn faster than their construction allows, which literally could cause them to burn up.

If you don't know the voltage rating of a motor, you can take a wild guess at it by trying various voltages and seeing which one provides the greatest power with the least amount of heat and mechanical noise. Let the motor run for several minutes, then feel the heat on the outside of the motor case. Listen to the motor; it should not seem as if it is straining under the stress of high speeds.

CURRENT DRAW

Current draw is the amount of current, in milliamps or amps, that the motor requires from the power supply. Current draw is more important when the specification describes *motor loading*, that is, when the motor is turning something or doing work. The current draw of a free-running (no-load) motor can be quite low. But have that same motor spin a wheel propelling a robot across the flow, and the current draw might increase several hundred percent.

Most DC motors use a permanent magnet inside. In these motors, which are the most common, current draw increases with load. You can see this visually in Figure 13-10. The more the motor has to work to turn the shaft, the more current is required. The load used by the manufacturer when testing the motor doesn't follow any kind of standard, so in your application the current draw may be more or less than that specified.

A point is reached when the motor does all the work it can do and no more current will flow through it. The shaft stops rotating; the motor has *stalled*. This is considered the worst-case condition. The motor will never draw more than this current unless it is shorted out. If your robot is designed to handle the stall current, then it can handle anything.

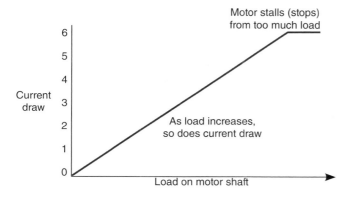

Figure 13-10 All motors draw current; the amount of current depends on the design of the motor and how much load is placed on the motor. As the load increases, so does the current. At some point, the load is too much for the motor and it stops turning, but it still consumes current. This is called the stall current.

When adding a motor to your robot you should always know the approximate current draw under load. All multimeters can be used to test current drawn by a motor. Learn how in the section "Testing Current Draw of a Motor," later in this chapter.

SPEED

The rotational *speed* of a motor is given in revolutions per minute, or *RPM*. Many continuous DC motors have a normal operating speed of 4000 to 7000 RPM. Certain special-purpose motors, such as those used in tape recorders and computer disk drives, operate as slowly as 2000 to 3000 RPM, and there are motors that operate at 12,000 RPM and higher.

You don't need a lot of speed for most robotic applications. In fact, for DC motors used to move your robot, the speed of the motor needs to be reduced to no more than 200 or 250 RPM—often even slower.

The easiest way to slow down a motor is to attach some gears to it. This also increases the turning force (called *torque*; see the next section), allowing the motor to push bigger robots or lift heavier objects. Discover more about gears and how they're used in motors for robots in Chapter 16, "Mounting Motors and Wheels."

TORQUE

Torque is the force the motor exerts upon its *load*—the load is whatever it's moving. The higher the torque, the larger the load can be and the faster the motor will spin. Reduce the torque, and the motor slows down, straining under the workload. Reduce the torque even more, and the load may prove too demanding for the motor. The motor will stall to a grinding halt and, in doing so, eat up current—not to mention, put out a lot of heat.

Torque is perhaps the most confusing design aspect of motors. This is not because there is anything inherently difficult about it, but because motor manufacturers have yet to settle on a standard means of measurement. Motors made for industry are rated one way; motors for the military, another. And most motors for consumer or hobby applications come with no torque ratings at all.

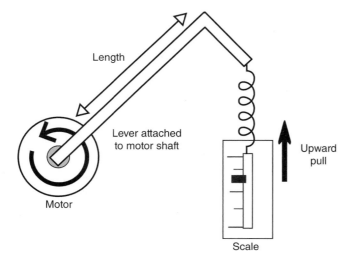

Figure 13-11 The power output or torque of a motor can be measured using a simple graduated spring scale (a fish-weighing scale will do). The motor is attached to the scale using a lever. The amount of pull with the length of the lever are used to indicate the torque rating of the motor.

At its most basic level, torque is measured by attaching a lever to the end of the motor shaft and a weight or gauge on the end of that lever, as depicted in Figure 13-11. The lever can be any number of units: 1 centimeter, 1 inch, or 1 foot. Remember this, because it plays an important role in torque measurement.

The weight can be either a hunk of lead or, more commonly, a spring-loaded scale. Turn the motor on, and it turns the lever. The amount of weight the motor lifts is its torque. There is more to motor testing than this, of course, but it'll do for the moment.

Now for the ratings game. Remember the length of the lever? That length is used in the torque specification.

- If the lever is 1 inch long, and the weight successfully lifted is 2 ounces, then the motor is said to have a torque of 2 ounce-inches, or *oz-in.* (Some people reverse the "ounce" and "inches" and come up with "inch-ounces." Whatever.)
- Or the torque may be stated in grams rather than ounces. In this case, a lever calibrated in centimeters may be used. This gives you grams-centimeter, or *gm-cm.*
- Torque for very large motors may be rated in pound-feet, or *lb-ft.*
- Becoming more popular is the newton-meter unit of torque. You may see it as *N m*, *N-m* or *Nm*. One N-m is equal to the torque that results from a force of 1 newton (no, not the fig kind) applied to a lever that is 1 meter long. (If you're interested, the newton is equal to the amount of force required to accelerate a mass that weighs 1 kilogram at the rate of 1 meter per second squared.)

STALL AND RUNNING TORQUE

The typical motor is rated by its *running torque—that is,* the force it exerts as long as the shaft continues to rotate. For robotic applications, it's the most important rating because it determines how large the load can be and still guarantee that the motor turns.

Manufacturers use a variety of techniques to measure running torque. The tests are impractical to duplicate in the home shop, unless you have an elaborate dynometer and sundry other tools. Instead, you can empirically determine if the motors are sufficient for the job by constructing a simple test platform, as described in the next section.

Another torque specification, *stall torque,* is sometimes provided by the manufacturer instead of or in addition to running torque. Stall torque is the force exerted by the motor when the shaft is clamped tight. The motor does not turn.

Testing Current Draw of a Motor

You can often just look at a motor and know it'll have enough torque for your robot. Less ensured is knowing how much current the motor demands when it's running. It's not possible, or even advisable, to infer the current draw of a motor just by its size, shape, or type.

Knowing the current draw is very important: you need to make sure the batteries of your robot have the right capacity and that any electronics you use can handle the current. Should the motors draw too much current, your electronics could overheat and be permanently damaged.

Many motors are provided with a current draw specification. As noted earlier in this chapter, the spec may be the current of the motor when it's free-running. That's helpful information, but it's not enough to know how the motor will behave when it's pulling weight—or worse, if the motors are loaded down so much they stop turning.

You can use a multimeter to accurately test the current draw of your motor when it's free-running, under normal load, and even stalled—completely stopped. There are two methods, each described below.

If you are new to the concept of multimeters, you'll want to read the manual that comes with yours, and see Chapter 21, "Robot Electronics—the Basics," for more information. What follows assumes you already have at least a basic understanding of how to operate your multimeter.

DIRECT MOTOR CURRENT MEASUREMENT

The steps that follow assume your multimeter has a special input for testing high currents. Many do, and the input is labeled **10A** (or similar), like the one in Figure 13-12. It's a rare motor for a desktop robot that draws more than 10 amps, even when stalled, but even so, check the manual for your meter to be sure the 10A input has a replaceable fuse. (The fuse is usually located in the battery compartment.)

If any of the following are true, skip this section and move to "Indirect Motor Current Measurement."

● My meter does not have a 10A (or higher) input.
● My meter doesn't have an input fuse.
● I suspect the motor may draw current in excess of 10 amps.

Follow the connection diagram in Figure 13-13. Be sure to connect the red (+, positive) lead of your meter in the 10A socket. Then, connect the test leads to your motor using jumper wires, the kind with heavy-duty alligator clips on the ends. You need two jumpers, one each for the red and black test leads. Then,

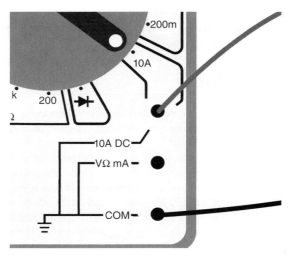

Figure 13-12 You may directly measure the current consumption of a motor by using a digital multimeter with a high amperage (10 amps or higher) input.

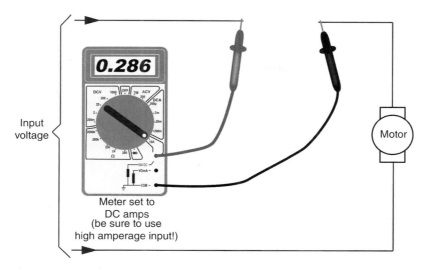

Input voltage

0.286

Meter set to
DC amps
(be sure to use
high amperage input!)

Motor

Figure 13-13 Test the current draw of a motor by interrupting the power line to the motor. Place the + and − leads of the multimeter between this power line, and take the reading. The meter must be dialed to read DC current.

1. Dial the meter to the 10A setting.
2. Apply power to the motor.
3. Observe the reading on the meter. It will be in amps. A reading of 0.30, for example, means 30 mA, or 0.3 of an amp. A reading of 1.75 means 1.75 amps.

This gives you the free-running torque of the motor. For other torque tests, you need to load down the motor.

Small hobby motors without a gearbox can be loaded down just by squeezing the shaft between your fingers. As you slow the motor down, watch the current increase. If you can stop the shaft from turning completely, the motor will be stalled. That's about as high as the current will ever get from the motor.

Small hobby motors with a gearbox can be loaded down by applying pressure to the output shaft of the motor or one of the gears in the gearbox nearest the shaft. This, of course, assumes the gears in the gearbox are accessible.

 With some gearbox motors you can temporarily remove the motor from the gearbox. Test it alone, then put the motor back into the gearbox.

Larger motors are harder to test just by stopping the shaft with your fingers. It's better to attach a wheel to the motor (the larger the diameter, the better), then try to apply load to the wheel. Don't use a pair of pliers or another tool to clamp down on the motor shaft or gears, as this may cause damage. At the very least, you can take the free-running measurement.

INDIRECT MOTOR CURRENT MEASUREMENT

You can still determine current draw from your motors if your multimeter lacks a high-current input or if the motor you're testing draws over 10 amps. The solution is to insert a low-value, high-wattage resistor inline between the + battery terminal and the motor.

The value of the resistor can be anything between 1 and 10 ohms, but it needs to be high wattage—10 or 20 watts, even more for motors that draw in excess of 10 amps.

1. Plug the red (+, positive) lead into the standard volts/ohm/current jack.
2   Use jumper clips to insert the resistor into the motor circuit, as shown in Figure 13-14.
3. Dial your multimeter to DC volts. If your meter is not autoranging, select a voltage at or just higher than the voltage you will apply to the motor.
4. Connect the red (+, positive) lead of your multimeter to the side of the resistor closest to the voltage source.
5. Connect the black (–, negative) lead of your multimeter to the other side of the resistor.
6. Apply power to the motor.

What you'll see is the voltage developed across the resistor. You then use one of the simple formulas of Ohm's law to determine the current flowing through the resistor. The formulas is

$$I = E/R$$

where I is current, E is voltage, and R is resistance.

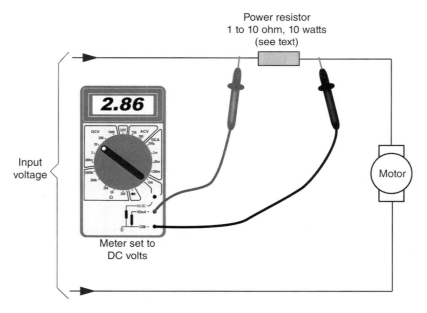

Figure 13-14 An alternative method for testing current draw is to place a low-ohm (1 to 10 Ω) high-wattage resistor inline between the power source and the motor. Change the meter to read voltage. See the text for how to correlate the voltage reading to current.

You are solving for *I*, or current, because you know the voltage developed across the resistor and you know the value of the resistor, in ohms. Let's suppose:

- The measured voltage is 2.86, and
- The resistor is rated at 10 ohms.

$$I = 2.86/10$$

or

$$I = 0.286, \text{ or } 286 \text{ milliamps}$$

Watch the voltage go up as you apply load to the motor shaft. See the previous section on how to test for currents under load and when stalled, using nongeared and geared motors.

Dealing with Voltage Drops

On most robots it's the motors that draw the most current. As the motors stop and start, the voltage provided by them can change. This happens because under heavy current draw, the voltage provided by the batteries can momentarily sag. The sag may be for only a fraction of a second, but it can be long enough to cause problems.

If your robot's control computer is connected to the same battery source as the motors, the voltage drop can cause what's known as a brownout; if the brownout causes the voltage to drop below a certain threshold (Figure 13-15), the control electronics may not operate correctly. Quite literally, your robot can go berserk!

Brownouts are particularly troublesome when using microcontrollers (see Part 7), which are small computers that run a program you devised. During a brownout, the microcontroller may spontaneously reset, causing it to rerun its programming from the beginning. This can actually occur several dozens or even hundreds of times a second.

During these brownouts your robot may become inactive (that's good), or it may lurch or spin or do something else unpredictable (that's bad).

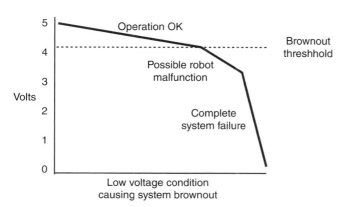

Figure 13-15 A brownout may occur when the system voltage of your robot falls below a certain minimum level. As the voltage falls under the brownout threshold, operation of the robot becomes unpredictable.

There are several ways of avoiding voltage drops caused by motors:

- Add more volts to your robot's batteries. If the batteries normally supply 6 volts, but may "sag" to 4.5 volts during heavy motor draw, add another cell to bring the voltage to 7.2 or 7.5. The extra margin might prevent a voltage sag that causes a brownout.
- Add bigger batteries with a higher capacity (amp hours). The volts may be the same, but the added current capability can provide a reserve against voltage drops.
- Add a second battery pack to operate the electronics. This is my preferred method, because it effectively isolates the power supplies for the motor and electronics. As the motors draw current, they pull it from their own batteries and not from the pack powering the electronics.

For small robots you can often use a AAA- to D-size battery pack for the motors, and a 9-volt battery for the robot control electronics. If you have lots of electronics—microcontroller, multiple sensors, maybe a video camera—you might need to beef up the second battery. Use a separate multicell battery pack selected to provide the current needed for the electronics.

In order for the dual power supply technique to work, the ground (–, negative) lead from both battery packs must be connected together. This topic is discussed in more detail in Chapter 24, "All About Robot Gray Matter."

14

Using DC Motors

DC motors are the mainstay of robotics. A surprisingly small motor, when connected to wheels through a gear reduction system, can power a hefty robot seemingly with ease. A flick of a switch, a click of a relay, or a tick of a transistor, and the motor stops in its tracks and turns the other way. A simple electronic circuit enables you to gain quick and easy control over speed—from a slow crawl to a fast sprint.

This chapter shows you how to apply continuous DC motors (as opposed to stepping or servo motors) to power your robots. The emphasis is on using motors to propel a robot across your living room floor, but you can use the same control techniques for any motor application, including gripper closure, elbow flexion, and sensor positioning. What follows applies to DC motors with and without gearboxes attached to them.

 This chapter discusses electronic components and circuits related to controlling DC motors. If you are brand-new to these subjects, be sure to see Chapter 21, "Robot Electronics— The Basics," and Chapter 22, "Common Electronic Components for Robotics."

The Fundamentals of DC Motors

There are a many ways to build a DC motor. By their nature, all DC motors are powered by direct current—hence the name *DC*—rather than the alternating current (AC) used by most motorized household appliances.

THE PERMANENT MAGNET MOTOR: AFFORDABLE, EASY TO USE

Perhaps the most common DC motor is the permanent magnet type, so-called because it uses two or more permanent magnet pole pieces that remain stationary. The turning shaft of the motor, the *rotor* (or *armature*), is composed of several sets of wires—called *windings*. These wires

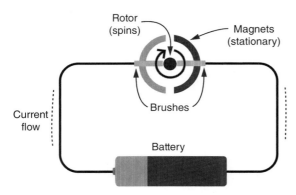

Figure 14-1   Simplified block diagram of a DC motor. Current flows from the battery to brushes, which are electrically connected to the commutator. Windings around the rotor are energized, causing the rotor to spin.

are wrapped around pieces of metal. At the end of the rotor is a *commutator*, which is used to alternately apply current to the windings.

Figure 14-1 shows a simplified diagram of current passing from a battery, through the commutator (shown here simplified), and energizing the rotor, which is in the middle. As the rotor spins, current is applied to the windings of the motor in such a way that the rotor is kept in motion. Only when current ceases to flow through the windings (or the motor shaft is physically blocked from turning) does the motor stop.

Note the two dark gray bars in the motor diagram shown in Figure 14-1. These are *brushes*, and they serve as terminals to apply the current from the battery to the commutator. On very inexpensive hobby motors, the brushes are often just pieces of copper wire, bent to a handy shape. On more expensive motors, brushes are made of conductive carbon. Both of these motors are known as *brushed motors*.

Both types of brushes can wear down over time, which can break the electrical connection between the battery and the commutator. This is why, when used long enough, a DC motor will just go kaput.

Brushless motors use electronics, not brushes, to alternate the current between windings. Brushless motors are used extensively in computer disk drives, "noiseless" fans, CD and DVD players, and precision electronics. You should know about them, as motors pulled from these components and sold as surplus may require additional electronics in order to operate.

You may also encounter brushless motors used in R/C servo motors. These motors tend to be quite expensive and are reserved for applications where motor failure could be catastrophic, like suddenly losing control of a $2000 R/C helicopter.

REVERSIBLE DIRECTION

One of the prime benefits of DC motors is that most (but not all) are inherently reversible. Apply current in one direction—the + and – on the battery terminals—and the motor spins clockwise. Apply current in the other direction, and the motor spins counterclockwise. This capability makes DC motors well suited for robotics, where it is often desirable to have the motors reverse direction. Use it to back a robot away from an obstacle or to raise or lower a mechanical arm.

If you're buying your DC motors surplus, you may encounter some that are not reversible. This could be due to the way the motor windings are constructed inside the motor or it could be due to an intentional mechanical design. Read the description for the motor carefully. It will usually indicate whether it's bidirectional; or at least, if it's not, the description will specify that the shaft turns CW (clockwise) or CCW (counterclockwise) only.

See the section later in this chapter under "Controlling a DC Motor" for various ways DC motors can be reversed.

Reviewing DC Motor Ratings

Motor ratings, such as voltage and current, were introduced in Chapter 12. Here's a quick recap of the main points of interest when selecting and using a DC motor for your robot:

- DC motors can often be operated at voltages above and below their specified rating. If the motor is rated for 12 volts and you run it at 6 volts, the odds are the motor will still turn but at reduced speed and torque. Conversely, if the motor is run at 18 volts, the motor will turn faster and will have increased torque.
- But this does not mean that you should intentionally underdrive or overdrive the motors you use. Significantly overdriving a motor may cause it to wear out much faster than normal. However, it's usually fairly safe to run a 10-volt motor at 12 volts or a 6-volt motor at 4 or 5 volts.
- DC motors draw the most current when they are *stalled*. Stalling occurs if the motor is supplied current but the shaft does not rotate. Any battery, control electronics, or drive circuitry you use with the motor must be able to deliver the current at stall, or major problems could result.
- The rotational speed of a DC motor is usually too fast to be directly applied in a robot. Gear reduction of some type is necessary to slow down the speed of the motor shaft. Gearing down the output speed has the positive side effect of increasing torque.

Controlling a DC Motor

As I've noted, it's pretty easy to change the rotational direction of a DC motor. Simply switch the power lead connections to the battery, and the motor turns in reverse. And when you want the motor to stop, merely remove the power leads to it.

That's fine for when you're playing around on your workbench, but what are the options when the motor is part of a robot? You have several, actually, and each has its place. The ones you'll read about in this book are:

- Switch
- Relay
- Bipolar transistors
- MOSFET transistors
- Motor bridge modules

Motor Control by Switch

You can manually operate your robot using switches. This is a good way to learn about robot control and experiment with different types of robot base designs. The switches attach to your robot by wires. You can control both the operation and the direction of the motors.

SIMPLE ON/OFF SWITCH CONTROL

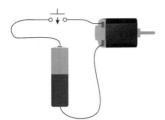

A very basic single-pole single-throw (SPST) *switch* can control whether a motor is on or off. Most robots use two motors, so having two switches lets you independently control each motor.

- To make the robot go forward, turn both switches on at the same time.
- To make the robot turn one direction or another, turn one switch on while leaving the other off.
- And, of course, to stop the robot, turn both switches off.

CONTROLLING DIRECTION USING SWITCHES

By using a double-pole, double-throw (DPDT) switch (read more about these in Chapter 22, "Common Electronic Components for Robotics"), you can control the direction of your motors— push the switch forward to have the motor spin one way; pull the switch back and the motor spins the other way.

Even if you don't plan on controlling your robot using switches, it's still handy to review the material that follows so that you understand how DPDT switches handle the reversal of the motor. The same basic technique of polarity reversal is used in all the other approaches.

Again, for the typical robot that uses two motors and two wheels, you use a pair of DPDT switches to control both motors. One switch operates the left motor, and the other switch operates the right motor. And if your DPDT switches have a *center-off* position, the same switches can be used to turn the motors off when you want to stop your robot.

See Figure 14-2 for an example of how to connect a DPDT switch from a battery pack to a motor. For two motors you'd use two switches. Put the batteries and switches in a project box. Use two sets of wire pairs (four wires total) to connect your switch control panel to the motors on the bot.

Remember to use DPDT switches with a center-off position. When they are in the center position, the motors receive no power, so the robot does not move.

Try to get momentary-contact switches so that when you release them they spring back to the center. This makes it a lot easier to control your robot by flipping the switch. Momentary-contact DPDT switches cost a little more, but they're much more convenient.

Figure 14-3 shows how the switches (the technique also applies to relays, discussed next) are wired in order to achieve motor reversal. It may look a little weird at first, but it's actually quite logical in how it works. The switches are wired so that in one position—say, Position A—current from the battery flows through the motor in one direction. Flipping to Position B changes the direction of battery flow to the other way around. This naturally makes the motor go the other way.

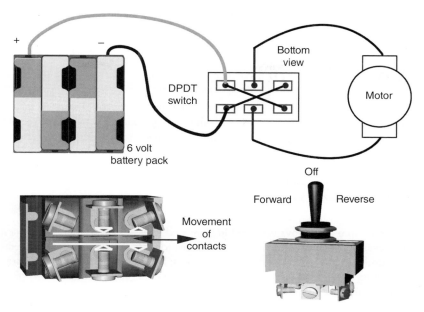

Figure 14-2 How to wire a double-pole, double-throw switch to control the direction of a DC motor. A switch with a center-off position allows you to stop the motors. Use two switches to manually control the direction of two motors.

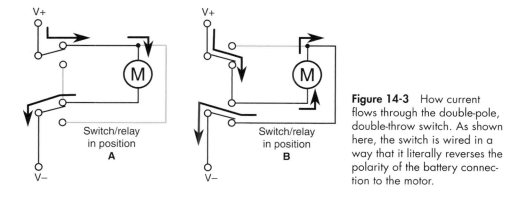

Figure 14-3 How current flows through the double-pole, double-throw switch. As shown here, the switch is wired in a way that it literally reverses the polarity of the battery connection to the motor.

Some things to try:

- Steer by turning on one motor only (leave the other one off). Note the speed of the turn.
- Now steer by making one motor go forward and the other motor go backward. Note the speed of the turn . . . it's faster. The robot actually spins, turning in place. This is how military tanks turn. It's referred to as tank steering, or, more commonly in robotics, *differential steering*.

Operating your robot with switches is great for learning how things work, but you'll soon want to graduate to hands-off methods, where your mechanical creations will steer themselves.

The remainder of the DC motor control methods concentrate on techniques that allow for fully autonomous robots.

Motor Control by Relay

Before getting to the all-electronic methods of motor control, I want to take a moment to talk about another, somewhat more old-fashioned approach: using *relays*. Yes, I know this is the twenty-first century, and what's all this about using something like a relay that was invented almost 200 years ago. Daft indeed!

But there are many good reasons to look at relays, if only because they're a natural stepping-stone from switch control to fully electronic control. But also, small relays for small robots are very cheap and very easy to use.

While it's true that relays wear out in time, and they're slower than electronic motor control, neither is a particularly relevant excuse to avoid them. Small *reed* relays for the average-size desktop amateur robot can switch several hundred thousand times before even beginning to act funny. As for their slowness, the motors are even slower—the slowest part of your robot.

INSIDE A RELAY

A relay is an electrically operated switch. It's just like the manual switches detailed in the previous section, but with the added feature of being controlled via electric signals.

The operation of a relay is simple: an electric coil is placed around or near a piece of metal that acts as a switch plate. When current is applied to the coil, the coil acts like an electromagnet. The magnetism pulls the metal switch plate closer to it. This engages the switch.

Nearly all relays self-reset; that is, when current is removed from the coil, a spring on the switch plate pulls it back into its original position. This disengages the switch.

SIMPLE ON/OFF RELAY CONTROL

You can accomplish basic on/off motor control with a single-pole relay, just as you did with manual switches. The relay is wired so that when it's inactive (OFF) current from the battery is not switched to the motor. When the relay is activated (ON) the circuit between battery and motor is complete, and the motor turns.

How you activate the relay is something you'll want to consider carefully. You *could* control it with a push-button switch, but that doesn't get you anything more than using manual switches alone. Relays can easily be driven by digital signals, the kind from a simple board or microcontroller on your robot.

See Figure 14-4 for the basic way of connecting a relay to any kind of controlling electronics. If you're new to electronics in general, a review of some terminology is in order.

- *Logical 0* (referred to as *LOW*) is digital terminology that means 0 volts is applied to the relay.
- *Logical 1* (referred to as *HIGH*) means that voltage (of some level) is applied to the relay. In most digital electronic circuits, this is 5 volts. In fact, we can just assume it's 5 volts, unless told otherwise.

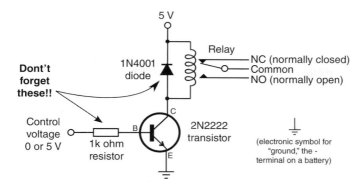

Figure 14-4 A relay can be used to electronically turn a motor on and off. A transistor and a resistor drive the relay; you control the operation of the relay by applying 0 or 5 volts as a control voltage.

- *Gates* are inputs or outputs on a digital circuit, such as from a computer or a microcontroller. As you might guess, you'd use an *output gate* when operating a relay via a digital circuit.

In this circuit, LOW turns the relay off and HIGH turns it on. The relay can be operated from most any digital gate. The chapters in Part 6 deal much more extensively with using electronic and computerized control.

CONTROLLING DIRECTION VIA RELAY

Changing the direction of the motor is only a little more difficult using relays than turning it on and off. As with the manual switches in the previous section, this requires a DPDT relay, wired in series after the on/off relay just described. Refer to Figure 14-5 for how these two relays are connected. With the switch contacts in the DPTD relay in one position, the motor turns clockwise. Activate the relay, and the contacts change positions, turning the motor counterclockwise.

Again, you easily control the direction relay with digital signals. Logical 0 makes the motor turn in one direction (let's say forward), and logical 1 makes the motor turn in the other direction.

You need two relays, not just one, to duplicate the functionality you enjoyed with the DPDT center-off switch. An SPST relay is used to turn the motor on and off, and a DPDT relay is used to control the direction of the motor.

You can see how to control the operation and direction of a motor using just two signals (*data bits*) from a digital circuit like a computer or microcontroller. Since most robot designs incorporate two drive motors, you can control the movement and direction of your robot with just four data bits. In fact, this is true for all the electronic motor control schemes in this chapter.

CURRENT SPECS FOR RELAYS

When selecting relays for your robot, make sure the contacts are rated for the motors you are using. All relays carry contact ratings, and they vary from a low of about 0.5 amp to over 50 amps, at 125 volts. Higher-capacity relays are larger and require more current to operate.

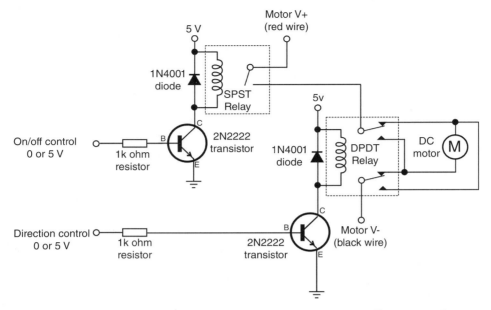

Figure 14-5   How to combine two relays to fully control a motor: turn it off and on, and reverse direction. The top relay is a single-pole type; the bottom relay is a double-pole type.

This means they need bigger transistors to trigger them and require much more care in selecting all the components in the system. Not very pretty.

As this book is primarily about creating amateur robots under about 5 or 10 pounds, you'll be using smaller motors, which means smaller relays. For the robots described in this book, you don't need a relay rated higher than 2 or 3 amps. If you plan on building bigger robots with much bigger motors, you should consider the motor bridge module, detailed later in this chapter.

SIMPLIFIED RELAY DRIVER ELECTRONICS

With two motors you need four relays, which means four transistors, four diodes, and four resistors. What are these extra components for?

- The transistors are used to boost the current from the digital gate (e.g., from your microcontroller), as not all gates have sufficient current to directly drive the relay. The transistors act as current amplifiers.
- The diodes protect the transistors from current that flows backward from the relay coil when it is switched off. This happens because, when the relay coil is deenergized, some of the current that was flowing through it is regurgitated back out. This *back EMF* (EMF stands for electromotive force) can damage the transistor; the diode prevents this mess from happening.
- The resistors control the amount of current flowing from the digital gate to the transistor. Without the resistor, the transistor would suck up too much current and possibly damage the gate.

A typical value for these resistors is between 1 kΩ (1000 ohms) and 4.7 kΩ (4700 ohms), assuming 5-volt circuitry. The lower the value of the resistor, the more current will flow from the gate and to the base of the transistor. Use the 1-kΩ value for larger relays that need more current; otherwise, select a higher value.

Use the highest-value resistor connected to the base of the transistor that allows for reliable operation of the relay. You might start with 4.7 kΩ and work downward until you find a resistor value that works best for your specific circuit.

See the RBB Support Site (see Appendix) for additional ways to drive relays, including from the handy ULN2003 Darlington array integrated circuit chip.

Motor Control by Bipolar Transistor

No, bipolar transistors don't exhibit manic-depressive behavior. In this case, *bipolar* is merely the term used to describe their internal construction—which we won't be getting into here since it's not particularly relevant (besides, there are like 10,000 books and Web sites that already talk all about it). There are other types of transistors, but the ones we're interested in for the time being are the bipolar variety.

Bipolar transistors are more accurately known as *bipolar junction transistors,* or *BJTs.* Same thing, slightly more words.

For robotics motor control, you use a bipolar transistor much as you would a switch. In fact, the transistor acts just like a switch: apply or remove current, and the transistor (switch) turns on or off. In order to do its work as a motor control switch, the transistor needs a few extra common electronic components, specifically, a resistor and a diode. The purpose of these are described shortly.

BASIC TRANSISTOR MOTOR CONTROL

See Figure 14-6 for the most basic implementation of using a transistor to operate a motor. A digital LOW or HIGH signal is applied to the input of the transistor circuit. Depending on whether the input is LOW or HIGH, the motor turns on or it turns off.

- When you connect the Motor control input to 5 volts (HIGH), the motor turns.
- When you connect the Motor control input to the ground connection (LOW), the motor stops.

Transistors of the type shown in Figure 14-6 exhibit an inherent voltage drop between its collector ("C") and emitter ("E") connections. The drop is usually about 0.7 volt. It's enough that you may notice your motor runs a bit slower than when it's directly connected to the battery.

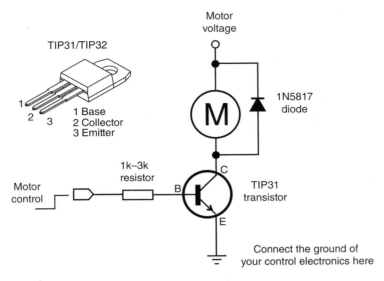

Figure 14-6 Fully electronic control of a motor is done using a transistor. This simple circuit starts and stops a motor, depending on the input signal. Select the value of the resistor so that the transistor switches fully on when the input signal is applied (the motor should turn at almost full speed).

FULL-BRIDGE TRANSISTOR CONTROL

A more elaborate form of transistor motor control uses the *full-bridge*, also called an *H-bridge*. The "H" comes from the way the motor is connected to its control circuitry, looking a wee bit like the letter H.

H-bridges require four transistors, along with associated components—resistors, flyback diodes, and perhaps more, depending on the design. They're harder to build, and harder to get right. Given the availability of other motor control circuitry, it makes more sense to use these other techniques, which tend to be easier and cheaper.

Instead, you can use MOSFET transistors, as described in the following section, or rely on one of the many low-cost motor bridge ICs or modules. Those are covered later in this chapter. Both provide more flexibility, and they are often cheaper and easier to build.

See the RBB Support Site for other ideas for creating H-bridge circuits with unijunction transistors.

Motor Control by Power MOSFET Transistor

MOSFET stands for "metal oxide semiconductor field effect transistor." *Metal oxide* indicates the process used to manufacture them, and *field effect* refers to the way the transistor conducts current. *Power MOSFET* is a further classification that indicates the device is intended to drive some kind of load, like a motor.

MOSFETs look like bipolar transistors (they come in the same type of packages), but internally there are some important differences. First, due to their MOS construction, they are more susceptible to being damaged by static electricity. Always keep the protective foam around the terminals of the device until you're ready to use it.

Second, the names of the terminals are different from what you find on bipolar transistors. These variations are discussed in more depth in Chapter 22, "Common Electronic Components for Robotics," but for now just know that if you connect a MOSFET to your circuit incorrectly, odds are great that it'll be destroyed the instant you turn on the power. Bipolar transistors aren't usually so sensitive.

BASIC MOSFET MOTOR CONTROL SWITCH

You can construct a hearty digital switch for powering motors and other high current devices with an N-channel power MOSFET transistor. Get the kind in the popular TO-220 style transistor case, such as the FQP30N06. This particular device can handle up to 32 amps at 60 volts—much higher than you'll ever have in a tabletop bot!

To get the full power-carrying ability of the MOSFET you should mount it on a metal heat sink, which draws heat away from the device. But for typical DC motors used with desktop robotics you can probably skip the heatsink.

A basic circuit that uses an N-channel power MOSFET (like the FQP30N06) is shown in Figure 14-7. It's a simple on/off switch to control a motor. Apply a 5-volt signal to the gate connection of the transistor, and the motor is turned on. Resistors are used for additional protection of the circuit controlling the transistor.

Notice the *Motor voltage* label. The voltage to your motor can be different from the voltage that controls the MOSFET transistor. Often, the control electronics in your robot are powered by 5 volts; the voltage to your motors can be 5 volts or over, up to the specified limit of the MOSFET, which is often at least 20 or 30 volts. In the case of the FQP30N06L, the voltage limit is 60 volts.

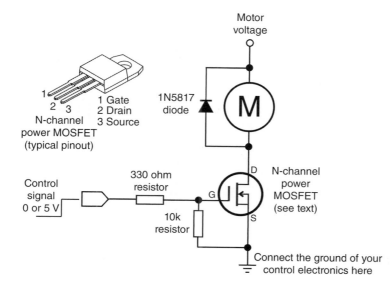

Figure 14-7
A MOSFET power transistor provides nearly the full voltage to the motor when it is turned on.

MOTOR H-BRIDGE USING MOSFET TRANSISTORS

Figure 14-8 shows the basic concept of the MOSFET transistor H-bridge. The gates of the transistors are connected to either ground or 5 volts. Turning on Q1/Q4 causes current to flow through the motor in one direction, making the motor spin clockwise. Turning on Q2/Q3 causes the current to flow through the motor in the opposite direction. The result: The motor spins counterclockwise.

The two types of MOSFET—N-channel and P-channel—refer to the microscopic conductive channel found inside the device. The two types differ in their chemical makeup, which in turn affects how the devices conduct electrons. The arrangement shown in the circuit takes advantage of N- and P-channel behavior to build an H-bridge that acts as close to a mechanical switch as possible.

Note that "turning on" an N-channel or P-channel MOSFET is relative and, for a P-channel transistor, may work opposite to what you think:

Gate Signal	N-channel	P-channel
0 volts (LOW, logical 0)	Turns off	Turns on
5 volts (HIGH, logical 1)	Turns on	Turns off

Unlike bipolar transistors, which exhibit a drop in voltage when current is passed through them, MOSFET transistors pass through nearly all of the volts to the motor.

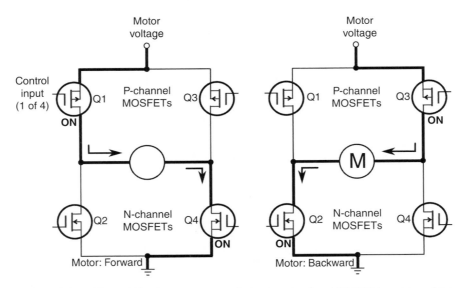

Figure 14-8 A (very) basic H-bridge motor control circuit, using four MOSFET transistors. Note that the two uppermost transistors are P-channel type; the two lower transistors are N-channel type. Don't get these crossed up.

The N- and P-channel MOSFET transistors you use should be complementary pairs, that is, transistors that share similar specifications. This provides a balance in current-carrying capability. For example, you might use the popular FQP30N06 and FQP27P06 transistors.

Motor Control by Bridge Module

Circuits for controlling motors are big business, and it shouldn't come as a surprise that dozens of companies offer all-in-one solutions for running motors through fully electronic means. These products range from inexpensive $2 integrated circuits (ICs) to sophisticated modules costing tens of thousands of dollars. Of course, we'll confine our discussion to the low end of this scale!

Motor control modules incorporate the H-bridge design you learned about in the previous section. The module may consist of just an IC, or it may be a premade circuit board with the H-bridge electronics on it. Either way, motor control bridges have two or more pins on them for connection to control electronics and, of course, connections to power and to the motors.

Typical functions for the pins are:

- *Motor power.* Connect these to the battery or other source powering the motors. I like to use a completely separate battery pack for the motors and the rest of the robot's electronics. Using the motor power pins on the motor bridge, this is very easy to do.
- *Motor enable.* When enabled, the motor turns on. When disabled, the motor turns off. Some bridges let the motor "float" when disabled; that is, the motor coasts to a stop. On other bridges, disabling the motor causes a full or partial short across the motor terminals, which acts as a brake to stop the motor very quickly.
- *PWM.* Most H-bridge motor control ICs are used to control not only the direction and power of the motor but its speed as well. The typical means used to vary the speed of a motor is with pulse width modulation, or PWM. This topic is described more fully under "Controlling the Speed of a DC Motor." On many H-bridge ICs, the motor enable and PWM input share the same pin, as shown in Figure 14-9.
- *Direction.* Setting the direction pin changes the direction of the motor.

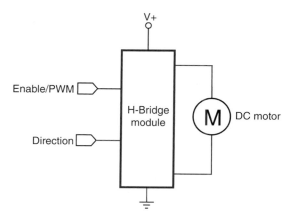

Figure 14-9 Basic hookup diagram of a motor control bridge module. Many modules are designed to operate two motors; these provide separate hookup pins for each motor.

- *Brake.* On bridges that allow the motor to float when the enable pin is disengaged, a separate brake input is often used to specifically control the braking action of the motor.
- *Motor out.* These are outputs for connecting to the motor.

The better motor control bridges incorporate overcurrent protection circuitry, which prevents them from being damaged if the motor pulls too much current and overheats the chip. Some even provide for *current sense,* useful when you want to determine if the robot has become stuck.

Recall from earlier in this chapter that DC motors will draw the most current when they are stalled. If the robot gets caught on something and can't budge, the motors will stop, and the current draw will spike.

USING THE L293D AND 754410 MOTOR DRIVER ICs

Among the most common—and least expensive—motor bridge ICs are the L293D and its close cousin, the 754410. Both come in small 16-pin IC packages, and their hookup is identical. The big different between the two is the maximum amount of current that the chip can handle.

- The L293D can supply up to 600 mA of current (continuous, per channel).
- The 754410 can supply up to 1.1 A of current (continuous, per channel).

On both chips, the supply voltage ranges from 4.5 to 36 volts, and they have connections for not just one motor but two.

From here on out I'll refer just to the L293D, but know the discussion applies to the 754410 as well.

When buying the L293D, be sure it has a *D* at the end of it. "D" denotes diodes. Recall the use of diodes used to protect components in the transistorized H-bridges described earlier in this chapter. The L293D (and the 754410) have these diodes built in. If you don't get the D version of the L293, you'll need to add these diodes yourself.

While we're on the subject of diodes: There is some disagreement among robo-builders about whether the diodes built into the 754410 are intended for flyback protection. The datasheet for the 754410 is not clear on the subject. Adding to the confusion is that several versions of the datasheet contain other errors. If you want to be sure that the 754410 is fully protected, you can add external diodes, such as 1N5817. But if playing safe is the key, this suggestion applies equally to the L293D and any other motor bridge IC.

Figure 14-10 shows the basic connection for the L293D to a pair of motors. Notice the two power pins for the L293D: one is to power the IC, and it should be 5 volts. The other is the power for the motor, and that can be up to 36 volts. The minimum allowed motor voltage is 4.5 volts.

Each motor is controlled by a set of three pins, also called input lines. The motors are referenced as *Motor1* and *Motor2*.

For Motor1, the control lines are labeled Input1, Input2, and Enable1. For Motor2 the lines are marked Input3, Input4, and Enable2.

In order to activate Motor1, the Enable1 line must be HIGH. You then control the motor and its direction by applying a LOW or HIGH signal to the Input1 and Input2 lines, as shown in this table.

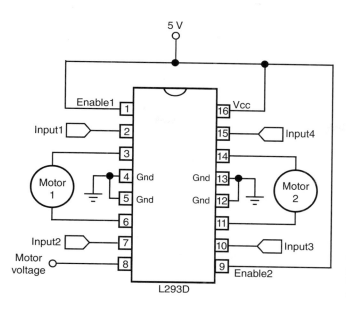

Figure 14-10 H-bridge integrated circuits incorporate all the necessary electronics to drive a motor.

Motor 1		
Input1	**Input2**	**Action**
LOW	LOW	Motor stops
LOW	HIGH	Motor turns forward
HIGH	LOW	Motor turns backward
HIGH	HIGH	Motor stops

Controlling Motor2 is the same, except it uses Input3 and Input4. The enable lines are also used to control the speed of the motor (Enable1 for Motor1, and Enable2 for Motor2). Instead of a constant HIGH signal, you can apply a rapid succession of LOW/HIGH pulses; the length of the pulses determines how fast the motor goes. For more, see the section "Controlling the Speed of a DC Motor," later in this chapter.

The L293 series and 754410 chips have a "fast motor stop" feature when the Enable line is HIGH and both Inputs are LOW or HIGH (either, doesn't matter). Depending on the motor, this may not produce much of a braking effect. If you need to quickly stop the motor, momentarily reverse its direction, then stop it. If you don't want the "fast motor stop" feature, disable the motor by bringing the Enable line LOW.

 The four center pins of the L293D serve as the IC's ground connectors and can also be used as part of a heat sink. In order for the chip to drive its maximum current, you should add a larger metal heat sink over the IC. You can buy these premade or solder on strips of bare (uncoated) copper metal pieces to the center pins in a kind of "wing" arrangement.

USING THE L298 MOTOR DRIVER IC

Another very popular and reasonably priced all-in-one H-bridge motor driver is the L298. Like the L293D it can control two motors, not just one. It can handle 2 amps per motor, though to get the maximum current be sure to add a heat sink. The L298 has a large cooling flange with a hole in it, making it easy to attach a homebrew metal heatsink to it.

If there's a downside to the L298 it's that it comes in a special "Multiwatt 15" package, with 15 offset pins that don't match the standard 0.100" spacing of breadboards. Rather than fuss with bending the pins to match a breadboard just get a ready-made *breakout board* for the L298. This is a small circuit board with holes drilled in it to accept the chip. You then plug the breakout board into your breadboard. Problem solved.

Figure 14-11 shows a basic connection diagram for controlling two motors using the L298 motor bridge IC. As with the L293D, there are three input pins for each motor: Input1, Input2, and Enable1 controls Motor1. Input3, Input4, and Enable2 controls Motor2. The motors connect to Output1/Output2, and Output3/Output4, as shown.

And, like the L293D, the L298 uses two different supply voltages. The voltage on pin 9 powers the chip itself, and should be 5 volts. The voltage on pin 4 supplies the motors, and it can be up to 46 volts.

The L298 has two additional inputs for sensing current on either motor. I won't be demonstrating these current sense inputs here, but know they are available should you wish to experiment with them. The L298 datasheet discusses how to use these inputs.

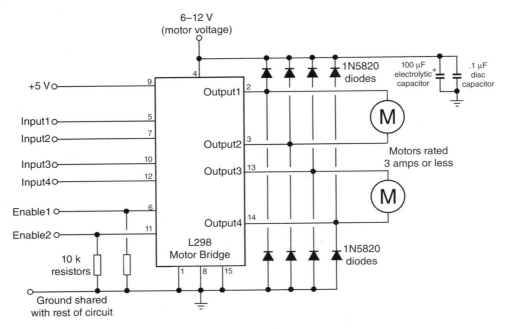

Figure 14-11 The L298 motor bridge IC can drive two motors up to 3 amps each. External protection diodes are needed. Motors are operated by applying appropriate control signals to the Enable and Input lines.

Let's look at how to control just one of the motors, Motor1. In order to activate the motor the Enable1 line must be HIGH. You then control the motor and its direction by applying a LOW or HIGH signal to the Input1 and Input2 lines, as shown in this table.

Input1	Input2	Action
LOW	LOW	Motor breaks and stops*
HIGH	LOW	Motor turns forward
LOW	HIGH	Motor turns backward
HIGH	HIGH	Motor breaks and stops*

*To coast a motor to a slower stop, apply a LOW signal to the Enable1 line.

Unlike the L293D, the L298 does not have built-in protection diodes, so you'll need to add those. (If you're using a ready-made module with the L298 these diodes are already included.)

The chip needs several other external parts for proper operation. Because of the added complexity of using the L298, frankly most folks opt to get a fully integrated module based on the chip. Ready-made and kit modules are available, and they end up being easier (and often cheaper) to use than collecting and assembling all the parts yourself. Several projects in Part 4 use a ready-to-go L298 module.

ALL-IN-ONE BRIDGE MODULES

As you've seen in this chapter, effective and efficient motor control requires careful selection and construction of various electronic parts. I've only demonstrated the most simple relay, transistor, and motor control IC circuits.

If you're wanting something more robust your best bet is get a ready-to-go bridge module. This all-in-one circuit board contains the bridge circuitry and all associated electronics. Just attach wires to the rest of your robot and you're set to go.

Over the years the all-in-one module has gained in popularity; these days you can get a dual-motor (controls two motors) module for under $15. Control connections from your robot's main brain comes in two forms:

- Digital on/off inputs for controlling motor enable, PWM, and direction. These can be controlled by an Arduino or similar microcontroller, or by simpler electronics, such as light sensors (see how in Chapter 37, "Make Light-Seeking Robots"). An example of H-bridge with digital inputs is shown in Figure 14-12.
- Serial inputs that feature "set and forget" operation. These require a microcontroller where a sequence of binary data effectively communicates a series of instructions to the module. A benefit of the serial control module is that you can order the motor to change speed, without having to worry about the complexities of motor speed control (described in the next section).

Figure 14-12 This modular motor control board is operated using digital inputs and controls up to two motors.

Many of these intelligent bridge modules are available in single- or dual-motor versions, with current capacities of 50, 75, even 100 amps—ideal if you're building a very large or combat robot.

 Also available are *ESC motor speed controllers,* originally intended for use with high-speed R/C racing vehicles. Find these at any R/C hobby store. Though ESC motor speed controllers are designed for use with R/C receivers, you can use an ordinary microcontroller to simulate the signals that it expects to see. Just treat it like a servo motor.

Controlling the Speed of a DC Motor

There will be plenty of times when you'll want the motors in your robot to go a little slower or perhaps track at a predefined speed. Speed control with continuous DC motors is a science in its own right, but the fundamentals are quite straightforward.

The proper way to control the speed of a motor is to feed it short on/off pulses of its normal voltage. The pulses are very fast, so fast that the motor doesn't have time to respond to each on/off change. What happens is that the motor ends up averaging the ons with the offs, so, effectively, less voltage gets to the motor.

As noted previously in this chapter, this system of motor speed control is called PWM. It is the basis of just about all motor speed control circuits. The longer the duration of the pulses, the faster the motor because it is getting full power for a longer period of time. The shorter the duration of the pulses, the slower the motor.

Check out Figure 14-13, which shows the on and off nature of PWM. The time between each pulse is called the *period,* and it's usually just a brief moment in time—microseconds. In the typical PWM system, there may be from 1000 to over 20,000 of these periods each second.

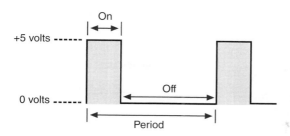

Figure 14-13 The speed of nearly all DC motors may be varied by changing the duty cycle (on versus off times) of its supply voltage. The longer the on time, the faster the motor will turn.

Notice that the power or voltage delivered to the motor does not change—it's always 5 volts (or 6 volts, or 12 volts, or whatever). The only thing that changes is the amount of time the motor is provided with this voltage. The longer the on time in relation to the off time, the more power the motors gets. Most motors function adequately at PWM ratios of 25 percent or higher. Depending on the motor and other factors, at lower PWM rates the motor may not receive enough power to turn its load.

Remember: The frequency of the pulses—how many occur in a second—does not change, just their relative on and off times. PWM frequencies of 1 kHz (1000 cycles per second) to over 20 kHz are commonly used, depending on the motor.

Unless you have a specification sheet from the manufacturer of the motor, you may have to do some experimentation to arrive at the "ideal" pulse frequency to use. You want to select the frequency that offers maximum power with minimum current draw.

Avoiding Electrical Noise

Electric motors generate lots of noise. In one form, the noise is like the static of a thunderstorm on an AM radio. It doesn't matter that the lightning bolt is miles away; the electrical charge travels through the atmosphere and into the radio's circuits, ending up as a loud snap, crackle, and pop in the speaker.

Fortunately, it's rather easy to cut down on the electrical noise produced by your robot's motors. The solution is to place one or more capacitors as near to the motor terminals as possible. The capacitor literally "soaks" up the electrical transients produced by the motors, which in turn reduces the noise they make.

- As part of good operating procedure, place a 0.1 µF ceramic disc capacitor across the motor terminals. Just solder the sucker right in there, along with the power leads to the motor.
- If the motor still generates too much electrical noise for the circuit, then as an extra precaution solder a 0.1 µF ceramic disc capacitor from each power terminal to the metal case of the motor. (This can be a bit tricky. To get the solder to stick to the motor case be sure it's clean. Rough up the metal with a small file. If needed, dab on some solder flux paste over the area and turn up the head of the soldering tool.)

Figure 14-14 shows the idea. The added capacitor(s) act to filter out the electrical noise from the motor, helping to reduce the amount of noise that travels back to the circuit.

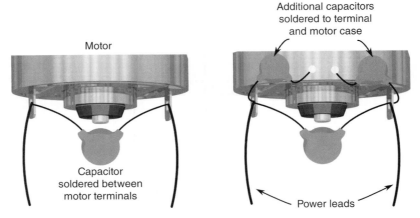

Figure 14-14 Electrical noise generated inside a DC motor can affect the operation of robot electronics. This noise can be suppressed by using small ceramic disc capacitors (0.01 to 0.1 μF) soldered to the motor terminals as shown.

Choices for Robot Motors

When I started out building robotics in the 1970s there were few ready-made motor solutions for small bots. What couldn't be scavenged from old toys had to be found in surplus catalogs or dusty attics. It was rare the motor matched the needs of the bot. Dark days indeed.

Times change fortunately, and the world is awash with choice of small motors for bots. You can still find old and used motors from toys, surplus stores, and attics, but you can also pick from a variety of motors specially built for DIY projects. Figure 14-15 shows a small selection of some of them.

Figure 14-15 A selection of motors for small robot projects. Motors with gearheads already attached can be directly coupled to wheels.

A bare motor lacks a gearbox. These are useful for building "brushbots" (the shaking of the motor makes the robot move), or can be attached to gears you provide.

Toy motors with attached gearheads. You can choose from assembled motors or kits. The higher-quality gear motors use metal gears for longer life.

When choosing a drive motor for your robot consider how you'll attach it to wheels or sprockets. Many of the motors made for DIY projects are made to interface with matching wheels and other mechanics. See Chapter 16, "Mounting Motors and Wheels," for more ideas.

CHAPTER 15

Using Servo Motors

DC motors are inherently an *open feedback* system—you give them juice, and they spin. How much they spin is not always known, at least not without additional mechanical and electronic parts.

Servo motors, on the other hand, are a *closed feedback* system. This means the output of the motor is coupled to a control circuit. As the motor turns, the control circuit monitors the position. The circuit won't stop the motor until the motor reaches its proper point. All without your having to do anything extra.

Servo motors have earned an important place in robotics. And fortunately for robot builders, another hobby—model radio control—has made these motors plentiful, easy to use, and quite inexpensive.

In this chapter you will learn what you need to know to use radio control (R/C) servos in your robot projects. While there are other types of servo motors, it is the R/C type that is commonly available and affordable, so I'll be sticking with those only.

Be sure to also check out Chapter 16, "Mounting Motors and Wheels," to learn how to attach servos to your bot, and the chapters in Part 6, "Robot Brains," on how to program servos to do various wonderful things.

How R/C Servos Work

Servo motors designed to be operated via a radio-controlled link are commonly referred to as *radio-controlled* (or *R/C*) servos, though, in fact, the servo motor itself is not what's radio-controlled. The motor is merely connected to a radio receiver on the model plane or car. The servo takes its signals from the receiver.

Figure 15-1   Three common sizes of servo motor: standard, mini, and micro. Flanges molded into the motor case allow easy mounting, and the output of the servo can be readily attached to wheels, linkages, and other mechanisms.

This means you don't need to control your robot via radio signals just to use an R/C servo (unless you want to, of course). You can control a servo with your PC, a microcontroller such as the Arduino.

Figure 15-1 shows three common sizes of R/C servo motors, which are used with model flyable airplanes and model racing cars. The largest is the "standard" motor; it measures about 1-1/2" × 3/4" × 1-3/8". For this style of servo, its size and the way it's mounted are the same, regardless of the manufacturer. That means that you have your pick of a variety of makers and can compare prices. There are other common sizes of servo motors besides that shown in the figure, however. I'll get to those in a bit.

A PEEK INSIDE

Inside the servo is a motor and various other components, neatly packaged (see Figure 15-2). While not all servos are exactly alike, all have these three major parts: motor, reduction gear, and control circuitry.

- *Motor.* A DC motor capable of reversing direction is at the heart of the servo.
- *Reduction gears.* The high-speed output of the motor is reduced by a gearing system. Many revolutions of the motor equal one revolution of the output gear and shaft of the servo.
- *Control circuitry.* The output gear is connected to a potentiometer, a common electronic device similar to the volume control on a radio. The position of the potentiometer indicates the position of the output gear.

The motor and potentiometer are connected to a control board, all three of which form a *closed feedback loop.* The servo is powered by 4.8 to 7.2 volts.

You can't run an R/C servo simply by connecting it to a battery. It needs special control signals to operate. As it turns out, it's not terribly hard to control a servo using simple programming. Numerous programming examples are provided in Part 6.

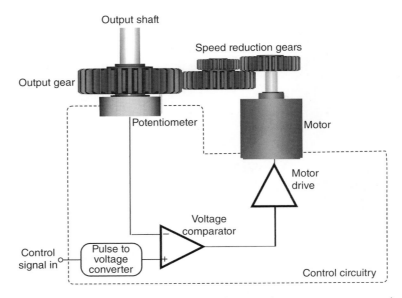

Figure 15-2 How an RC servo works. A control signal causes the motor to turn in one direction or the other, depending on the current position of the output shaft.

LIMITING ROTATION

As you can surmise, servo motors are designed for limited rotation rather than for continuous rotation (CR) like a DC motor. While there are servos that rotate continuously, and you can modify one to freely rotate (see later in this chapter), the primary use of the R/C servo is to achieve accurate rotational positioning over a range of up to 180°.

While 180°—half a circle—may not sound like much, in actuality such control can be used to steer a robot, move legs up and down, rotate a sensor to scan the room, and more. The precise angular rotation of a servo in response to a specific digital signal has enormous uses in all fields of robotics.

Control Signals for R/C Servos

The control signal that commands a servo to move to a specific point is in the form of a steady stream of electrical pulses. The exact duration of the pulses, in fractions of a millisecond (one-thousandths of a second), determines the position of the servo, as shown in Figure 15-3.

Note that it is not the number of pulses per second that controls the servo, but the *duration* of the pulses that matters. This is very important to fully understand how servos work and how to control them with a microcontroller or other circuit.

Specifically, the servo is set at its center point if the duration of the control pulse is 1.5 milliseconds. Durations longer or shorter command the servo to turn in one direction or the other.

- A duration of 1.0 milliseconds (ms) causes the servo to turn all the way in one direction.
- A duration of 2.0 ms causes the servo to turn all the way in the other direction.
- And to recap, a duration of 1.5 ms causes the servo to return to its midpoint.

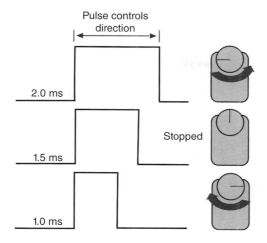

Figure 15-3   The length of the control pulses determines the angular position of the servo shaft. The pulse range is from 1.0 ms (or 1000 μs) to 2.0 ms (2000 μs). A pulse of 1.5 ms (1500 μs) positions the servo shaft in the middle.

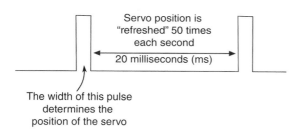

Figure 15-4   Control pulses are repeated ("refreshed") at roughly 50 Hz (50 times each second).

The servo needs about 30 to 50 of these pulses per second, as shown in Figure 15-4. This is referred to as the *refresh* (or *frame*) rate; if the refresh rate is too low, the accuracy and holding power of the servo are reduced. If there are way too many pulses per second, the servo may jitter and fail to work properly.

PULSES ALSO CONTROL SPEED

As mentioned, the angular position of the servo is determined by the duration of the pulse. This technique has gone by many names over the years. One you may have heard is *digital proportional*—the movement of the servo is proportional to the digital signal being fed into it.

The power delivered to the motor inside the servo is also proportional to the difference between where the output shaft is and where it's supposed to be. If the servo has only a little way to move to its new location, then the motor is driven at a fairly low speed. This ensures that the motor doesn't "overshoot" its intended position.

But if the servo has a long way to move to its new location, then it's driven at full speed in order to get it there as fast as possible. As the output of the servo approaches its desired new position, the motor slows down.

People often refer to the pulses used to control an R/C servo as *pulse width modulation*, or PWM. That's okay, but it can lead to some confusion.

Technically speaking, R/C servos employ what might be better termed *pulse duration modulation*. With PWM (detailed in Chapter 14, "Using DC Motors"), it's the duty cycle of the pulses—the ratio that each pulse is on versus off—that matters. R/C servos don't care about duty cycles or ratios. All they care about is how long the pulse is. As long as the servo receives at least 20 of these pulses per second (50 is better), it's happy.

You can call whatever goes on inside a servo anything you like, just as long as you remember that a PWM signal for a DC motor bears no relation to the "PWM" signal used to control an R/C servo. In fact, trying to use a PWM signal intended for a DC motor will likely overheat and damage a standard R/C servo.

VARIATION IN PULSE WIDTH RANGES

Most standard servos are designed to rotate back and forth by 90° to 180°, given the full range of timing pulses. You'll find the majority of servos will be able to turn a full 180°, or very nearly so.

The actual length of the pulses used to position a servo to its full left or right positions varies among servo brands, and sometimes even among different models by the same manufacturer. You need to do some experimenting to find the optimum pulse width ranges for the servos you use. This is just part of what makes robot experimenting so much fun!

The 1-to-2-ms range has built-in safety margins to prevent possible damage to the servo. Using this range provides only about 100° of turning, which is fine for many tasks.

But if you want a full stop-to-stop rotation, you need to apply pulses shorter and longer than 1 to 2 ms. Exactly how long depends entirely on your specific servo. Full rotation (to the stop) for one given make and model of servo might be 0.730 ms in one direction and 2.45 ms in the other direction.

You *must* be very careful when using shorter or longer pulses than the recommended 1-to-2-ms range. Should you attempt to command a servo beyond its mechanical limits, the output shaft of the motor will hit an internal stop, which could cause gears to grind or chatter. If left this way for more than a few seconds, the gears may be permanently damaged.

The 1.5-ms "in-between" pulse may also not precisely center all makes and models of servos. Slight electrical differences even in servos of the same model may produce minute differences in the centering location.

Timing signals for R/C servos are often stated in milliseconds, but a more accurate unit of measure is the *microsecond*—or millionth of a second. In the programming chapters that follow you'll more often see timing pulses for servos stated in microseconds.

To convert milliseconds to microseconds, just move the decimal point to the right three digits. For example, if a pulse is 0.840 milliseconds, move the decimal point over three digits and you have 0840, or 840 microseconds (lop off the leading zero; it's not needed).

The Role of the Potentiometer

The potentiometer of the servo plays a key role in allowing the motor to set the position of its output shaft, so it deserves a short explanation of its own.

The potentiometer is mechanically attached to the output shaft of the servo (in some servo models, the potentiometer *is* the output shaft). In this way, the position of the potentiometer very accurately reflects the position of the output shaft of the servo.

The control circuit in the servo compares the position of the potentiometer with the pulses you feed into the servo. The result of this comparison is an *error signal*. The control circuitry compensates by moving the motor inside the servo one way or the other. When the potentiometer reaches its final proper position, the error signal disappears and the motor stops.

Special-Purpose Servo Types and Sizes

While the standard-size servo is the one most commonly used in both robotics and radio-controlled models, other R/C servo types, styles, and sizes also exist.

- *Quarter-scale* (or *large-scale*) servos are about twice the size of standard servos and are significantly more powerful. Quarter-scale servos make perfect power motors for a robot arm.
- *Mini-* and *micro* servos are diminutive versions of standard servos and are designed to be used in tight spaces in a model airplane or car—or robot. They aren't as strong as standard servos.
- *Sail winch* servos are designed with maximum strength in mind and are primarily intended to move the jib and mainsail sheets on a model sailboat.

R/C servo motors enjoy some standardization. This sameness applies primarily to standardized servos, which measure approximately 1.6″ × 0.8″ × 1.4″. For other servo types the size varies somewhat among makers, as these are designed for specialized tasks.

Table 15-1 outlines typical specifications for several types of servos, including dimensions, weight, torque, and transit time. Of course, except for the size of standard servos, these specifications can vary according to brand and model. Keep in mind that there are variations on the standard themes for all R/C servo classes.

Table 15-1   Typical Servo Specifications

Servo Type	Length	Width	Height	Weight	Torque	Transit Time
Standard	1.6″	0.8″	1.4″	1.3 oz	42 oz-in	0.23 sec/60°
1/4-scale	2.3″	1.1″	2.0″	3.4 oz	130 oz-in	0.21 sec/60°
Mini-micro	0.85″	0.4″	0.8″	0.3 oz	15 oz-in	0.11 sec/60°
Low profile	1.6″	0.8″	1.0″	1.6 oz	60 oz-in.	0.16 sec/60°
Sail winch small	1.8″	1.0″	1.7″	2.9 oz	135 oz-in	0.16 sec/60° 1 sec/360°
Sail winch large	2.3″	1.1″	2.0″	3.8 oz	195 oz-in	0.22 sec/60° 1.3 sec/360°

A couple of the terms used in the specs require extra discussion.

- As explained in Chapter 13, "How to Move Your Robot," the *torque* of the motor is the amount of force it exerts. Servos exhibit very high torque thanks to their internal gearing.
- The *transit time* (also called *slew rate*) is the approximate time it takes for the servo to rotate the shaft a certain number of degrees, usually 60°. The faster the transit time, the faster-acting the servo will be.

You can calculate equivalent RPM by multiplying the 60° transit time by 6 (to get full 360° rotation), then dividing the result into 60. For example, if a servo motor has a 60° transit time of 0.20 seconds, that's one revolution in 1.2 seconds (0.2 × 6 = 1.2), or 50 RPM (60/1.2 = 50).

Gear Trains and Power Drives

The motor inside an R/C servo turns at several thousand RPMs. This is way too fast to be used directly; all servos employ a gear train that reduces the output of the motor to the equivalent of about 50 to 100 RPM. Servo gears can be made of nylon, metal, or a proprietary material.

- *Nylon* gears are the lightest and least expensive to make. They're fine for general-purpose servos.
- *Metal* gears are much stronger than nylon and are used where brute-force power is needed, but they significantly raise the cost of the servo. They're recommended for heavier walking robots or large robotic arms. On many servos only some of the gears are metal; the rest are a heavy-duty nylon or other plastic material.
- Gears made of *proprietary* materials include Karbonite, found on Hitec servos. These materials are offered as stronger alternatives to plastic.

Replacement gear sets are available for many servos, particularly the medium- to high-priced ones ($20+). Should one or more gears fail, the servo can be disassembled and the gears replaced.

Output Shaft Bushings and Bearings

Besides the drive gears, the output shaft of the servo receives the most wear and tear. On the least expensive servos this shaft is supported by a plastic or resin bushing, which obviously can wear out very quickly if the servo is used heavily. A *bushing* is a one-piece sleeve or collar that supports the shaft against the casing of the servo.

Metal bushings, typically made from lubricant-impregnated brass (sometimes referred to as *Oilite*, a trade name), last longer but add to the cost of the servo. The better servos come equipped with *ball bearings*, which provide longest life.

When looking at servos, you'll often see a notion regarding the bearing type, either bushing or bearing, and whether it's metal or plastic. You also may see a notation for "Top" or "Bottom"; this refers to having a bushing or bearing on the top and/or bottom of the output gear (top *and* bottom is best), like that shown in Figure 15-5.

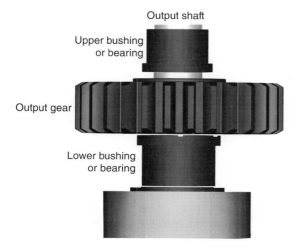

Output shaft

Upper bushing
or bearing

Output gear

Lower bushing
or bearing

Figure 15-5 Ball bearings or bushings may be placed at the bottom and/or top of the servo output gear, in order to prolong the life of the servo motor.

Connector Style, Wiring

While some servo specifications may vary from one model to another, one thing that's much more standardized is the connectors used to plug servos into their receivers.

CONNECTOR TYPE

There are three primary connector types found on R/C servos:

- "J" or Futaba style
- "S" or Hitec/JR style
- "A" or Airtronics style

Servos made by the principal servo manufacturers—Futaba, Airtronics, Hitec, and JR—employ the connector style popularized by that manufacturer. In addition, servos made by competing manufacturers are usually available in a variety of connector styles, and connector adapters are available.

PINOUT

The physical shape of the connector is just one consideration. The wiring of the connectors (called the *pinout*) is also critical. Fortunately, all but the "old-style" Airtronics servos (and the occasional oddball four-wire servo) use the same pinout, as shown in Figure 15-6.

With very few exceptions, R/C servo connectors use three wires, providing ground, DC power (+V), and signal. The +V DC power pin is virtually always in the center; that way, if you manage to plug the servo in backward, there's less chance of damage. (An exception to this is what's referred to as "old-style Airtronics," a wiring scheme no longer in use. You may encounter it if you have some older-model Airtronics servos.)

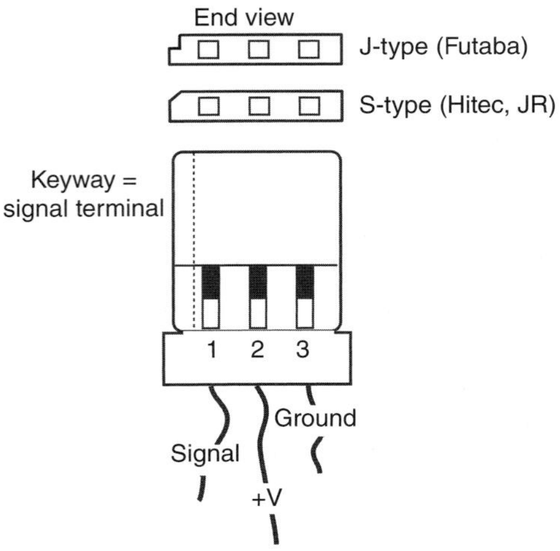
End view
J-type (Futaba)
S-type (Hitec, JR)
Keyway =
signal terminal
1 2 3
Ground
Signal
+V

Figure 15-6 Standard three-pin connector used on the vast majority of RC servo motors. The connector may or may not be "keyed" using a groove or notch.

COLOR CODING

Most servos use color coding to indicate the function of each connection wire, but the actual colors used for the wires vary among servo makers. Some of the more common wire color coding is:

- White, orange, or yellow: Signal
- Red: +V (DC power)
- Black or brown: Ground

USING SNAP-OFF HEADERS FOR MATED CONNECTORS

R/C servos and their mating receivers use polarized connectors to help prevent plugging things in backward. These polarized connectors are fairly expensive, and most folks instead use 0.100″ "snap-off" *pin headers,* common in electronics. You can buy these things at any local or online electronic parts outlet, and they're pennies a piece.

For a servo, snap off three pins, then solder them to your circuit board. Since these header pins lack any kind of polarization, it's possible to plug servos in backward. You'll want to mark how the servo connector should attach to the header, to help prevent this.

Fortunately, reversing the connector *probably* won't cause any damage to either the servo or the electronics, since reversing the connector merely exchanges the signal and ground wires. This is *not* true of the "old-style" Airtronics connector: if you reverse this connector, the signal and DC power (+V) lines are swapped. In this case, both servo and control electronics can be irreparably damaged.

> ⚠ Damage may occur if you wire up the pin headers wrong. Mix up the Ground and +V pins, and within seconds your servo will be *permanently ruined*—perhaps even die a violent death. I've seen the bottoms of servos blown out when they have been connected backward in a circuit.

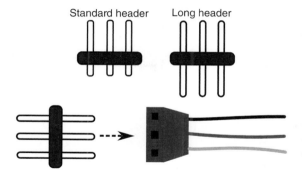

Figure 15-7   A standard male header, versus one that uses long pins on both ends. You want the long-pin version to connect your RC servos to a solderless breadboard.

You can also use headers with solderless breadboards to easily and quickly connect servo motors to the rest of your robot electronics, but it works best if you use the kind of header where the pins are long on both ends (Figure 15-7).

Analog versus Digital Servos

The most common, and most affordable, R/C servos are analog, meaning their control electronics uses traditional circuitry for controlling the motor. Digital servos use onboard microcontrollers to enhance their operation.

Among the added features of digital servos include higher-power and programmable behavior. With the proper external programmer (available separately, and at extra cost) it's possible to control the maximum speed of the servo, for example, or make the servo always start from power-up in a specific position.

Except for higher torques, from an applications standpoint there is little difference between analog and digital servos. You control both the same way. However, the higher torque of a digital servo means the motor is drawing more current from its power source, which means batteries tend not to last as long between charges.

For most robotics applications, digital servos are not required; an exception is when building a walking robot, where the extra torque of digital servos comes in handy. Six-legged walking bots may use 12 and even 18 servos just for the legs. The higher torque helps to offset the added weight of all those servos.

Electronics for Controlling a Servo

Unlike a DC motor, which runs if you simply attach a battery to it, a servo motor requires proper interface electronics in order to rotate its output shaft. While the need for interface electronics may complicate to some degree your use of servos, the electronics are actually rather simple. And if you plan on operating your servos with a PC or microcontroller, all you need for the job is a few lines of software.

A DC motor typically needs power transistors, MOSFETs, or relays if it is interfaced to a computer. A servo, on the other hand, can be directly coupled to a circuit or microcontroller with *no additional electronics*. All of the power-handling needs are taken care of by the control circuitry in the servo.

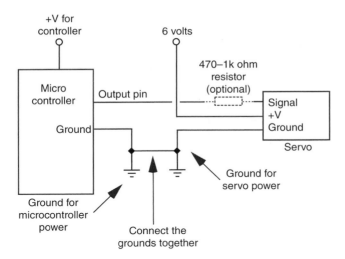

Figure 15-8 General connection diagram for attaching a microcontroller to a servo motor. The 470 to 1kΩ resistor is optional, and is included to prevent damage should a defective servo try to draw excessive current.

CONTROLLING A SERVO VIA A MICROCONTROLLER

All microcontrollers can be used to control an R/C servo. The basic connection scheme is shown in Figure 15-8.

- The microcontroller and servo can share the same power source, assuming the controller has an onboard regulator, but it's much better to use a separate source for the servo. Why? Servos draw a lot of current when they're first turned on or are in motion. By using separate sources—a 9-volt battery for the controller, for example, and a set of 4 AA cells for the servo—you avoid messy power line problems.
- When using separate power sources, be sure to connect the grounds from these sources together. Otherwise, your server will not work properly.
- Only one input/output (I/O) line from the microcontroller is needed to operate the servo. You may insert an optional 470-Ω (ohm) 1/8-watt resistor inline between the controller and the Signal input of the servo. This helps protect the microcontroller in case the servo has an (unlikely) internal electrical problem.

Be sure to power the servos using a separate battery pack. Connect as shown. You might also wish to add a 1-μF to 22-μF tantalum capacitor between the +V and ground of the servo power source to help kill any noise that may be induced into the electronics when the servo turns on and off.

Tantalum capacitors are polarized. Be sure to connect the + lead of the capacitor to the +V pin of the servo. For more information about tantalum and other types of capacitors, see Chapter 22, "Common Electronic Components for Robotics."

See the chapters in Part 6, "Robot Brains," for specific hookup diagrams and programming code for using servos with the Arduino and other microcontrollers. Be sure to also visit the RBB Support Site (see Appendix) for additional programming examples.

Figure 15-9   A popular servo tester, typical of the models available for under $10, operates up to three motors at a time. Control the rotation of the servo by turning the dial.

USING A SERIAL SERVO CONTROLLER

Even if your microcontroller is a speed demon and has no trouble creating pulses for 12, 18, or even 24 servos, you may not want to use it for that task. Instead, you may wish to use a dedicated serial servo controller. It acts as a pulse-making coprocessor.

Serial servo controllers connect to your microcontroller via a serial communications line. Using programming code (examples are provided with the unit you buy), you then send commands to each servo, telling it to move to a certain position. The work of producing the properly timed pulses is the job of the servo controller, thus freeing up your microcontroller to do other, more important jobs.

USING A SERVO TESTER

You may not always want to connect your servo to a microcontroller or serial servo controller just to play, or see if the servo works. Available online are a number of self-contained battery-operated servo testers, like the one in Figure 15-9. Connect a 6-volt battery supply, plug in a servo, and spin the control knob to operate the servo.

Most testers provide three operating modes:

- Manual—turn the dial to move the servo output shaft.
- Automatic—the servo rotates back and forth on its own. This is both a handy test mode, and a nifty bare-bones robot control mechanism.
- Neutral—the tester sends a continuous stream of 1500 μs pulses to center the servo at its middle point.

You can make your own servo tester using an inexpensive LM555 timer IC. See the RBB Support Site for a handy schematic for using the LM555 for testing and controlling and servo.

USING GREATER THAN 6 VOLTS

Servos are designed to be used with rechargeable model R/C battery packs, which put out from 4.8 to 6 volts, depending on the number of cells they have. Servos allow a fairly wide latitude in input voltage, and 6 volts from a four-pack of AAs provide more than enough juice. As the batteries drain, however, the voltage will drop, and you will notice your servos won't be as fast or strong as they used to be.

But what about going beyond the voltage of typical rechargeable batteries used for R/C models? Indeed, *some* servos can be operated at 7.2 volts, but always check the datasheet that comes with the servos you are using. Unless you need the extra torque or speed, it's best to keep the supply voltage to your servos at no more than 6 volts, and preferably between the rated 4.8- to 6-volt range specified in the manufacturer's literature.

Using Continuously Rotating Servos

So far I've only talked about servos that are meant to turn a portion of a circle. These are used when precise angular positioning is required, such as scanning a sensor from side to side.

But R/C servos can also rotate continuously, either by design or via a modification that you perform yourself. R/C servos make terrific drive motors for your robot. One benefit of CR servos as drive motors for a robot: They tend to be less expensive than comparable DC gear motors of the same specification, and they come with their own driver electronics. They're definitely worth considering for your next robot.

Servos that rotate continuously act like an ordinary geared DC motor, except they're still controlled by sending the motor pulses.

- To make the motor go in one direction, send it 1-ms pulses.
- To make the motor go in the other direction, send it 2-ms pulses.
- To make the motor slow to a stop, send it 1.5-ms pulses.
- To make the motor stop altogether, stop sending it pulses.

Stopping the motor by ceasing the pulses works for all except digital servos. With most digital servos, when pulses are stopped the servo will merely continue with the last good position information it received.

You're not likely to use digital servos for continuous rotation, so this problem seldom comes up in real life. But keep it in mind just in case.

Not long ago, you'd have to retrofit a standard R/C servo to make it turn continuously. Though the process of converting a regular servo to a continuously rotating one, it is time consuming and there's always the risk of damaging the servo.

These days there are many servo manufacturers that offer "pre-converted" servos; these turn continuously in either direction right out of the box. No extra work needed. Cost is the same or only a dollar or two over the price of a standard limited-rotation servo. Manufacturers include Parallax, SpringRC, PowerHD, Feetech, and GWS.

Figure 15-10 Many continuous rotation servos provide a means to adjust the center (also called null) point of rotation, where 1500 µs pulses will cause the motor to completely stop.

Many (though not all) pre-made CR servos provide an adjustment control to "dial in" the center point of the servo where it will not rotate when fed 1500 µs pulses. The adjustment control is frequently located on the side of the servo, as shown in Figure 15-10. See Chapter 24, "All About Robot Gray Matter," for additional information.

Check out Chapter 17, "Build Robots with Wheels" and Chapter 18, "Build Robots with Tracks," for numerous robot projects using CR servos. You'll find even more examples in Part 9, "Online Robot Projects."

 For the true DIY'ers out there, see the RBB Support Site for step-by-step instructions on how to modify a typical standard servo for continuous rotation!

Using Servos for Sensor Turrets

After all this talk of using R/C servos as robot drive motors it's easy to forget what they were created for in the first place: for precise angular position. A common application in robotics is the sensor turret, so-called because it acts as a rotating turret (like a cannon gun turret) for one or more robot sensors. Typical sensors for turrets include ultrasonic and infrared proximity detection—these are detailed in Chapter 31, "Proximity and Distance."

The concept of the rotating sensor turret is simple: put a sensor on top of the servo, and then "scan" it back and forth by alternating the position of the servo left and right. You can mount the servo to the robot using a variety of methods. Foam tape and glue are always options, but I prefer mechanical fasteners that can be easily removed and reattached; the fasteners make it easier to build and change the robot.

Using Servos for Legs, Arms, and Hands

Another useful application for regular (non-CR) servos is as joints in walking robots and robotic arms and hands. By commanding the servo to turn to a certain angular position, you can precisely change the bend of the joint, allowing the bot to walk, move an arm into position, or grasp and object.

Figure 15-11   Regular servos used in a robotic arm provide precisely controlled movable joints. This arm contains four servos: fingers open/close, wrist rotation, wrist flex, and elbow.

Figure 15-11 shows several standard size non-CR servos used in a homebrew robot forearm and hand. Each servo operates a joint, or in case of the hand, opens and closes the fingers.

Several servo-based projects using joints are provided in Chapter 19, "Build Robots with Legs" and Chapter 20, "Build Robotic Arms and Grippers."

CHAPTER 16

Mounting Motors and Wheels

You've got two motors. You've got a robot body. What comes next isn't always simple: you have to somehow mount the two motors onto the robot body, hopefully without making the thing look like a junkyard reject! Then there's the problem of attaching the wheels.

DC motors and R/C servos each have their own means of mounting to a robot platform or frame. Some are easier than others. In this chapter you'll learn ways to mount both common and not-so-common motors to robot frames and platforms. And you'll learn about attaching wheels to those motors.

To round out the chapter, you'll find helpful information about using standard drivetrain components—things like gears, chain, belts, and couplers.

Mounting DC Motors

There are no hard standards in the design of a DC motor. Depending on how the motor was meant to be used, it may be a snap to mount, or it could be cumbersome, requiring a hodgepodge of hardware.

- Generally speaking, motors meant for use in a variety of applications tend to have holes, brackets, or flanges that make mounting easier.
- Those motors engineered to work with just a specific product rely on the design of that product for secure mounting. There are no holes, no brackets, no flanges.

Motors meant for robotics (or at least home-shop tinkering) are made with mounting in mind. These include the various Tamiya motor kits, all of which have flanges or other means for secure mounts. Solarbotics, Pololu, and several other robotics-centric online retailers import motors that have been especially selected because they offer mounting ease. Or else they provide their own mounting solutions, especially crafted for the motors they offer.

These and other motors like them are the easiest to work with, and if you're just starting out in robotics, these should be the ones you choose. It'll make your life much easier!

USING BUILT-IN FLANGES

Oh, happy joy—a motor with its own mounting flanges! It doesn't get much better than this, so enjoy while you can. You know what you need to do here: just find some nuts and screws that fit, mark where the holes should go on your robot, and attach.

When mounting with flanges, I like to insert the head of the screw through the flange on the motor side (adding a flat washer if the head might come through the flange). A nut secures everything on the other side (Figure 16-1).

USING MOUNTING HOLES

Some motors have already-threaded holes on their faceplate, the part with the shaft sticking out. The threads may be imperial (inch) or metric. For imperial, 2-56 and 4-40 threads are the most common. For metric, you'll often find holes tapped for 3mm or 4mm machine screws.

If you're mounting the motor perpendicular to the body of the robot, you may need to come up with a bracket to secure the motor to. More about brackets later in this chapter.

 It's important that the machine screws you use don't go too far into the motor, or else they may obstruct the rotating parts inside.

USING BRACKETS

You need brackets if you must attach a vertically mounted motor to a horizontal surface. Metal brackets are common finds at any home improvement store. They tend to be a bit bulky, being made for general household applications. For smaller bots I prefer lightweight plastic brackets available at specialty online stores and miniature metal hardware available from the larger electronic parts outlets.

Figure 16-1 On motors with a mounting flange, insert the fastener from the flange side, when possible. In this way the exact length of the screw is not as critical, and you can use self-tapping screws that secure into the base material (wood, plastic, etc.).

You can also fashion your own mounting brackets using metal or plastic. Cut the bracket to the size you need, and drill mounting holes. This technique works well when you are using servo motors for radio-controlled model cars and airplanes.

Brackets from Metal

You can make your own L-shaped metal brackets by bending a strip of aluminum, brass, or stainless steel metal. You can find these metal strips at many hobby stores. Aluminum and brass are easier to work with. They're best for smaller motors. For larger and heavier motors, opt for stainless steel.

To make a metal bracket:

1. Cut the strip to the desired length—for example, make the strip slightly more than 2″ long for an L-shaped bracket that is 1″ × 1″. Each side of the L is a "leg" of the bracket, and it's 1″ long.
2. Mark the midpoint of the bracket with a pencil. For that 2″ bracket, for instance, place a mark 1″ in.
3. With the metal strip held securely in a vise, drill at least one hole in each leg (see Figure 16-2). For larger brackets, drilling two holes is even better. Stay away from the midpoint mark you made in step 2.
4. For this next part, use a vise secured to your workbench. Put the strip into the vise so that the midpoint mark is just visible, and tighten the vise.
5. Use a hammer and a block of wood to fold over the exposed leg of the bracket. You want a neat 90° angle.

When making small brackets, drill the holes first, then cut the strip to length. If you need to make lots of brackets, draw the layout on a piece of paper and use it as a template for placing the holes and cuts.

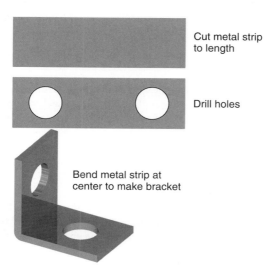

Cut metal strip to length

Drill holes

Bend metal strip at center to make bracket

Figure 16-2 Make your own metal angle brackets by cutting strips of metal to length, then drilling holes. Bend at the middle to finish the bracket. Use copper, brass, or aluminum; steel is much harder to work with.

Bracket Blocks from Wood and Plastic

You can make convenient block-style mounting brackets out of small pieces of 1/2" or thicker wood or plastic. Start with a minimum 1.5"-long block. On one face of the block drill two holes close to the outside edges. On the other face drill at least one hole toward the center part of the block. You can then use the block as a type of L bracket.

For larger motors, use bigger and longer blocks. You may need two and even three mounting holes per face. Drill larger holes for bigger machine screws.

For plastic materials, visit your local neighborhood plastics retailer (check the Yellow Pages for one near you). Most have a discard bin in the front showroom with odds and ends. Look for small scraps of thick nylon, acetal resin (Delrin), and ABS plastic. The scrap is usually sold by the pound; enough material for a half dozen blocks should cost only a few dollars.

Using Clamps

If the motor lacks mounting holes, you can use clamps to hold it in place. U-bolts, available at the hardware store, are excellent solutions. Choose a U-bolt that is large enough to fit around the motor.

A technique that works with smaller motors is to use hold-down straps designed for EMT (electrical) conduit pipes. The straps are available in various sizes to hold down pipes of different diameters. These pipe straps are available at your local home improvement and hardware stores, in both metal and plastic. The plastic ones are easier to work with and lighter.

Mounting and Aligning Motors with Aluminum Channel

The same (but somewhat larger) aluminum channel used to construct robot frames can be used to mount and align DC motors. Find a channel that's large enough inside for your motors to drop in place. A snug fit is best.

Cut the channel to length and place the motors within the channel, end to end. If the motors protrude from the channel, you might be able to secure them in place using nothing more than a cable tie—for demonstration purposes, the illustration shows one tie, but you'll probably need several to hold the motor in place. Cinch up the tie so that it firmly holds the motor.

Mounting R/C Servos

The world is quite a different place when working with R/C servos. By their nature, servos have mounting flanges; what's more, servo sizes and mounting configurations are fairly standardized, giving you the option of premade mounting solutions if you don't want to "roll your own." Whatever the method, servos should be securely mounted to the robot so the motors don't fall off while the thing is in motion.

ATTACHING SERVOS WITH HOT MELT GLUE

Gluing is a quick and easy way to mount servos on most any robot body material, including heavy cardboard and plastic. For a permanent bond, use only a strong glue, such as two-part epoxy. For a somewhat less permanent construction, I prefer hot melt glue because it doesn't emit the fumes that epoxy does and it sets much faster.

When gluing, it is important that all surfaces are clean. Rough up the surfaces with a file or heavy-duty sandpaper for better adhesion. If you're gluing servos to LEGO parts, apply a generous amount so the extra adequately fills between the "nubs." LEGO plastic is hard and smooth, so be sure to rough it up first.

If you don't need or want to permanently attach the servos, use less hot melt glue or use a dab of ordinary white household glue. The bond is weaker, allowing you to more easily pry the servo off and reuse it for something else.

ATTACHING SERVOS WITH SCREWS

Unless you have a good reason not to, the best way to mount R/C servo motors to your robots is with screw fasteners. You have two kinds of screws to choose from:

- Self-tapping metal or wood screws don't need a nut on the other end to hold things in place. Drill a small pilot hole to start, then insert the screw. The threads of the screw dig into the material and hold it into place.
- Machine screws and nuts are ideal if you need to disassemble your creation and rebuild it, or use parts for something else.

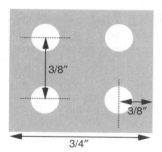

Servo "Tab" Mounting Brackets

For R/C servo motors of any size you can create simple mounts using a pair of "tab brackets." Start with a strip of 1/8"-thick aircraft plywood or plastic, and drill a pair of holes to match those on the flange of the servo motor. No closer than 3/8" from these holes, drill one or two holes for a metal or plastic angle bracket.

After drilling, cut the strip near the pair of servo flange holes. Repeat this process again for the second bracket. Once you get the hang of it you can drill and cut a pair of servo brackets in just minutes. Figure 16-3 shows a pair of tab mounting brackets on a servo.

Figure 16-3 A pair of servo mounting tabs, shown with servo and 3/8" metal angle brackets for securing to a robot base.

Figure 16-4   The basic mount accepts a standard size servo, and provides numerous holes for attaching to the robot base.

Specialty Stamped Metal Servo Mounts

With the popularity of servos for robotics applications there's no shortage of metal mounts for all types and sizes of servos. These are mass produced using a stamping process: the mount is punched out of a piece of aluminum with various strategic holes, and then bent to form flanges and ridges.

Figure 16-4 shows a common type of metal servo mount. Although it's a basic and fairly simple mount, it has a wide variety of uses. The mount accepts a standard servo using screws—often included, but check with the seller. The servo can then be securely mounted to the body of the robot, or even attached to another mount to create a compound linkage.

Additional types of stamped metal are used in concert with the mounts, providing useful hinges for arms, legs, and other jointed linkages. You can read more about these mounts in Chapter 19, "Build Robots with Legs" and Chapter 20, "Build Robotic Arms and Grippers." Stamped metal mounts are available from online sources and are not likely to be something you'll find at a local store; see Sources in the RBB Support Site for a list of selected online retailers who offer a variety of these premade metal mounts.

On the Web: Making Your Own Mounts

You can also construct your own servo mounting brackets using 1/8"-thick aluminum or plastic using layout and drilling plans on the RBB Support Site (see Appendix).

Mounting Drivetrain Components to Shafts

Drivetrain components are things like wheels, gears, sprockets, and other stuff that attach to your robot's motors. Unless the parts are specifically designed for one another, connecting the shaft of the motor to drivetrain components is one of the more difficult tasks in building a robot.

Still, it's not impossible, and robotics wouldn't be as much fun without the occasional challenge. What follows are the most common methods used to connect a motor shaft to a wheel, gear, or other drivetrain component.

PRESSFITTING

For smaller drivechains, it's not uncommon to use pressfit parts, where the shaft fits very tightly into the wheel or other component. Usually these are manufactured to fit this way, and the parts are assembled with heavy-duty hydraulic presses. You probably don't have such a press in your garage; you'll need to instead use a small hammer, a bit of spit for lubrication, and a few well-chosen curse words to knock the shaft into position.

SETSCREWS

A setscrew physically clamps the wheel or other part directly to the shaft. Setscrews are convenient and elegant, but they're not common in amateur robotics because drivetrain parts that use them tend to be more expensive. You're more likely to encounter setscrews when using small metal R/C parts like gears or locking collars (see the section that follows, on making your own shaft couplers).

Setscrews usually have hex socket heads, so you need a hex wrench in order to remove or tighten the screw. When purchasing R/C parts that use setscrews, the wrench is usually included.

PROPRIETARY INTERLOCK

Some motors and wheels (and other drivechain components) are made to go together. Good examples are the various DC gearbox motors and wheels from Tamiya. Many of the motors and wheels are designed for interchangeability. The wheels fit onto the motor shaft using nuts, pins, or other means.

SERVO HORNS

The output shaft of R/C servo motors has about two dozen tiny splined teeth carved around its circumference. These splines, and a metal screw that's inserted into the bore of the shaft, are made to secure various styles and sizes of *servo horns*. Most servos come with at least one of these horns—typically a round or small, X-shaped jobbie.

Servo horns are made of plastic or metal (when of metal, it's usually aluminum). You can drill holes for attaching things to the horns. This is one of the primary ways of securing wheels to a servo. You can also glue parts to the horns, though this works best when bonding plastic pieces to plastic horns. Hot melt glue and epoxy are good choices.

ADHESIVES

Some motor shafts, wheels, and other drivetrain components can be bonded with an adhesive. This technique works best for plastic parts—don't try to glue a metal shaft to a plastic wheel.

An example of using adhesives is to cement a LEGO axle into a non-LEGO plastic wheel. You can then use the axle/wheel combo in a LEGO creation. Drill out the hub of the wheel so it's just smaller than the axle. Then gently tap the axle into the hole using a small hammer. You may set the axle in place using epoxy or household adhesive.

Mounting Wheels to DC Gear Motors

In the world of amateur robotics, you can now find motors that match the wheels made for them. Those are the easiest to use when building a bot, but there are many other options, too. With some effort, you can adapt a wheel to most any kind of motor, using either a direct connection—as described in this section—or a coupler, as detailed later in the chapter.

USING MATCHING MOTORS AND WHEELS

When I first started in robotics I used to spend an inordinate amount of time combing through various surplus catalogs, looking for motors and wheels that could go together. It wasn't always easy to find matches. Today there are numerous specialty online robotics retailers that offer low-cost DC motors and wheels that are designed to complement one another.

And when I say low-cost, I do mean low-cost—it's easy to find a motor-and-wheel set that's priced under $10 each. You need two for a basic bot. Add some batteries, wire, some simple electronics (or even a microcontroller like the PICAXE), and you have a fully functional autonomous robot for under $30. Not bad.

BUILDING CUSTOM WHEELS

While matching motors and wheels are handy, selection isn't always great. You may want a smaller or larger wheel than what's offered, or you may not like the width of the rubber tracks on the wheel. That's when you need to come up with your own motor-to-wheel solution.

 Be sure to also see the section "Using Rigid and Flexible Couplers" for more ideas on how to match wheels to motor shafts. Using couplers requires a bit more work than the methods outlined here, but sometimes you have no other choice.

Wheels with Setscrews

If your wheel already has a hub with a setscrew, you're in business . . . assuming the wheel hub is the right size for the motor shaft. If it is, you're ready to go. But if it's not, there are a couple of things you can do to solve the problem:

- If the wheel hub is too small, drill it out to fit the shaft. Obviously, this works only if the hole you drill isn't so large that it destroys the wheel hub.
- Use a reducing bushing, which is basically a short length of hollow tube. The inside of the bushing is sized to match the motor shaft; the outside, the wheel hub. You then carefully drill a hole through the bushing. Use a longer setscrew, if needed, to make it extend through the bushing and make contact against the shaft.

Using Flanges

But what if the wheel lacks a setscrew? One solution is to use a *flange* that does have a setscrew, and a hub with a properly sized bore that matches the motor shaft. You attach the flange to the wheel using fasteners, then mount the flange to the shaft.

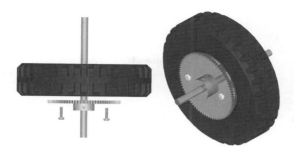

Figure 16-5 The concept behind using a metal or plastic gear as a flange for mounting a wheel to a motor shaft. Attach the face of the gear against the wheel hub; the gear provides a hub that can then be secured to the motor shaft.

A flange is anything smaller than the diameter of the wheel and that has a setscrew and hub compatible with the motor shaft you want to connect it to. Specially made flanges are available through Lynxmotion, Jameco, Servo City, and other online retailers. The flanges come in different bore sizes. Cost is modest.

A convenient ready-made flange is a surplus gear. Drill two or three holes through the face of the gear and matching holes into the wheel, being sure to keep the gear and wheel concentric. Mount the flat part of the gear against the wheel (see Figure 16-5), then attach the assembly to the motor shaft.

Mounting Wheels to R/C Servos

Servos reengineered for full rotation are most often used for robot locomotion and are outfitted with wheels. Since servos are best suited for small- to medium-size robots (under about 3 pounds), the wheels for the robot should ideally be between 2″ and 5″ in diameter, and lightweight.

WHEELS ENGINEERED FOR R/C SERVOS

By far the easiest way to attach wheels to servos is to use a wheel that's specially engineered for the job. Many specialty robotics retailers sell wheels meant for use with the standard-size Hitec and Futaba servos.

Now for the bad news: Your choice of wheel diameters is pretty limited. You'll find just a few sizes, with 2-1/2″ (give or take a few fractions of an inch) the most common. If you need a smaller or larger size, you can always make your own wheels, as detailed next.

Servos differ in the type of spline used on their output gear. The three most common spline types are noted simply by the servo manufacturer that popularized them: Hitec, Futaba, and Airtronics. If you are purchasing wheels for your servos, make sure the wheels use a matching hub spline. Wheels made for standard-size Futaba servos will also work on any other brand that uses the same Futaba-style spline, such as Parallax.

Figure 16-6 Most any kind of wheel can be converted for use with an R/C servo by attaching a servo horn to the side of the wheel. If your servo didn't come with an assortment of horns, you can purchase them separately.

MAKING YOUR OWN WHEELS FOR SERVOS

The general approach for attaching wheels to servos is to use the round servo horn that comes with the servo and secure it to the wheel using screws or glue (see Figure 16-6). The underside of the horn fits snugly over the output shaft of the servo. Here are some ideas:

Lightweight foam tires, popular for model airplanes, can be glued or screwed to the servo horn. Popular brands are Dave Brown and Du-Bro, and these can be found at most any well-stocked R/C hobby store. The tires are available in a variety of diameters, with the 2″, 2-1/2″, and 3″ diameters the best for small bots.

Large LEGO "balloon" tires have a recessed hub that exactly fits the small, round servo horn included with Hitec and many other servos. You can simply glue the horn into the rim of the tire.

A gear glued or screwed into the servo horn can be used as an ersatz wheel or as a gear that drives a wheel mounted on another shaft.

Homemade O-ring wheels can be constructed out of two plastic discs, cut to any diameter you like—though about 3-1/2″ is a practical maximum. The O-ring is the rubber tire of the wheel. At the center of the discs, mount a large, round servo horn, then fasten the pieces together using miniature machine screws and nuts.

Pulley horns look like three discs cemented together with a space in between. You can turn the pulley horn into a unique wheel by adding a pair of rubber O-rings as tire treads.

Urethane skateboard/inline roller-skate wheels make for incredibly "grippy" wheels for robots. The trick, if it can be called that, is to find metal or plastic discs that just fit into the hub of the wheel. Most skate and inline blade wheels use metric sizes, with a hub diameter of 22mm. A 0.625″-diameter fender washer fits into the bore of the wheel but is (usually) large enough to stop against the ridge that's molded into the center of the wheel. Hold all the pieces together with a 4-40 machine screw that goes from the outside of the wheel and directly into the servo motor output shaft.

Attaching Mechanical Linkages to Servos

A key benefit of using R/C servos is the variety of ways you can connect stuff to them. In model airplane and car applications, servos are most often connected to a push/pull linkage (called a *pushrod*). As the servo rotates, the pushrod draws back and forth, like that in Figure 16-7.

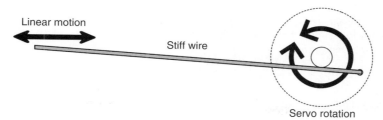

Figure 16-7   Rotary motion can be converted to linear motion by using a pushrod linkage. As the servo rotates, the end of the linkage moves back and forth.

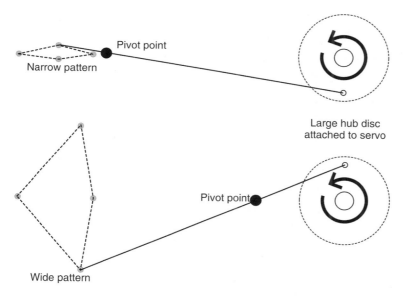

Figure 16-8   Control the angular displacement of the pushrod by changing the location of the pivot point, which is simply a mechanical conduit (small tube, hole, plastic grommet) that restricts side-to-side motion. Avoid placing the pivot point too close to either end of the pushrod.

You can use the exact same hardware designed for model cars and airplanes with your servo-equipped robots. Visit the neighborhood hobby store and scout for possible parts you can use. Look for pushrods and clevis ends.

CONTROLLING LINEAR MOVEMENT

So you can see how pushrods allow you to convert the rotation of a servo to linear movement. There are two key ways of controlling the amount of linear movement you get:

- Use a larger or smaller horn on the servo. The larger the horn, the larger the movement of the pushrod.
- Use a specific pivot point for the pushrod, a mechanical constriction like an eyelet, channel, or hole that limits the movement of the pushrod to just back-and-forth motion. As shown in Figure 16-8, the closer the pivot is to the servo, the wider the movement pattern; conversely, the farther away from the servo, the narrower the pattern.

There will always be some side-to-side motion in the linear movement. For this reason, it's a good idea to design a bit of "slop" in the pushrod system, such as a loose clevis end on either end of the pushrod.

ADDING LUBRICATION

Parts that slide against each other should be kept lubricated to prevent them from binding up. Use a thick grease, not oil. Grease for hobby R/C is available at any store that carries these parts. It comes in a small tube; you want the white or clear stuff. My favorite is white lithium grease.

Drivetrain Components for Robotics

Pushrod mechanical links are one form of drivetrain—get the power of a motor from one place to another. There are hundreds of drivetrain components you might use in your robot creations, but the following table summarizes the most commonly used. These drivetrain parts go between the motor and the wheel or other driven element, such as legs, tracks, or arm segments.

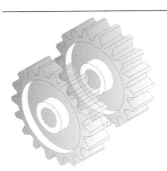

Gears

Gears are a principal component of drivechains and are primarily used in robotics to reduce the speed, and increase the torque, of the wheel drive motors. They are also used to share the power between several wheels or other components.

Because of the mechanical precision required to properly mesh gears, most amateur robot builders do not construct their own gear assemblies. More about gears later in the chapter.

Timing Belts

Timing belts are also called *synchronization* or *toothed* belts, and they can be used to make tracked robots, as well as substitute for more complex gear systems.

Widths range from 1/8" to 5/8", and lengths from just a few inches to several feet in diameter. Material is usually rubber. Belts are rated by the pitch between "nubs" or "cogs," which are located on the inside of the belt. The cogs interface with matching rollers.

Endless Round Belts

Endless round belts are used to transfer low-torque motion. The belt looks like an overgrown O-ring and, in fact, is often manufactured in the same manner.

Grooved pulleys are used with round belts. The diameter of the pulleys can be altered to change torque and speed. Like timing belts, you can use round belts as the tire material on wheels and for unusual treads in a tracked robot.

Roller Chain

Roller chain is exactly the same kind used on bicycles, only in most robotics applications it's smaller.

Roller chain is available in miniature sizes, down to 0.1227" pitch (distance from link to link). More common is the #25 roller chain, which has a 0.250" pitch. For reference, most bicycle chain is #50, or 0.50" pitch. The chain engages a sprocket that has teeth of the same pitch.

Idler Wheels

Idler wheels (also called idler pulleys or idlers) take up slack in belt- and chain-driven mechanisms. The idler is placed along the length of the belt or chain and is positioned so that any slack is pulled away from the belt or chain loop. Not only does this allow more latitude in design, it also quiets the mechanism.

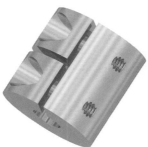

Couplers

Couplers come in two basic styles: rigid and flexible. They are used to directly connect two shafts together, so you don't need a gear or belt to transfer the power. Couplers are common when connecting wheels and motors that aren't otherwise designed for one another.

Bearings

Bearings are used to reduce the friction of a spinning component, such as a wheel or idler, around a shaft. There are lots of types of bearings, but ball bearings are the most common. The ball bearing is composed of two concentric rings; between each ring is a row of metal balls. The rings—and the ball bearings—are held in place by a flange.

Bushings

Bushings and bearings serve the same general purpose, except a bushing has no moving parts. (Note: Some people call these *dry bearings*.) The bushing is made of metal or plastic and is engineered to be self-lubricating. Bushings are used instead of bearings to reduce cost, size, and weight, and are adequate when friction between the moving parts can be kept relatively low.

Using Rigid and Flexible Couplers

Couplers are used to connect two drive shafts together end to end. A common application is to use a coupler to connect the drive shaft of a motor with the axle of a wheel. Couplers can be rigid or flexible.

Rigid couplers are best used when the torque of the motor is low, as it would be in a small tabletop robot. Conversely, flexible couplers are advised for higher-torque applications, as they are more forgiving of errors in alignment. Why is this? A rigid coupler may shear off or damage the motors or shafts when misaligned.

PURCHASING READY-MADE COUPLERS

There are many types of commercially available rigid and flexible couplers, and cost varies from a few dollars to well over $50, depending on materials and sizes. Common flexible couplers include helical, universal joint (similar to the U-joint in the driveshafts in older cars), and three-piece jaw.

The couplers attach to the shafts either with a press fit, by a clamping action, by setscrews, or by keyway. Press fit and clamp are common on smaller couplers for low-torque applications; setscrews and keyways are used on larger couplers.

Three-piece jaw couplers consist of two metal or plastic pieces that fit over the shafts. These are the "jaws." A third, rubberized piece, the spider, fits between the jaws and acts as a flexible cushion. One advantage of three-piece couplers is that because each piece of the jaw is sold separately, you can readily "mix and match" shaft sizes—couple a 1/8" motor shaft to a 1/4" wheel shaft.

MAKING YOUR OWN RIGID COUPLERS

To save money, you can make your own rigid couplers using metal tubing, metal or plastic standoffs, or threaded couplers.

Couplers from Tubing

You can get brass, steel, and aluminum tubing of various diameters at hobby stores; larger aluminum tubing can be found at home improvement outlets. Most tubing is sold by its inner diameter—1/8" tubing measures an eighth of an inch inside, for example. The thickness of the tubing determines its outside diameter. Get tubing that fits over the motor or wheel shaft you're attaching to.

When matching up tubing, note its I.D. (inside diameter). Tubing at the hobby store is meant to fit into the next size larger, as in a folding telescope. You can use this feature to match one shaft diameter to another. Typical thickness of the tubing is between 0.014" and 0.049", though this varies somewhat by brand.

For lightweight robots with small motors, look at 1/8" or 3/16" I.D. If possible, bring your motor or wheel (or just the shaft) to the store with you so you can test the pieces for proper fit.

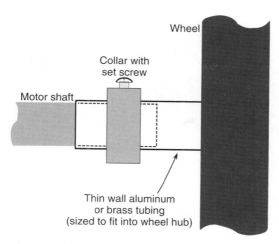

Figure 16-9 Thin tubing or plastic doesn't have enough "bite" for a setscrew. But you can use a metal collar around the tubing to provide compression against a solid motor shaft. Tighten the setscrew in the collar for a solid fit.

To use, cut the tubing to length. Use a tubing cutter instead of a saw. You can then secure the tubing to the shaft in several different ways, including the following:

- Crimp it on with an appropriately sized metal collar. The collar comes with a setscrew. Carefully tighten the setscrew over the tubing (see Figure 16-9). This type of metal collar is available at any R/C hobby store and often goes by the name Dura Collar. If you can't find a collar just the right size for the tubing, you may need to select the next size smaller, then drill it out. The collar is typically plated brass, so it's not as hard to drill as it may look.
- Use large wire crimper pliers to squeeze the tubing around the shaft. This works only for thin-walled tubing and when working with smaller motors that don't develop a lot of torque.

When using the crimp-on metal collar trick, you can help cinch up the fit by drilling a small (1/16″) hole in the tubing where the setscrew for the collar will go. This works well when using thicker-walled tubing. When assembling the collar over the tubing, carefully line up the hole with the setscrew, then tighten the screw.

 Don't forget that some wheels have outsized hubs, like the one in Figure 16-10. You may not even need to use tubing to make a coupler. Just attach the collar around the existing hub of the wheel, and tighten the set screw. With plastic wheels the hub will deform, cinching around the shaft for a fairly tight fit.

Couplings from Standoffs and Threaded Couplers

Commercially available rigid couplers often use setscrews to hold the coupler to the shaft. Thin-wall tubing is too thin for a setscrew. You need a thicker-walled coupler.

There are two ready-made sources of short metal and plastic begging to be turned into shaft couplers: standoffs and something called threaded couplers.

Figure 16-10 Metal collars may be used on wheels with an outsized hub (the hub protrudes from the wheel, giving more surface area to the shaft). Place the collar around the hub of the wheel. The compression tightens the hub around the motor shaft.

Standoffs are commonly used in electronics projects to keep two circuit boards or other pieces separate from one another. I regularly use metal and plastic standoffs to add additional "decks" to my robots. The decks are like tiers in a wedding cake. The standoffs are typically 1″ or longer and are made of nylon, steel, or aluminum. I like working with aluminum over steel, as aluminum is easier to drill and cut. You can get either the threaded or the nonthreaded kind, but the unthreaded kind is easier to work with.

Threaded couplers are just like standoffs but are almost always made of heavier steel. You get them at the local hardware or home improvement store. They come in thread sizes from 6-32 to 5/16″ and larger. The 6-32 thread size is about right for 1/8″ shafts.

To make shaft couplers, you should have a vise and a drill press in order to make accurate holes. You also need a tap set with drill bits. The bits are already correctly sized for the taps.

 Nylon can't hold a thread as well as aluminum or steel. If you go with a nylon standoff, use it only with shafts that have a flatted side to them. Round shafts require more pressure to keep them from turning, but this added pressure of the setscrew against the nylon just strips out the threads.

1. Secure the standoff or coupler into the vise. Make it nice and snug.
2. Use a center punch to mark where you will be drilling. You want two holes, no closer than about 1/4″ from the ends. The punch makes a mark in the plastic or metal that serves as a pilot for keeping the bit on track.
3. For metal standoffs and couplers, apply some light machine or cutting oil on the mark. Slowly drill the holes for the setscrews all the way through one side of the standoff or coupler. Don't drill through to the other side unless you want to put setscrews on that side as well.
4. Turn the standoff or coupler on end and tighten in the vise. Change the drill bit to the size of the shaft you're using. For example, if the shaft is 1/8″ in diameter, use a 1/8″ drill bit.
5. Drill halfway down the length of the standoff or coupler, then turn it upside down. Drill through the second half.

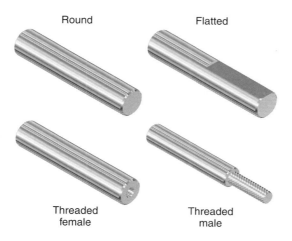

Round   Flatted

Threaded
female

Threaded
male

Figure 16-11   Common shaft types you will encounter: round, flatted, and threaded (both male and female). The flatted type is also called a D-shaft, because it resembles the letter D.

6. Now to tap the holes for the setscrews. Change the bit (if needed) to the correct size for the setscrew you wish to use.

7. Put a drop or two of machine or cutting oil into the holes you've drilled (metal standoff or coupler only), then carefully tap them to make threads. You can skip the cutting oil when tapping into plastic.

 I'm not going to give a lot of detailed explanation about how to use taps to thread a hole, as this information is everywhere on the Internet. Do a quick Web search for phrases like *how to drill and tap*.

On the Web: Making Your Own Flexible Couplers

Flexible couplers use a soft material, like aquarium tubing, to join two shafts together. Read more about how to build them on the RBB Support Site.

Working with Different Shaft Types

Motor shafts come in several shapes and forms. A few of the more common ones are shown in Figure 16-11. Most motors use a simple round shaft; most secure to a gear or wheel hub using a tight friction fit. A flatted or "D" shaft is best when using a setscrew, as the tip of the screw can settle into the flat depression. Flatted shafts may also be used for friction fit. The "D" helps prevent the shaft from spinning inside the wheel or gear hub.

Some motors have threaded shafts. For example, several motors in the Tamiya educational motor lineup have a short male-threaded shaft end. Using locking nuts you can secure wheels and other components onto the end. R/C servo motors use a female-threaded shaft to secure a servo horn or other accessory to the motor. On servo motors, the shaft is also splined to help prevent slippage.

Other shaft types you may encounter include the hex and square, so-called because of their hexagonal or square shape. They're used with wheels, gears, and other parts that have matching-shaped hubs.

PART 4

Make Your First Robot

CHAPTER 17

Build Robots with Wheels

In previous chapters you've seen how you can construct practical and functional mobile robots using only common tools and readily available materials. You can use wood, plastic, or metal—or, for quick prototypes, heavy-duty cardboard or foamboard intended for art projects.

This chapter extends what you've discovered in previous pages, offering numerous plans and design concepts for building robots that run about on wheels. The choice of construction material is up to you, and you're free to experiment with different sizes and assembly techniques.

See Chapter 18 if you're wanting to build a robot that uses tank tracks, and see Chapter 19 if you have a hankering to make a robot that walks on legs.

See also Chapters 12 through 16 for information on general robot design, as well as information on powering your robot with motors. And be sure to check out the free bonus projects on the RBB Support Site (see Appendix).

Basic Design Principles of Rolling Robots

With few exceptions, bots that roll use wheels or tank treads to get from one place to another. As you read in Chapter 13, "How to Move Your Robot," wheeled robots use a number of steering techniques. The most common—for wheels or treads— is two motors on each side of the vehicle.

DRIVE MOTOR ARRANGEMENTS

The most popular mobile robot designs uses two identical motors to spin two wheels on opposite sides of the base. These wheels provide forward and backward locomotion, as well as left and right steering. If you stop the left motor, the robot turns to the left. By reversing the motors relative to one another, the robot turns by spinning on its wheel axis ("turns in place"). You use this forward-reverse movement to make "hard" or sharp right and left turns.

Centerline Drive Motor Mount

You can place the wheels—and, hence, the motors—just about anywhere along the length of the platform. If they are placed in the middle, as shown in Figure 17-1, you should add casters to either end of the platform to provide stability. Since the motors are in the center of the platform, the weight is more evenly distributed across it.

A benefit of centerline mounting is that the robot has no "front" or "back," at least as far as the drive system is concerned. Therefore, you can create a kind of *multidirectional* robot that can move forward and backward with the same ease. Of course, this approach also complicates the sensor arrangement of your robot. Instead of having bump switches only in the front of your robot, you'll need to add some in the back in case the robot is reversing direction when it strikes an object.

 Depending on the size of the robot and its weight distribution, you may be able to get by with just one caster. By placing slightly more weight over the caster, the bot will favor tipping to that side.

Front-Drive Motor Mount

You can also position the wheels on one end of the platform. In this case, you add one caster on the other end to provide stability and a pivot for turning, also shown in Figure 17-1. Obviously, the weight is now concentrated more on the motor side of the platform. Even out the weight distribution by putting the batteries in the center of the platform.

One advantage of front-drive mounting is that it simplifies the construction of the robot. Its "steering circle," the diameter of the circle in which the robot can be steered, is still the 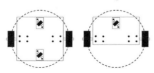 same diameter as the centerline drive robot. However, it extends beyond the front/back dimension of the robot. This may or may not be a problem, depending on the overall size of your robot and how you plan to use it.

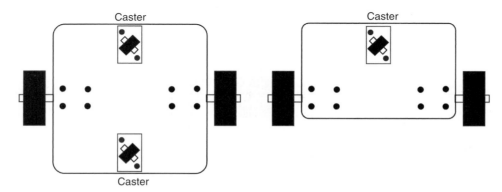

Figure 17-1 Center- versus front-drive motor mounting. With the center-drive arrangement, you typically need a caster on each end, though you can use just one if the robot is properly balanced.

PICKING THE RIGHT WHEELS

Wheels are made up of tires (tyres for those of you who speak British English) mounted on hubs (Figure 17-2). A tire is rubber, plastic, metal, or some other material, and the hub is the portion that attaches to the shaft of the axle or motor. The hub usually has a hole for an axle, which is most often the driveshaft of a motor.

Some wheels for robots are molded into one piece. Others, such as the Dave Brown Lite-Flight wheels, are composed of separate pieces assembled at the plant. The Lite-Flight wheels use a plastic hub that attaches to the motor shaft or axle, and onto the hub is mounted a foam tire.

Wheel Materials

The first order of consideration is the materials used for the wheel. The least expensive wheels, like those used on low-cost toys, are molded in one piece, usually a hard plastic. The wheel doesn't have a separate tire and hub. While these wheels are acceptable for some robots, you probably want a softer tire surface. This requires a softer rubber or foam, over a rigid hub.

Rubber over plastic: The hardness of the rubber greatly influences traction. One common measure of hardness is *durometer,* tested by a device called (get this!) the durometer. There are several durometer scales, each labeled with a letter, such as A or D. A durometer of 55A is relatively soft and pliable; 75A is medium; and 95A is quite hard.

Rubber over metal: Typical of wheels made for R/C racing, these are heavier and sturdier, and are well suited to bigger robots. You can also get small rubber tires mounted on aluminum hubs. These are typically sold at hobby stores as tail wheels for model airplanes.

Foam over plastic: Foam wheels are also a mainstay in the R/C racing field. Like their rubber counterpart, hardness varies.

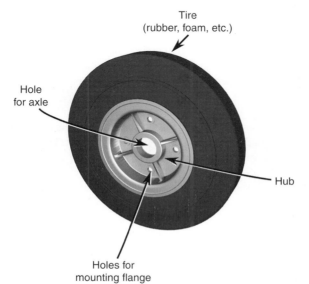

Tire
(rubber, foam, etc.)

Hole
for axle

Hub

Holes for
mounting flange

Figure 17-2 The construction of the typical wheel is a rubber (or other material) tire mounted on a hub.

Rubber/foam over spoked wheels: As the size of the wheel increases, so does its weight. Spokes are used to reduce the weight of very large wheels. Smaller bicycle or wheelchair wheels are suitable for larger robots.

Pneumatic wheels: Traditional foam and rubber tires are merely fitted over their hubs. In a pneumatic wheel, the tire is filled with air, which gives the wheel more bounce, but with added rigidity. Wheels for wheelbarrows and some wheelchairs are pneumatic.

Airless tires: Similar in concept to the pneumatic wheel, airless tires are hollow and filled not with air but with a rubber or foam compound. They are common on wheelchairs and heavy-duty materials-handling carts. They're great for larger bots that have to carry a lot of weight.

Wheel Diameter and Width

There are no standards among wheel sizes. They vary by their diameter, as well as their tread width (the tread is the plastic, foam, or rubber material that contacts the ground).

- The larger the diameter of the wheel, the faster the robot will travel for each revolution of the motor shaft. You can quickly calculate linear speed if you know the speed, in revolutions per minute or second, of the motor. Simply multiply the diameter of the wheel by *pi,* or 3.14, then multiply that result by the speed of the motor. See the section "Using Wheel Diameter to Calculate the Speed of Robot Travel" for more details.
- The larger the diameter of the wheel, the lower the torque from the motor. Wheels follow the laws of levers, fulcrums, and gears. As the diameter of the wheel increases, the amount of torque delivered by the wheel decreases.
- Wider wheels provide a greater contact area for the wheel, and therefore traction (from friction) is increased.
- The wider the wheels, the more those wheels may cause mis-steering (called *tracking*). A heavy robot with wide wheels, especially one where the motors are low torque, may have a greater tendency to steer off course. Very wide wheels can cause extra friction that resists smooth turning. Very narrow wheels can also cause tracking issues if there is little traction between the tire material and the ground.

When selecting the wheel diameter and width, match the wheel to the job. A robot with modest-size wheels of fairly narrow proportions (say, 1/4″ wide for a wheel of 2.5″ to 3″ in diameter) will be more agile than if it were equipped with much wider wheels.

Wheel Placement and Turning Circle

Where the wheels are located on the robot base affects the turning circle of the robot. Whenever possible, locate the wheels within the body of the base, rather than outside it. This decreases the effective size of the robot and allows it to turn in a tighter circle. Figure 17-3 shows wheels mounted both within the area of the base and outside it.

UNDERSTANDING WHEEL TRACTION

As in a car, wheels on your robot are meant to grip the driving surface. This provides traction and allows it to move forward. Yet, oddly enough, with robots both too little and too much grip can be a bad thing.

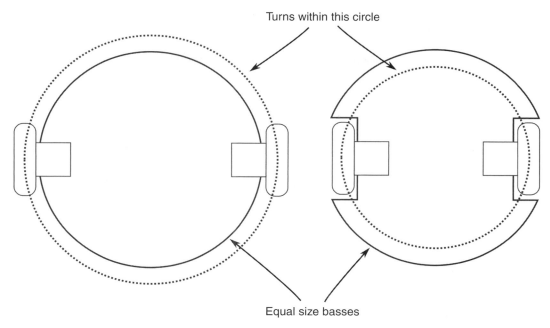

Turns within this circle

Equal size basses

Figure 17-3   The turning circle describes the area occupied by a robot when it makes a spinning turn. Given the same-diameter base, wheels on the inside of the base circumference have a smaller turning circle.

Picking up from Chapter 13, "How to Move Your Robot," let's look at how a differentially steered robot is designed. It has two motors and wheels mounted on opposite sides. Traction going straight ahead is simple: when both motors are activated in the same direction, the robot moves forward or backward in a straight line.

Wheel traction becomes an issue in turns. There are two ways to turn a differentially steered robot: by reversing the motors relative to one another (hard turn) or by stopping one motor and activating the other (soft turn). In both kinds of turns,

- Inadequate traction causes the wheels to slip, so it's anyone's guess where the robot will be heading afterward.
- Excessive traction can cause "chatter"—the wheels grip the road surface so well, they have to bounce in order to negotiate the turn. The effect is most pronounced in soft turns and is compounded in 4- and 6-wheel designs.
- Four or more driven wheels, mounted in sets on each side, will function much like tank tracks. In tight turns, the wheels will experience significant friction and skidding. If you choose this design, position the wheel sets closer together.

Most wheels for robotics use a rubber tire material. The softness of the rubber and its surface help determine its *compliance*. A very soft and mushy tire material—like that found on some model racing cars—may cause too much traction, hindering proper steering. A very hard tire material, such as a hard plastic, may not provide enough grip.

The effectiveness of any tire material is determined by the surface it rolls over. A hard tire on a hardwood floor can be a bad combination; a moderately soft tire on Berber carpet is a much better combination.

USING WHEEL DIAMETER TO CALCULATE THE SPEED OF ROBOT TRAVEL

The speed of the drive motors is one of two elements that determine the travel speed of your robot. The other is the diameter of the wheels. For most applications, the speed of the drive motors should be under 130 RPM (under load). With wheels of average size, the resultant travel speed will be approximately 4 feet per second. That's actually pretty fast. A better travel speed is 1 to 2 feet per second (approximately 65 RPM), which requires smaller-diameter wheels, a slower motor, or both.

How do you calculate the travel speed of your robot? Follow these steps:

1. Divide the RPM speed of the motor by 60. The result is the revolutions of the motor per second (RPS). A 100 RPM motor runs at 1.66 RPS.
2. Multiply the diameter of the drive wheel by *pi,* or approximately 3.14. This yields the circumference of the wheel. A 7″ wheel has a circumference of about 21.98″.
3. Multiply the speed of the motor (in RPS) by the circumference of the wheel. The result is the number of linear inches covered by the wheel in 1 second.

With a 100 RPM motor and 7″ wheel, the robot will travel at a top speed of 35.17″ per second, or just under 3 feet. That's about 2 miles per hour! You can slow down a robot by decreasing the diameter of the wheel. By reducing the wheel to 5″ instead of 7″, the same 100 RPM motor will propel the robot at about 25″ per second.

Bear in mind that the actual travel speed of your robot when it's fully accessorized may be less than this. The heavier the robot, the larger the load on the motors, so the slower they will turn.

For your reference, here is handy table comparing travel speed, in inches per second, against a variety of motor RPM and several common small wheel sizes.

 See the Robot Speed Calculator on the RBB Support Site. Enter the motor RPM and the diameter of the wheels you're using, and the calculator tells you the travel speed in inches per second.

SUPPORT CASTERS AND SKIDS FOR WHEELED ROBOTS

Differentially steered robots need something on the front and/or the back to prevent them from tipping over. There are several common approaches, listed below.

Nonrotating Skid

The purpose of a skid is to glide over the ground without using any moving parts. The skid is rounded (a cap or acorn nut works well) to facilitate a smooth ride. Polished metal, hard plastic, or Teflon are common choices.

For obvious reasons, skids are not suitable for robots that may travel over uneven surfaces or when there are many obstructions, like cables and old socks.

Swivel Caster

Swivel casters are available with wheel diameters from 1″ to over 4″. Match the size of the caster with the size of the robot. You'll find the common 1-1/4″- to 2″-diameter caster wheel is suitable for most medium robots. For larger bases you can opt for the 3″ and even 4″ casters.

Swivel casters are commonly available with plate or stem mounting and in the following wheel styles:

- Single wheel
- Dual wheel ("twin wheel")
- Ball style

The ball style is used with furniture and tends to be heavy. If you use it at all, reserve it for heavier robots. Single-wheel casters are the most common and easiest to find. Look for a caster that swivels easily.

Ball Caster

Ball casters act as omnidirectional rollers. Unlike swivel casters, which must rotate to point in the direction of travel, ball casters are ready to move in any direction at any time. This makes them ideal for use as support casters in robots.

The size of the ball varies from pea-sized to over 3″ in diameter, and they are available in steel, stainless steel, or plastic. Pololu sells a variety of small ball casters for desktop robots; industrial supply outlets such as Grainger, McMaster-Carr, and Reid Tool and Supply offer the bigger ones.

Omnidirectional Wheel

Omnidirectional wheels are basically rollers mounted on the tread of a tire. The tire turns on an axis like any other, but the rollers allow for movement in any direction. For what they do as casters, omnidirectional wheels mean extra cost, size, and weight. I've not found that they work any better than a ball transfer or even a well-made swivel caster.

Tail Wheel

One alternative to the swivel caster is the tail wheel, used on R/C model airplanes (and, therefore, available at most hobby stores). The wheels come in sizes ranging from about 3/4″ to over 2″ and are used with specific mounting hardware.

In summary: For a small robot, under a couple of pounds and measuring 7″ in diameter or less, a nonrotating skid is usually acceptable. For centerline motor mounting, use two skids: one each in the front and rear of the bot. For larger or heavier robots, a skid may dig into soft surfaces, or it may snag on bumps, cables, and other obstructions. For these, use a swivel caster or a ball caster.

Successful Use of Casters

The casters on your robot must not impede the direction or speed of the machine's travel. Cheap swivel casters can catch and not swivel properly when the robot changes direction.

Keep these points in mind when selecting and using casters with your robots:

- Test them for smooth swivel action. Casters with ball bearings tend to give better results.
- In most cases, since the caster is provided only for support and not traction, the caster wheel should be a hard material to reduce friction.
- When using two casters on either end of the base, there's a possibility of the robot becoming trapped if the casters touch ground but the drive wheels do not. You can fix this by using only one caster instead of two, or place slightly more weight over the end with the caster. You may also reduce the height of the casters.

Two-Motor BasicBot

The BasicBot is a simple differentially steered base that's easy to construct of wood, plastic, cardboard, picture mat, or foamboard. It's an ideal first robot, and its round shape makes it well suited for use as a wall follower, maze solver, or other robot that works in confined spaces. The base measures 5″ in diameter; many craft and hobby stores sell 1/8″- or 1/4″-thick wood and plastic already cut into this size (or close to it), saving you from cutting out a circle using a saw or mat-cutting blade. A finished BasicBot is shown in Figure 17-4.

Figure 17-4 The BasicBot uses one or two decks, a Tamiya Twin Motor gearbox kit, a set of small tires, and a ball caster for balance. Construction takes less than an hour.

The BasicBot uses the following motors and mechanical parts, all of which are available at Tower Hobby and many other online hobby stores.

- Tamiya Twin-Motor Gearbox, #70097: The motor comes as a kit and is assembled in about 20 minutes using a screwdriver and small needle-nose pliers.
- Tamiya Ball Caster, #70144: You get two ball caster units; you need only one for the BasicBot, so save the second caster for another project. Construction takes about 5 minutes.
- Tamiya Truck Tire Set, #70101: You get four tires; you need only two.

Tamiya also offers the model #70168 Double Gearbox kit. It is functionally identical to the Twin-Motor kit, except its dimensions are slightly different. This means that if you use the double gearbox you'll need to adjust the drilling pattern in order to properly mount the motor to your robot base.

CONSTRUCTING THE BASICBOT

Refer to Figure 17-5 for the cutting and drilling layout of the bottom deck. Use a 1/8″ bit to drill the holes. The location of the four holes on the base isn't supercritical, but the spacing is. You may wish to use the constructed motor and ball caster to mark off the holes.

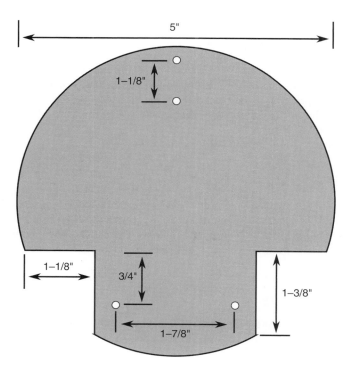

Figure 17-5 Cutting and drilling layout for the BasicBot. All holes are 1/8″.

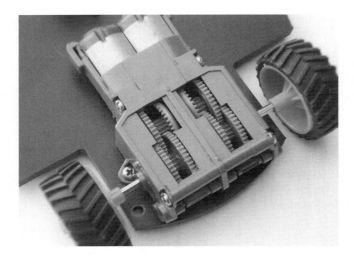

Figure 17-6 Underside of the BasicBot, showing the Twin-Motor Gearbox and ball caster.

1. Begin by constructing the twin-motor gearbox according to the instructions that come with it. You have the choice of building the motors with a 58:1 or 203:1 gear ratio. Opt for the 58:1 ratio if you'd like a faster robot. For maze following and other tasks you're better off with the slower, 203:1 ratio.
2. Before inserting the motors into the gearbox, solder wires to them and connect the wires to a set of switches or control electronics. See Chapter 14, "Using DC Motors," for ways to control small motors.
3. Construct the ball caster according to the instructions that come with it. The caster comes with various pieces to alter its height. Your finished caster should measure about 1″ from the base to the ball socket.
4. Mount the motor box and ball caster. Assuming a 1/4″-thick base, use 4-40 × 1/2″ machine screws and nuts. You can use 3/8″-long screws if the base is 1/8″ thick.
5. Mount rubber tires onto two of the truck tires, then insert the wheels over the motor shaft. Figure 17-6 shows the underside of the BasicBot with the Twin-Motor Gearbox assembled and attached.

ADDING A SECOND DECK

The one-deck BasicBot has room for a small microcontroller board, mini solderless breadboard, and a four-cell AAA battery holder. You can increase the area of the BasicBot if you need more space.

Make a second deck (Figure 17-7) with another 5″-diameter circle. Drill matching holes in both the bottom and the second deck. Use 1″- or 1-1/2″-long metal or plastic standoffs to act as "risers" to separate the two decks.

USING MORE EFFICIENT MOTORS

The Tamiya Twin-Motor kit makes for an affordable and easy-to-use motor set for small desktop robots, but these motors are not terribly efficient, and they consume a lot of current.

Figure 17-7 Top and bottom decks of the BasicBot, showing placement of the motor and ball caster.

You can replace the DC motors used in the twin-motor kit with more efficient ones. The gearbox kit uses a pair of Mabuchi FA-130-size motors. You may substitute for a more efficient motor, as long as it conforms to the "130" motor size. Pololu and several other online sources offer replacement of 130-size motors that consume much less current than the stock units that come with the Tamiya kit.

Bonus Project: Double-Decker RoverBot

Find more project plans for two-wheeled robots on the RBB Support Site. One is the RoverBot, which requires only rudimentary construction skills (straight cuts only) and can be made using any number of materials, including 1/4″-thick aircraft-grade plywood, 1/4″-thick expanded PVC sheet, or 1/8″ acrylic plastic. It's designed to be easy to build and easy to expand.

On the support site you'll find complete construction plans, including cutting and drilling layouts, parts list and sources, and assembly instructions.

Building 4WD Robots

What's better than a two-wheeled robot? Why, four wheels, of course! Four-wheel-drive (4WD) bots are able to traverse more kinds of terrain than their two-wheeled cousins, moving from indoors to out with ease. They're the preferred method of exploring grassy or dirt areas. And because they have four wheels, 4WD robots are statically balanced and have no need (or use) for a caster. Like a 2WD robot, 4WD bases use differential steering to explore their world.

Alas, 4WD robots are a bit harder to construct, and, depending on how they're designed, they cost more. But the advantages of a 4WD base often outweigh the disadvantages of a higher price tag and extra time in the shop.

What applies to the typical 4WD robot also applies to those using six (or more) wheels. As you add drive wheels, the complexity, weight, and cost of the robot can skyrocket. You might consider instead a tracked base, which functions like a multi-wheel-drive system, with an infinite number of wheels. See the following section for more on robots that use tracks.

SEPARATE MOTORS OR LINKED DRIVE?

While there are 4WD systems where two of the wheels are unpowered, in the typical 4WD all four wheels provide oomph to the robot. That means you either must use separate motors for each wheel or somehow link the wheels together so they're driven by the same motors. Figure 17-8 shows the basic concept.

- Separate motors cost more but require less mechanical complexity. You merely add two more motors and wheels to the base. With separate motors you can also control them individually, which offers some benefits over loose terrain like unpacked dirt.
- Linked drive saves the added cost of two extra motors, but requires you to develop a system where each motor powers two wheels at the same time.

With either method, the wheels of the 4WD robot are the same diameter and are placed toward the center of the base. The farther apart the wheels on each side, the more difficult the steering. On many 4WD bots, the wheels are placed with only minimal separation.

 Of the two, 4WD systems with separate motors provide the most power, simply because each motor is dedicated to a single wheel. On linked-drive systems, the one motor is shared between wheels, so the overall power of the base is less.

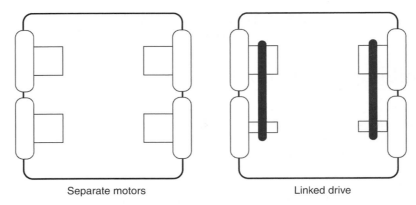

Separate motors   Linked drive

Figure 17-8 4WD robots take two general forms: those with four independent motors and those with two motors. The wheels on each side of the bot are linked to a single motor.

Constructing a Separate-Motor 4WD Robot

A simple yet fully functional 4WD robot can be built using three pieces of wood, plastic, or metal. The motors (let's say, R/C servos) are mounted in what I call *side rails*. They're just overgrown motor mounts, with cutouts for the body of the motor and holes for the screws. The side rails attach to a deck piece by way of any kind of corner angle bracket.

Figure 17-9 shows the basic concept of a 4WD base. The spacing between each pair of servo motors depends on the diameter of your wheels and the way the servos are mounted in the side rails (i.e., output shafts facing one another provides less space than if they face outward). In all, it's a pretty straightforward design: mount the servos in side rails; stick the side rails to the base.

The choice of corner angle brackets is up to you. I used some plastic brackets I had lying around the shop, but the common 3/4″ × 1/2″-wide corner angle bracket available at any hardware store works as well.

On my prototype I used a set of six 4-40 × 1-1/2″ machine screws, along with some 5/8″ nylon standoffs I bought surplus, as risers between the bottom deck of the bot and an optional second deck. Holes are cut in the bottom deck to feed wires through for connecting to batteries, microcontroller, and other electronics.

Constructing a Linked-Drive 4WD Robot

A 4WD linked-drive system uses a single motor on each side of the robot and a power train coupling to connect each motor to its wheels. The most common techniques for coupling the motor and wheels are gear, chain, or belt drive.

- Belt and chain drive are probably the easiest methods, because both offer a bit of "slop" in aligning all the parts. A central motor (on each side of the bot) powers both wheels using either a flexible belt or a segmented chain; see Figure 17-10 for details. For best results, the belt should be *cogged*; that is, it should have nubs molded in that assertively mesh with teeth in the sprockets used on the motor and wheel shafts.
- Gear drive uses a main drive gear from the motor to mesh with subgears attached to the wheels (Figure 17-11). This is the method most commercially made 4WD toys use, but it requires extra precision when constructing a homebrew solution.

Figure 17-9   A 4WD robot, using servo motors and some oversized plastic wheels ripped off from a toy. The base is constructed by attaching the servos in side rails, and the side rails are attached to a top plate.

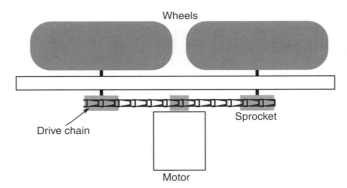

Figure 17-10   Chains (shown here) and belts can be used to link the wheels of a 4WD robot that uses only two motors. The chain allows for less precise construction requirements.

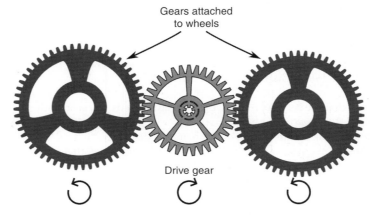

Figure 17-11   Gears used to transfer power from a central motor to a pair of wheels. The gears must be carefully positioned or else the teeth may not mesh properly.

 You must be very careful to align the gears with the proper spacing, or else they won't mesh properly. The mechanism will bind if the gears are too close together, or slip and chatter if they're too far apart.

For small robots (let's say, under about 8″ to 10″), Tamiya makes several products that can be used to create belt- or chain-drive 4WD bases. First is the Tamiya Ladder-Chain Sprocket set (#70142), which consists of plastic molded sprockets and individual links of chain that you connect together. Add or remove links to make the chain the exact length you wish. Because of the parts assortment you get, you will need two complete sets for one robot base.

Another method is the Tamiya Track and Wheel Set (#70100), which comes with molded plastic sprockets and rubber 1″-wide tracks. This set is normally used to make a tracked vehicle (see Chapter 18), but it also works well as a belt-drive system for a 4WD vehicle.

Two Quickie Wheeled Bot Platforms

Here are a couple of examples of wheeled platforms that you can build in less than 30 minutes each. They use my easy, famous, and patented "no-cut/no-drill" building technique—okay, not patented and probably not famous, but definitely easy!

2WD PLATFORM WITH DC GEARMOTORS

Figure 17-12 shows a quick-build robot that uses two commonly available DC gearmotors and wide tires, plus a couple of bits and pieces from the home improvement store. The robot is assembled using only hot melt glue. See Table 17-1 for a parts list.

Follow these steps to build:

1. Gather the parts as shown in Figure 17-13A. Using a metal file to lightly chamfer the corners of the TP15 plate (to avoid snagging on things). For ease of construction, solder wires to the motor terminals before assembling the robot. Apply some electrical tape to the terminals to prevent the wires from pulling loose.

Figure 17-12   This 2WD QuickBot is constructed using a pair of commonly available (and inexpensive) DG gearmotors and wheels, along with ready-made metal parts from the home improvement store.

Table 17-1 2WD QuickBot Parts List	
1	Simpson Strong-Tie TP15 steel nail plate
1	Simpson Strong-Tie RTR rigid tie connector
2	DC gearmotors, 1:48 gear ratio, with wide tires (motor and wheel often sold together, and in pairs)
1	Tail skid assembly consisting of: 1 6/32 × 1″ machine screw 2 6/32 nylon acorn nut 2 6/32 standard hex nut

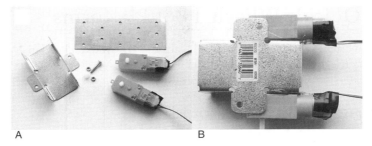

A   B

Figure 17-13   (A) Parts assortment for the 2WD QuickBot. (B) Motor placement for gluing.

Figure 17-14   The finished 4WD platform, using premade/predrilled materials for quick-assembly. Construction time is a matter of minutes.

2. Apply glue to each motor so that you can attach it to the RTR tie connector as shown in Figure 17-13B. Apply only enough glue to hold the parts.
3. Glue the TP15 nail plate to the underside of the RTR rigid tie connector. Alignment isn't critical, but try to center the parts width-wise.

Assemble the tail skid using 6/32 hardware. The nylon acorn (cap) nut provides a smooth gliding surface for most floors. Use the nuts to adjust the height of the skid. Tighten the nuts against each other to keep the screw from working loose.

Having the motors and wheels on one end makes the tail of the robot very light. Even out the weight distribution by placing the batteries on the tail end.

4WD PLATFORM WITH SERVO MOTORS

Figure 17-14 shows a completed 4WD bot built using a galvanized steel nail plate, first discussed in Chapter 9, "Robots of Metal." The parts list is provided in Table 17-2.

Table 17-2 4WD QuickBot Parts List	
1	Simpson Strong-Tie TP37 steel nail plate
4	Standard size continuous rotation motors, Futaba spline; e.g., Parallax 900-00008
4	2.5"- to 2.7"-diameter wheels, with Futaba spline hubs; e.g., Parallax 28109
4	Stamped aluminum servo brackets, with hardware

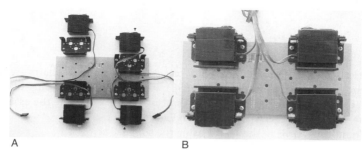

Figure 17-15 (A) Secure the servo mounts to the nail plate using 4-40 screws and nuts. (B) The output shafts of the servos should be aligned toward the center of the base.

The specific model of nail plate I used is a Simpson Strong-Tie TP37, which measures about 3" × 7". The plate has predrilled holes that *almost* match the holes in the stamped servo brackets. By "almost" I mean the alignment is close enough to assemble without needing to drill out the holes. You may need to apply a gentle force to push the fasteners through the holes.

Differences in products may affect the alignment of the predrilled holes. If you find your nail plate and servo brackets don't quite match up you may either need to pull out the trusty drill—use a 1/8" bit—or use an alternative assembly method, such as hot melt glue.

1. Using a metal file to lightly chamfer the corners of the nail plate. This helps keep the base from marring furniture, or scratching ankles of innocent bystanders.
2. Attach all four servo brackets to the nail plate using the predrilled holes, as shown in Figure 17-15A. Use 4-40 × 1/4" machine screws and 4-40 nuts. Be sure the fasteners are on tight. You may wish to use 3/8" screws and add a split washer to help keep the nuts and screws in place.

Figure 17-16   Attach the wheels with the screws provided with each servos.

3. Insert the four servos into the brackets so that the output shafts are offset toward the center of the base (see Figure 17-15B). Each servo requires only two screws and nuts placed in opposite corners of the mounting flange. Complete the assembly by attaching the wheels to the servos, as shown in Figure 17-16.
4. Mount the wheels to the servos using the supplied screw.

Avoid electrical shorts by covering the base with a square of lightweight foam board. Secure the board to the base using a light dab of hot melt glue.

CHAPTER 18

Build Robots with Tracks

In Chapter 17 you read how to make robots using wheels. But wheels aren't the only way to roll across the ground. *Tracks* add another dimension to robot locomotion. This chapter extends the idea of rolling robots to cover those that chug across the living room floor on tank-like tracks.

Be sure to check out Chapter 17 for information on wheeled robot designs, and Chapter 19 for bots that walk on their own two (or four, six, or eight) legs.

Refer to Chapters 12 through 16 for information on general robot design, as well as information on powering your robot with motors. And be sure to check out the free bonus projects on the RBB Support Site.

The Art and Science of Tank-Style Robots

Another popular form of the rolling robot is the tank track design, so-called because it uses tracks (also called *treads*) similar to those on military tanks. Like 4WD robots, tank track bots don't need a balancing caster or skid. They use differential steering like 2WD and 4WD bases and are expressly designed for use over uneven terrain. Figure 18-1 shows a representative home-brew tracked-drive robot, made with a rubber track stolen from a 1/8-scale tank toy and refitted over a plastic base.

Throughout this chapter I used the terms *track* and *tread* somewhat interchangeably, though in actual use the *tread* is considered to be the rubberized part of the *track* that contacts the ground.

Figure 18-1 Tank-style robots use rubber, plastic, or metal tracks, along with a drive motor and a series of unpowered idler wheels that keep the tracks in place. Low-cost toys are a common source of useful rubber tracks.

FINDING THE RIGHT TANK TRACKS

The first order of business is to locate a suitable track for the tank-style robot. Common track materials are rubber, plastic, and metal.

- Rubber tracks are perhaps the most common, found (for example) on many tank and earth-mover toys. You can rob the toy of its tracks and other parts and use them on your robots.
- Plastic tracks are made of rigid segments, linked together using pins or rivets. Several companies (Lynxmotion, Vex) make plastic segments for the express use as robot tracks.
- Metal tracks are found on high-end die-cast toys, as well as snowmobiles. These are heavy and expensive and are ideally suited for larger bots.

Regardless of the material, the tracks work in the same way: a drive sprocket positively engages with matching teeth or indentations in the track. The track is laid out along more or less the full length of the robot. Nonpowered idler sprockets or untoothed wheels keep the track in place.

Each kind of track has its own unique method of engaging with its drive sprocket. Whenever possible, always purchase (or rob from a toy) a track with its corresponding sprockets and idlers.

BOTS WITH FLEXIBLE RUBBER TRACKS

As noted, one of the best sources for inexpensive all-rubber tracks is toy tanks. These are sold in different scales, from about 1:64 (miniature) to upward of 1:10 or even 1:6. (The scale is the ratio of the size of the model to its original. A scale of 1:24, for example, means the model is 1/24 the size of the original. Most toy tanks are in the range of 1:24 to 1:32 scale.)

Look for a toy where the track is not too elastic and where, at a minimum, the drive sprocket and idler rollers can be removed and placed on your own custom base. Some toy tanks offer

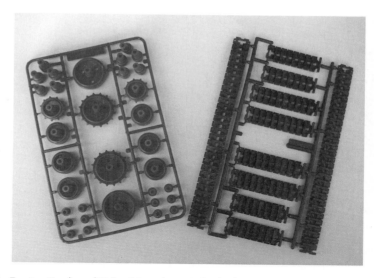

Figure 18-2   Tamiya Track and Wheel Set, showing both plastic parts and rubber track. You connect the track pieces in various lengths to suit the size of your robot.

easier hacking, where you can simply remove the turret and top of the vehicle, and replace the electronics with your own microcontroller and H-bridge. For these, you don't need to build a body for your robot, as you have one ready-made in the toy itself.

 A good source for smaller rubber tracks is LEGO Technic sets. The sets come with suitable sprockets that are made to engage with the teeth on the inside of the tracks.

USING THE TAMIYA TRACK AND WHEEL SET

A commonly used track for robots is made by Tamiya and sold by itself as Tamiya Track and Wheel Set (item #70100). A number of online sources, such as Tower Hobbies and Hobbylinc, offer this set. The track is also included in a few other Tamiya products, as the Tamiya Tracked Vehicle Chassis Kit and the Tamiya Remote Control Bulldozer Kit. These also come with motors.

The Tamiya track is rubber and comes in segments of various lengths. You put the segments together to build a track. Sprocket and idler rollers are included (see Figure 18-2). Pick out the parts you need. The segments connect using a little nub on the edges of the track. Despite how it sounds—or even looks—the tracks are fairly robust and seldom break apart unless forced.

 In a pinch, you can glue the pieces together with a flexible adhesive, such as silicone caulk. Make sure the adhesive doesn't seep into the part of the track that engages with the sprocket and that the seam is smooth.

Figure 18-3   The TrackBot, using a couple of Tamiya gear motor kits and a Tamiya Track and Wheel Set. Construction is easy, and the whole thing costs under $25.

Build a TrackBot

You can construct a practical and sturdy tracked robot base using two motors (DC gear or R/C servo) and a Tamiya Track and Wheel Set (#70100). The finished base is shown in Figure 18-3. Ideal construction materials are 1/4″ aircraft-grade birch plywood or 6mm expanded PVC.

In addition to the base material you need:

1	Tamiya Track and Wheel Set (#70100)
2	Tamiya 3-Speed Crank-Axle Gearbox kit (#70093), or equivalent
8	4-40 × 1″ machine screws
8	4-40 nylon insert locking nuts
18	#4 washers
6	4-40 1/2″ machine screws, nuts
4	3/4″ × 1/2″ wide corner angle brackets

Begin by cutting out the base deck and side rail pieces as shown in Figure 18-4. Refer to Figure 18-5 for the drilling details for the two side rails (make two of these). Drill the holes using a 1/8″ bit. Hole placement is fairly critical, especially the distance between the two mounting holes for each motor, so use care.

Note the alternative mounting holes. These holes are for providing a higher tension for the track. See the section "Track Assembly" for more info.

I've used a pair of Tamiya 3-Speed Crank-Axle Gearbox kits (#70093), as they are inexpensive and offer a choice of three different gear ratios: 17:1, 58:1, and 204:1. Normal speed is 204:1, but if you want a really fast tank, pick the 58:1 ratio.

Before construction, use a hacksaw or heavy-duty lineman's pliers (be sure to wear eye protection!!) to cut the hex shaft to a length of 2-1/4″. Complete the assembly of the motors as described in the instruction sheet that comes with them. You'll need a small Phillips screwdriver. Be sure to use the same gear ratio for both motors.

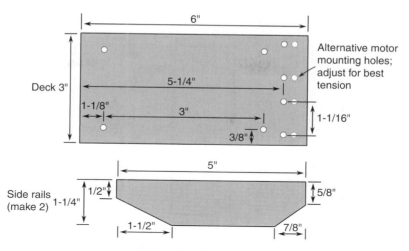

Figure 18-4   Cutting and drilling guide for the TrackBot. All holes are 1/8".

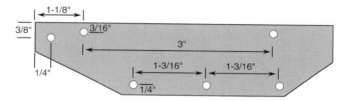

Figure 18-5   Drilling guide for the side rails for the TrackBot. All holes are 1/8". Hole placement is fairly critical in order to maintain proper tension of the track. (Even so, you may need to re-tension the Track as detailed in the "Track Assembly" section.)

Prior to mounting the motors, use the included hex wrench to lightly tighten the setscrew that secures the driveshaft to its output gear. You need to make a "right" and a "left" motor, as shown in Figure 18-6. Position the driveshaft so there's no more than 1/4" protruding from the side of the motor.

Assembling the Side Rails
Refer to Figure 18-7 for the following. The side rails are what the tracks are attached to. Let's begin with the left side rail.

1. Use a 4-40" × 1" screw, washer, and locking nut to attach the large idler wheel to the hole on the far left side of the rail. The wheel should face you. Tighten the locking nut so that the wheel just begins to stop turning freely, then back off about 1/8 of a turn. When you rotate the wheel there should be no drag. Neither should there be any kind of excessive wobble.
2. Use three more screws, washers, and nuts to attach the small idler wheels to the three holes along the bottom of the rail. Tighten them as you did the large wheel.
3. Using 4-40 × 1/2" screws, washers, and nuts, attach two corner angle brackets as shown in the illustrations.

Construct the right side rail in the same manner, but in mirror image.

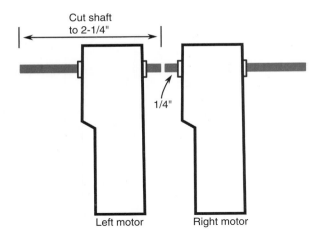

Figure 18-6   Motor construction and drive axle shaft placement for the TrackBot. You need to cut the axle shaft to length, using a saw or pair of heavy-duty lineman's pliers (be sure to wear eye protection!).

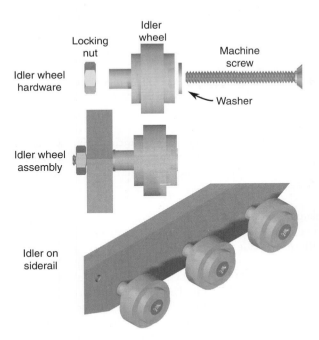

Figure 18-7   Construction detail for the idler wheels. Tighten the locking nut so the wheel just starts to drag, then back off the screw 1/8 turn.

Track Assembly

Assemble the tracks from the Track and Wheel Set using the following lengths for each side:

- 1 each 30-links segment
- 2 each 10-links segments
- 1 each 8-links segment

The segments interlock into one another. Use only your fingers; avoid using mechanical force or tools to link the segments together, or else the rubber may get torn. It can take a few tries to get the hang of it. Refer to the instructions that come with the set for details.

To install the track, loop over the four idler wheels. Wrap the track around the teeth of the drive sprocket, and carefully push the sprocket onto the motor shaft. Note: Do not overstretch the track or it may become unlinked.

The design of the TrackBot allows for adjusting the tension of the track—important, because if the tracks are too loose, they'll pop off easily. Remove the motors, and drill new holes about 3/16″ away from the original set. Remount the motors.

You can create a slot for adjusting the track tension by drilling several holes close together in line. Use a thin rat-tail file to remove the material between the holes, thus making a slot.

Attach the Side Rails to the Base

Assemble the side rails to the base, using 4-40 × 1/2″ screws, washers, and nuts. The motors are located at the rear of the base; the front should be flush with the leading edge of the right and left side rails.

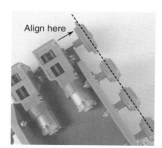

Align here

Attaching the DC Gear Motors

Use 4-40 × 1/2″ screws and nuts to attach the two motors to the base. The short ends of the driveshafts should face each other as shown in Figure 18-8.

Temporarily attach the large sprockets to the driveshafts. Using the hex key included with the motor, slightly loosen the setscrew that holds the driveshaft in place. As needed, move the driveshaft in the motor so that the sprocket lines up with the idler wheels in the side rail. Even a

Figure 18-8 Attach the two motors to the base with screws so that the short ends of the driveshafts face each other.

Figure 18-9 Edge view of the idler wheels attached to the TrackBot.

slight misalignment can cause the track to pop off when the robot turns, so try to be as accurate as possible. When done, tighten the setscrew on each motor.

Figure 18-9 shows how the rails are attached to the base.

Replacement FA-130 Motors

The miniature Mabuchi FA-130 motor included with the 3-Speed Crank-Axle Gearbox kit is not very efficient, consuming several amps when the motor is stalled (prevented from turning, but still powered). If you use electronic control to operate the motors, be sure it is rated for at least 2 amps, if not more.

Optionally, you can substitute the FA-130 for a higher-voltage, lower-current version. Pololu and several other online resources offer 130-size replacements that consume much less current and can be used with small motor bridge circuits.

Optional: Using Servo Motors

If you prefer, you may use an R/C servo to drive the tracked base. Mount the motor to the robot base using your favorite technique (see Chapter 15 for some easy-to-use mounts). Then, interface the sprocket to the motor with these steps:

1. Cut off the molded-in cap on the side of the drive sprocket.
2. Drill out the center using a 1/4″ bit.
3. Drill out two holes in a small round servo disc to match two of the holes in the drive sprocket. It turns out the holes match exactly when using the small round servo disc that comes with most Futaba and Futaba-style servos.
4. Attach the servo disc to the sprocket using a pair of 4-40 × 3/8″ machine screws and nuts.
5. Use the screw included with the servo to mount the drive sprocket to the motor.

Adding a Second Deck

Need more space? You can add a second deck to the tracked base by removing the nuts on the top of the corner angle brackets and replacing them with 4-40 threaded hex standoffs. Cut out the wood or plastic for the top deck, and secure it in place over the hex standoffs using 4-40 × 1/2″ screws.

Figure 18-10 Plastic track segments are connected using stainless steel pins. Because there is no stretch to the tracks they don't easily pop off (see the section "Dealing with Detracking").

BEST STEERING FOR TRACKED ROBOTS

Because a tank track exposes a considerable amount of its surface onto the ground at any one time, in a turn the tracks must actually slip, or skid, over the earth. The part of the track farthest from the midpoint of the vehicle skids the most.

Unlike a differentially steered two-wheel bot, where it is possible to turn by simply stopping one wheel, this is not advisable with a tracked drive. The track exposes too much surface area to the ground, which greatly increases friction. The stopped track will skitter over the ground (and possibly come off), and turning is harder to control.

SPECIAL CONSIDERATIONS FOR ALL-RUBBER TRACKS

Because of size, cost, and weight concerns, the track material on most robot tanks is rubber. Rubber has a higher compliance than plastic or metal. If the robot is operated over a surface that is also fairly *compliant* (means having resiliency or "give"), turning may be difficult for the little tank.

Another potential issue of using a rubber track is static friction, or *stiction*. (There may be other frictional components involved, but we'll bypass them for this discussion.) With stiction, a rubber track may have difficulty skidding over a highly polished material, such as a glass table-top or hardwood floor.

There are numerous techniques to reduce the steering problems inherent in all tracked vehicles. One is to use a less compliant track material. Not all rubber compounds are equally elastic. A good rubber track for a tank design exhibits only limited elasticity (stretch). The surface of the rubber is smooth and may have molded-in "cleats" that reduce the surface area of rubber touching the ground at any one time. With less surface area, there is less rubber to skid.

USING PLASTIC TRACKS

An alternative to rubber tracks is tracks of hard plastic. An example is the track for the Vex Robotics Design System. The kit, which is designed for the Vex line of robots but can be adapted to other applications, consists of a series of plastic links that you put together.

Another example of hard plastic track is available from Solarbotics, Robotshop, and other specialty online retailers. These tracks are composed of ABS plastic links (see Figure 18-10), connected by miniature stainless steel rods. You connect the links together to make a track of any size you want.

Plastic sprockets and idlers are also available to make a complete tracked system. Drive sprockets are available for Futaba and Futaba-style servo motors, Solorbotics, and other DC motors that use 7mm double-flatted driveshafts. Figure 18-10 also shows a complete tracked drive subsystem that uses the plastic links, ready to be mounted on a base. Of course, you need two of these drive subsystems to make a complete robot.

If there is a disadvantage to hard plastic tracks, it's that the plastic may slip over hard surfaces—the exact opposite of rubber tracks. Depending on the design of the track, you may be able to overcome this by applying small pieces of rubber material over the track segments. This provides enough compliancy to improve locomotion and steering.

DEALING WITH DETRACKING

Rubber and plastic tracks (or metal for that matter) differ in their resistance to *detracking*—also called *derailing,* or "throwing a track." Detracking occurs mostly when negotiating a turn. This is when the frictional forces acting against the track are at their highest. As the vehicle attempts to turn, heavy sideways pressure is exerted at the front and back of the track. If the pressure is great enough, the track may come off its drive sprocket or guide rollers.

Detracking rears its ugly head the most when using highly elastic rubber tracks. The more elastic the track material, the more readily it will stretch during a turn. The problem is magnified if the tank is loaded down with weight. The heavier the vehicle, the more likely you'll have a thrown track. To limit this problem:

- Reduce the weight on the vehicle.
- Make slower turns.
- Try to find a rubber track that doesn't stretch as much. The lower the elasticity, the less likely the track will pop off.
- As necessary, tighten the track by adjusting the distance between the drive sprocket on one end and the idler roller on the opposite end. This limits the track from stretching too much more. Avoid overtightening, which can deform the track and place excessive stress on the drive components.
- Decrease the surface area of the track on the ground. You may do this by changing the elevation of the idlers toward the front and back.
- Experiment with the width between the tracks. Longer, narrower track widths resist turning more than shorter, fatter widths.
- Add "keepers" to the idlers that don't touch the ground. The keepers are like oversized rims that keep the track in place.

By their nature, plastic and metal tracks don't stretch, so, assuming they are placed snugly onto the sprocket and idlers, detracking is rare.

CHAPTER 19

Build Robots with Legs

Legs are biologically inspired solutions to that old problem of how to get your robot from here to there. And not only as a means to mobilize your creation, but to step over common obstacles like yesterday's lunch bag, your dirty socks, and the family turtle—stuff that might confound the typical robot with wheels or treads.

In this chapter you'll learn about the role of legs in creating mobile robots, and the special requirements and limitations they impose. You'll read about ready-made solutions, as well as several designs for making your own six-legged bots on a budget and from scratch.

Be sure to check out Chapter 17 for information on wheeled robot designs, and Chapter 18 for bots that roll on tank tracks.

 Refer to Chapters 12 through 16 for information on general robot design, as well as information on powering your robot with motors. And be sure to check out the free bonus projects on the RBB Support Site.

An Overview of Leggy Robots

With legs, a robot can live among humans, ideally without any kind of special adjustments or alterations to the environment. Ramps, curbs, steps, stairs, and cracks in the sidewalk all pose no more of a problem for the robot than they do for any other ambulatory human being.

As of this writing, human-size legged robots that can go anywhere and do anything are still the province of either science fiction or well-funded government research. Less ambitious are the full-scale walking robots of industrial and educational research that show promise but are still too cost prohibitive.

Small-scale legged robots are another matter. With a modest inventory of servo motors and some specially crafted brackets, you can construct walking bots with two, four, six, and even more legs. A programmable walkerbot is well within the reach of the garage-shop tinkerer.

Legged robots aren't for beginners. If you're just starting out, hone your skills by constructing a couple of wheeled bots. This applies whether you buy a ready-made kit of parts or build it from scratch. Walking robots require greater precision and attention to detail—when improperly constructed, they rattle apart, may simply not work, or fall over in a crashing heap.

NUMBER OF LEGS

The most common forms of legged bots are (see Figure 19-1):

Bipeds, which walk on two legs. In a true bipedal robot, one leg lifts up to make a step. Balance can be tricky when the robot has just one foot on the ground, making this type among the hardest to master. The alternative is the "shuffling" walking robot common in toys whose legs don't actually lift up and down.

Quadrupeds, meaning four legs. The more sophisticated four-legged bots demonstrate a variety of walking styles, some mimicking animals. These can be tricky to build, because each leg needs three separate joints—otherwise known as degrees of freedom, or DOF— in order to keep balance and make turns.

Hexapods, for which six legs provide excellent balance and mobility. Among walking robots, they're the most common and, despite the extra legs, the easiest to build.

Octapods are the arachnids of the robot world with eight legs. In many designs the robot may walk predominately on six legs, and use the additional two as a kind of manipulator.

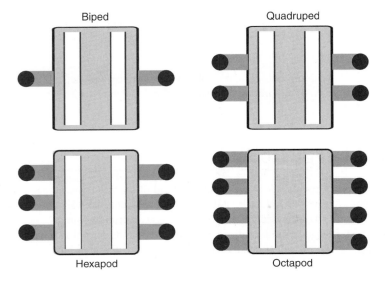

Figure 19-1 Four major types of walking robots, defined by their number of legs.

And, of course, there are examples of robots with other appendage counts. One-legged robots, for example, look like pogo sticks without a rider. They move around by bouncing. Then there are bots with 10 or more legs, caterpillars, snakes, and more. Pretty rad stuff.

STATIC VERSUS DYNAMIC BALANCE

Balance is the ability of the robot to remain upright when it's standing on its legs. Balance can be static or dynamic.

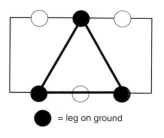

= leg on ground

- *Static* balance means the legs provide a natural stability, using any (or a combination) of several techniques. In four- and six-legged bots, the most common static balance comes from always having at least three legs on the ground at the same time, forming a tripod stance. With all types, static balance is improved by having a low center of gravity; much of the weight is at least 50 percent below the overall height of the robot.
- *Dynamic* balance means the robot uses sensors to keep itself upright. When the bot feels it's starting to tip over, the sensors activate one or more motors to shift the robot's weight one way or another. As the weight shifts, the tilt is negated, and the robot is kept upright.

BIPEDS: ANDROID OR HUMANOID

The terms android and humanoid are used to describe robots that are modeled after the human form. An *android* is a robot designed to look as much like a human being as possible, either male, female, or an *androgynous* mix of the two (androgyny is where we get the term android, meaning "like male and female").

Conversely—and somewhat confusingly—a *humanoid* robot is one that shares the basic architecture of a human. It has two legs, with a head at the top, and two arms at the side. But rather than duplicate the appearance of people, it's meant more to replicate human ability, such as being able to walk through a hallway meant for humans or sit down in a chair meant for humans.

NUMBER OF JOINTS/DEGREES OF FREEDOM

Leg motion is provided by a series of joints. Some joints allow the leg to swing back and forth, while others permit side-to-side motion. The number of joints that provide motion is called the *degrees of freedom*, or *DOF*. For the most part, the more DOF, the more agile the robot. For amateur robots the 2- and 3-DOF bot designs are more common.

Except for some unique designs that use clever mechanical linkages, each leg DOF needs a separate motor to control the joint. The more motors, the more expensive the robot; it's heavier and larger, too. So consider the number of leg DOF when picking a walking robot design. That hexapod with 3-DOF legs looks mighty impressive; just remember it needs 18 motors. It's not unusual for these kinds of robots to cost upward of $600, and that's before any microcontrollers or other electronics.

Figure 19-2 Commercially manufactured hexapod kit featuring 3-DOF legs. The extra degree of freedom in the legs provides a more defined stepping action. (*Photo courtesy Lynxmotion.*)

OPERATING TERRAIN

Bots with 2 DOF per leg work best on smooth, unobstructed surfaces—a kitchen floor or a hallway with wood laminate are good choices. The surfaces shouldn't be too slick, unless you add rubber pads on the robot's feet. Otherwise, the contraption will tend to skitter on the floor as it frantically moves its limbs. It's comical but not too productive.

The more sophisticated 3-DOF robots with four- or six-legs can be used on more challenging terrain. This is because they're endowed with an additional articulation that permits the legs to literally step up and down rather than just swing in and out, which is typical of the simpler 2-DOF designs. The full leg-lift helps the robot clear low objects and even thick carpet nap.

Many of the more elaborate ready-made walking robot kits are designed with this feature. Figure 19-2 shows an example: its third DOF permits the legs to lift up vertically, clearing obstacles at least 1/2" to 1" in height. Of course, these designs are more expensive because they require three servos per leg and additional leg hardware.

Selecting the Best Construction Material

Walking robots need to be strong yet lightweight. The heavier the robot, the less likely it can stand on its own two feet, so to speak. Pine lumber and soft plywood are pretty much out, because they're too bulky and heavy for the strength they offer.

There's lots more about wood, plastic, and metal in the chapters in Part 2 of this book. Be sure to see Chapter 7, "Robots of Wood," Chapter 8, "Robots of Plastic," and Chapter 9, "Robots of Metal."

Materials choice, in order of preference, is:

- *Aluminum,* cut and drilled to shape. For walking robots under 1 foot high, thickness ranges from about 20 gauge (0.0320″) to 8 gauge (0.128″). Premade parts are often cut by a powerful waterjet machine controlled by a computer, then bent using special jigs. When building your own, the likely tools are hacksaw, drill press, and bench vise.
- *Polycarbonate plastic* is a tough, scratch-resistant material commonly used as a substitute for glass. You can cut it with hand or power tools. A thickness of 1/8″ is ideal for robot making.
- *PVC plastic,* in 6mm (about 1/4″) thickness. While not as strong as polycarbonate, PVC is lots easier to work with, requiring nothing more than regular woodworking tools.
- *ABS plastic,* in 1/8″ or 1/4″ thickness. It's a bit easier to cut and drill than polycarbonate, and parts can be glued using a common and inexpensive solvent cement.
- *Wood,* but not just any wood, specifically aircraft-grade birch or other hardwood plywood. The 1/4″ or 3/8″ sheets provide adequate strength, and parts may be glued using a quality wood glue.
- *Acrylic plastic* is one of the least desirable of the materials commonly used to construct walking robots. Though similar in appearance to polycarbonate, it's not quite as strong and is susceptible to cracking under stress. The repetitive bending of the plastic can permit hairline fractures and "crazes" to form over time.

Scratch Build or Parts Kits

Perhaps the hardest aspect of building a walking robot is fabricating the leg pieces. So before starting any legged robot project, take an honest look at your tools, skills, and budget, and decide whether you want to build your own from scratch (i.e., from raw materials) or whether you want to assemble a walkerbot from premade parts. Because of the growing popularity of amateur robotics, a number of online sources offer parts specifically designed for constructing legged robots, anything from two legs on up.

BUILD YOUR OWN FROM SCRATCH

If you have reasonably good shop skills, you can consider making your own walking bot from scratch, using your choice of wood, plastic, or metal. The most common construction in a legged robot is the X-Y joint, so-called because a pair of motors produces a linear movement in both the X (right/left) and Y (up/down) planes. Shown in Figure 19-3 is an X-Y joint created using 6mm PVC plastic. A pair of R/C servo motors is attached to the joint using miniature fasteners.

Plans for the joint parts appear on the RBB Support Site (see the Appendix), and are used for a number of bonus projects you might want to try, like the robust six-legged walking robot and the articulated robot wrist.

BUILD FROM PREMADE PARTS

Numerous sources such as Lynxmotion and Pitsco provide premade parts for X-Y joints. Prefabbed aluminum brackets like the one in Figure 19-4 come cut and drilled to work with most

Figure 19-3 Homemade X-Y joint components for constructing a robotic leg (among other things). The parts are fashioned out of wood or plastic.

Figure 19-4 A precut and preformed aluminum X-Y joint, available from a number of online specialty robotics retailers.

any standard-size servo motor. The bracket weighs about the same as the PVC X-Y joint (half an ounce), but the PVC version is considerably cheaper to build.

Various types of these aluminum brackets can be combined to create different arrangements. For example, you can create an X-Y bracket with an L bracket in order to reconfigure the orientation of the servo motors used for the joint. Among other things, this can be used to provide a more streamlined shape or to create mechanisms for 3-DOF joints.

Some kits with premade parts use modular metal and/or plastic construction and are designed as development systems for building walking and rolling bots. Example: The Bioloid robot sets offer numerous permutations including sophisticated bipedal designs. The kits, which are designed for advanced study, are available in different parts assortments.

Figure 19-5   A yoke mechanism distributes weight or force so that it's not all against the shaft of the motor. This helps the motor work more efficiently and last longer.

BUILD FROM A MIXTURE OF PARTS

There's nothing stopping you from combining your own homemade parts with prefabbed ones, mixing and matching as needed. For example, you might combine a PVC X-Y joint with a pre-made L bracket. You might also provide your own chassis for the robot, leaving just the leg pieces as store-bought.

MOTOR SHAFT SUPPORT AND YOKE BRACKETS

The simplest of servo brackets, like the one in Figure 19-3, lack a means to support the load placed on the servo motor by the weight of the leg pieces. This isn't a major problem for robots that see only occasional use and aren't too heavy. But it's something you'll want to consider as your walkerbots gain weight or are frequently used for show-and-tell.

   The most common way to alleviate the strain on the motor is by using a yoke or double-sided bracket (see Figure 19-5). Here, the load on the motor is distributed evenly between the motor shaft and a secondary (and passive) shaft. The two shafts are in line with one another. The second passive shaft can be a steel or aluminum rod, or even a machine screw. The shaft may use ball bearings or bushings to ensure unimpeded rotation.

Leg Power

When it comes to the muscles of a walking robot, radio-controlled servos are the preferred choice; they're compact and widely available, and they require no special interfacing or drive electronics connected to your robot's central computer.

Where rolling robots can make do with just about any standard-size R/C servo, legged robots need a bit more attention to the details. Specifically, you need servo motors that provide enough torque to lift the legs and move the robot. It's not unusual for a six-legged hexapod to weigh several pounds. With underpowered servos, the robot will just sag to the ground, unable to sustain its own weight.

CONTROLLING SERVOS

Recall from Chapter 15, "Using Servo Motors," that servos require a special pulse signal in order to operate. Typical operating voltage for servos is between 4.8 and 6.0 volts, though many robot builders push to the edges of the envelope and supply 7.2 volts—this increases the torque of the motor by as much as 30 or 40 percent.

Overvolting an R/C servo motor has the potential of burning it out, so you must exercise care if you wish to experiment with this technique. Some brands, and even models within each brand, are more tolerant of the higher voltages than others, so you need to experiment. When using a higher-than-normal voltage with your servos, give them a periodic touch test to make sure they aren't overheating. Immediately disconnect the servo if it seems to be getting too hot or is putting out an unusual smell.

Because of the heavy demands of the motors in a walking robot, you need to run them from a separate battery supply. When using this arrangement, be absolutely sure that the ground connections of the two battery supplies are connected, or your robot will not function properly.

USING A DEDICATED SERVO CONTROLLER

Legged robots require lots of servos. Rather than hogtie the robot's main processor with the job of running them all, you can hand it off to a coprocessor instead. That's the idea of the *serial servo controller,* or SSC.

These compact circuit boards are designed to receive one-time instructions from your robot's microcontroller on which servos to activate and what position to move them to. The SSC then independently controls the servos without any further intervention by the robot's controller.

It's called a *serial* servo controller because the microcontroller communicates with the SSC via a simple serial communications link. The link can be one-way or two-way. Most microcontrollers provide a simple command structure for sending serial data to another device. For example, the Arduino has a SoftwareSerial library that can use any I/O pin as a serial line.

USING HIGHER TORQUE SERVOS

The weight of the robot (body, electronics, batteries) can place a strain on motors used for moving leg joints. The typical standard servo provides under 50 or 60 oz-in of torque, which is acceptable for desktop robots and smaller/lighter walking robots.

But for anything larger, you want a servo that delivers higher torque, 90 oz-in or above. To deliver this torque the servo may use more ruggedized constructing, including metal gears and ball bearings. It may also use an all-digital circuitry that can enhance torque. When purchasing servos for your walking robots be sure to pay extra attention to the torque rating.

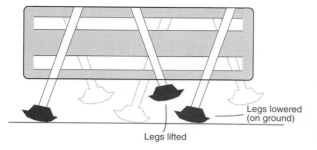

Legs lowered
(on ground)

Legs lifted

Figure 19-6 The alternating tripod gait common in hexapod robots. Three legs always touch the ground; the robot moves by alternating the side of the triangle that's on the ground and "sweeping" those legs in the opposite direction of intended travel.

Walking Gaits for Legged Robots

Gait refers to the pattern of leg movement as an animal or insect walks. Gaits can and do differ depending on the speed of travel—a running gait is wholly different from a walking gait. The leg motions are distinctive in each one. So, too, for robots, though almost all are restricted to walking gaits. There aren't too many legged robots that can run, at least not without falling over and making fools of themselves.

Figure 19-6 shows most common gait of a six-legged robot that has independent control of each leg. This gait is often referred to as an "alternating tripod gait," because at least three legs are always touching the ground at the same time, providing static balance. (A similar gait, also an alternating tripod, is used when the robot uses legs that are linked together. This gait is described in more detail in the section "Build a 3-Servo Hexapod," next.)

The alternating tripod gait goes through a number of sequences: For each side of the tripod, the legs are lifted, lowered, and swept forward or backward in unison. This provides for a reasonably fast walk while still maintaining good static balance as the weight of the robot is shifted from one side to the other.

The legs act either to lift or to power.

- Legs are *lifted* to orient them into a new position. During lift, the leg does not provide propulsion, nor does it contribute to the balance of the robot.
- Legs that *power* propel the robot in the opposite direction of the movement.

To be sure, there are other gaits for six-legged creatures, living or robotic. They include metachronal (or wave) and ripple. Space forbids me from detailing each one of these, but you can learn more about these and others with a Web search.

Build a 3-Servo Hexapod

Using simple linkages, you can construct a fully functional hexapod walking robot powered by just three servos. Figure 19-7 shows the completed Hex3Bot, which measures 7″ by 10″, and stands 3-1/2″ tall. Weight is only 11.5 ounces when constructed out of 6mm expanded PVC, the recommended material. (You can also use 1/4″ aircraft-grade plywood, but this will make the robot weigh a bit more.)

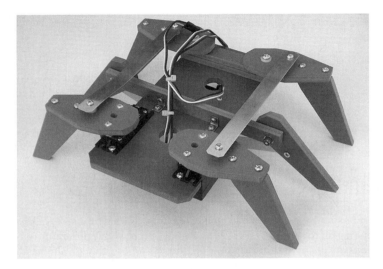

Figure 19-7   The completed Hex3Bot walking robot, which uses three R/C servo motors to get around. The Hex3Bot is an example of a hexapod design that uses linked legs.

Like any hexapod robot with static balance, the Hex3Bot keeps from falling over by always having at least three legs—in a tripod arrangement—on the ground at any time. Walking is accomplished by sweeping the front and rear legs on each side in alternating cycles. The middle legs, which only lift up and down, act to "rock" the robot from side to side, providing the third leg of the tripod.

The walking gait is composed of a three-step sequence:

1. Lift right or left middle leg. Only one leg is down at any time. The robot tilts to the side opposite the middle leg that is down.
2. Power sweep the front/rear legs that are touching the ground. The robot propels forward.
3. Non-power sweep the front/rear legs that are not touching the ground. The robot doesn't move in this step; it merely positions the legs for the next sequence.

PARTS YOU NEED

In addition to fastener hardware (see the section "Assemble to Complete the Hex3Bot" for a list), you need the following stuff to build this robot:

- 1 piece of 12" × 12" 6mm-thick expanded PCV (preferred), or 1/4" birch or other hardwood aircraft-grade plywood
- 3 standard-size servo motors.
- 2 12" lengths of 1/2"-wide by 0.016"-thick brass strips (available at hobby and craft stores, such as K&S Metals, #1412110231)
- 2 six-arm ("star") servo hubs (they usually come with the servo)
- 1 large round servo hub (usually comes with the servo)

You need the following fasteners to complete the Hex3Bot (get extras in case you lose any during construction):

4	3/8″ × 3/8″ miniature L bracket, Keystone 633
8	#4 × 1/2″ sheet metal screws
18	4-40 × 1/2″ machine screws
8	4-40 × 7/16″ machine screws (okay to substitute 1/2″ length)
2	4-40 × 7/16″ flathead machine screws (okay to substitute 1/2″-length pan head)
6	4-40 × 3/4″ machine screws
2	4-40 × 3/8″ machine screws
2	6-32 × 1-1/2″ machine screws
2	#6 nylon washers (okay to substitute #6 metal washers)
8	#6 nylon spacer, 1/8″ thick (check the special parts drawers at the better hardware stores, or you may substitute a stack of two or three nylon washers to build a spacer that is a total of 1/4″ long)
6	#6 metal washers
6	#4 metal washers
2	6-32 locking nuts
10	4-40 locking nuts
24	4-40 hex nuts

Machine screws are pan or round head, unless otherwise noted.

CUT AND CONSTRUCT THE LEGS

All holes are 1/8″ unless otherwise noted.

Begin by constructing the leg pieces; a cutting and drilling template is provided in Figure 19-8. The four front and rear legs are composed of two pieces—upper and lower. The middle legs have just one piece. The "feet" (bottoms) of all the leg pieces are cut at a 30° angle. Use a protractor to mark the angle. It's okay if it's a degree or two off, but try to be as accurate as possible to match all cuts.

Assemble the upper leg and lower leg pieces using #4 sheet metal (*not* wood) screws (see Figure 19-9). Using a pencil and the top leg as a template, mark the placement for the mounting in the edge of the bottom leg. Prepare a pilot hole for the screw with a small 1/16″ bit. It doesn't need to be a deep hole. Fasten the leg pieces using the sheet metal screws. They'll self-tap as you tighten them.

Notice that there are two styles of legs: the rear legs, which include a hub for connecting to a servo, and the front legs, which attach to the robot using a screw and lock nut. You will make two of each style. I've set the hole spacing for a six-arm "star" servo hub, common on Futaba and Futaba-style standard-size servos. If you use a different hub, you'll need to adjust the spacing of the holes. Drill out the holes in the hub to accommodate the 4-40 machine screws. Use two 4-40 × 1/2″ machine screws and nuts to attach the servo hub to the underside of the upper leg piece.

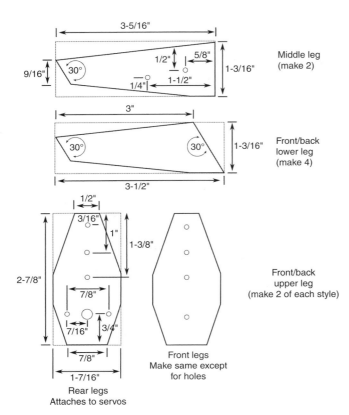

Middle leg
(make 2)

Front/back
lower leg
(make 4)

Front/back
upper leg
(make 2 of each style)

Front legs
Make same except
for holes

Rear legs
Attaches to servos

Figure 19-8   Cutting and drilling template for the Hex3Bot's legs. All holes are 1/8" except as noted.

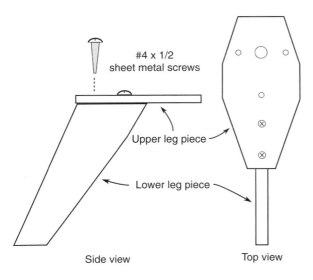

#4 x 1/2
sheet metal screws

Upper leg piece

Lower leg piece

Side view

Top view

Figure 19-9   The legs are assembled by attaching the upper leg piece to the lower piece, using #4 sheet metal screws.

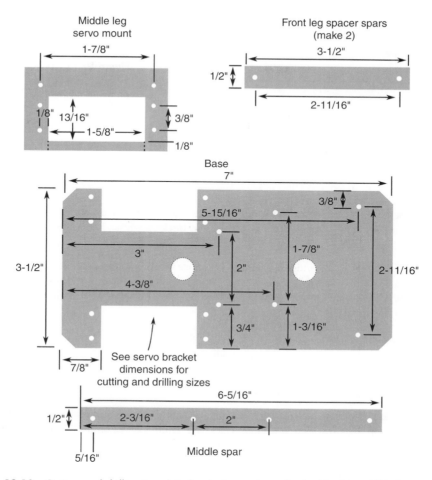

Figure 19-10 Cutting and drilling template for the base pieces for the Hex3Bot. All holes are 1/8″ except as noted in the text.

CUT THE BASE PIECES

Follow the cutting and drilling guide in Figure 19-10 to prepare the base, middle spar, middle leg servo mount, and the two front leg spacer spars. Be careful when drilling the holes for mounting the servos; these require a fair amount of accuracy.

The two large holes down the middle of the base are for passing wires through. The holes can be most any size and shape—1/2″ is handy—as long as they don't interfere with the construction of the bot.

Use a 9/16″ bit (to accommodate a 6-32 screw) for the following:

- Two holes on the far right of the base
- Two holes on the front leg spacer spars

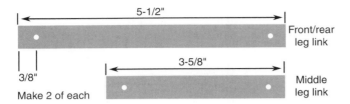

Figure 19-11 Linkages are made from 1/2" brass strips, cut to length. The holes on either end are 1/8". Make two of each length.

PREPARE THE LINKAGES

The Hex3Bot uses locking nuts for its moving links. The nuts should not be overtightened on their screws, or the robot may have difficulty in moving; nor should they be too loose. Use a screwdriver and wrench (or nut driver) to tighten the locking nuts so that they just begin to impede free movement of the linkage. Then back off about 1/8 to 1/4 turn.

Using two 1/2"-wide by 0.016"-thick brass strips, cut and drill the four linkages as shown in Figure 19-11. From each of the 12" lengths of brass, cut one strip to 5-1/2" and another to 3-5/8". Use a hacksaw with an 18- or 24-tooth-per-inch blade. Use a file to smooth the cut edges, and file down the sharp corners.

- The longer links connect the front and rear legs together.
- The shorter links connect the middle leg servo to the middle legs.

BUILD THE LEG ASSEMBLIES

Prepare the left front/rear leg assembly using one rear leg and one front leg, plus one long linkage strip. Use 4-40 × 3/4" machine screws, nylon spacers, and 4-40 locking nuts, as shown in Figure 19-12A. Be sure not to overtighten the locking nuts. Repeat for the right front/rear leg assembly. Make this one in mirror image to the first.

Prepare the middle leg assembly by attaching the legs as shown in Figure 19-12B. Use the middle spar, large round servo hub, short linkage strips, and fasteners as noted. You will need to drill out two of the holes in the servo hub to accommodate the screws. The other end of the linkages connects to the large servo hub. Drill out two holes at the 5 and 7 o'clock positions on the hub. Figure 19-13 shows the completed leg assemblies.

ASSEMBLE TO COMPLETE THE HEX3BOT

Remember the note about the locking nuts. The nuts should not be overtightened on their screws, or the robot may have difficulty in moving. Use a screwdriver and wrench to tighten the nuts so that they just begin to impede free movement. Then back off about 1/8 to 1/4 turn.

1. Begin with the servo that operates the middle legs. Attach the servo mount to the underside of the base using a pair of 3/8" × 3/8" miniature L brackets and 4-40 × 7/16" machine screws and 4-40 nuts.
2. Fasten the servo into the mount using two or four 4-40 × 1/2" machine screws and 4-40 nuts. (When using only two screws, mount them on opposite corners of the servo.) See Figure 19-14A.

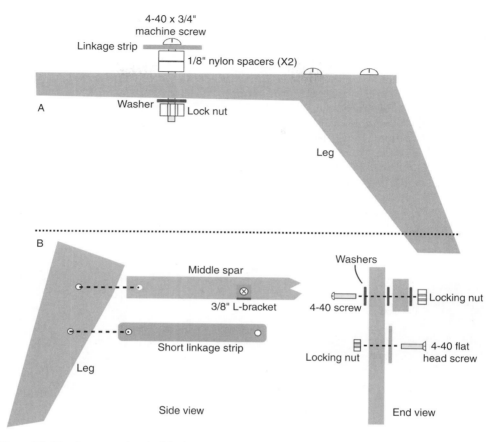

Figure 19-12 Construct detail of the leg assemblies showing the hardware used. The locking nut should not be so tight that it prevents the linkage from moving freely.

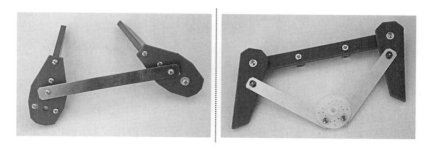

Figure 19-13 How the completed leg assemblies should look. Left: front/rear legs. Right: middle legs.

3. Mount two servos to the base using 4-40 × 1/2″ machine screws and nuts (see Figure 19-14B). The driveshaft of the motors should be closest to the rear of the base and should be on the side opposite the middle leg servo.
4. Attach the middle leg assembly to the top of the base using a pair of 3/8″ × 3/8″ miniature L brackets and 4-40 × 7/16″ machine screws and 4-40 nuts. See Figure 19-14C.
5. Connect each servo to your microcontroller (or similar circuit), and apply 1.5-millisecond pulses. This centers the servo to its middle, or neutral, position.
6. Attach the large round servo horn to the middle leg servo using the screw included.
7. Attach the left leg assembly to the left servo motor using the screw included with the servo.
8. Use a 6/32 × 1-1/2″ machine screw, washers, and 6/32 locking nut to attach the front left leg to the base. Place the two front leg spacer bars between the bottom of the leg piece and the base. Don't forget the washers, as shown in Figure 19-15.
9. Repeat steps 7 and 8 for the right leg.

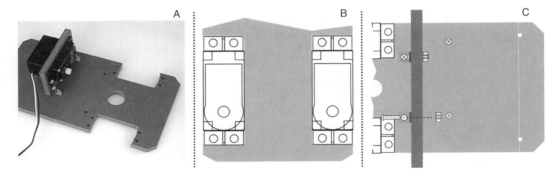

Figure 19-14 Assembly detail for attaching motors and middle leg assembly.

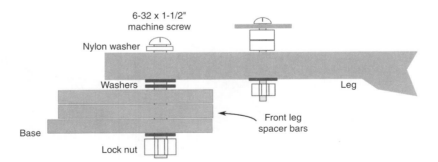

Figure 19-15 Assembly detail for attaching the front legs to the Hex3Bot base. Assemble both the right and the left front legs in the same fashion.

OPERATING THE HEX3BOT

You need a microcontroller or serial servo controller to run the Hex3Bot's servos. The walking sequence is straightforward:

1. Begin by rotating the middle leg servo so that one of the center legs (let's say the left leg) is on the ground. Experiment to find a rotation angle that lifts the front/rear legs on the same side completely off the ground.

 In the prototype Hex3Bot this angle was about 60° from center—the exact angle on your Hex3Bot may vary slightly depending on the particular measurements of the parts you built. You should expect some variance, and be ready to compensate for it in your program code.

2. Sweep the right front/rear legs back—toward the end with the servos—to push the robot forward. (It will actually move forward and turn to the left at the same time; this "waddling" is common with this form of hexabot.)
3. Sweep the left front/rear legs forward to move them into position.
4. Rotate the middle leg servo so that the opposite (right) leg is on the ground and the right front/rear legs are lifted.
5. Sweep the left front/rear legs backward to propel the robot forward.
6. Sweep the right front/rear legs forward to move them into position.
7. Repeat these steps to continue moving.

- To make the robot go in reverse, have the front/rear legs push in the opposite direction to that described previously.
- To make the robot turn, have only one side of the front/rear legs push.

MOUNTING ELECTRONICS AND ACCESSORIES

There's very little room on the top of the Hex3Bot for mounting batteries or electronics. To make space you'll probably want to add a second deck, separating the deck with the base using standoffs or risers of some type or another. The standoffs should be at least 1-1/4″ long, in order to clear the mechanical movement of the legs.

In lieu of standoffs you can construct your own risers using a suitably long machine screw, some aquarium tubing, and a couple of nuts. Start by drilling matching holes in the Hex3Bot base and the second deck. For whatever riser length you want, add 3/4″ to accommodate the thickness of the base and deck, plus extra for securing a nut on top. In the case of a 1-3/4″ riser, select a 2-1/2″ screw, and cut the aquarium tubing to 1-3/4″.

Bonus Project: Build a 12-Servo Hexapod

See the RBB Support Site on how to create the Hex12Bot, a six-legged robot with 2 DOF per leg. The robot measures 11-1/2″ by 7″, and stands over 6″. Provided are full construction plans, with cutting and drilling templates, programming examples, and parts lists and sources.

CHAPTER 20

Build Robotic Arms and Grippers

Robots without arms can't "reach out and touch someone." Arms extend the reach of robots and make them more like humans. For all the extra capabilities arms provide a robot, it's interesting that they aren't difficult to build. Your arm designs can be used for factory-style, stationary "pick-and-place" robots, or they can be attached to a mobile robot as an appendage.

This chapter deals with the concept and design theory of robotic arms and hands. You'll learn about the various types of robotic arms, and how to build several kinds. You'll also learn ways to construct hands, more commonly called grippers or end effectors.

The Human Arm

Take a close look at your own arms for a moment. You'll quickly notice several important points. First, your arms are amazingly adept mechanisms, no doubt about it. Each arm has two major joints: the shoulder and the elbow (the wrist, as far as robotics is concerned, is usually considered part of the gripper mechanism). Your shoulder can move in two planes, both up and down and back and forth. The elbow joint is capable of moving in two planes as well: back and forth and up and down.

The joints in your arm, and your ability to move them, are called *degrees of freedom* (DOF). Your shoulder provides 2 DOF in itself: shoulder rotation and shoulder flexion/extension (shoulder flexion is motion upward to the front; shoulder extension is motion downward to the rear). The elbow joint adds a third and fourth degree of freedom: elbow flexion/extension and elbow rotation.

Robotic arms also have degrees of freedom. But instead of muscles, tendons, ball-and-socket joints, and bones, robot arms are made from metal, plastic, wood, motors, solenoids, gears, pulleys, and a variety of other mechanical components. Some robot arms provide but 1 DOF; others provide 3, 4, and even 5 separate DOF.

Degrees of Freedom in a Typical Robotic Arm

Human anatomy offers an inexact comparison with robotic arm systems. Our bone and muscle structure provides movement in a way that is seldom duplicated in robot arms. For example, out of simplicity, most robotic arms don't use a ball joint for the shoulder. In the human arm, this joint provides multiple degrees of freedom. In the robot version, shoulder motion is duplicated with two and sometimes three separate joints.

A basic robot arm has three degrees of freedom (additional freedoms are provided by the "wrist" and gripper):

- DOF #1 is rotation of the arm at its base. The base may rotate up to 360°, depending on design, though it's more common to limit rotation to about 180°. This represents the mechanical extents of the typical R/C servo, which is often used to move the joints in a low-cost robotic arm.
- DOF #2 and DOF #3 are essentially the shoulder and elbow joints, respectively. Together they allow the arm to lift and lower, and to position its gripper at the height and distance to grasp an object in front of it.

All the joints of the arm, including those in the gripper section, work in tandem to locate, grasp, and move objects. By pitching the shoulder, elbow, and wrist joints (if available) forward, the arm can reach down and pick up something on the ground.

Arm Types

Robot arms are classified by the shape of the area that the end of the arm (where the gripper is) can reach. This accessible area is called the *work envelope*. For simplicity's sake, the work envelope does not take into consideration motion by the robot's body, just the arm mechanics.

The human arm has a nearly spherical work envelope. We can reach just about anything, as long as it is within arm's length, within the inside of about three-quarters of a sphere. Imagine being inside a hollowed-out orange. You stand by one edge. When you reach out, you can touch the inside walls of about three-quarters of the orange peel.

In a robot, such a robot arm would be said to have revolute coordinates. The three other main robot arm designs are polar coordinate, cylindrical coordinate, and cartesian coordinate. You'll note that there are 3 DOF in all four basic types of arm designs. Let's take a closer look at each one. See Figure 20-1 for depictions of all four.

1. **Revolute coordinate** arms are modeled after the human arm, so they have many of the same capabilities. Revolute coordinate arms are a favorite design choice for hobby robots. They provide a great deal of flexibility, and, besides, they actually *look* like arms. See later in this chapter for details on how to construct a revolute coordinate arm.
2. **Polar coordinate** arms have a half-sphere work envelope. A mechanical turntable rotates the entire arm; the polar coordinate arm lacks a means for flexing or bending its shoulder, however. The second degree of freedom is the elbow joint, which moves the forearm up and down. The third degree of freedom is accomplished by varying the reach of the forearm.

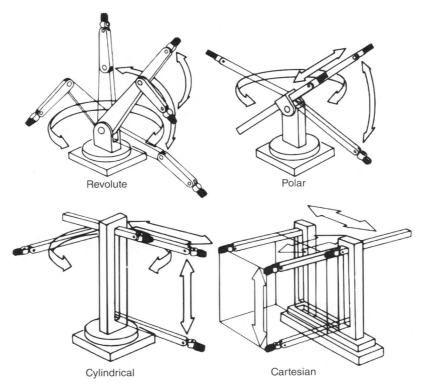

Figure 20-1   The four main classes of robot arms: revolute, polar, cylindrical, and cartesian.

3. A **cylindrical coordinate** arm looks a little like a robotic forklift. Its work envelope resembles a thick cylinder, hence its name. Shoulder rotation is accomplished by a revolving base. The forearm is attached to an elevator-like lift mechanism. The forearm moves up and down this elevator to grasp objects at various heights.

4. The work envelope of a **cartesian coordinate arm** resembles a box. It is the arm most unlike the human arm as it has no rotating parts. The base consists of a gantry that moves an elevator column. The forearm moves up and down the column and has an inner arm that extends the reach closer to or farther away from the robot.

Actuation Techniques

Any part of a robot that moves something is called an actuator. Motors are actuators, and so are arms. Actuation is moving the joints in the arm. There are three principle methods for this: electrical, hydraulic, and pneumatic. For small robots, electrical actuation is the method of choice; it's the least expensive and the easiest to implement.

- *Electrical actuation* is done with motors, solenoids, and other electromechanical devices. It is the most common and easiest arm type to implement. The motors for elbow flexion/extension, as well as the motors for the gripper mechanism, can be placed in or near the base. Cables, chains, or belts connect the motors to the joints they serve.
- *Hydraulic actuation* uses fluid-reservoir pressure cylinders, similar to the kind used in earth-moving equipment and automobile brake systems. Though industrial robots use a noncorrosive fluid that inhibits rust, a homebrew bot can use water and plastic parts.
- *Pneumatic actuation* uses pressurized air instead of fluid. Valves open and close to let air pressure into cylinders. The cylinders are connected to the arm joints to move them.

Build a 3 DOF Robotic Wrist

Sometimes a whole arm isn't required. All the bot really needs is a hand (gripper) on the end of a wrist-like mechanism that provides basic movement. The human wrist has 3 degrees of freedom: it can twist (rotate) on the forearm, it can bend up and down, and it can rock from side to side. You can add some or all of these degrees of freedom to a robotic hand.

Here are plans for making a 3 DOF wrist using either of two methods: homebrew plastic or wood parts you cut yourself, and metal brackets.

BUILD THE WRIST FROM METAL BRACKETS

You can construct a working robotic wrist using four servos and three metal servo brackets. You can get this type of bracket online through a number of sources, including Lynxmotion, Amazon, and eBay. Look around for a good deal.

Try to find brackets that come with machine screws and nuts; this saves you the hassle of providing these yourself. If yours don't come with hardware, get a set of about two dozen 4-40 × 3/8″ panhead machine screws and matching 4-40 nuts.

 Your servos should include a variety of servo horns so that you can best match the hole pattern in the metal bracket. The servo should also include a screw for attaching the horn on the output shaft of the motor.

Prior to assembly use a servo tester to set all servos to their neutral centered position. Connect the servo to the tester and apply power to it. Use a 4 × AA or AAA battery holder, with either nonrechargeable (6 volts) or rechargeable (4.8 volts) cells. Be mindful of plugging in the batteries the correct way—don't reverse the polarity of the batteries, or else the tester and the servo may be damaged.

Figure 20-2 Construct the robotic wrist in two sub-assemblies, making note of the orientation of each bracket in relation to its neighbors.

1. For each servo, visually align a servo horn so that two of its holes match (or approximately match) mounting the holes on the bottom of the bracket. Depending on the servo horn you use the holes may not precisely match; that's okay as long as they are close. (For my servos I found the double-arm horn to provide the best overall fit.) Mark which holes to drill with a pencil.
2. Using a 1/8" bit and small hand or motorized drill, drill out the marked holes in each servo horn. Careful! Don't hold the horn in your hand while you drill. Use a drill vice or a pair of heavy pliers.
3. For each servo attach a servo horn to the underside of the bracket bottom. Refer to the diagram for placement on the bracket. Using two screws and nuts to secure servo horn. The nuts should be located on the side of the horn.
4. Attach each horn to its servo using the servo horn screw provided. The screw should be tight, but don't overdo it and strip the plastic of the servo output shaft.
5. Assemble the servos and brackets as shown in Figure 20-2. Note the orientation of each bracket; servos attach to its next neighbor at right angles.
   Note: You can mount each servo into its bracket using two screws positioned at opposite corners of the servo flange. Be sure the nuts are on tight.
6. Combine the two servo sub-assemblies as shown in Figure 20-3.

Each servo has a natural limitation of rotation based on the position of its neighbors. At some angular positions the brackets may nudge one another. Keep this in mind when programming the various wrist movements and avoid going over these limits.

Figure 20-3   The fully assembled robotic wrist, with all four servos attached.

Figure 20-4   The completed homebrew robotic wrist.

Having trouble using screws to mount the servo horns to the brackets? Consider using hot glue instead. Apply glue to the horn, then immediately press it into position on the bracket. Be sure not to ooze glue into the center of the horn.

I recommend high-temperature hot glue to make the bond as strong as possible. After applying the glue wait 5 or 10 minutes to let it set up and strengthen.

ON THE WEB: BUILD THE HOMEBREW ROBOT WRIST

Perhaps you prefer to make your robotic wrist out of homemade parts. With a saw and drill you can construct your own servo brackets using wood or plastic. Assemble the brackets, then mount the servos into them.

Refer to the RBB Support Site for instructions on making a homebrew robot wrist. The functionality is similar even though the orientation of the servos is a bit different. You're free to experiment with different joint styles and orientations. Figure 20-4 shows the completed homebrew wrist.

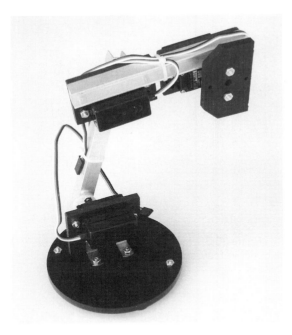

Figure 20-5   Completed robotic revolute coordinate arm. Attach the arm to a base or robot and add a gripper mechanism.

Build a Functional Revolute Coordinate Arm

You can build your own revolute coordinate robotic arm with about $20 in parts—not including servo motors. The size of the arm is scalable; the prototype shown in Figure 20-5 stands about 8″ tall and has a reach (without gripper) of over 9-1/2″. It provides 4 degrees of freedom, including a rotating base and shoulder, elbow, and wrist joints.

You will need four standard-size R/C servo motors. For added strength, select servos with a torque of no less than 45 oz-in; 65 to 85 oz-in is preferable. On most servos, the higher the torque, the slower the servo turns, so bear this in mind when selecting the motors for your arm.

To complete the arm, you'll need:

- 1 12″ length of 3/8″ U-channel extruded aluminum
- 1 3″ ball bearing turntable
- 1 small piece (about 12″ × 8″) 1/4″ aircraft-grade plywood or PVC plastic
- 1 pair of small 3/4″ corner angle brackets
- Small assortment of 4-40 machine screws and nuts

SET SERVOS TO NEUTRAL (CENTERED) POSITION

Prior to assembly use a servo tester to set all servos to their neutral centered position. First set the tester to standard mode, and use the dial to move the servo horn to its clockwise and counterclockwise extents. This checks the servo for proper operation.

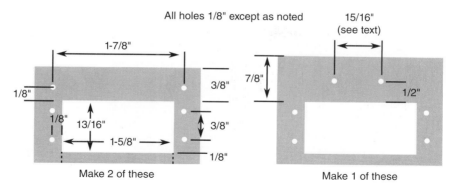

Figure 20-6 Cutting and layout guides for the servo mounts used in the revolute coordinate arm.

Next set the tester to produce constant 1500 μs pulses, and wait until the servo is centered. Disconnect the servo from the tester. When building try not to move the output shaft of the servo, but if you do, simply connect it back to the tester and re-center it.

MAKE THE SERVO MOUNTS

The robotic arm uses two types of servo mounts and one general-purpose (and optional) solid plate for attaching to the shaft of the servo. See Figure 20-6 for the basic cutting and drilling guide. You need to make a total of three mounts. You can construct these parts with 1/4″ air-craft-grade plywood or PVC plastic.

The spacing of the two holes in the top flange of the larger mount depends on the horns that come with your servos. The dimensions shown are for the large circular horn that comes with most Futaba and Futaba-style servos. If you use a different servo and horn, the hole spacing may be slightly different; adjust the spacing of the holes in the top flange accordingly.

These require some precision in cutting, so no one would blame you if you opted for the store-bought alternative stamped metal bracket, the same kind used in the wrist; see Figure 20-2.

BUILD THE BASE

The base of the arm consists of a 3″ ball bearing turntable ("lazy Susan") mounted on 4-1/2″-diameter round plastic, which serves as a bottom plate (this plate can be square if that's easier for you to cut). You can get this size turntable at better stocked hardware stores and home improvement outlets, or via mail order. Search for *3″ lazy susan*—I found mine for under $2 sold through Amazon.

1. Begin by drilling four holes to mount the turntable. Use the turntable itself to mark the holes in the center of the base. Once marked, drill the holes with a 1/8″ bit. Don't mount the turntable just yet.
2. Using the largest horn that comes with your servos, find two holes opposite one another on the horn that are about 3/4″ to 1-1/4″ apart. Use a 1/8″ bit to drill these out.

Figure 20-7 Arm base construction.

Figure 20-8 Assemble the servos to the servo mounts using machine screws and nuts.

3. Place the horn in the center of the 4″ bottom plate. Insert two 4-40 × 1/2″ machine screws through the horn and plate. See Figure 20-7A.

4. On the other side of the bottom plate, thread the screws through a pair of 3/4″ metal corner angle brackets. Secure the screws with 4-40 nuts. See Figure 20-7B.

5. Flip the bottom plate over again and mount the turntable using at least two screws on opposite corners. To insert the screws you'll need to spin the turntable so that all the flanges (top and bottom) are exposed. See Figure 20-7C.

These construction plans don't include mounting the fourth servo under the bottom plate. I'm leaving that up to you, based on where you want to put your arm. You can mount the arm on top of a mobile robot (it should be at least 8″ to 10″ in size), or you can build a stationary arm that operates within a confined space.

MOUNT THE JOINT SERVOS

You need to attach three servos into their mounts. Two sizes of mounts are used, as mentioned in "Building the Servo Mounts": two regular-size mounts and one with a larger flange. Examples are shown in Figure 20-8. Secure the servos to their mounts using at least two screws, one on each corner of the motor.

Figure 20-9   Orientation of the servo motors in the mounts. Attach the motors to the mounts with their output shafts oriented as shown.

Each of the servos needs to be oriented a certain way within its mount. Refer to Figure 20-9 for a guide. When viewed from the front of the servo, the output shaft of the motor should be located as shown. Secure the servo within its mount using at least two 4-40 × 1/2″ machine screws and nuts.

CONSTRUCT UPPER ARM

Make an upper arm (the part of the arm between shoulder and elbow) from a 6″ length of 3/8″ aluminum U-channel. Cut with a hacksaw, and file down the ends to remove any burrs. Then:

1. Using the small round or double-arm horn that comes with your servos, use a 1/8″ bit to drill out two holes on opposite sides for mounting screws. Once drilled out, mark three holes on each end of the upper arm: the two holes you just drilled and the center hole for the servo screw. Place the marks on the flat ("bottom") of the U in the U-channel.
2. Use a 1/8″ bit to drill each of the marked holes (for best results, use a center punch to begin the hole—see Chapter 11, "Putting Things Together" for more tips and tricks).
3. Use a 1/4″ bit to drill out the center hole on each end of the upper arm. This makes the hole large enough for you to insert the servo horn screw.
4. Use 4-40 × 3/8″ or 4-40 × 1/2″ machine screws and nuts to secure the servo horns to the upper arm.

MOUNT SHOULDER SERVO TO BASE

Use the servo in the larger of the two mounts for the shoulder joint. This servo attaches to the two 3/4″ corner angle brackets using 4-40 × 1/2″ machine screws and nuts (see Figure 20-10A). The servo is oriented so that its mounting flange points toward the base. The backside of the motor is over the centerline of the base.

ATTACH UPPER ARM TO SHOULDER

With the servo already set to its centered position, use the screw supplied with the servo to attach the shoulder joint motor to the bottom of the upper arm.

Point the arm piece straight up when attaching the motor, so that the joint will rotate equally in both directions. Don't overtighten the screw or else it might strip the output shaft of the servo. See Figure 20-10B.

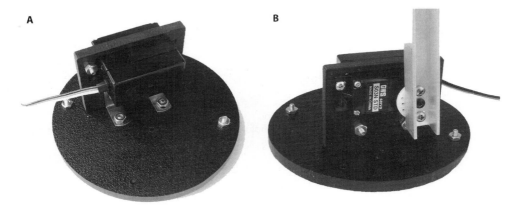

Figure 20-10 Secure the shoulder motor (on tallest of servo mounts) to the rotating base.

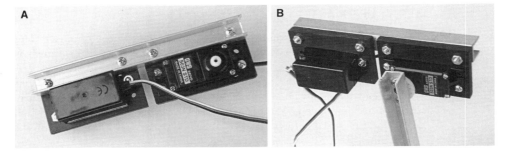

Figure 20-11 Attach servos to the forearm, mount upper arm to elbow.

ATTACH SERVOS TO FOREARM

Using 4-40 × 1/2″ machine screws and nuts, attach the two servos (the ones in the smaller mounts) to the forearm. Be sure to orient the servos so that one faces the "front" of the arm, and the other the back.

- The output shaft of the servo that faces the front (same side as the open part of the U-channel) should be on the right side (Figure 20-11A).
- The output shaft of the servo that faces the back (flat part of the U-channel) should be on the left side.

MOUNT UPPER ARM TO ELBOW

With the servo already set to its centered position, use the screw supplied with the servo to attach the elbow joint motor to the top of the upper arm. See Figure 20-11B.

Don't overtighten the screw or else it might strip the output shaft of the servo. As you did with the shoulder motor, mount the arm piece pointing straight up. That way the joint will rotate equally in both directions.

ATTACH WRIST

Finish the arm by attaching a wrist mechanism. Shown in Figure 20-5 is a simple plate for mounting grippers of different designs.

You'll need to extend the length of the wires from at least the two servos mounted on the forearm in order to reach the control electronics. You can get servo extensions in 12″ lengths (and longer) at hobby stores specializing in radio control parts, and through online and mail-order outlets that sell servos.

Build a Robotic Arm from a Kit

The revolute coordinate arm described in the preceding section is designed around simple components that can be constructed using ordinary tools and limited precision. If you want something more elaborate but don't want to build it from scratch, you can choose from among a number of specialty robotic arm kits or use custom construction set parts.

ARMS FROM ROBOTICS CONSTRUCTION SET PARTS

Think of these as Erector sets for grown-ups—they're parts of various shapes and sizes made of stamped metal, with holes already drilled in them for attaching to R/C servos, motors, and other robotic components. One popular robotics construction set is the TET-RIX Robotic Design System from online educational outlet Pitsco. Several sets are available, each with a different assortment of parts. You can also purchase individual parts as needed. Most construction sets are made with stamped aluminum, though they may also contain plastic pieces.

Even with the variety of parts that come in the typical robotics construction set, you're not limited to using just those components. Unless there's a restriction otherwise (you're building a robot for a school competition, for example), you should feel free to add your own bits and pieces.

See Chapter 40, "Make Robot Arms," for more info on using stamped parts for making a cool appendage for your robo-buddy.

ARMS FROM SPECIALTY KITS

Thanks to the popularity of amateur and educational robotics, there are plenty of specialty kits that are constructed just for the purpose of building a robot arm. Some are low-cost and meant for casual experimentation using manual switch control, but others, like the arm in Figure 20-12, use R/C servo motors and microcontrollers to precisely position the arm components. Such arms may or may not come with a gripper; you can add one you made yourself or get a gripper kit and build it from premade parts.

This particular arm has 4 degrees of freedom:

- Shoulder rotation, using a rotating base. This allows the arm to pick up objects in a circular arc of about 90° to each side.
- Shoulder flexion/extension, using a servo mounted near the base.
- Elbow flexion/extension, using a servo mounted off a yoke from the shoulder joint.
- Wrist flexion/extension, using a servo mounted at the end of the forearm.

Figure 20-12   Robot arm kit, shown with control electronics and gripper.

In almost all cases, arm kits are designed for use with standard- and mini-size R/C servos. The kits are available with and without servo motors, in case you already have some in your parts bin. If you supply your own servos, take note of any special requirements that are listed in the documentation for the arm. Some or all of the servos may need to have a minimum torque, for example.

Coming to Grips with Grippers

Arms aren't much good without hands. In the robotics world, hands are usually called *grippers* (also *end effectors*) because the word more closely describes their function. Few robotic hands can manipulate objects with the fine motor control of a human hand; they simply grasp or grip the object, hence the name gripper. Never sticklers for semantics, I'll use the term *gripper* from here on out.

Gripper designs are numerous, and no single design is ideal for all applications. Each gripper technique has unique advantages over the others. Let's take a look at a number of useful gripper designs you can use for your robots.

CONCEPT OF THE BASIC GRIPPER

In the world of robotics there are hundreds of ways to make a gripper. The designs tend to be application-specific: like Captain Hook in *Peter Pan,* a metal hook offers advantages a normal hand cannot. A perfectly useful gripper might be designed for a single task, like collecting Ping-Pong balls or picking up chess pieces—whatever the robot is made to do.

Figure 20-13 shows a typical robotic gripper, depicted in three different states: all the way open, halfway open, and all the way closed. This style of gripper uses "fingers" that stay parallel as they open and close. The design allows the gripper to apply even pressure on either side of an object and close in on the object without pushing it away. The mechanism is not difficult, but you can see that it adds a bit of complexity to the construction of the gripper.

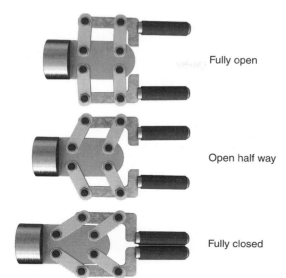

Fully open

Open half way

Fully closed

Figure 20-13   Robotic gripper in three states: all the way open, closed, and midpoint. In each case the "fingers" of the gripper remain parallel.

To be useful, most grippers are attached to the end of an arm. If the arm is on a wheelbase, the bot can steer around the room looking for things to pick up and examine. By orienting the gripper vertically or horizontally via a rotating wrist mechanism, the robot can grasp all manner of objects.

There's no rule that says the gripper must be attached to an arm; it can be connected directly to the robot itself. Figure 20-14 shows a "two-finger" gripper at the front of a robot base that collects balls or other objects.

BUILD A TWO-PINCHER GRIPPER

The two-pincher gripper consists of two movable fingers, somewhat like the claw of a lobster. For ease of construction, let's use extra Erector set parts or similar construction kit. Cut two metal girders to 4-1/2" (since this is a standard Erector set size, you may not have to do any cutting). Cut a length of angle girder to 3-1/2". Use 4-40 or 6-32 × 1/2" machine screws and nuts to make two pivoting joints.

Refer to Figure 20-15A for an overview of assembly. Cut two 3" lengths and mount them. Nibble the corner off both pieces to prevent the two from touching one another. Nibble or cut through two or three holes on one end to make a slot. Use 4-40 or 6-32 × 1/2" machine screws and nuts to make pivoting joints in the fingers (Figure 20-15B).

The basic gripper is finished. You can actuate it in a number of ways. One way is to mount a small eyelet between the two pivot joints on the angle girder. Thread two small cables or wire through the eyelet and attach the cables (see Figure 20-15C). Connect the other end of the cables to a solenoid or a motor shaft. Use a light-compression spring to force the fingers apart when the solenoid or motor is not activated.

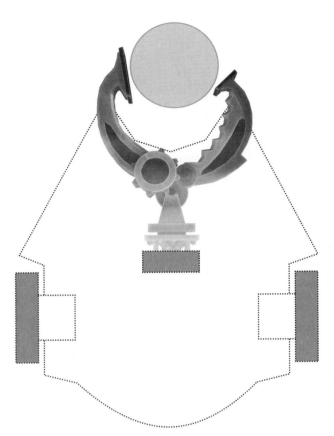

Figure 20-14   Grippers may be mounted directly on the base of a robot and used in specialty applications, such as collecting and holding balls, cans, or eggs.

You can add pads to the fingers by using the corner braces included in most Erector set kits and then attaching weather stripping or rubber feet to the brace.

TIPS FOR MAKING PARALLEL FINGER GRIPPERS

Figure 20-16 show another approach to constructing two-pincher grippers. By adding a second linkage to the fingers and allowing a pivot for both, the fingertips remain parallel to one another as the fingers open and close.

The downside: constructing a parallel gripper requires more parts and precision, so this may not be something you want to do in your home shop. Fortunately there are a number of ready-made grippers that use one mechanism or another to provide for parallel finger action. Figure 20-17 shows a low-cost model popular on the Web (I got mine through Amazon). It works with most any standard size servo. Build time is about 5 minutes. Add a bracket to attach it to the rest of the arm.

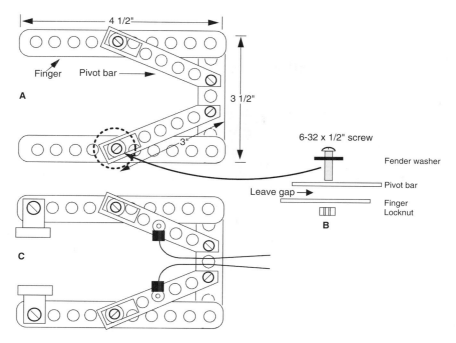

Figure 20-15 Construction detail of the basic two-pincher gripper, made with Erector set (or similar) parts.

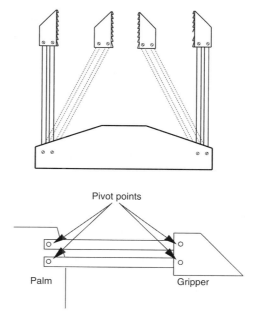

Figure 20-16 Adding a second linkage to the fingers and allowing the points to freely pivot causes the fingertips to remain parallel to one another.

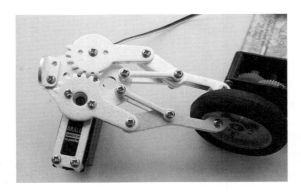

Figure 20-17 A commercially made gripper provides a quick and easy way to hand fingers/hands to your robot—but of course they're not quite as fun to build!

TOOL CLAMP GRIPPER

I figure if someone else has gone to the trouble and expense of making a product that just happens to work in robot projects, the least we can do is take advantage of their kindness!

In this project you use an ordinary plastic tool clamp, available at dollar stores and discount tool outlets, as a motorized robotic gripper. An R/C servo motor opens and closes the clamp. This gripper is simple to build and surprisingly strong.

Plastic tool clamps come in a variety of shapes and sizes. You want one that's as close as possible to the clamp shown here. The clamp measures 5″ in overall length and is about 1-1/2″ wide when the clamp is closed. The clamp is fitted with a plastic locking mechanism, which you'll remove as part of the construction steps outlined as follows.

Cut Gripper Mount

Using 6mm expanded PVC or 1/4″ hobby or aircraft plywood, cut the mount of the gripper as detailed in Figure 20-18. The only true dimension-sensitive cut is the inside of the servo mount.

This cut is best made by drilling a hole at one corner, then threading a thin (woodworking) coping saw blade through the hole. Attach the blade to the coping saw, and carefully cut the rectangular hole for the servo. Be sure not to cut away too much, or you won't have room for the servo mounting holes.

Attach Servo to the Mount

Using 4-40 × 1/2″ machine screws and nuts, attach a standard-size servo to the gripper mount, as shown in Figure 20-19. The angled portion of the mount should face toward the right. The servo shaft should be closest to the angled portion of the mount.

If you'd like to add a side-to-side wrist joint to the gripper, attach a double-arm servo horn (it should come with the servo) to the angled portion of the mount, also using 4-40 × 1/2″ machine screws and nuts. The nuts should be on the "top" of the mount (same side as the servo output shaft), and the horn should be on the bottom.

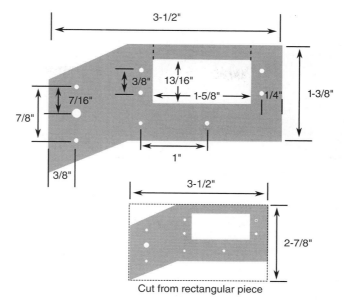

Figure 20-18   Cutting and drilling layout for the mount for both R/C servo motor and tool clamp.

Figure 20-19   R/C servo motor attached to the mount. Note the orientation of the output shaft of the servo motor.

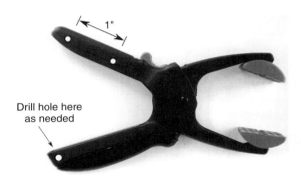

Drill hole here
as needed

Figure 20-20 Drill three holes for mounting the tool clamp and for attaching the activating rod from the servo motor.

Prepare and Mount the Clamp

Prepare the clamp by first removing the plastic locking mechanism. I've found the best way is to simply use brute force: start by releasing the lock and opening the clamp all the way. Using a pair of heavy-duty pliers, grip the locking mechanism and literally tear it out. If any pieces of the locking mechanism remain inside, nudge them out with a small flat-bladed screwdriver.

> You may ruin the first clamp you try to modify this way, so better get two at the store just in case. They're not that expensive, and, besides, in many instances they sell a pair of clamps in one package. That's how I've always bought them.

Drill three 1/8"-diameter holes, as noted in Figure 20-20. On one handle you drill two holes; the spacing must match the holes in the mount. The other hole is for the servo linkage, and its exact position isn't critical. Make sure you drill all the holes in the approximate middle of the clamp handles.

Drilling complete:

1. Attach the clamp to the gripper mount using 4-40 × 3/4" machine screws and nuts.
2. Attach a double-arm horn or adjustable arm horn to the servo. (When using a double-arm horn, cut off the opposite arm.)
3. Use a servo tester to move the servo to its centered (neutral) midpoint.
4. Use heavy lineman's pliers to cut a large safety pin as shown in Figure 20-21. You can get a dozen for just a dollar or two at the local discount store.
5. Using pliers, bend 1/4" or so at the ends of the wire.
6. Place one bent end into a hole in the servo horn and the other into the empty hole you drilled in the clamp handle.
7. Once you're sure the wire is the right length (move the servo back and forth a few times), use a pair of pliers to cinch the wire securely in place. You don't need or want to clamp the wire shut, just close it up a bit so it won't easily fall out.

The completed tool clamp gripper is shown in Figure 20-22. At the base of the gripper is another double-arm servo horn, for attaching to a servo that's mounted on the robot. This servo lets the gripper scan back and forth for its prey. You don't need to use this servo horn if you're planning on bolting the gripper directly to your robot.

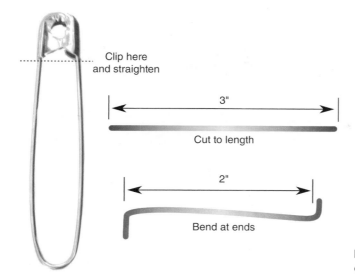

Figure 20-21 Use part of a safety pin as a connecting rod.

Clip here and straighten

3"

Cut to length

2"

Bend at ends

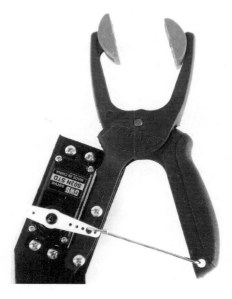

Figure 20-22 Finished tool clamp gripper, showing the activating rod between the clamp and the R/C servo motor.

MAKE A GRIPPER FROM A ROBOT TOY ARM

You've probably seen those colorful plastic "robot arms" at the toy store. Squeeze the handle on one end and a pincher gripper closes on the other. It's all done with simple mechanics. With just a bit of artistic surgery and a standard R/C servo you can turn the toy arm into a workable robot gripper!

Figure 20-23 shows the robot arm before and after conversion. What follows are plans for modifying a commonly available toy arm that's available online and at many toy and discount stores. Yours might differ in some ways; feel free to adapt the plans accordingly.

Figure 20-23   Before and after views of turning a toy arm into a robot gripper.

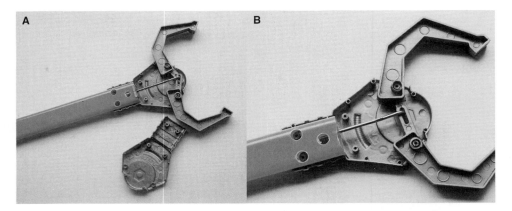

Figure 20-24   Use a screwdriver to open the gripper and remove the return spring.

1. Start by removing the handle. After removing the screws holding the top and bottom halves of the handle in place I used a pair of pliers to break out the plastic from the recesses. That's okay—the handle isn't used in the final gripper design. Use some masking or electrical tape to keep the connecting rod from moving too much.

2. Use a screwdriver to disassemble the gripper. Catch all the little screws in a bowl for safekeeping. Gently lift the gripper cover off to expose the mechanism inside (see Figure 24A).

3. You don't need (or want) the spring that keeps the gripper open when you're not squeezing on the handle. Use pliers to pry out the little piece of retainer plastic that acts as backstop for the spring (see Figure 24B).

4. Dab some light machine oil to any moving part, in the gripper mechanism, and reassemble. Don't overtighten the screws, or it might cause the gripper to bind up.

5. Connect servo tester to a standard size servo. Set the tester to provide 1500 µs pulses to move the servo to its centered neutral position.

6. Use hot melt glue to attach the servo to the handle so that the output shaft of the servo faces opposite the gripper end.

A   B

Figure 20-25   Glue a standard R/C servo to the handle of the toy arm, and attach the connecting rod to it.

 On my toy arm there was just enough of a flange to secure the servo. Depending on the design of the arm you use you might need to make a little mounting plate for the servo out of foamboard.

7. Secure the connecting rod to a double arm or X horn by first drilling a 1/8″ hole in the horn, and slipping the rod through. Use heavy-duty pliers to bend the rod to keep it from coming back out (see Figure 20-25A).
8. Use the servo tester to slowly move the servo back and forth, watching that nothing is binding up (see Figure 20-25B). If all looks good apply extra glue around the edges of the servo to help keep it in place.

ON THE WEB: MORE GRIPPER PLANS

Visit the RBB Support Site for bonus gripper plans, including a solenoid-driven "clapper" two-pincher gripper, fingers that open and close using a homemade worm gear system, and ideas on how to construct flexible fingers that have a human-like *compliant* grasp.

PART 5

Robot Electronics

CHAPTER 21

Robot Electronics—The Basics

In previous chapters you learned all about the mechanics of robots, including their construction, motors and wheels, and power systems. In this chapter you'll discover the electronics that endow bots with the appearance of life. In this and future chapters you'll find out ways to use modern (but still inexpensive) advances in electronics to create fully programmable robots able to truly think on their own. It's an exciting endeavor, so let's get started.

Tools for Electronics You Should Have

Compared to mechanical construction, you need relatively few tools to build the electronic centerpieces of your robots. You can always spend lots and lots of cash for all kinds of testing equipment and specialized electronic gear, but what follows are the basics that will get you started.

MULTIMETER

A *multimeter*, also called a *volt-ohm meter, VOM,* or *multitester,* is used to test voltage levels and the resistance of circuits—among other things. This moderately priced tool (see Figure 21-1) is the basic prerequisite for working with electronic circuits of any kind. If you don't already own a multimeter, you should seriously consider buying one. The cost is minimal, considering its usefulness.

There are many multimeters on the market today. A meter of intermediate features and quality is more than adequate. Meters are available at RadioShack and most online electronics outlets. Shop around and compare features and prices.

Digital or Analog
There are two general types of multimeters available today: digital and analog. The difference is not in the kinds of circuits they test, but in how they display the results. *Digital* multimeters,

Figure 21-1 A digital multimeter checks resistance, voltage, and current. This model also performs simple checks of a number of common electronic components, including capacitors, diodes, and transistors.

which are now the most common, use a numeric display not unlike a digital clock. *Analog* meters use the more old-fashioned—but still useful—mechanical movement with a needle that points to a set of graduated scales.

Automatic or Manual Ranging

Many multimeters require you to select the range before it can make an accurate measurement. For example, if you are measuring the voltage of a 9-volt battery, you set the range to the setting closest to, but above, 9 volts. With most meters it is the 21- or 50-volt range. Because you need to select the range, there are lots of options on the dial, but in reality these options are really just variations on a theme. The meter is easier to use than it looks.

Autoranging meters don't require you to do this, so they are inherently quicker to use. When you want to measure voltage, for example, you set the meter to volts (either AC or DC) and take the reading. The meter displays the results in the readout panel.

For the sake of completeness, the examples in this book that explain how to use a meter assume you have a manual (*non*automatic) ranging model. If yours has automatic ranging, then just skip the step that says to dial in the upper range of your expected measurement.

Accuracy

The accuracy of a meter is the minimum amount of error that can occur when taking a specific measurement. For example, the meter may be accurate to 2000 volts, plus or minus 1 percent. A 1 percent error at the kinds of voltages used in robots—typically, 5 to 12 volts DC—is only 0.1 volts. Not enough to quibble about.

Digital meters have another kind of accuracy: the number of digits in the display determines the maximum resolution of the measurements. Most digital meters have three and a half digits, so they can display a value as small as 0.001—the half digit is a "1" on the left side of the display.

Functions

Digital multimeters vary greatly in the number and type of functions they provide. These functions are selectable by rotating a dial on the front of the meter. At the very least, all standard multimeters let you measure AC and DC voltage, DC amperage, and resistance.

The maximum ratings of the meter when measuring volts, milliamps, and resistance also vary. For most applications, the following maximum ratings are more than adequate:

DC voltage	1000 volts
AC voltage	500 volts
DC amperage	200 milliamps
Resistance	2 megohms (2,000,000 ohms)

One *very important* exception to this is when you are testing the amount of current draw from motors. Many DC motors draw in excess of 200 milliamps.

Better multimeters have a separate DC amperage input that allows readings of up to 10 amps (sometimes as high as 20 amps). If you have the budget for it, I highly recommend that you get a meter with this feature. In many cases, it's a separate input on the front of the meter and is clearly labeled, like that in Figure 21-2.

The high-amperage input may or may not be fuse-protected; if it is fuse-protected and you exceed the current rating for the input, a fuse will blow and you'll have to get it replaced. On some inexpensive multimeters the inputs are not fused, and exceeding the maximum ratings could result in permanent damage to the device. So be careful!

Meter Supplies

Multimeters come with a pair of test leads, one black and one red. Each is equipped with a pointed metal probe. The quality of the test leads included with the multimeter is usually minimal, so you may want to purchase a separate set that's better. The coiled kind is handy; the test leads stretch out to several feet, yet recoil to a manageable length when not in use.

Standard point-tip leads are fine for most routine testing, but some measurements may require that you use a *clip lead*. These attach to the end of the regular test leads and have a spring-loaded clip on the end. You can clip the lead in place so your hands are free to do other things. The clips are insulated with plastic to prevent short circuits.

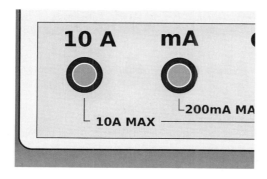

Figure 21-2 For robotics work you'll want a multimeter with a high-amperage (10 amps or higher) input. You use this to easily test the current draw of motors, among other tasks.

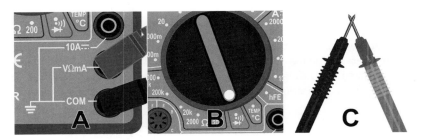

Figure 21-3 Steps in reading continuity using a digital meter.

Using the Meter: The Basics

To use your multimeter, first set it next to whatever circuit you're testing. Make sure it's close enough so the test leads reach the circuit without any risk of pulling either the meter or the circuit into your lap.

Then start with this:

1. Plug in the test leads; the black lead of your meter goes into the – or *COM* jack, and the red lead of your meter goes into the + or labeled function jack (the label may be something like *VΩmA*, which represents the kinds of tests you can do when the lead is inserted into that jack—in this case, voltage (V), resistance (Ω), and low-milliamp (mA) current testing (Figure 21-3A).

2. Check for proper meter operation by doing a *continuity test*. This involves selecting one of the following operating modes. Depending on the features of your meter, choose Resistance (Ω), Diode check, or Continuity. If using Resistance and the meter is not autoranging, choose the lowest Ω setting (Figure 21-3B).

Touch the metal tips of the test probe together. If the meter is functioning properly—the battery is good, the test leads are not broken—the results should be as shown (Figure 21-3C).

Meter Setting	Good	No Good
Resistance (Ω)	Zero or nearly zero resistance	Infinite* or very high resistance
Diode check[†]	Good	Infinite value*
Continuity[†]	Good	Infinite value*

* Meters show infinite value differently, but most display a blinking "1" on the left side of the display.

† Many digital multimeters provide an audible beep when using the Diode check or Continuity settings, and the continuity test is good.

Once the meter has checked out, select the desired function and range, and apply the leads to the circuit under test.

 What?!? The meter doesn't pass its simple continuity test? The reasons could be simple and easy to fix: check that the internal battery is good. Replace as needed. Inspect the test leads for breaks. If the metal prongs of the leads are old, they could be corroded or rusted. Clean or replace. And finally, if the meter is internally fused, the fuse could be blown. Try the spare.

Using the Meter: Testing a Battery

You can use your multimeter to test batteries and other low-voltage DC power sources. Merely as an example to get you started, here are the steps:

1. With the meter on, dial to the DCV (DC volts) setting. If your meter is not autoranging, select a range that is one step higher than the expected voltage; for most robot tasks this will be 20 volts or less.
2. Ensure the test leads are plugged into the proper jacks, as detailed in the previous section "Using the Meter: The Basics."
3. Touch the black lead to the negative (−) terminal of the battery or pack; touch the red lead to the positive (+) terminal.
4. Note the value displayed by the meter. If the battery (or pack) is good, the voltage should be close to the expected value—for example, an AA alkaline battery should be about 1.5 volts, give or take 0.1 or 0.2 volts.
5. For grins, switch the test leads so that black goes to the + battery terminal and red goes to −. Note the voltage again; it should now be a negative value, indicating that the polarity of the test connection to the battery is reversed.

Using the Meter: Verifying the Value of a Resistor

Another common use of a multimeter is verifying the value of a resistor. Here's how.

1. From your parts bin, select any four-banded resistor with the color brown as its third band. This will ensure the resistor is between 100 and 990 ohms. (Resistors and their values and markings are discussed in more detail in Chapter 22, "Common Electronic Components for Robotics.")
2. Refer to the resistor color code table in the RBB Online Support Site to look up the value of the resistor. For example, if the first three color bands are orange-orange-brown, the indicated value of the resistor is 330 Ω.

3. With the meter on, dial to the Ω setting. If your meter is not autoranging, select a range that is one step higher than the expected reading. For instance, if the ranges are 2, 20, 200, 2000, and so forth, select 2000 (2k), as it is one step higher than the expected value of 330 Ω.
4. Ensure the test leads are plugged into the proper jacks, as detailed in the previous section "Using the Meter: The Basics."
5. With the resistor resting on the table or workbench, apply the test leads to either side of the resistor. Be sure not to touch the metal of the test leads, or else the natural conductivity of your skin will influence the result.
6. Read the value on the meter. See Figure 21-4 for an example.

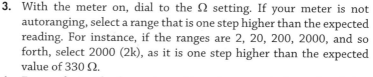

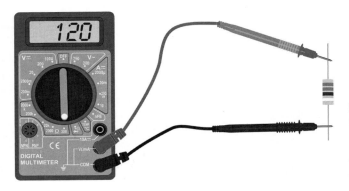

Figure 21-4 Reading the value of a resistor. Be sure not to touch the test probes, or the resistance of your skin will influence the result. When using a meter without autoranging, adjust the range dial just above the expected value.

So the resistor doesn't read exactly what it should? There's no cause for alarm. The fourth band on a resistor with four color bands indicates its tolerance, or how far off the printed value it can be from the actual value. A gold band indicates 5 percent tolerance; a silver band, 10 percent tolerance. If the resistor is 330 Ω with 10 percent tolerance, its reading on the multimeter can range from 300 to 360 Ω.

Meter Safeguards, Good Operating Habits

When using a meter, even on low-voltage circuits, observe these best operating habits:

- Never hold the test probes by their metal part. Not only can this give you a nasty shock when testing AC household current, but your skin resistance can alter the test results.
- On each use of the meter, test its basic operating status with the continuity check detailed previously.
- None of the projects in this book involve working directly with AC household current—everything is battery-powered. Should you wish to use your multimeter with an AC circuit, be sure to consult the instruction manual that came with the meter for important safety precautions.
- Be very careful when selecting the operating mode of the meter. Never accidentally set the meter to read resistance and then test a voltage source. Your meter could be damaged otherwise or, at the least, burn out a fuse if it is so equipped.
- Turn off the meter when you're done with it. This preserves battery life.

ON THE WEB: USING A LOGIC PROBE

Another handy tool for testing electronic circuits is the logic probe, so-called because it verifies signals used in logic circuits (anything that deals with digital 0s and 1s, LOWs and HIGHs). These kinds of circuits include microcontrollers. I've prepared a free "Logic Probe 101" article on the RBB Support site (see Appendix) that provides more information.

SOLDERING IRON PENCIL

You can build robots without owning a *soldering iron pencil* (or simply soldering iron), but it's darned difficult to do anything more advanced than just put together basic kits. Even if you

Figure 21-5 Soldering station with adjustable heat output and integral tip cleaner. (*Photo courtesy American Hakko Products Inc.*)

never plan to make your own circuit boards for your robots, you still need a soldering pencil for basic electronic chores, such as attaching wires to motors. A soldering station, like that in Figure 21-5, combines a heat-regulated iron plus a stand. In more elaborate models you can adjust the temperature.

Not only do you want a soldering iron, you really want the kind that lets you change the tip and heating element. Why? These are the parts that, over time, wear out. Rather than buy a whole new soldering pencil, you only have to buy replacement parts.

For routine electronic work, you should get a soldering pencil with a 25- to 30-watt heating element. Anything higher may damage electronic components. You can use a 40- or 50-watt element for wiring switches, relays, and power transistors. If you can afford it, opt for a model with a temperature dial. They cost a bit more, but they're far more flexible.

See the section "How to Solder," later in this chapter, for a step-by-step guide on soldering.

HAND TOOLS FOR ELECTRONIC CONSTRUCTION

You need just a few hand tools for electronic construction. The ones described here will fill your electronic toolbox quite nicely. All of these tools are inexpensive. The first three on the list, which are among the most used tools for electronic construction, are shown in Figure 21-6.

- *Flush wire cutters,* sometimes referred to as "nippy" cutters. These let you cut off wire flush with the surface of a circuit board.
- *Wire strippers* for smaller-gauge wire. Be sure it can handle between 18- and 26-gauge wire (the higher the number, the smaller the diameter of the wire). Most electronic hookup wire is 22 gauge. The stripper should have a dial that lets you select the gauge of wire you are using.
- *Solder clamp or vise.* The clamp or vise serves as a "third hand," holding together pieces to be soldered, so you are free to work the soldering pencil and feed the solder. One with a built-in magnifying glass is nice. The ones with simple alligator clips are the least expensive, but they do the job.
- Set of *flat-bladed* and *Phillips screwdrivers,* including sizes #1 and #0.
- Small *needle-nose pliers.*
- *Dental picks.* These are ideal for scraping, cutting, forming, and gouging into things. You can buy these surplus.

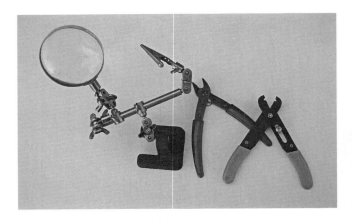

Figure 21-6 Trio of the most used tools for working with electronics: a vise or "third hand" (magnifier optional, but handy), flush wire cutters, and wire strippers.

 Always wear eye protection when using flush wire cutters or, for that matter, any wire-cutting tool. It's quite common for wires to literally shoot out at high velocity when cut. You don't want anything flying into your eyes.

Making Electronic Circuits—The Basics

You have at your disposal numerous ways to construct the electronic circuits for your robots. Those designs involving only switches and batteries and motors can simply be wired together, one to the other, and there is no need to centralize the components in a single place. Options include:

- *Solderless breadboard.* Quickly and easily construct circuits by plugging components into sockets on a plastic board. No soldering necessary. See Chapter 23, "Making Circuits" for more information.
- *Permanent circuit board.* Select from among several methods for soldering parts to build a permanent circuit. You can use generic boards that accept common components, or design your own printed circuit board (PCB). Also see Chapter 23 for details.
- *Wire wrapping.* Use a low-cost tool to interconnect electronic components with very fine wire. See the related App Note on the RBB Support Site for more information.

Understanding Wires and Wiring

Almost every electronic circuit uses wire of one kind or another. Wiring is a science all to itself, but we'll concentrate just on three main aspects: insulation, gauge, and conductor type.

INSULATION

Most of the wire used in building robot electronics is insulated with a plastic covering. This keeps one wire from touching another and causing a short circuit. Apart from esoteric aspects about insulation, the most important is its color. Get into the habit of using different-color wiring to denote what it's being used for in your circuit. For example, red wire is often used for the + (positive) battery connection; black wire for the – (negative) connection.

GAUGE

The thickness, or *gauge,* of the wire determines its current-carrying capabilities. Generally, the larger the wire, the more current it can pass without overheating and burning up.

See "Electronic Reference" on the RBB Support Site for common wire gauges and the maximum accepted current capacity, assuming reasonable wire lengths of 5 feet or less.

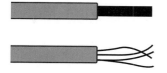

CONDUCTOR TYPE

Wire is made of one or more strands of metal.
- Single-strand or *solid* wire has just one metal conductor. Solid wire is commonly used when building circuits using a solderless breadboard.
- Multiple-strand wire has many conductors and is said to be *stranded.* For any given wire gauge the conductors in stranded wire are small. Stranded wire is more flexible and doesn't break as easily when it's repeatedly flexed.

How to Solder

Few electronic projects can be assembled without soldering wires together. Soldering sounds and looks simple enough, but there's a bit of science to it. If you are unfamiliar with soldering, or you need a quick refresher course, read the primer on soldering fundamentals provided in this section.

SOLDERING TOOLS YOU NEED

Good soldering means having the proper tools. If you don't have them already, you can purchase them at RadioShack or most any electronics store. Let's review the soldering-related tools you need.

I've already introduced the soldering iron earlier in the chapter. See Figure 21-7 for a description of the pencil's main parts. Try to get one with a three-prong power cord. This provides important grounding of the tool, which is needed for safety.

Stand

If your soldering pencil doesn't come with a *stand,* be sure to get one. They're used to keep the soldering pencil in a safe, upright position. You should *never* simply lay a hot soldering pencil down on your work table.

Sponge

Keep a damp (never dry) sponge by the soldering station. Be sure to keep it wet. Use the sponge to wipe off globs of solder that may remain on the tip. Otherwise, the glob may come off while

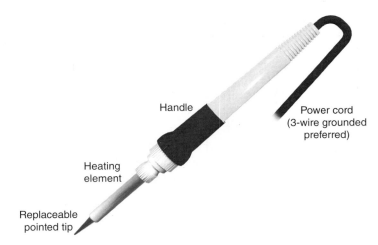

Handle

Power cord
(3-wire grounded
preferred)

Heating
element

Replaceable
pointed tip

Figure 21-7   Parts of the soldering pencil. Be sure yours is a three-wire grounded model. Don't use an ungrounded soldering tool for electronics work. The replaceable tip is a good feature. When it's worn, you can get a new tip rather than a whole new soldering pencil.

you're soldering and ruin the connection. In a pinch, you can substitute a wetted and folded-up paper towel or napkin. Be sure it stays wet—you don't want it to catch fire when you try to clean the soldering tip against it!

Solder

Use only rosin core solder approved for use in electronic circuits. It comes in different thicknesses. For best results, use the thin type (0.031″) for most of your electronics. Don't use acid core or silver solder on electronic equipment. (Note: Certain "silver-bearing" solders are available for specialty electronics work, and they are acceptable to use.)

For noncommercial applications, you have the option of lead-bearing or lead-free solders. Both are safe when handled properly. Lead-bearing solder is a bit cheaper and easier to work with as it has a lower melting temperature. The choice of which type of solder to use is yours.

Miscellaneous Soldering Tools

And there are a few more soldering tools worth mentioning:

- A *heat sink* looks like a small metal clamp. It's used to draw heat away from components during soldering.
- Ordinary household *isopropyl alcohol* makes a good, all-around soldering cleaner. After soldering, and when the components and board are cool, clean the board with *rosin flux remover*.
- A *solder vacuum* (or "solder sucker") is a suction device used to remove excess solder. It is often used when desoldering—that is, removing a wire or component from the board, so that you can fix a mistake.

CLEANING THINGS PRIOR TO SOLDERING

Before soldering, make sure all parts of the connection are clean. If you're soldering a component onto a printed circuit board, clean the board first with warm water, a kitchen scouring powder, and a nonmetallic scrubbing pad. Rinse thoroughly, and let dry.

Next, wet a cotton ball with normal household isopropyl alcohol and wipe off all the connection points. Wait a minute for the alcohol to *completely* evaporate, then start soldering.

SETTING THE CORRECT SOLDERING TEMPERATURE

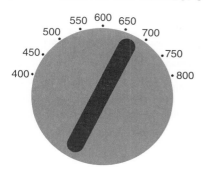

A soldering tool with a temperature control is often referred to as a *soldering station*. If that's what you've got, and you're using traditional lead-bearing solder, dial the station to between 665° and 680°F (352° to 360°C). This provides maximum heat while posing the minimum danger of damage to the electronic components.

If your soldering iron/station has just the control and lacks a heat readout, initially set it to low. Wait a few minutes for it to heat up, then try one or two test connections. Adjust the heat control so that solder flows onto the connection in under 5 seconds.

To do its job, your soldering tool needs to be significantly hotter than the melting point of solder. Most lead-bearing solders have a melting point of about 362°F (183°C). For lead-free solder, the range is much wider, but in general their melting point is 40° to 70°F higher. Increase the temperature of the soldering tool accordingly.

STEPS FOR SUCCESSFUL SOLDERING

The idea behind successful soldering is to use the soldering tool to heat up the work—whether it is a component lead, a wire, or whatever. You then *apply the solder to the work*. Don't apply solder directly to the soldering pencil. If you take that shortcut, you might end up with a "cold" solder joint. A cold joint doesn't adhere well to the metal surfaces of the part or board, so electrical connection is impaired.

Here are the steps to solder a component onto a circuit board:

1. Use small needle-nose pliers to bend the leads of the component to match the spacing of the holes in the circuit board (see Figure 21-8). Eyeball the correct distance; you'll get more accurate at judging where to place the bends as you gain experience.
2. Insert the component leads through the holes in the circuit board. The component should rest fully against the board, or be very close to it.
3. With your fingers, gently bend the leads of the component to the sides to prevent the component from falling out.
4. Place the board so that the side you are soldering faces you. Apply the tip of the soldering pencil against both the component lead and the "pad" around the hole.
5. Wait 3 to 5 seconds, and then touch the end of the solder against the opposite side of the solder pad. The solder should begin to flow into the pad and around the component lead (Figure 21-9). Allow just enough solder to flow into the joint, then immediately remove both the solder and soldering pencil.
6. If the solder does not flow, wait a few seconds more and try again. If it's still not flowing, the soldering pencil may not yet be hot enough. Remove the tip from the work, and wait another minute for the soldering pencil to reach proper operating temperature.
7. It takes several seconds for the solder to cool. During this time, be absolutely sure not to disturb the solder joint, or it could be ruined.

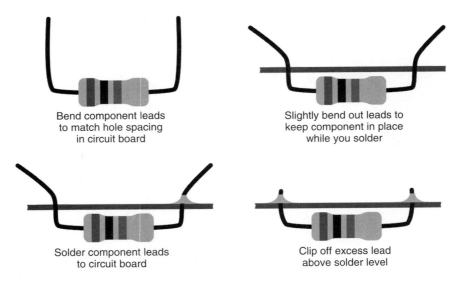

Bend component leads
to match hole spacing
in circuit board

Slightly bend out leads to
keep component in place
while you solder

Solder component leads
to circuit board

Clip off excess lead
above solder level

Figure 21-8   Process of soldering a component to a circuit board. Begin by bending the wire leads. Apply solder to one pad and lead at a time. Clip off any excess lead above the solder level. (A short "whisker" jutting out is fine.)

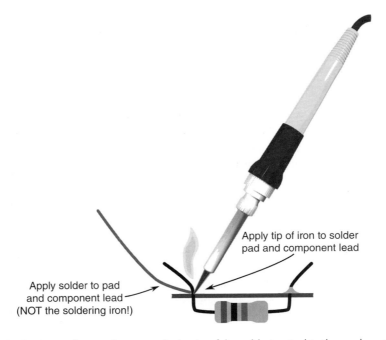

Apply tip of iron to solder
pad and component lead

Apply solder to pad
and component lead
(NOT the soldering iron!)

Figure 21-9   Remember to always apply the tip of the soldering tool to the work, not the solder. (Exception: You may apply the solder directly to the tool when tinning the tip. Do this periodically after wiping off the tip against a damp sponge.)

Don't apply heat any longer than necessary. Prolonged heat can permanently ruin electronic components. A good rule of thumb is that if the soldering tool is on any one spot for more than 5 seconds, it's too long.

Finishing Up

When soldering on printed circuit boards, you'll need to clip off the excess leads that protrude beyond the solder joint. Use a pair of diagonal or flush cutters for this task. You don't need (or want) to cut off any part of the solder, so don't be too aggressive with the nippers; it's okay if a little stubble of wire still sticks out.

Be sure to protect your eyes when cutting the lead; a bit of metal could fly off and lodge in an eye. Not fun.

Tips for Better Soldering

- If at all possible, you should keep the tool at a 30° to 40° angle for best results. Most tips are beveled for this purpose.
- Apply only as much solder to the joint as is required to coat the lead and circuit board pad. A heavy-handed soldering job may lead to solder bridges, which is when one joint melds with joints around it. That can cause a short circuit.
- When the joint is complete and has cooled, test it to make sure it is secure. Wiggle the component to see if the joint is solid.
- Be sure to insert the component completely through the circuit board, and check that it isn't crooked (see Figure 21-10). Excessive component lead length can cause short circuits if they touch other exposed parts of the board.

SOLDER TIP MAINTENANCE AND CLEANUP

As the soldering tool comes to temperature, clean off the tip by wiping it against a damp sponge. As you work, periodically repeat this step to keep the tip free of excess solder.

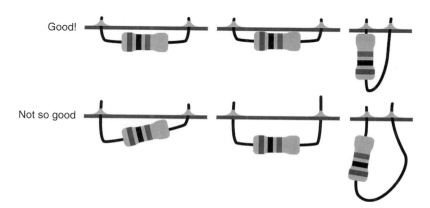

Figure 21-10   Good and not-so-good examples of soldering. When soldering components to a circuit board, be sure the parts are in fully and straight.

The tip should always be *tinned,* meaning it should have a very light coat of solder on it. Tinning involves cleaning off the tip with the damp sponge, then directly applying a bit of solder to the tip (this is one instance when applying the soldering tool directly to the solder is allowed). Remove any excess solder by wiping again with the damp sponge.

When the job is done let the tool cool down for at least 10 minutes before putting it away. After many hours of use, the soldering tip will become old, pitted, and deformed. This is a good time to replace the tip. Old or damaged tips impair the transfer of heat, and that can lead to poor soldering joints. Be sure to replace the tip with one made specifically for your soldering tool. Different brands of tips are generally not interchangeable.

CHAPTER 22

Common Electronic Components for Robotics

Components are the things that make your electronic projects tick. Any given robot project might contain a dozen or more electronic components of varying types, including resistors, capacitors, integrated circuits, and light-emitting diodes.

In this chapter, you'll learn about the components commonly found in electronics for robotics, and the roles they play in making the circuits work. We've got a lot of ground to cover, so let's get started.

But First, a Word About Electronics Symbols

Electronics use a kind of specialized road map to tell you what components are being used in a circuit and how they are connected together. This pictorial road map is called the *schematic*; it is a blueprint that tells you just about everything you need to know to build the circuit.

Schematics are composed of special symbols that are connected with intersecting lines. The symbols represent individual components, and the lines are the wires that connect these components together. The language of schematics, while far from universal, is intended to enable most anyone to duplicate the construction of a circuit with little more information than a picture.

Forget what you might have read in some of the other books—learning how to read a schematic isn't hard. Especially as it relates to robotics, all it takes is learning the meaning of a few basic symbols.

In this chapter you're introduced to the symbol for each of the primary electronic components. These symbols are combined with those shown in Figure 22-1 that show how the various components are wired together. Purely for example purposes, Figure 22-2 shows how some symbols are interconnected to build a simple circuit—push the switch and the light-emitting diode turns on.

There are hundreds of different electronic components, but only a small handful see regular use in electronic circuits for robotics. So instead of listing them all, I'll cover only those that you're most likely to encounter and use.

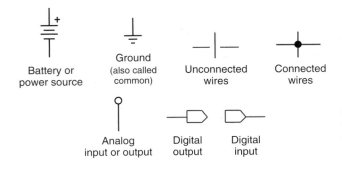

Battery or power source

Ground (also called common)

Unconnected wires

Connected wires

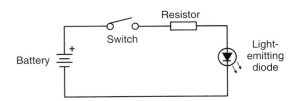

Analog input or output

Digital output

Digital input

Figure 22-1 Basic schematic symbols used for wiring circuits. There are variations of those shown here, but these are the wiring symbols used throughout this book.

Resistor

Switch

Battery

Light-emitting diode

Figure 22-2 Example of a basic circuit, shown in schematic form. It consists of a battery, switch, resistor, and light-emitting diode. In many instances, the symbols are self-explanatory or pictorially descriptive.

Fixed Resistors

Besides wire, *resistors* are the most basic of all electronic components. A resistor opposes the passing of current through it. It's like squeezing down a rubber hose to keep water from flowing through. Fixed resistors (there are also variable resistors, discussed next) apply a predetermined resistance to a circuit.

By using resistors of different values in a circuit, different parts get varying amounts of current. The careful balance of current is what makes the circuit work.

HOW RESISTORS ARE RATED

The standard unit of value of a resistor is the *ohm*, represented by the symbol Ω. The higher the ohm value, the more resistance the component provides to the circuit. The value on most fixed resistors is identified by color-coded bands, as shown in Figure 22-3 (you can't see the colors here, but you get the idea). The color coding starts near the edge of the resistor and comprises four, five, and sometimes six bands of different colors. Most off-the-shelf resistors for amateur projects use standard four-band color coding. They're the easiest to find, and the least expensive.

To read the value, start with the band closest to the edge and decode the colors using Table 22-1.

The first and second colors are the first and second *digits* of the value; the third color is the *multiplier,* the number of zeros you need to add. For example, if the first color is brown, the second color is red, and the third color is orange:

> Brown = 1
> Red = 2
> Orange = Multiply by 1000 (the same as simply adding three zeros)

which gives you 12,000.

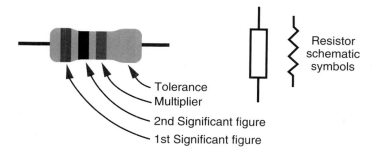

Resistor schematic symbols

- Tolerance
- Multiplier
- 2nd Significant figure
- 1st Significant figure

Figure 22-3 Component outline and schematic symbols for a resistor. In this book the hollow rectangle is used for resistors, but you may encounter the "sawtooth" symbol used in schematics elsewhere.

Table 22-1	Resistor Color Code Chart			
Color	**1st Digit**	**2nd Digit**	**Multiplier**	**Tolerance**
Black	0	0	1	—
Brown	1	1	10	±1%
Red	2	2	100	±2%
Orange	3	3	1,000	±3%
Yellow	4	4	10,000	±4%
Green	5	5	100,000	±0.5%
Blue	6	6	1,000,000	±0.25%
Violet	7	7	10,000,000	±0.1%
Gray	8	8	100,000,000	±0.05%
White	9	9	—	—
Gold			0.1	±5%
Silver			0.01	±10%
None				±20%

The fourth band in a four-band resistor is the *tolerance*, which is the amount (in percentage form) that the resistor may actually vary from its printed value. For example, a silver band means the resistor has a 10 percent tolerance; assuming a value of 12,000, 10 percent is 1200. That means the actual value of the resistor can be anything between 10,800 and 13,200.

Note the ± in front of the tolerance values in Table 22-1. That means + or – the indicated percentage. It can go both ways.

So-called precision resistors have a tolerance of less than 1 percent. Generally, these have five or more bands. The color coding is the same, but the extra bands provide more accurate numbers—for most projects they aren't needed.

The ohm value of resistors can vary from very low to very high. To make it easier to notate higher resistance values, resistors employ a common shorthand.

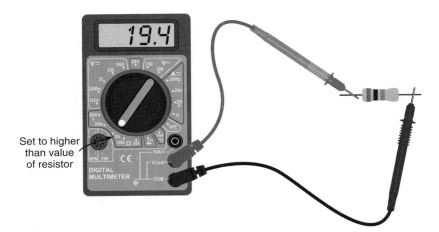

Set to higher
than value
of resistor

Figure 22-4 How to check the value of a resistor using a multimeter. Dial the meter to read ohms, and, if the meter is not autoranging, select a range just higher than the expected resistance value.

- The letter k (or K) is used to denote 1000. So a resistor with a value of 5 kΩ is the same as 5000 Ω. Sometimes the Ω symbol is dropped, because it's understood. So you may also see the resistor noted with just *5k*.
- The letter M is used to denote 1 million, 1,000,000. A resistor with a value of 2 MΩ has a value of 2 million (2,000,000) ohms, or 2 *megohms* for short.
- Resistor notations for decimal values can vary depending on the country of origin. In the United States, a 4.7-ohm resistor is notated simply as 4.7 Ω. But some countries, like the United Kingdom and Australia, often use a different system, where the letter *R* replaces the decimal point, as in *4R7*. Similarly, a 4.7-kΩ resistor is shown as *4k7*.

Resistors are also rated by their *wattage*. The wattage of a resistor indicates the amount of power it can safely pass through its body without burning up—the correct term for this is *power dissipation*. Resistors used in high-load, high-current applications, like motor control, require higher wattages than those used in low-current applications. The majority of resistors you'll use for hobby electronics will be rated at 1/4 or even 1/8 of a watt.

 Check out the RBB Support Site for a colorized resistor color chart, as well as a handy color code calculator.

TESTING THE VALUE OF A RESISTOR

You can readily test the value of any resistor, as shown in Figure 22-4.

1. Dial the multimeter to read ohms. If your meter is not autoranging, select a maximum range just above the marked value of the resistor. (If you don't know the value, select a high-resistance range to start.)
2. Connect the black (– or COM) lead to one end of the resistor; connect the red lead to the other end.

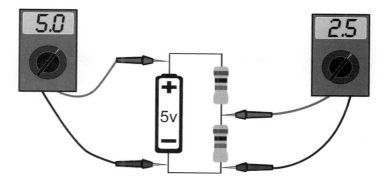

Figure 22-5   Resistors are commonly used to alter the voltage levels in a circuit. Two resistors connected in series as shown form a voltage divider. The actual voltage between the resistors depends on the values of the resistors. The readings shown here assume two resistors of the same value.

3. Read the result on the multimeter. If not using an autoranging meter, try a lower-resistance range to improve the accuracy of the measurement. If the meter shows over range (*Over range* indication, or the meter flashes 1.---) go back up one range.

COMMON APPLICATIONS FOR RESISTORS

Of the myriad uses of resistors in electronic circuits, two stand out as among the most common. We'll concentrate on those.

Using Resistors to Divide a Voltage

Remember that a resistor literally resists current flowing through it. In a working circuit this feature can be used to control voltage, since current, voltage, and resistance are all related—learn more about this later in the section "Understanding Ohm's Law."

Picture two resistors strung together like that in Figure 22-5. This is called a *series* connection, because the two resistors are in series with one another. (If they were side by side in the circuit, they'd be said to be in *parallel*. We don't need to investigate this connection scheme right now, so we'll move on.)

The circuit shown is powered by a 5-volt supply. The voltage at the point where the two resistors are connected in the middle will be somewhere between 0 and 5 volts. Exactly what that voltage is depends on the values of the resistors.

- If both resistors are of equal value, the voltage at the center is exactly one-half the supply voltage, or 2.5 volts.
- If the resistors are not the same value, the voltage is a ratio of the difference of their resistance. For example, if the top resistor is 5 kΩ and the bottom resistor is 10 kΩ, the voltage at the center is 3.33 volts.

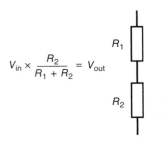

$$V_{in} \times \frac{R_2}{R_1 + R_2} = V_{out}$$

Figure 22-6 The basic (and simple) formula for calculating the divided voltage, when two resistors of unequal value are wired in series.

How do you come up with these voltages, other than testing them with a multimeter each time? All it takes is some simple math. See Figure 22-6; the top resistor is referred to as R_1, and the bottom is R_2.

$$V_{in} \times \frac{R_2}{R_1 + R_2} = V_{out}$$

Let's test this formula by plugging in the 5-kΩ and 10-kΩ values of the resistors and the 5 volts from the power supply. The formula becomes:

$$5 \times \frac{10,000}{15,000} = 3.33$$

or, to simplify:

$$5 \times 0.66 = 3.3$$

Using Resistors to Limit Current

Many electronic components, notably light-emitting diodes and transistors, will suck up as much current as the power supply will provide. This is bad because these components will burn out if they receive too much current. They're made to handle only a certain amount of current, and beyond that they are permanently damaged.

By stringing a resistor in series with these other components you can limit the amount of current they receive. This, after all, is the main purpose of a resistor . . . to resist current.

Figure 22-7 shows a very typical wiring diagram of a battery illuminating a light-emitting diode, also known as an LED. To prevent the LED from frying because it's consuming too much current, a resistor is placed between it and the positive side of the battery. The circuit uses a resistor to limit the current. But how do we arrive at its value?

Again, all it takes is a little bit of math, plus knowing some things about the typical LED.

- First, most LEDs will burn out if they consume—also referred to as *draw*—more than about 30 milliamps (30 mA). So we want to make sure the LED gets less, and perhaps substantially less, than this amount of current. For example purposes, we wish to have the LED receive no more than 15 mA.
- Second, you need to know the *forward voltage drop* across the LED. This is literally the amount of voltage that is lost when current is passed through the component. The typical voltage drop of an LED is 1.5 to 3.0 volts, though this can vary pretty widely when you start using some of the specialty LEDs. For our purposes, we'll assume 2.0 volts for the drop.

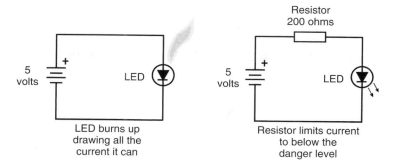

Figure 22-7   Another common use of resistors is to limit current; such a resistor is used to prevent a light-emitting diode (LED) from drawing too much current from its power source. Without a resistor to limit current, the LED will quickly burn out.

Apply this simple formula to determine the value of the resistor:

$$R = \frac{V_{in} - V_{drop}}{mA}$$

V_{in} is 5, and V_{drop} (the voltage drop) is 2.0. We want to limit current to 15 milliamps, so the formula becomes

$$200 = \frac{5 - 2.0}{0.015}$$

or

$$200 = \frac{3.0}{0.015}$$

Notice that the current draw for the LED is a decimal fraction; gotta do it this way because the formula assumes amps, not milliamps (there are 1000 milliamps in 1 amp).

Resistors with four color bands come in only specific standard values, and, as it happens, 200 ohms is a standard value (though not one stocked by all resellers). When your calculation results in a nonstandard resistor value, pick the next-higher standard value, such as 220 Ω. This way, the worst thing that'll happen is that the LED won't glow quite as brightly.

 In addition to calculating the resistance value, you often need to come up with the wattage value, too. For most circuits you'll be fine with the standard 1/8- and 1/4-watt resistors, but how do you know if you need a bigger wattage? It's easy when you use Ohm's law, presented next.

UNDERSTANDING OHM'S LAW

In the early 1800s, German physicist Georg Ohm experimented with the relationships between voltage, current, and resistance. He came up with a method of accurately calculating these relationships, and this became Ohm's law.

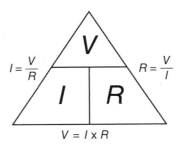

Figure 22-8 The Ohms law triangle, a mnemonic diagram that shows how to calculate the missing value when two other values are known. You can calculate for resistance, voltage, or current (current is referenced as the letter I).

Figure 22-8 shows the basic triad that makes up the law—it's called the Ohm's law triangle. It indicates the math you do to calculate one value when you know the other two. In all cases, you use either multiplication or division, also shown in the figure.

V stands for voltage (note: in some texts describing Ohm's law, voltage is shown as E).
R stands for resistance.
I stands for current (it's not C, as that stands for the capacitance of a capacitor, as in flux capacitor).

Example Ohm's Law Calculation

Let's just take one of the formulas, the one for calculating V (voltage). For that, you need to know two of the other elements of the triangle, I (current, in amps) and R (resistance). The formula is:

$$V = I \times R$$

Suppose the current is 1.2 amps, and the resistance is 50 ohms. Simply multiply 1.2 times 50; the result is 60, for 60 volts.

Calculating Power (Watts)

You can use an extension of Ohm's law to calculate power dissipation in a circuit. This is helpful to ensure that the wattage of the resistors you choose is high enough. Higher-wattage resistors are bigger and can handle more power passing through them.

The extension isn't part of the simple Ohm's law triangle, but is a more complicated variation involving a wheel.

Let's return to the example of selecting a resistor to limit current to an LED. The formula to calculate power is very simple:

$$V \times I = W$$

- V = voltage through the LED. This includes the voltage drop through the LED; from the previous example this voltage is 3.0.
- I = current to the LED. From the previous example this is 15 mA, or .015.
- W is the power dissipation, in watts. The resistor needs to be rated to dissipate at least this amount of power.

Substituting the actual values, you get

$$3.0 \times 0.015 = 0.045$$

The answer is in whole watts; 0.045 is less than 1/20 of a watt, so the standard 1/8-watt resistor is more than enough.

Potentiometers

Potentiometers are technically variable resistors. They let you "dial in" a specific resistance. The actual range of resistance is determined by the upward value of the potentiometer, and this upward value is how the potentiometer is marked. As with fixed resistors, the values are in ohms. For example, a 50-kΩ potentiometer will let you dial in any resistance from 0 to 50,000 ohms.

Potentiometers (or *pots* for short) are of either the dial or the slide type. The dial type is the most familiar and is used in such applications as radio volume and electric blanket thermostat controls. The rotation of the dial is nearly 360°. In one extreme, the resistance through the potentiometer is zero; in the other extreme, the resistance is the maximum value of the component.

LINEAR OR AUDIO TAPER

As you turn the dial of a potentiometer, the resistance varies from the lower extreme—usually 0 ohms, or very close to it—to the indicated value of the pot. The scale of the change is dependent on the internal construction of the component. There are two scales: linear and audio. The scale is referred to as the *taper* of the potentiometer.

- With *linear taper,* the most common, the value changes in proportion to the setting of the dial. For example, with a 10-kΩ pot, turning it a quarter of the way will yield 1/4 of the full scale, or 2.5 kΩ. For nearly everything you do in robotics you'll want a linear taper potentiometer.
- With audio taper (also called *log* or *logarithmic taper*), the value of the potentiometer is a logarithmic function of the position of the dial. Given a 10-kΩ pot, the component still varies from 0 Ω to 10 kΩ; however, the change is not a straight line but a curve that's especially steep. Audio taper pots are a fairly common find in the surplus market. You don't want one of these unless you're working on an audio project.

USING A POTENTIOMETER

Most pots have three connections (see Figure 22-9), which basically form two resistors in series. In fact, potentiometers behave just like two resistors in series, and they can be used for the same kinds of things; for example, as voltage dividers. The ratio of the values of the two resistors used in the divider determine the voltage.

As shown in Figure 22-10, the two terminals on either side of the potentiometer function like the top of the fixed resistor R_1 in Figure 22-6, and the bottom of the fixed resistor R_2. The center terminal, called the *wiper*, is the connection between R_1 and R_2. As you turn the dial of the pot, you vary the ratio between the two resistances.

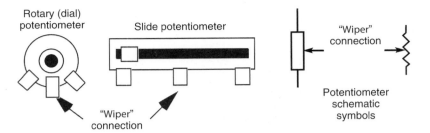

Figure 22-9 Component outline and schematic symbols for a potentiometer (or "pot" for short). The wiper is the center connection of the pot. It's the wiper connection that provides the varying resistance value.

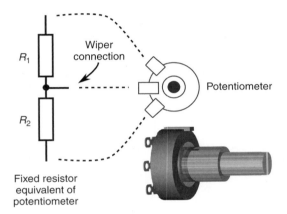

Figure 22-10 A potentiometer is basically two resistors wired in series, like that in Figure 22-6. Except in a pot, the values of the two resistors are constantly changing as you rotate the dial.

You can quickly test the operation of a potentiometer by connecting it to a multimeter (see Figure 22-11).

1. Dial the multimeter to read ohms. If your meter is not autoranging, selecting a maximum Ω just above the marked value of the potentiometer.

2. Connect the black (– or COM) lead to the center wiper terminal of the pot, and connect the red lead to either of the end terminals.

3. Slowly rotate the pot in one direction or the other, and watch the resistance go up and down.

OTHER TYPES OF VARIABLE RESISTORS

Potentiometers are the most common type of variable resistor, but there are several other kinds you'll encounter in your robot-building lifetime. The two most often used are photoresistors and the force-sensitive resistor.

● *Photoresistors* are sensitive to light. Their value changes as the intensity of the light varies. Photoresistors often go by the names *photocell* and *CdS cell*; the CdS comes from cadmium sulfide, the mixture of chemicals that make the component sensitive to light.

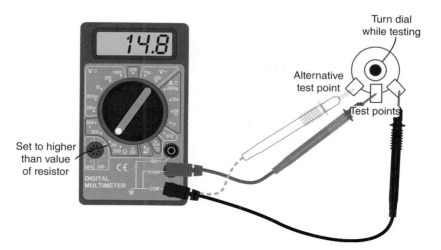

Figure 22-11   How to check the value of a potentiometer using a multimeter. The concept is the same as when testing a resistor. Turn the dial as you test to watch the resistance value change.

- *Force-sensitive resistors,* or FSRs, register force or pressure against them. The resistance of the component goes up or down as the force/pressure changes. There are numerous kinds of FSRs used for various kinds of sensing jobs. For example, a flex sensor provides a varying resistance as it's twisted or bent. For others, resistance changes as any part of the membrane of the component is pressed against.

 Read more about photoresistors in Chapter 33, "Environment," where you'll learn how to use CdS cells to make your robot respond to light. Be sure to check out Chapter 30, "Touch," for additional details about FSRs.

READING POTENTIOMETER MARKINGS

Unlike fixed resistors, potentiometers aren't marked with a color code. Instead, their value is marked either directly—for example, 10,000 or 10K—or indirectly using a *decade* numbering system. In this system the value is a three-digit number such as *503,* which means 50, followed by three 0s, or 50,000. The number is given in ohms, meaning the pot is 50 kΩ.

Capacitors

After resistors, *capacitors* are the second most common component found in the average robotics electronic project. Capacitors serve many purposes. They can be used to delay the action of some portion of the circuit or to remove bothersome electrical noise within a circuit. These and other applications depend on the ability of the capacitor to hold an electrical charge for a predetermined period of time.

Capacitors come in many more sizes, shapes, and varieties than resistors, though only a small handful are truly common. However, most all capacitors are made of the same basic stuff: two or more conductive elements are separated by an insulating material called the *dielectric* (see Figure 22-12).

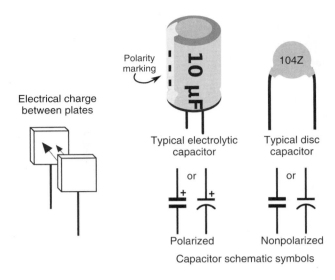

Polarity marking

Electrical charge between plates

Typical electrolytic capacitor

Typical disc capacitor

Polarized   Nonpolarized

Capacitor schematic symbols

Figure 22-12   Component outline and schematic symbols for two popular styles of capacitors. Capacitors may be polarized or nonpolarized. When using polarized capacitors, there will be a polarity marking on the body of the component. Be sure to properly orient the capacitor in the circuit.

This dielectric can be composed of many materials, including air, paper foil, epoxy, plastic, even oil. When you select a capacitor for a particular job, you must generally also specify the dielectric. The most common are summarized in Table 22-3, later in this chapter, along with their common uses.

HOW CAPACITORS ARE RATED

Capacitors have two important ratings:

- *Capacitance.* Capacitance is the ability of the component to hold a charge. The larger the capacitance, the longer the charge is retained.
- *Dielectric breakdown voltage.* At higher voltages the dielectric becomes partially or completely electrically conductive and the capacitor no longer functions as it should. The capacitor must be used below this voltage.

Capacitance is measured in *farads*. The farad is a large unit of measurement, so the bulk of capacitors available today are rated in microfarads; one microfarad is a millionth of a farad.

When the capacitor is under 1 microfarad, its value may be shown as a decimal point number—for example, 0.1 for one-tenth of a microfarad. Or it may be shown as a nanofarad. A nanofarad is a thousandth of a microfarad—that 0.1-microfarad capacitor is instead listed as 100-nanofarad. Same value, different way of expressing it. An even smaller unit of measure is the picofarad, or a millionth of a microfarad.

The "micro-" in the term *microfarad* is most often represented by the Greek "mu" (μ) character, as in 10 μF, or 10 microfarads. Keeping up with the shorthand, the nanofarad is nF, and the picofarad is pF.

In older books and magazines on electronics you may see microfarad shorted to *mfd*. Example: *10 mfd* is 10 microfarads (10 μF). Mfd isn't used as much today, but it's good to know what it means in case you want to try out an older circuit you find at the library.

HOW CAPACITORS ARE MARKED

Capacitors are routinely marked with at least their capacitance value; many capacitors are also marked with value tolerance notation and a breakdown voltage. Additionally, capacitors that are polarized—they have a + and a − lead—also carry a polarization marking.

Capacitance Value

The capacitance values for some capacitors are printed directly on the component. This is true of larger capacitors with values of 1 µF or higher, if for no other reason than that their larger physical size allows the manufacturer to directly print the value on the component.

But for other capacitors, things aren't always so simple. Smaller capacitors often use a common three-digit *decade* marking system to denote capacitance. The numbering system is easy to use if you remember it's based on picofarads, not microfarads.

A number such as 104 means 10, followed by four zeros, as in

100,000

or 100,000 picofarads. To make the conversion, move the decimal point to the left six spaces: 100,000 becomes 0.1. Therefore, that 104 capacitor is 0.1µF, or 100 nF. Note that values under 1000 picofarads do not use this numbering system. Instead, the actual value, in picofarads, is listed, such as 10 (for 10 pF).

Table 22-2 provides a quick glance at how several common capacitor number markings convert to their µF (microfarad) equivalents.

 As with resistors, what's printed on a capacitor may not exactly match its real value. There are several ways to indicate capacitor tolerance. Rather than take up the space here, I've uploaded a guide on capacitor selection, "Understanding Capacitor Markings," to the RBB Support Site.

Dielectric Breakdown Voltage Value

The dielectric breakdown voltage is specified only for certain capacitors. For those that have it, the voltage is marked directly, such as "35" or "35V." Sometimes, the letters *WV* are used after

Table 22-2 Capacitor Value Reference			
Marking	**Value (µF)**	**Marking**	**Value (µF)**
xx (number from 01 to 99)	xx pF		
101	0.0001	331	0.00033
102	0.001	332	0.0033
103	0.01	333	0.033
104	0.1	334	0.33
221	0.00022	471	0.00047
222	0.0022	472	0.0047
223	0.022	473	0.047
224	0.22	474	0.47

the voltage rating. This indicates the *working voltage*. You should not use the capacitor in a circuit with a voltage that exceeds this value.

On capacitors without a breakdown voltage printed on them you must estimate the value based on the type of dielectric it uses. This is an advanced topic and not covered in this book; nevertheless, it seldom comes up in electronics for robotics because most circuits use 12 volts or less, and most capacitors have a rated breakdown voltage of 25 to 35 volts. As a safety margin, select a capacitor with a breakdown voltage at least double the operating voltage of the circuit.

Polarization Marking

Some capacitors are polarized. Markings on the capacitor indicate the + or the − terminal.

If a capacitor is polarized, it is *extremely* important that you follow the proper orientation when you install the capacitor in the circuit. If you reverse the leads to the capacitor—connecting the + side to the ground rail, for example—the capacitor may be ruined. Other components in the circuit could also be damaged.

 By convention, the polarizing mark on aluminum electrolytic capacitors is typically the − (negative) lead. The polarizing mark on tantalum electrolytic capacitors is typically the + (positive) lead.

UNDERSTANDING CAPACITOR DIELECTRIC MATERIAL

Capacitors are classified by the dielectric material they use. The most common dielectric materials are listed in Table 22-3. The dielectric material used in a capacitor partly determines which applications it should be used for.

Table 22-3   Capacitor Dielectric Materials (and Their Uses)

Dielectric	Typical Applications	Polarized*
Aluminum electrolytic	Power supply filtering; electrical noise filtering (decoupling)	Yes
Tantalum electrolytic	Same as aluminum electrolytic, but smaller and more compact; more expensive; breakdown voltages may not be as high	Yes
Ceramic	Most common; inexpensive, general purpose; often referred to simply as a "disc capacitor" due to its typical shape	No
Silver mica	High precision, typically used for high-frequency applications; expensive	No
Polyester (or Mylar) polypropylene	Low-power, low-frequency signal applications where ceramic capacitors are not adequate; more stable than ceramic capacitors	No
Paper foil	Your grandpa's Heathkit radio	No (but better ask Grandpa)

* Typical. There are exceptions to everything, including whether a capacitor is polarized. So be on the lookout for any polarization markings, or the lack thereof. For example, some aluminum electrolytic capacitors are nonpolarized. These are usually marked NP, and are designed for special applications, such as use in stereo speaker systems.

COMMON APPLICATIONS FOR CAPACITORS

Unlike resistors, for which it's easy to demonstrate practical applications in a circuit, capacitors are a bit more nebulous. They tend to work by interacting with other components, rather than doing things just on their own.

Capacitors are often used to filter, or remove, the rapid variations of an input voltage, leaving only a steady voltage. This is quite useful in all kinds of electronics, because some components produce large, instantaneous "glitches" of voltage. These glitches, referred to as power line noise, may disrupt neighboring components, especially integrated circuits.

Capacitors may also be used with resistors as part of a timing circuit. The value of the capacitor determines how long an event lasts—timing is controlled by how long it takes for the capacitor to charge or discharge its current. In all of the applications in this book involving the ubiquitous LM555 timer IC, for example, there's always a capacitor that—when combined with at least one resistor—determines the duration of the timer.

Diodes

The diode is a rudimentary form of semiconductor. Diodes (see Figure 22-13) are used in a variety of applications, and there are numerous subtypes. Here is a list of the most common that are used in the field of robotics:

- *Rectifier.* Let's call this the "average" diode. It's so called because one of its principal uses is to rectify AC current to provide DC only. It's used for many other things, too, and is frequently employed in robotics circuits for motor control applications.
- *Schottky.* Special kind of diode that has improved performance. Used for applications where speed, low voltage drop, and "snap action" are needed.
- *Zener.* This diode limits voltage to a predetermined level. Zeners are used for low-cost voltage regulation.
- *Photo.* All semiconductors are sensitive to light, but photodiodes are made especially for the task. Photodiodes and their close cousin, the phototransistor, are frequently used as light sensors in robotics projects.
- *Light-emitting.* These diodes emit infrared or visible light when current is applied.
- *Laser.* The now-common penlight lasers use a specially constructed diode that emits single-color laser light. They can be used for visual effects and some sensor applications.

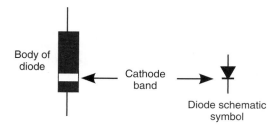

Body of diode

Cathode band

Diode schematic symbol

Figure 22-13 Component outline and schematic symbols for a diode. Diodes are polarized, as noted by a colored band. The band denotes the cathode, which is the negative (–) connection.

HOW DIODES ARE RATED

Diodes carry three important ratings: peak inverse voltage, current, and forward voltage drop (there are others, but these are the main ones).

- **The peak inverse voltage (PIV) rating** roughly indicates the maximum working voltage for the diode. In the case of the common rectifier diodes, this maximum voltage is 50, 100, even 1000 volts, depending on the component. These voltages are well beyond what you'll typically find in electronics for a robot, giving you a wide variety of components to choose from.
- **The current rating** is the maximum amount of current that can be passed through the diode without risk of overheating and eventual self-destruction. Current ratings are in amps or, in the case of small diodes, milliamps. This rating is very critical in robotics applications, where it's common to have circuits that require more current than what the average diode can handle.
- **Forward voltage drop** is the amount of voltage that is essentially lost when it passes through the diode. The voltage drop affects the overall behavior of the diode; for example, with a very low drop, the diode is faster-acting.

UNDERSTANDING DIODE POLARIZATION

All diodes are polarized, and they have positive and negative terminals. The positive terminal is called the *anode,* and the negative terminal is called the *cathode.*

You can readily identify the cathode end of a diode by looking for a colored band or stripe near one of the leads. Figure 22-13 shows a diode that has a band at the cathode end. Note how the band corresponds with the heavy line in the schematic symbol for the diode.

EXPLORING THE COMPOSITION OF DIODES

Among other variations, diodes are available in two basic flavors, germanium and silicon, which indicate the material used to manufacture the active junction (the part that conducts current) within the diode. The two types of materials also have an effect on the forward voltage drop of the device: about 0.7 volts for silicon and 0.3 volts for germanium.

Since the voltage drop can change the behavior of the circuit, you should always be careful which of these two kinds you use. If the circuit doesn't say, a silicon diode is probably fine; if a germanium component is called for, be sure not to substitute a silicon type.

COMMON APPLICATIONS FOR DIODES

Diodes are used for a fantastically wide array of applications. The most typical applications rely on a diode acting as a kind of "electronic check value," allowing voltage to flow through the device in one direction only. Two such applications are blocking an AC voltage and providing protection against reverse polarity.

Block AC Voltage

Diodes pass voltage in one direction only: from anode to cathode. This means if a voltage contains an alternating current (AC), the diode will block the current trying to sneak through the diode from the other direction. In an AC signal the polarity reverses from positive to negative, and back again.

In electronics terms this is called rectification. One application of rectification is to prevent an AC signal from passing through a circuit, a common requirement of power supplies for electronics.

Reverse Polarity Protection

Connecting a battery backward to an electronic circuit can easily damage the circuit. You can prevent such damage by putting a diode in series with the positive power supply connection. This technique works because the diode passes current in one direction only: from anode to cathode. It won't allow current to flow the other way.

Light-Emitting Diodes (LEDs)

All semiconductors emit light when an electric current is applied to them. This light is generally very dim and only in the infrared region of the electromagnetic spectrum. The *light-emitting diode* (LED; see Figure 22-14) is a special type of semiconductor that is expressly designed to emit copious amounts of light. Most LEDs are engineered to produce a specific color of light, as well as infrared and ultraviolet. Red, yellow, and green LEDs are among the most common, but blue, violet, and even white light (all-color) LEDs are available.

LEDs carry the same specifications as any other diode. The *typical* LED has a maximum current rating of 30 milliamps or less, though this varies greatly, and depends on size, type, even color. And like a diode, all LEDs exhibit a forward voltage drop—only it's often much higher than that of a standard diode. Depending on the LED (and often related to its color), expect a voltage drop of between 1.5 and 3.5 volts. Some specialty high-brightness LEDs have even higher drops.

And, as with a standard diode, the terminals on an LED are its anode (+) and cathode (–). Rather than a white or colored stripe, most LEDs distinguish the two using other methods as noted in Figure 22-14. Not all LEDs follow the same marking conventions, so you may need to experiment. (Usually nothing bad happens if you connect an LED backward—that is, if you switch the anode and the cathode—but the LED will not light.)

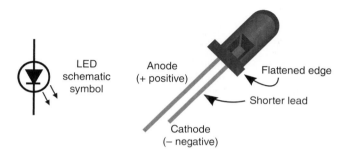

Figure 22-14   Component outline and schematic symbols for a light-emitting diode (LED). Diodes are polarized, and, depending on the component, the polarization may be noted by a flattened edge and/or a shorter lead. These denote the cathode, which is the negative (–) connection.

POWERING LEDs

LEDs are most often used in low-power DC circuits and are powered with 12 volts or less. Always remember that the component can be ruthlessly damaged if you expose it to currents exceeding its maximum rating. So, unless the LED has a built-in resistor, you always need to add one to limit the amount of current that flows through the LED. See the application examples for the resistor, mentioned previously in the chapter, which demonstrate this process.

SHAPE, SIZE, AND LIGHT OUTPUT

Light-emitting diodes come in all kinds of shapes and sizes. The most common are cylindrical and shaped with a domed top. Popular sizes are:

- T1, or miniature: 3mm in diameter
- T1-3/4, or standard: 5mm
- Jumbo: 10mm

While the most common LED is round, there are also square, rectangular, even triangular LEDs. The shapes are handy for different kinds of applications—the triangles look like arrows, so they can be used to show direction.

LED Colors

Most LEDs emit a single color, but others are designed to produce two or three colors. You can control which color is shown by applying current to various terminals on the LED (Figure 22-15).

- A two color (or *bicolor*) LED contains red and green LED elements (other colors are possible too, but the red/green combination is the most common). You control which color is shown by reversing the voltage to the LED. You can also produce a yellowish-orange by quickly alternating the voltage polarity.
- A *tricolor* LED is functionally identical to the bicolor LED, except that it has separate connections for the red and green diodes. You can produce the intermediate color of yellowish-orange by turning on both the red and green diodes at the same time.
- A *multicolor* LED contains red, green, and blue LED elements. You control which color to show by individually applying current to separate terminals on the LED. These are also called RGB LEDs.

There's a bit of confusion as to what, exactly, constitutes bicolor and tricolor LEDs. Both can produce three colors: red, green, and yellowish-orange. The way they do it differs. Compounding the confusion is that some sources call a multicolor LED a tricolor LED, because it (rightly) contains three colors.

Common Anode or Cathode

To reduce the number of terminals coming out of the multidiode LED, all of the anodes, or all of the cathodes, of each diode in the device are wired together. When all the anodes are combined, the LED is said to be *common anode*. And when all of the cathodes are combined, it's *common cathode*.

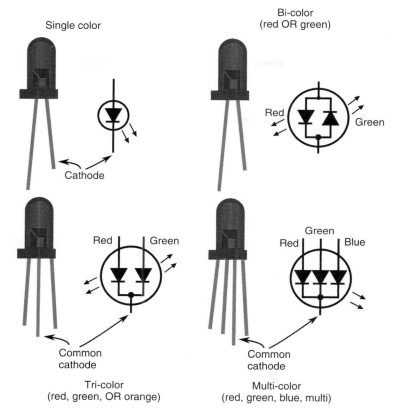

Figure 22-15 LEDs that display more than one color have additional connection points. The wiring of the connections depends on the type of LED. Shown here are several of the most common.

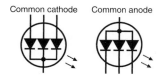

Which one you choose depends on what's available and how the circuit is designed. Common cathode is more common among multiple-color LEDs; both common anode and common cathode are used in 7-segment and other LED displays. Be sure to match the device with the circuit plans you're using.

MULTI-LED DISPLAYS

LEDs can come one in a package or as part of a larger package with other LEDs. Each individual LED in the package can be individually lit. Three common variations on the multi-LED theme are:

> **7-segment display.** Has seven individual LEDs in special shapes to form a large numeral. By controlling which LEDs are on and off, the display can show numbers 0 through 9.
> **Bar graph display.** Typically contains 10 miniature rectangular LEDs.
> **Dot matrix display.** Contains rows and columns of dots; any number, letter, or special character can be created by lighting up certain dots.

Small signal
transistor

Transistor
schematic symbol
(NPN shown here)

Figure 22-16 Component outline and schematic symbols for a transistor. The outline view is of a small signal transistor.

ADDRESSABLE MULTICOLOR LEDs

There's a new genre of light-emitting diode (LED) that makes child play out of outfitting your robot with pulsating special effects. You get all the colors of the rainbow, with varying intensity no less. What's more, you can string multiple LEDs on a single microcontroller pin, and operate all of them at once.

The LEDs are *addressable,* meaning that in your program code you specify which light in the string you want to control. Only the intended recipient of the message obeys the lighting command; all the others ignore it, and continue with their last instruction. Read more information on addressable LEDs in Chapter 36, "Visual Feedback from Your Robot."

Transistors

Transistors (see Figure 22-16) were designed as an alternative to the old vacuum tube. They're used for many of the same things, either to amplify a signal or to switch a signal on and off. At last count there were *several thousand* different types of transistors available.

Transistors are divided into two broad categories: signal and power. You can usually tell the difference between the two merely by size.

- *Signal.* These transistors are used with relatively low-current circuits, like radios, telephones, and some hobby electronics projects. They are available in either plastic or retro-looking metal cases. The plastic kind is suitable for most uses, but some precision applications require the metal variety. No projects in this book require great precision, so the cheap plastic signal transistor is good enough.
- *Power.* These transistors are used with high-current circuits, like motor drivers and power supplies. Power transistors come in metal cases, though a portion of the case (the back or sides) may be made of plastic.

Figure 22-17 shows several common varieties of transistor cases and how they are referred to, such as *TO-92* and *TO-220.* The signal transistor is rarely larger than a pea and uses slender wire leads. The power transistor has a large metal case, to help dissipate heat, and heavy, spoke-like connection leads.

TO-92   TO-5 or TO-18

Signal

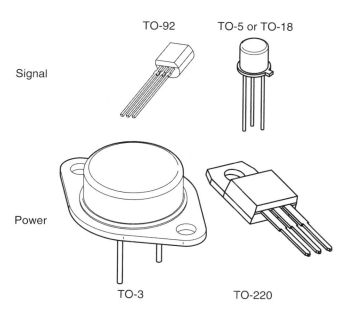

Power

TO-3   TO-220

Figure 22-17   There are many shapes and sizes of transistors, but a common trait of most of them is that they have three connection wires. Larger transistors are generally used for applications involving high currents, such as powering motors. Shown are four popular transistor cases.

HOW TO IDENTIFY A TRANSISTOR

Transistors are identified by a unique code, such as 2N2222 or MPS6519. Refer to a data book to ascertain the characteristics and ratings of the particular transistor you are interested in. Transistors are rated by a number of criteria, which are beyond the scope of this book. None of these ratings are printed directly on the transistor.

Transistors have three- or four-wire leads. The leads in the typical three-lead transistor are *base, emitter,* and *collector*. (A few transistors, most notably certain types of the field-effect transistor—or FET, for short—have a fourth lead. None of the circuits in this book use this type, which are not common anyway.)

NPN, PNP—TWO SIDES OF THE SAME COIN

Transistors can be either NPN or PNP devices. This nomenclature refers to the sandwiching of semiconductor materials inside the device. You can't tell the difference between an NPN and a PNP transistor just by looking at them.

NPN
transistor

PNP
transistor

However, the difference is indicated in the catalog specifications sheet as well as by the schematic symbol for the transistor. In an NPN device, the arrow is shown leaving the transistor; in a PNP device, it's the opposite. This differentiation helps you to quickly tell whether you should use an NPN or a PNP type transistor for the circuit.

ENTER THE MOSFET

So far I've been talking about a class of transistor called the *bipolar junction transistor,* or *BJT*. These are by far the most common. Another form of transistor is the MOSFET. This collection of alphabet soup stands for *metal-oxide semiconductor field-effect transistor*. It's often used in circuits that demand high current and high precision.

N-channel   P-channel
MOSFET   MOSFET

MOSFET transistors don't use the standard base-emitter-collector connections you just read about. Instead, they call them *gate, drain,* and *source* connections. And note, too, that the schematic diagram for the MOSFET is different from that of the standard transistor.

Like the bipolar junction transistor, MOSFETs come in two varieties: N-channel and P-channel. And, as before, you can't tell the difference between an N-channel and a P-channel MOSFET just by looking at it.

 The schematic symbols for the two types of MOSFETs are only ever-so-slightly different, and not all schematic diagrams bother to show which one is used in the circuit. You *cannot* substitute N-channel for P-channel, so when using these devices, be absolutely sure you've got the right ones in your hands.

Integrated Circuits

The *integrated circuit,* colloquially referred to as an *IC* or *chip,* forms the backbone of the electronics revolution. The typical integrated circuit comprises many transistors and other components. As its name implies, the integrated circuit is a discrete and wholly functioning circuit in its own right. ICs are the building blocks of larger circuits. By merely stringing them together you can form just about any project you envision.

Integrated circuits are enclosed in a variety of *packages.* The actual integrated circuit itself is just a tiny sliver inside this package. For the hobbyist, the easiest-to-use IC package is the *dual in-line package* (or *DIP*), like the one in Figure 22-18. The illustration shows an 8-pin DIP, but other sizes are common, too.

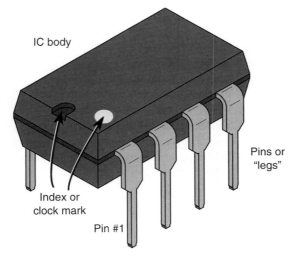

IC body

Pins or "legs"

Index or clock mark

Pin #1

Figure 22-18   Integrated circuits come in a plastic rectangular body, with four or more connection pins, or legs. A printed index mark or notch shows the "top" of the component. Pin 1 is always the first pin when going counterclockwise.

IDENTIFYING INTEGRATED CIRCUITS

As with transistors, ICs are identified by a unique code printed on the top. Codes may be a simple number sequence such as 7400 or 4017. This code indicates a type of device that is made by many different manufacturers. You can use this code to look up the specifications of the IC in a reference book.

More than likely, the number identifying the IC will contain letters that further distinguish it from other ICs of the same family, that is, ICs that do pretty much the same thing but have different operating characteristics or manufacturing technologies. The differences among these IC families are quite complex and well outside of what this book is about, but to get you started, let's take the 7400 as an example.

The 7400, which dates back to the swingin' 1960s, contains four digital NAND gates; NAND gates are one of several common forms of logic circuits used to create computers. Variations of this chip include:

- *7400*—Base chip as originally manufactured.
- *74ALS00*—Advanced low-power Schottky; enhanced lower-power version.
- *74HCT00*—CMOS version of the 7400 compatible with older devices.

And there are *many* others. When a circuit specifies just the base chip—the 7400 rather than a specific family member—it usually means that the circuit isn't particularly picky, and you can use most any IC in that family.

Many ICs also contain other written information, including manufacturer catalog number and date code. Do not confuse the date code or catalog number with the code used to identify the device. Date codes can look like part numbers—9206 might mean the IC was made in June 1992.

MICROCONTROLLERS AND OTHER SPECIALTY ICs

Chips like the 7400 (and all its kin) are considered standard building blocks and are available from a variety of manufacturers. Add to these the thousands of specialty ICs that are unique to specific chip makers and perform unique tasks.

For robotics, the most common of these specialty ICs is the microcontroller, a type of all-in-one computer that is designed to directly connect with external inputs and outputs. Microcontrollers are more fully detailed in Chapters 24 through 28.

Still other specialty ICs might perform any of the following tasks:

- Sense the tilt of your robot to let you know if it's fallen over (accelerometer)
- Detect small changes in temperature and convert this information to a signal that can be read by a microcontroller (temperature sensor)
- Convert one voltage to another, with very high efficiency (switching mode regulator)
- Create sound effects and intelligible speech (sound and speech synthesizer)
- Control the operation and speed of a DC motor (H-bridge motor driver)

and, of course, many others.

As with standard-building-block ICs, specialty chips carry markings that indicate the manufacturer name and part number, date code, and other important information.

Switches

Most robots use at least one switch—to turn it off or to reset the thing when its programming goes bonkers. A lot of robots use multiple switches, for things like setting operating modes or detecting when the bot has bumped into a wall, chair leg, or person.

Switches come in a variety of shapes and sizes, but their overall functionality is universal: switches are composed of two or more electrical *contacts*. In one position of the switch, the contacts are not connected, and current doesn't flow through the device. The switch is said to be *open*. In the other position, the contacts are connected together, and current flows. The switch is said to be *closed* (see Figure 22-19).

Poles. The most basic switch has a single pair—or *pole*—of contacts. But some switches are composed of several poles, so there are extra pairs of contacts—one switch functions as multiple switches all connected together. Multiple poles allow the switch to control multiple circuits at the same time.

Throws. The basic switch has two positions, on or off. But some switches have more positions. Each position is called a *throw*. Instead of just on or off, a two-throw switch is on-on. Each "on" can connect to a different part of the circuit. Most often, multiple throws are combined with multiple poles to provide all manner of choices.

Momentary position. Switches that are spring-loaded are called *momentary*, because when you release the switch it automatically goes back to its normal position. The most common type of momentary switch is the push button: press it, and the switch closes; release it, and the switch returns to open all on its own.

Switches are further defined by how they are operated. There are four principal types, and each is self-explanatory by its name: toggle, push-button, slide, and rotary.

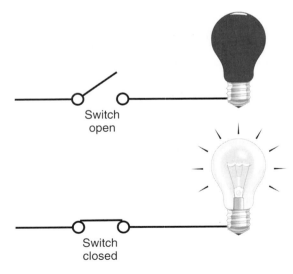

Switch
open

Switch
closed

Figure 22-19   Basic function of a switch: When open, the light is off (unpowered). When closed, the switch completes the circuit, and the light turns on.

MIXING AND MATCHING

Poles, throws, and momentary position can all be mixed together in various combinations. Switches follow a common nomenclature to describe how they function. Here are just a few of the combinations you'll likely encounter (see Figure 22-20).

- Single-pole, single-throw (SPST) has one set of poles and one throw. It's the basic, no-frills switch.
- Single-pole, double-throw (SPDT) has one set of poles and two throws.
- Double-pole, single-throw (DPST) has two set of poles and just one throw. You can think of it as two SPST switches combined into one.
- Double-pole, double-throw (DPDT) has two set of poles and two throws.

MOMENTARY AND CENTER-OFF

Add to the three main pole/throw combinations spring-loaded momentary and center-off positions. A DPDT switch with a center-off position is said to be on-off-on. One or both of the *on* positions may be momentary, that is, spring-loaded. Release the switch, and it returns to its off position.

When you read specifications about a switch, you may see something like this:

(on)-off-(on)

It describes a switch with a center-off position. The parentheses mean that the position is momentary.

NORMALLY OPEN, NORMALLY CLOSED

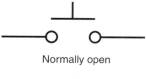

Normally open

Normally closed

The contacts of a momentary switch can be either normally open or normally closed. The "normally" has to do with the position of the contacts when the switch is not being depressed.

In a normally open (*NO* for short) switch, the circuit is normally broken. Depressing the push button closes the contacts. It's just the opposite with a normally closed (*NC*) switch: depressing the button opens the contacts.

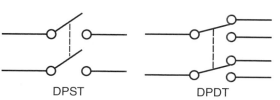

Figure 22-20   Common variations in switches indicate the number of poles and throws (positions). Common switches for use in robotics include single-pole, single-throw (SPST), and double-pole, double-throw (DPDT).

Relays

Relays work like switches, but they are changeable under electronic control. Instead of a human (or dog, or whatever) pressing a switch to change the contact settings, in a relay it's all done via electrical signals. Figure 22-21 shows the basic construction of a relay and how it works. As shown, the two basic parts of the relay are the *coil* and the *contacts*. In operation, when the coil is energized, the magnetic field that is created activates the switch contacts.

Because relays are essentially switches, they are defined in the same way: the poles and throws indicate the number of contacts. But unlike switches, most (but not all) relays are by their nature momentary. The switch contacts in a relay are activated when the relay is energized. Remove the juice from the relay, and the spring-loaded contacts go back to their original position.

COMMON RELAY TYPES

There are lots of different kinds and styles of relays out there, but most fall into one of the following three types (see Figure 22-22):

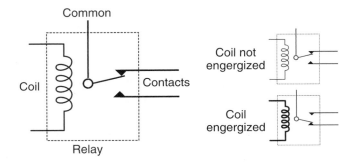

Figure 22-21   Relays are electrically controlled switches. Energizing a coil with current activates the switch contacts. Many relays have two contacts, in addition to a common connection. These contacts are marked normally open (NO) and normally closed (NC). The notation refers to when the relay is not energized.

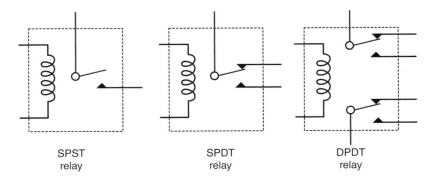

Figure 22-22   Three typical relay types used in robotics: single-pole, single-throw; single-pole, double-throw; and double-pole, double-throw.

- The SPST (single-pole, single-throw) relay has four connections: two for the coil and two for the switch contacts. The contacts may be normally open (the most common) or normally closed.
- The SPDT (single-pole, double-throw) relay has five connections: two for the coil, as usual, and three for the switch contacts. The contacts are often labeled as Common, NC, and NO. You wire your circuit to Common and either the NC or the NO connection.
- The DPDT (double-pole, double-throw) relays are an extension of the SPDT. They have a separate set of pole contacts, for a total of eight connections.

SIZES OF RELAYS

Relays are "rated" by the amount of current that can be passed through their switch contacts. The more current, the larger the relay. For the typical desktop robot application, the smallest of relays are ideal: these are about the same size and shape as an integrated circuit.

Larger currents (loads) require bigger contact points, so if you're operating a robot with a huge motor, you'll need a relay that can handle higher current. The current rating of the contacts are often noted on the body of the relay in amps and volts: *10 amps @ 125V,* for example. Such relays are often the domain of the big-brute combat robot, which is beyond the scope of this book, but there are guides and online sites that cover this topic. Do a Web search and visit your local library.

. . . And the Rest

Of course there are more kinds of electronic parts for your robots than the ones covered so far. But these are best left to examples in other chapters that show them in actual use. They include:

Speakers, to better hear your robot.

Microphones, to better your robot's hearing you (is that even a proper sentence?). See Chapter 35, "Producing Sound," for more on speakers and microphones.

Light-sensitive resistors, transistors, and diodes, to give your robot the gift of simple sight. These components are detailed more fully in Chapter 33, "Environment."

LCD displays, to let your robot communicate with you in words. Read more about how to use these in Chapter 36, "Visual Feedback from Your Robot."

. . . and more.

On the Web: Stocking Up on Parts

So what parts should you get for your robotics lab? How many resistors, and what values, should you stock up on? That can be a tough question if you're just starting out. The details would kill too many trees, so as a bonus to the readers of this book, I've outlined some suggestions (with suppliers and parts numbers) on the RBB Support Site.

23

Making Circuits

Y ou've built your robot, and it's ready to rumble. Well, almost! You still need to complete its electronics: brain, sensors, interface, display, and all the other things that make a full, complete bot.

Robot circuits take many forms, but they come down to two basic types: solderless breadboard and soldered circuit board. In this chapter you'll learn about both. You'll also learn how to make your own interconnecting cables to wire everything up, and tips on keeping things neat and tidy.

Using Solderless Breadboards

Solderless breadboards let you experiment with circuits without having to solder together components. You literally plug in the parts and string some wires to complete the connections. You save time, and you get to play with your creation that much sooner.

And, just as important, you can easily change the value of components to see if another one works better. Just unplug the old, and switch in the new.

The breadboard (see Figure 23-1) is made of plastic and is composed of many columns and rows. The columns are wired inside the board so that anything attached to that column shares a single electrical connection (see Figure 23-2). To use the board, you start by plugging in integrated circuits (ICs), resistors, and other components. You then establish how they're interconnected by stringing wires from one column to another.

While solderless breadboards are intended for experimentation and testing, there's no law that says you must eventually solder your circuit together to make it permanent. Many robot builders construct the solderless version only and use it that way for weeks, months, even years.

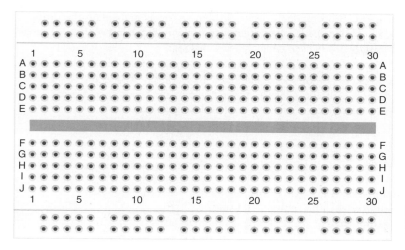

Figure 23-1 Solderless breadboard with 400 tie points.

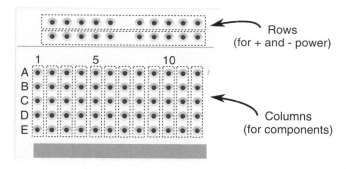

Rows
(for + and - power)

Columns
(for components)

Figure 23-2 Underneath the holes in the breadboard are strips of conductive metal with spring contacts. Plugging a wire into a hole pushes the wire into the contacts.

SIZES AND LAYOUTS

All solderless breadboards consist of holes with shared internal contacts. These contacts are spaced one-tenth of an inch apart—just right for ICs, most transistors, and discrete components like capacitors and resistors. The holes, called contact points or tie points, are the proper size for 23-gauge solid conductor wire. On each column you can connect together as many as five wires or components.

In addition to the column connections, many solderless breadboards have long rows on the top and bottom. These provide common tie points for the positive and negative power supply.

But from here, the size and layout of breadboards can differ greatly. Some measure a few inches square and contain only 170 holes. The 170-tie-point boards are made for simple circuits with one or two small ICs or a handful of small components. For bigger circuits you need a bigger breadboard. You can pick among models with 400, 800, 1200, even 3200 tie points.

Which size should you get? I believe in starting out small, the extra benefit being that small breadboards cost less. They're usually more than adequate for testing the typical robot control or sensor circuit. You can always get a larger board as your needs grow.

MAKING CONTACT

The contact points are usually made from a springy metal coated with nickel plating. The plating prevents the contacts from oxidizing, and the springiness of the metal allows you to use different-diameter wires and component leads without seriously deforming the contacts. However, the contacts can be damaged if you attempt to use wire that's larger than 20 gauge. (Remember: Wire gets bigger with *smaller* gauges.)

 The more you use your solderless breadboard, the faster it'll wear out. After a while, the springy metal isn't so springy. Dust can settle inside the contact points and decrease the electrical contact. This is another reason not to invest in a large and expensive breadboard. The cheap ones don't cost as much when they need to be replaced.

Things you should *not* try to plug into a solderless breadboard contact point include:

Stranded conductor wire, which won't work, even if you twist the strands.
Larger than 20-gauge wire or fat component leads.
Smaller than 26-gauge wire or really skinny component leads (the electrical contact will be iffy, at best).
High-voltage sources of any kind—these include wires from an AC wall socket or any circuit that uses high voltage. *Solderless breadboards are for low-voltage DC circuits only.*

CONNECTING WIRES FOR YOUR BREADBOARD

No solderless breadboard is complete without wires, but you can't use just any wire. The best wire for solderless breadboards is:

- 23-gauge
- Solid conductor
- Plastic insulated

You want wires of different lengths, with about 1/2" of the insulation stripped off each end. These *jumper wires* are available premade, or you can make them yourself. I prefer the premade kind. Get an assortment kit like the one in Figure 23-3, which contains jumper wires in various useful lengths.

Or if you decide to make your own set of breadboard wires, look for wire spool assortments with different colors. For starters, cut the wires into the following lengths:

Total Wire Length	Jumper Length	Quantity
1-1/4"	3/4"	10
1-1/2"	1"	15
1-3/4"	1-1/4"	15
2"	1-1/2"	15
2-1/2"	2"	10
3"	2-1/2"	10
3-1/2"	3"	5
4-1/2"	4"	5
6-1/2"	6"	5

Note: Jumper length assumes 1/2" of insulation is stripped off the wire on each end.

1. Start by cutting the wires to the total wire length, as indicated.
2. Use a pair of wire strippers to remove 1/2" of insulation off each end, as shown in Figure 23-4. While stripping the insulation, insert one end of the wire into the stripping tool (if it's adjustable, dial it for 23-gauge) and hold the other end with a pair of needle-nose pliers—the kind *without* serrated jaws is the best.
3. After stripping the insulation, use your needle-nose pliers to bend the exposed ends of the wire at 90°, as shown.

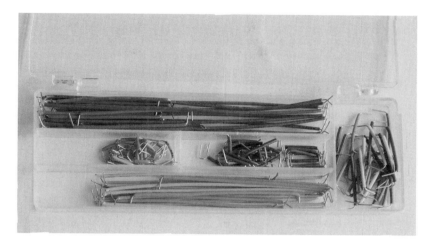

Figure 23-3   Save yourself some time and trouble by getting an assortment of premade breadboard jumper wires.

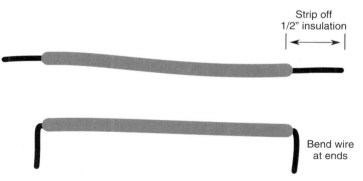

Figure 23-4   To make your own breadboard jumpers, begin by stripping off 1/2" of insulation from both ends, being careful not to nick the wire (that weakens it and can cause the wire to break off easily). Finish by bending the wires at the ends.

Steps in Constructing a Solderless Breadboard Circuit

Building a solderless breadboard project is merely a matter of placing the electronic components into the tie points and then completing the circuit using jumper wires. Start with a schematic or other plan, as shown in Figure 23-5.

1. Identify the components of the circuit and collect them before you begin. There's nothing worse than starting to build a circuit, only to discover you're missing some important part.
2. Visualize the best placement for the components. On simple circuits this is easy to do, as you'll have plenty of room. On more complex projects you'll want to carefully plan the best spot for each part, so that everything will fit.
3. Place the parts into the tie points of the breadboard starting from one end of the schematic and proceeding to the other. For the example circuit, the "ends" are the + and – power leads, but on others it could be a signal input or a sensor output. I've elected to start with the LED, as it's the largest of the components.
4. Finish the construction by adding any jumper wires that are needed, including the power leads.

 Though the breadboard is constructed with columns and rows at right angles, there's no rule that says you must place parts in such rigid rank-and-file order. Feel free to turn parts this way and that, to take best advantage of the tie points.

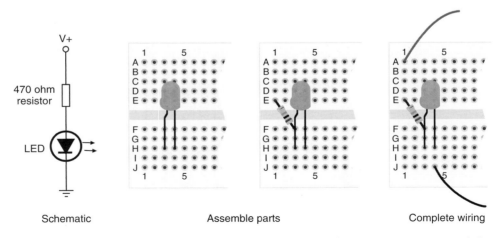

Schematic   Assemble parts   Complete wiring

Figure 23-5   Steps for assembling a solderless breadboard circuit from a schematic. Start with the largest components, and insert them into the breadboard. Add smaller components, then attach connecting wires.

Making Long-Lasting Solderless Circuits

Of course, you already know solderless breadboards are not "intended" for permanent circuits. But if your aim is to build a circuit on a solderless breadboard and keep it that way, you'll want to ensure a solid construction. Bird's-nest wiring and loose components just won't cut it.

Use a new board, one where the spring contacts aren't already getting loose. The low price of a small breadboard means you can keep a stock of these and use them as needed.

Firmly seat all components, including resistors, capacitors, and diodes. This means you need to trim the lead lengths so that the body of the component sits flush with the surface of the board.

Cut jumpers to length, and carefully route them around components and other jumpers so that none will accidentally pull out. Push the jumpers flush against the board.

 To help prevent parts and wires from coming loose, strap them in using some rubber bands. Avoid the use of things like cling plastic wrap for storing food. This generates lots of static electricity when you apply and remove it, possibly ruining your circuit.

Mounting the Breadboard to Your Robot

If you're using the breadboard directly on your robot, you'll want to securely mount it to keep it in place. Otherwise it may fall off the next time your bot does a pirouette turn, and all your hard work will be ruined as parts scatter onto the floor.

If the breadboard is readily accessible—so you can work on it—you can simply use double-sided foam tape to secure it to the robot (see Figure 23-6). Otherwise, use Velcro or other hook-and-loop material, so you can peel the board off to work on it.

When double-sided tape and Velcro won't work, use long cable ties to hold the board in place. Use two wraps, one at each end. Cinch the tie for a secure fit. If you need to later remove the board (so you can work on it on your bench), just cut the ties and discard. Use new cable ties when you're ready to remount the board on your robot.

Figure 23-6 A breadboard mounted on top of a robot, alongside a microcontroller. Jumper wires connect the microcontroller to the breadboard, which serves as a way to quickly and easily experiment with different robot sensors.

Some breadboards, such as the Global Specialties EXP-350, have mounting holes in the corners. The holes are usually pretty small, so you need 4-40 or maybe 2-56 miniature screws and nuts. The EXP-350 accepts 4-40 machine screws and nuts; use flathead screws for a flush look. Or it accepts 6-32 screws from the back, which tap straight into the plastic.

Tips for Using a Solderless Breadboard

Here are some tips for solderless breadboard success:

- When using static-sensitive components like CMOS integrated circuits or MOSFET transistors, build the rest of the circuit first, using "dummy" parts to make sure you wire everything properly. When you are ready to test the circuit, remove the dummy chip and replace it with the CMOS parts.
- When inserting wire, use a small needle-nose pliers (no serrated jaws please) to plug the end of the wire into the contact hole. Use the pliers to gently pull it out of the hole when you are done with the breadboard.
- Never expose a breadboard to high heat, or the plastic could be permanently damaged. ICs and other components that become very hot (because of a short circuit or excess current, for example) may melt the plastic underneath them. Check the temperature of all components while the circuit is under power.
- Use a chip inserter and extractor to implant and remove ICs. This reduces the chance of damaging the IC during handling.
- Avoid building a bird's nest, where the connection wires are routed carelessly. This makes it harder to debug the circuit and it greatly increases the chance of mistakes.
- Remember! Solderless breadboards are designed for low-voltage DC circuits *only*. They are not designed, nor are they safe, for carrying AC house current.

USING PIN JUMPERS

Solid conductor wire can break when it's used too much. Due to metal fatigue, the wire just snaps off—sometimes right inside the breadboard contact point (use a small needle-nose pliers to remove these).

For longer-lasting wires, you want to purchase or make a set of jumpers made with stranded conductor wire, with soldered-on machine pin ends. These last much longer than regular solid conductor jumpers, though they're a lot more expensive if you buy them ready-made.

 Check out the RBB Support Site (see Appendix) for a step-by-step pictorial on making your own pin jumpers.

MAKING PIN HEADERS FOR OFF-BOARD COMPONENTS

There are times when you want to use components—speakers, switches, potentiometers—that just won't fit into the holes of the breadboard. You can make pin jumpers for these, too. But instead of soldering a pin to each end of the wire, you solder a pin to one end, and on the other end you connect to the component, as shown in Figure 23-7.

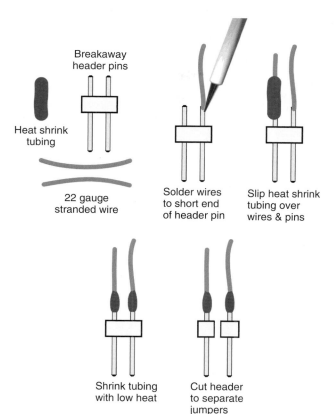

Breakaway
header pins

Heat shrink
tubing

22 gauge
stranded wire

Solder wires
to short end
of header pin

Slip heat shrink
tubing over
wires & pins

Shrink tubing
with low heat

Cut header
to separate
jumpers

Figure 23-7 Make your own handy plug-in connectors for components that can't directly plug into the solderless breadboard. Solder stranded (solid) wire to male pin headers, and insulate the ends with heat shrink tubing.

Figure 23-8 shows the general idea of using a custom header with a speaker. Just plug in the header whenever you want to use a speaker in your circuit designs. By making the hookup solid and strong you can avoid the frustration of loose wires and intermittent connections.

USING CLIP-ON JUMPERS

Another kind of wire is used when experimenting with and testing your robot electronic circuits. These are clip-on jumpers. The jumper is made with flexible insulated wire, where both ends have some kind of spring-loaded clip. You attach the clips to wires, components, or another part of the circuit to make temporary connections.

You should get at least one set of clip-on jumpers for routine testing and experimenting. Jumpers are available with push-pin hooks and alligator clips. Jumpers are available with clips on both ends, or just on one end; on the other end is a male or female header for attaching to a microcontroller or other board—see Figure 23-9.

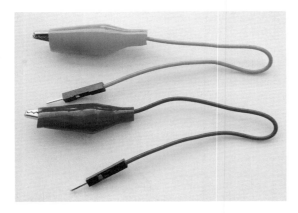

Figure 23-8 Example of a pin header permanently soldered to a component (a speaker shown here) that cannot by itself plug into the breadboard. Use stranded (as opposed to solid) conductor wire.

Figure 23-9 Use alligator clips with male or female headers for easy connections to microcontrollers and other circuit parts.

Making Circuit Boards

Sometimes a solderless breadboard won't cut it. Maybe your circuit needs to take up less space, or you want to create a longer-lasting version.

When it comes to making circuit boards, you've got plenty of choices. These include using a predrilled perforated board with metallic "traces" for the wiring already built-in, to fully custom boards that are specially made for the exact components in your circuit.

Here's a quick overview of your best options for building a soldered circuit board:

- *Solder breadboards.* These mimic the solderless breadboards you read about earlier in this chapter, but these are made for permanent soldering.
- *Unplated perforated boards.* Old-fashioned but still useful for very simple circuits, these are boards with holes already drilled in them. You wire up components directly.

- *Plated stripboards. Same idea as above, but these are plated in different grid styles to avoid wiring directly to component leads. Easy to work with and cheap. Lots of variety.*
- *Quick-turn printed circuit boards. Design a printed circuit board (PCB) on your computer, then send it out for manufacturing. Less expensive than you may think.*
- *Home-etched PCBs. Using a strong chemical, you can etch the pattern of your own PCB onto a copper-plated board. Time-consuming and messy, but can be a good learning experience.*
- *Custom prototyping boards. Some components with "rock star" status (most notably, microcontrollers like the PICAXE, AVR, and PICMicro) have various custom prototyping boards available for them. Take your pick.*
- *Wire wrapping. Semipermanent construction using very thin wire between components. You can undo the wires to make changes. You need special tools and wiring to make it work. See the RBB Support Site if you're interested in the circuit building process.*

So you know there are other methods that I'll be skipping here. Many are out of date, are specific to one manufacturer, require hard-to-find parts, or may require special tools.

CLEAN IT FIRST!

Circuit boards use a thin layer of copper to form *traces,* the wires of the circuit. Copper can oxidize and get dirty over time, both of which can lead to a poor solder joint. No matter which circuit-board-making method you use, prior to any soldering *be sure to thoroughly clean the board* using warm water and ordinary kitchen cleansing powder. You can scrub the board using your fingers, a folded-up paper towel, or a nonmetallic scouring pad.

After cleaning, rinse all of the cleanser off the board. Pat it dry with a paper towel, then let it air-dry—it takes a couple of minutes. For a superclean board, wet a cotton ball with household isopropyl alcohol and give the metal a final wipedown. Allow all of the alcohol to evaporate.

MAKING PERMANENT CIRCUITS ON SOLDER BREADBOARDS

Akin to the solderless breadboard is the *solder breadboard,* where *you can* make permanent any design you create on a solderless breadboard.

The solderboard comes pre-etched with 300 or more tie points. Circuits may be designed on a solderless breadboard, then transferred to the solder board when you're sure everything is working to your liking. Simply solder the components into place following the same design you used on the solderless board. Use jumper wires to connect components that can't be directly tied together. Figure 23-10 shows "half size" solder breadboard.

Small circuits take up only a portion of the solder breadboard. You can cut off the extra using a hacksaw or razor saw. (But beware of the "sawdust" from these boards; it's not healthy for you, so don't inhale or ingest any of it.) Leave space in the corners of the board to drill new mounting holes, so that you can secure the board inside whatever enclosure you are using. Alternatively, you can secure the board to a frame or inside an enclosure using double-sided foam tape.

USING POINT-TO-POINT PERFORATED BOARD CONSTRUCTION

An alternative to the solder breadboard is *point-to-point perforated* (or "perf") circuit board construction. This technique refers to mounting the components on a predrilled board and connecting the leads together directly with solder. Perf boards are basically just blank pieces of phenolic

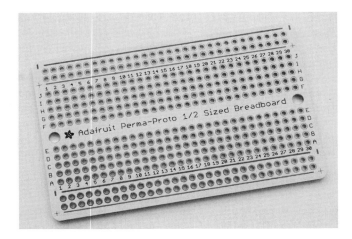

Figure 23-10   Test your circuits on the solderless breadboard, and once the design is finished, transfer the components to the solderboard following the same arrangement.

or other plastic, with holes drilled every 0.1″. This is the correct spacing for standard ICs, and it works well for other components.

For robot electronics, point-to-point perf board construction is best used—if at all—for very simple circuits containing just a few components. You use the board as a kind of structure for the electronic parts.

Building a Circuit on a Perf Board

Figure 23-11 shows the concept behind using a perf board—again, it's merely a board with holes already drilled into it. Basic construction goes like this:

1. Cut the board to the size you'll need for the circuit. You'll need to estimate the amount of board space.
2. Insert one component at a time through the holes of the perf board. Bend the leads on the opposite side to keep the component from falling out.
3. For components that are right next to one another, wrap leads that connect. One or two turns is enough.
4. For connecting components that are more than an inch or two from one another, use 30-gauge wire meant for wire wrapping (see the section on wire wrapping, later in the chapter), strip off about 1/2″ of the insulation from the end of the wire, and hook it around the component lead.
5. Use a soldering pencil to flow some solder to the wires.
6. Repeat until all the components are mounted and wired together.
7. Trim off any excess wire lengths to avoid short circuits. Double-check your work.

Perf board wiring is often referred to as *bird's-nest* construction—you can guess why. For circuits using more than a couple of components, use a copper layout board, detailed next. Since circuits on point-to-point perf boards are delicate, mount them on your robot using one or two small fasteners or a cable tie. Avoid having the board just flop around loose.

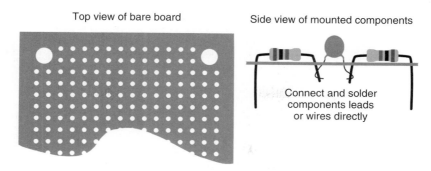

Top view of bare board   Side view of mounted components

Connect and solder
components leads
or wires directly

Figure 23-11   When using a perforated (perf) board, you directly connect components together or use solder component leads with insulated jumper wires.

Alternatives to Perf Board Mounting

A variation on the theme of point-to-point wiring on a perf board is to simply do away with the perf board. Over the years, a variety of techniques have been developed; here are some of them:

- *Dead bug* wiring is sometimes used with a single IC connecting to just a few components. It's called "dead bug" because when the chip is turned upside down, and things are soldered to its pins, it looks a bit like a little, black, dead bug. When soldering on resistors, capacitors, and similar components, cut the lead to length, then bend the end to make a U shape. Solder the U directly to the IC pin.
- *Wire wrap* sockets have extra-long pins for attaching to wires. The IC itself plugs into the socket after construction is complete. As with dead bug wiring, you can form fairly strong solder joints by bending the end of the component lead into a U and hooking it around the socket pin.
- *Lead-to-lead* construction is suitable when you're soldering the leads of one discrete component to another—for example, a resistor to an LED. Prepare the leads of both components by cutting them to the desired length, then form small hooks to make a good mechanical joint. Solder the two together.

USING PREDRILLED STRIPBOARDS

A stripboard has holes drilled in it, just like a perforated board. But on at least one side of the board is a series of copper metallic pads and/or strips that run through the center of the holes. These boards come in a variety of sizes and styles. All are designed for use with ICs and other modern-day electronic components. One application of the stripboard is circuit construction using wire wrapping, as detailed later in this chapter. But many circuits can be soldered directly onto the board.

Choose the style of grid board depending on the type of circuit you are building. Figure 23-12 shows a selection of basic grid layouts. Variations include:

Plated holes without strips: These are used like perforated boards, previously described, but you can solder component leads to the copper plating around the holes. You then complete the board by attaching wires to the component leads.

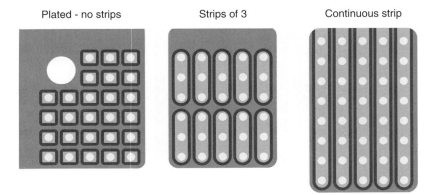

Plated - no strips   **Strips of 3**   **Continuous strip**

Figure 23-12   Example grid layouts for perforated stripboard. Select the layout that best matches the circuit you're building and your construction preferences.

Continuous strips that run the length of the board: To make a circuit, solder in components and wires, then use a sharp knife (or a specialty *spot face cutting* tool) to break the strip after all the connections are made for that particular part of the circuit.

Strips in groups of three to five holes: Rather than have you cut a long strip into smaller segments, these boards have done it for you. Each segment spans from three to five holes (sometimes more). Some boards have additional strips running perpendicular to the segments; these are common *busses* for easy connection to positive and negative terminals from the power supply.

Personally, I prefer boards with the three-hole segment plan. Components are tied together using three-point contacts. If you need to connect more components to a single point, you can link multiple segments together by bridging them with a piece of wire. Those boards with the extra strips for positive and negative terminals are ideal for circuits that use many ICs, as it simplifies connection to power. The interleaving of the power supply rails also helps reduce electrical noise.

Always use either the plated copper segments to bridge components together or else a length of bare wire. Avoid the temptation of using a big blob of solder to bridge segments together.

Using Premade Prototyping Boards

The popularity of several brands of microcontrollers has spawned a kind of cottage industry in prototyping (or *proto*) boards custom-made for them. Proto boards are empty PCBs with layouts and predrilled holes to accommodate a wide variety of projects. The prototyping board is designed to fit over or beside a microcontroller.

Start by soldering the microcontroller into its spot (better yet, use an IC socket, then plug the controller into the socket), then complete the board by adding other components. Many prototyping boards have a reserved section for general experimenting, where you can add your own circuits to the basic one already there. Some proto boards are even made with mini solderless breadboards stuck to them.

Some proto boards already have the microcontroller soldered to them. In this case, the controller may be in a more compact format than the typical dual in-line package (DIP) you're most familiar with. This makes using the board very convenient, as the controller is already built in.

The disadvantage is that if the MCU is not in a socket, any damage to the chip means the entire prototype board is a loss. My preference is to use a prototyping board that accepts the standard DIP IC packages, and then use sockets to allow easy swapping of controllers.

ON THE WEB: HOW TO MAKE YOUR OWN PRINTED CIRCUIT BOARDS

For an extra bit of finesse you can make your own custom printed circuit boards, either completely by hand, with the help of PCB software and service. Check out the RBB Support Site (see Appendix) for these and other illustrated topics:

- Creating Electronic Circuit Boards with PCB CAD
- Producing Arduino-Specific Boards with Fritzing
- Etching Your Own Printed Circuit Board

Using Headers with Your Circuits

Most robots are constructed from subsystems that may not be located on the same circuit board. So you need to connect these subsystems together using some kind of wiring system. For very simple connections, you can directly solder wires between boards and other components. But as the electronic systems of your bot get more complex, such direct connections make it harder to experiment.

The solution: Use connectors whenever possible. In this approach, you connect the various subsystems of your robot together using wires that are terminated with a connector of some type or another. The connectors attach to mating pins on each circuit board.

These three methods are the most popular:

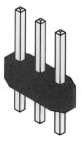

Use male header pins on the board, and attach female connectors to the incoming wires. This works best when you already have wires with the requisite connectors on them; otherwise, you need to make the connectors, which requires a special crimping tool.

Male headers come in several types: Standard headers are a solid block and you need to select the number of pins to match your application. *Breakaway* headers allow you to snap off whatever number of pins you want to use.

Use female header pins on the board, and attach male connectors to the incoming wires. With this method you can make all kinds of connectors for plugging in just about anything.

Terminal blocks are specifically designed to carry more current than pin headers. Solder the block to the board, then securely attach wires to the block using the included screws. You can get blocks with one or more terminals on them. Use these for attaching to large batteries, heavy-duty motors, and any other component that requires the use of heavier-gauge wires.

MAKING YOUR OWN MALE CONNECTORS

You don't need fancy cables and cable connectors for your robots. In fact, these can add significant weight to your bot. Instead, use ordinary 20- to 26-gauge wire, terminated with single- or double-row plastic headers.

You can make these yourself using breakaway header pins. You buy them in lengths of 10 or more pins, then break off as many as you need. Solder wires to the underside of the pins. Insert the top side into sockets on your circuit boards. Figure 23-13 shows how. The optional heat shrink tubing is applied by cutting a small piece of the tube and slipping it over the soldered wire. Shrink the tube using a heat gun or hair dryer set on high.

PREPARING FEMALE CONNECTORS

If you're connecting to a circuit board that already uses male pins, you'll need to use a female connector to hook things up. You can make these by purchasing a connector set that's designed for the pins you're plugging into. These vary by the number of pins and the pin spacing, so be sure to get the right set. Most pin headers use 0.100″ spacing. There are smaller and larger spacings, but these are obvious just by looking at them.

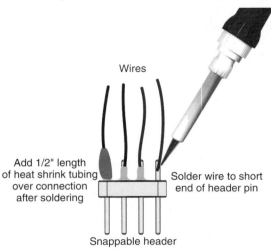

Wires

Add 1/2″ length of heat shrink tubing over connection after soldering

Solder wire to short end of header pin

Snappable header

Referring to Figure 23-14 as a guide, to make a connector prepare the end of a wire by removing about 1/4″ to 3/8″ of insulation. When using stranded conductor wire, twist the strands together. Insert the end of the wire into a crimp-on pin end. Secure the wire in the pin by crimping it; a crimping tool made for the job works best. You then insert the pin into the plastic *shell*; the shell forms the insulated connector. The pin simply snaps into place in the shell.

Figure 23-13 Make your own male header connectors by soldering wires to the short end of the header pins. For a professional look use heat shrink tubing over the solder joint.

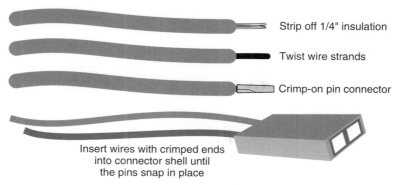

Figure 23-14 Make your own female connectors using a connector shell and crimp-on pins.

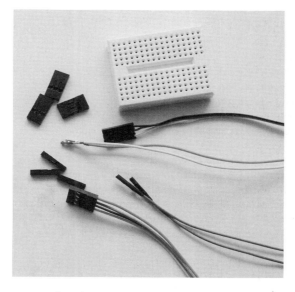

Figure 23-15 Make your own female connectors using a connector shell and precrimped pins.

USING PRECRIMPED WIRES AND SHELLS

If you ask me, making connector wires is a thankless job. Rather than soldering and crimping, I prefer to use wires where the female or male pin ends are already crimped on. These ends then insert into the plastic shell. The precrimped wires are available in different popular lengths and in various colors.

Figure 23-15 shows an assortment of precrimped wires lengths and connector shells. You can get the shells in various blocks: a 1×2 block is one row with two columns for pins.

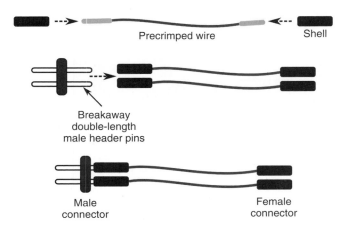

Figure 23-16   Use double-length breakaway male header pins with the female wire connectors. This method is stronger, and you don't have to keep so many types of precrimped wiring in your workshop.

1×1—Any single connection
1×2—Power, motors
1×3—Servos and most sensors
1×6—More advanced sensors
1×8—Multiple components

Combine blocks as needed: one each 1×2 and 1×3 blocks make a 1×5 block. You don't need every type of shell; a selection of just a few of the more common blocks will do you just fine.

I prefer to use only female pin ends. This not only reduces the stock necessary for building, but I find that the male ends bend easily. Whenever I need a male end, I use breakaway male headers that are *double-length*. These are about 1/4″ on both ends, long enough to plus into a solderless breadboard or other board, and into the wire connector. Figure 23-16 shows the simple concept.

Best Connections

Robots are known for becoming bird's nests of wiring. Tamp down the mess by using plastic ties to bundle the wires together. This keeps them all tidy. Or, instead of using individual lengths of insulated wire, use ribbon cable, which is made of many pieces of wire bonded together as a unit.

When making interconnecting cables, cut the wires to length so there is a modest amount of slack between subsystems. You don't want the wire lengths so short that the components are put under stress when you connect them together. But don't go overboard; you also don't want, or need, gobs of excess wire.

PART 6

Robot Brains

CHAPTER 24

All About Robot Gray Matter

"Brain and brain! What is brain!?"

In a now-iconic episode of the original *Star Trek* TV show, a race of alien women steal Spock's brain, and of course Captain Kirk wants it back. That's understandable: it's the *brains* that differentiate any mechanism—Vulcan, human, or robot—from a can opener.

You'll be glad to know Spock eventually got his brain back. While it may be a stretch to aim for logical Vulcan brains in all your bots, you'll likely still want some sort of *thinking center* in your creation. The brains of the robot processes outside influences, such as sonar sensors or bumper switches; then, based on their programming or wiring, determine the proper course of action.

A computer in one form or another is the most common brain found on a robot. A robot control computer is seldom like the PC on your desk, though robots can certainly be operated by most any personal computer. And, of course, not all robot brains are computerized. A simple assortment of electronic components—a few transistors, resistors, and capacitors—are all that's really needed to make a rather intelligent robot.

In this chapter let's review the different kinds of "brains" found on the typical homebrew robot, including microcontrollers—computers that are specially made to interact with (control) hardware. Endowing your robot with smarts is a big topic, so additional material is provided in Chapters 25 through 29.

Brains for the Brawn

We'll start by reviewing the six principal ways to endow your Scarecrow robot with a brain—no *Wizard of Oz* required here; it's all done with bits of wire and other parts.

Human control: Some very basic robots are controlled by human interaction. Switches on the robot, or in a wired control box, let you operate your creation by hand. Human control is an ideal way to learn about how robots operate. Examples include remotely controlled combat robots you see on television.

Discrete components: In years past, the typical robot brains used basic electronic components like the resistor and transistor (and even before that, tubes!). Now, with inexpensive microcontrollers, these kinds of circuits aren't seeing as much use, but they're still ideal for simple robots with simple jobs to do.

Microcontrollers: A microcontroller is computer-on-a-chip, with a "thinking" processing unit, memory, and means to connect with the outside world. Microcontrollers are the ideal form of robot brain because they're simple, cheap, and easy to use. You program them from your personal computer.

Onboard computers: Some bots need more computational power than microcontrollers can provide. Here low weight is important so laptops, netbooks, and computers with compact main boards are among the best choices as Robo Brainiac.

Remote computers: While the typical amateur bot is self-contained, there's no technical reason it can't be controlled by a computer located someplace else. This is typical of "teleoperated" robots, like those used by the military or in police bomb disposal. The computer is connected to the robot via wire, radio waves, or some other means.

Smartphones, tablets, and PDAs: Some consumer gadgets like smartphones and tablets can be pressed into service as robot controllers. The ideal device has a Bluetooth transmitter that can send operating commands to your robot over the air.

Start Out Simple!

I'm a believer in starting out simple. And it doesn't get any simpler than using manually operated switches to control a bot. While this is not a true "robot" in the formal sense of the word, it's a useful way to discover how robots work. By manually operating the robot with your own hands, you learn how it has to be done via fully electronic control.

Adding switches to operate a basic robot is easy. I prefer putting the switches and battery power in the same handheld remote—fewer wires that way. Use a piece of wood, plastic, or even picture frame mat board to hold two switches and a standard battery holder, as shown in Figure 24-1.

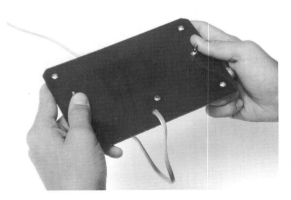

Figure 24-1 Two double-pole, double-throw (DPDT) switches can be used to control the motors of a robot. Read more about motor control using switches in Chapter 14, "Using DC Motors."

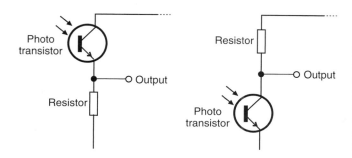

Figure 24-2 A few parts from basic electronic components form a workable robotic brain. Depending on how the parts are configured, the robot can display different "behaviors." In one variation of this light-detection circuit, the relay triggers in light; in the other variation, the relay triggers in dark.

In operation, the switches are wired so you can start and stop the motors and control their direction. By changing the direction of one or both motors, you learn how to maneuver the robot around a room.

Brains from Discrete Components

In the world of electronics, *discrete components* are parts like transistors, resistors, and basic-building-block integrated circuits (ICs). These components, used in some clever combination, can produce a working, thinking brain of a basic robot.

Figure 24-2 shows a common form of robot brain made from simple parts. Wired one way, this brain makes the robot reverse direction when it sees a bright light. The circuit is simple, as is the functionality of the robot: light shining on the photodetector turns on a relay.

You could add additional simple circuitry to extend the functionality of robots that use discrete components for brains. For instance, you could use an LM555 timer as a time delay: trigger the timer and it runs for 5 or 6 seconds, then stops. You could wire the LM555 to a relay so it applies juice only for a specific amount of time. In a two-motor robot, and using two LM555 timers with different time delays, your robot can be made to steer around things.

Programmed Brains

Perhaps the biggest downside of making robot brains from discrete components is that because the brains are hardwired as circuitry, changing the behavior of the machine requires considerable work.

Conversely, you can "rewire" a robot controlled by a computer simply by changing the *software* running on the computer. For example, if your robot has two light sensors and two motors, you don't need to do much more than change a few lines of *programming code* to make the robot come toward a light source rather than move away from it.

TYPES OF PROGRAMMABLE GRAY MATTER

An almost endless variety of computers and computer-like devices can be used as a robot brain. The five most common are:

- *Microcontroller.* A microcontroller is programmed in either assembly language or a high-level language such as C, Python, or Basic. Figure 24-3 shows a Parallax Activity Bot robot kit, which is driven by a Propeller 8-core microcontroller. Considering their importance in the art of robot building, the bulk of this chapter is devoted to microcontrollers.

- *Single-board computer.* This is also programmed but it generally offers more processing power than a microcontroller. It is more like a miniature personal computer.
- *Motherboard for a compact personal computer.* These are the main boards (or *motherboard*) in desktop computers, but selected for their small size.
- *Smartphone and tablets.* If you already have the processing power in a device you use every day, like your mobile phone, why not use it to make a robot? That's the idea behind powering bots using personal consumer electronics.

As noted, several popular microcontrollers are covered in their own chapters; let's take a closer look at the other options.

SINGLE-BOARD COMPUTERS

Single-board computers (SBCs) are a lot like "junior PCs" but on a single circuit board. While there are some that can run versions of Windows (such as Windows 10 IoT), many are engineered for an operating system that consumes less disk and memory space, such as Linux or even old-fashioned PC-DOS.

SBCs come in a variety of shapes and forms. A standard form factor supported by many manufacturers is *PC/104*, which measures about 4" square. PC/104 gets its name from "Personal Computer" and the number of pins (104) used to connect two or more PC/104-compatible boards together.

Among the most popular non-PC/104 SBCs is the Raspberry Pi, developed by a non-profit organization formed to help promote computer programming literacy in schools. Figure 24-4 shows the standard Raspberry Pi model which has the width and height of a credit card. Other versions of the Raspberry Pi are available; see Chapter 27 for more details.

PERSONAL COMPUTERS

Having your personal computer control your robot is an admirable use of available resources, but it's not always practical if you're planning on mounting the thing on top of good old Tobor (that's

Figure 24-3   A programmable controller uses software rather than specific wiring to determine how the robot reacts. This Activity Bot robot kit from Parallax uses the Propeller, which reads the value of sensors and applies power to the robot's two motors accordingly. (*Photo courtesy Parallax Inc.*)

Figure 24-4 The Raspberry Pi 3 Model B+ measures about 3 1/2" by 2 1/2".

robot spelled backward). The old-style desktop PC is simply too heavy, bulky, and power hungry to be an effective source of brains for your bot.

There are two ways to use a PC to control your robot:

- *Brains on bot.* Mount the computer on the robot. For a laptop, you can rely on its internal battery. But for a desktop PC meant to be plugged into the wall, you'll need to either run the computer using a large 12-volt battery and car power inverter or retrofit the computer with a power supply that can be juiced directly from the battery. Laptops and small mini- and pico-ITX PC motherboards from VIA are popular choices.
- *Brains off bot.* You use any kind of computer and link it to your robot via wires, radio frequency (RF) link, or optical link. This is common practice when using tabletop robotic arms; since the arm doesn't scoot around the floor, you can place it beside your PC and tether the two via wire. A USB connection is a favorite tethering technology—not to mention inexpensive and easy to use. There's also Bluetooth, Zigbee, and other types of radio links if you don't want the wires.

Regardless of form factor, the motherboard relies nearly entirely on USB ports for external communication. For data storage a solid-state drive (SSD) is the safest. Traditional hard drives may suffer damage if the robot is subjected to sudden and violent shock.

SMARTPHONES AND TABLETS

Rounding out the discussion of brains for your robot are smartphones and mobile tablets. These are typically used as remote control devices that communicate to a microcontroller on your robot. To be useful, the device:

- Should be user programmable, or at least have available a generic app that permits you to control your Arduino, BBC micro:bit, or other Bluetooth-enabled device.
- Provides some kind of communications link between itself and the robot electronics. On many devices this is through Bluetooth, but on others you might use Wi-Fi or even wired USB.

Of Inputs and Outputs

The architecture of robots requires inputs—things like sensors and bumper switches. And then there are outputs, such as motor control, light, and sound. The basic input and output of a computer or microcontroller is a two-state voltage level (i.e., off and on), which usually equates to either 0 and 5 volts, or 0 and 3.3 volts—the upper voltage is determined by the architecture of the computer or microcontroller you are using.

For example, to place an output of a computer or microcontroller to HIGH, the voltage on that output is placed, under software control, to 3.3 or 5 volts, depending on the system.

In programming, LOW is equivalent to off, or binary 0. HIGH is equivalent to on, or binary 1. The LOWs and HIGHs are *bits*.

Inputs and outputs are colloquially referred to as *I/O*. In addition to standard LOW/HIGH inputs and outputs, there are several other forms of I/O found on SBCs and microcontrollers. The more common are listed in the following sections, organized by type. Several of these are discussed in more detail in Chapter 29, "Interfacing Hardware with Your Microcontroller."

SERIAL COMMUNICATIONS

Robot subsystems need a way to talk to one another. This is often done with a serial communications interface. With serial communications, data is sent one bit at a time, as shown in Figure 24-5.

The most common types of serial communications include the following:

I2C (inter-integrated circuit): also shown as I^2C. This is a two-wire serial network scheme that allows ICs to communicate with one another. With I2C you can install two or more microcontrollers in a robot and have them talk to one another.

SPI (serial peripheral interface): This is a popular serial communications standard used by many electronic devices. SPI is most often used to interface with microcontrollers or microprocessor support electronics.

Synchronous serial port: This is a generic term for most any serial data link where information is transmitted one bit at a time, using (at least) two wires. One wire contains the transmitted data, and the other wire contains a clock signal. The term *synchronous* means the clock serves as a timing reference for the transmitted data. This is different from asynchronous serial communication (discussed next), which does not use a separate clock signal.

UART (universal asynchronous receiver transmitter): UARTs are more common in desktop computers, but they have applications in microcontrollers as well. *Asynchronous* means there's no separate synchronizing system for the data. Instead, the data itself is embedded with special bits (called start and stop bits) to ensure proper communication.

Ethernet: Famous for its use in connecting to local area networks, Ethernet allows very fast communications speed over reasonably long distances. It supports both wired and wireless transmission; the wireless version is more commonly known as Wi-Fi.

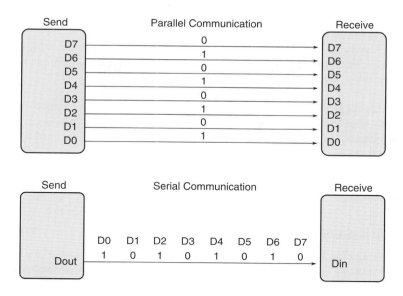

Figure 24-5   With serial communication data is sent one bit (0 or 1) at a time. With parallel communication, several bits are transmitted at once. Bits are commonly transmitted and received in multiples of eight—8 bits equals 1 byte.

> **Microwire:** This is a serial communications scheme used in National Semiconductor products, which is popular for use with the PICMicro line of microcontrollers from Microchip Technologies. It's similar to SPI. Most Microwire-compatible components are used for interfacing with microcontrollers.
>
> **1-Wire/MicroLAN:** This is a specialty serial communications link that requires just a single connection wire. It's used primary by components from Maxim and Dallas Semiconductor.

PARALLEL COMMUNICATION

Parallel data communication is more straightforward than serial, but it's not necessarily easier to implement. With parallel data, you combine the values of two or more I/O lines of the computer or microcontroller. With eight I/O lines you can communicate 256 different messages; this is because there are 256 different ways to set the two possible states (0 and 1) of the lines.

An example of using parallel communication is displaying text on a liquid-crystal display (LCD) panel. While you can purchase an LCD panel that connects to your microcontroller via serial data, these are much more expensive. With just six I/O lines you can directly control a run-of-the-mill LCD panel.

ANALOG AND DIGITAL CONVERSION

The command circuitry of your robot is *digital*; the world around us is analog (see Figure 24-6). Sometimes the two need to be mixed and matched, and that's the purpose of conversion. There are two principal types of data conversion:

ADC: Analog-to-digital conversion transforms analog voltage charges to binary (digital). ADCs can be outboard, contained in a single IC, or included as part of a microcontroller. Multiple inputs on an ADC chip allow a single IC to be used with several signal sources.

DAC: Digital-to-analog conversion transforms binary (digital) signals to analog voltage levels. DACs are not as commonly employed in robots, but that doesn't mean you can't be clever and think of a nifty way to use one.

PULSE AND SIGNAL DETECTION

Digital data are composed of electrical *pulses,* and these pulses may occur at a more or less even rate. Pulses and pulse rate (or frequency) are commonly used in robotics for such things as reading the value of sensors or controlling the speed of motors. The four major types of pulse and signal detect are:

External interrupt: An interrupt signals the attention of the processor that an event has taken place. It literally "interrupts" the normal program flow so the processor has a chance to deal with the event. Interrupts are often used to detect fleeting signals such as bumper switch closures.

Input capture: This is an input to a timer that determines the frequency (number of times per second) of an incoming digital signal. With this information, for example, a robot can differentiate between inputs, such as two different locator beacons in a room. Input capture is similar in concept to a tunable radio.

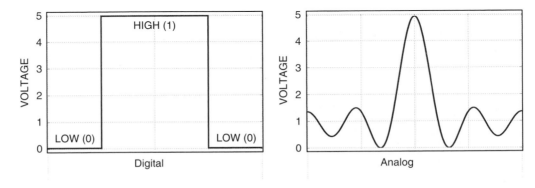

Figure 24-6 Comparison of digital and analog signals. Digital signals are in discrete steps and equate to numeric values. Analog signals are continuously variable.

PMW: Pulse width modulation is a digital output where pulses have a varying duty cycle (i.e., the "on" time for the waveform is longer or shorter than the "off" time). PMW is often used to control the speed of a DC motor.

Pulse accumulator: This is an automatic counter that counts the number of pulses received on an input over a period of time. The pulse accumulator is part of the architecture of the microcontroller and can be programmed to operate autonomously. This means the accumulator can be collecting data even when the rest of the microcontroller is busy doing something else.

Understanding Microcontrollers

Microcontrollers have become the favorite method for endowing a robot with smarts. And there's good reason: microcontrollers are inexpensive, have simple power requirements—usually just 3.3 or 5 volts—and most can be programmed using software on your PC. Once programmed, the microcontroller is disconnected from the PC and operates on its own.

A microcontroller is a computer-on-a-chip that is also made to connect to outside things. They come in all sizes, styles, and categories. Some are for esoteric applications, like running a car's engine or controlling the operation of industrial furnaces. Others are general-purpose, designed for a wide variety of applications. It's these that we're most interested in.

Figure 24-7 shows a low-cost modular microcontroller designed for experimentation—it contains the MCU itself, built-in buttons for user input, plus grid of 25 LEDs for displaying messages.

Here are other ways microcontrollers differ and the benefits of each type.

8-, 16-, OR 32-BIT ARCHITECTURE

As with your desktop computer, microcontrollers come with different data-processing architectures; these are denoted by the number of bits they process at one time.

Figure 24-7 Example of a low-cost modular microcontroller, designed for learning and experimentation. It's programmed from a PC or tablet.

- The oldest and simplest of microcontrollers used 4-bit processing. They aren't of much use today, and, besides, they're no longer common in the consumer chip-selling marketplaces.
- On the other end of the spectrum are MCUs that handle 64 bits at a time. These are specialty chips used for high-end applications and, as such, tend to be more expensive and difficult to program.
- Middle-of-the-road microcontrollers handle 8 or 32 bits at a time. Common for amateur robotics is the 8-bit variety, which, despite handling "only" 8 bits at a time, is ideally suited to the vast majority of robot programming tasks. Eight-bit MCUs are the least expensive and among the most widely available. An example of 8-bit microcontroller is the Arduino Uno. Examples of 32-bit controllers include the BBC micro:bit and the Raspberry Pi.

Recall from computer science class that the *bit* is the smallest value that can be stored in the memory of a microcontroller. A bit is either off or on. In most literature (and this book), those two values are denoted as 0 and 1 or alternatively as LOW and HIGH.

LANGUAGE PROGRAMMING

All microcontrollers need to be programmed for the task you want them to do. None "just work" out of the box (exception: they come preprogrammed with a demo). Microcontrollers are programmed by loading code into its main memory, where the program is either

Stored permanently—it cannot be revised or erased

Stored semipermanently—the program will remain even when the device is powered off, but can be revised or erased.

There are three general methods of programming a microcontroller, which, for the lack of better terms, can be grouped into four broad categories discussed below. These loosely defined terms relate to how the programming is stored and executed in the controller.

Compiled Low-Level Code

Low-level code is the generic term for the raw instructions that are stored inside the permanent (or semipermanent) memory of a microcontroller. The code commands every step of the microcontroller from the moment it is switched on.

Low-level code is written on another computer, and compiled from a more human-producible form, either assembly language or a high-level programming language like C or Pascal. The compilation process convert the human-written code to something that only the microcontroller can understand.

Compiled low-level code requires purpose-built programming hardware that loads the compiled code into the memory of the microcontroller. Because of this need of extra-cost hardware, this form of programming is not as popular, at least not for amateur and educational uses. The other methods detailed below are more common.

Programmer hardware is designed specifically for a given microcontroller or family of controllers. If you use different MCUs, you need a separate programmer for each one.

Bootloader with Hex Code

First, a definition: A *bootloader* is a special program that resides inside the microcontroller at all times, and was stored there by the maker of the MCU. The bootloader allows you to load your own code into the microcontroller without the need for special programming hardware. A large majority of microcontrollers today—Arduino, BBC micro:bit, Parallax Propeller, and many others—use this method.

Another definition: A *hex* file is a compiled version of your program for use by the microcontroller. The "hex" name comes from the way the data is typically stored in the file, as a series of *hexadecimal* (base-16) values.

You don't write the hex file yourself. It's generated for you as part of the process of compiling your user code into assembly language. The code you write, using software developed for your microcontroller, is usually in some high-level language: common examples are C, Basic, Python, and JavaScript.

Assembly language (or just "assembly") is a type of low-level computer programming language that you *can* write yourself, if you know how. Each system has its own assembly language; code for one type or family of microcontroller will not work on another. Despite the arcane appearance of assembly language code, some users prefer writing their programs with it. The learning curve can be steep.

When you're ready to program your microcontroller, the code is compiled into assembly language then transferred to the MCU. Figure 24-8 shows the basic process.

Not all microcontrollers programmed with a bootloader use hex files, but they use something similar. The important point is that the microcontroller itself can serve as its own programmer. No separate hardware is required other than a USB cable.

Bootloader with Interpreted Language Code

Some microcontrollers, such as the PICAXE and Parallax BASIC Stamp, use a language interpreter that's permanently built into the memory of the device. Programs you write on your PC are converted to intermediate "tokenized" code that is then sent to the microcontroller. Each token represents something the microcontroller is supposed to do. The interpreter on the MCU reads the tokens and takes the appropriate action.

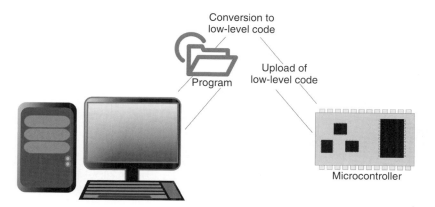

Figure 24-8   Programming cycle of a microcontroller that uses a bootloader. Programs are developed on a PC, where they are compiled to a machine-readable format, and then uploaded (usually via USB cable) to the microcontroller.

Like using a bootloader and hex file, controllers that use interpreted language programming are easier to set up, and they don't require special hardware apart from a USB connection. On the downside, a common complaint is that interpreted language code takes longer to process, so microcontrollers that use this technique may be slower.

With both types of bootloader-based programming, the factory-installed bootloader and your program share storage memory on the microcontroller. See Figure 24-9 for a simple overview of how this works.

In-Place Code

Many SBCs, and this includes the Raspberry Pi, use a different approach to programming and loading programs. Rather than code on one system and then transfer it to another via a cable, you write your programs directly within the device itself. Let's call this *in-place coding.*

With in-place code you use the computer like any other: connect it to a display so you can see what's going on, write text using a keyboard, and perhaps navigate through its operating system with a mouse. When you're done writing you run (or *execute*) the program.

Choice of Programming Language

A low-level programmable microcontroller is basically a blank canvas; it's up to you what you put in it and how you do it. With these controllers you have an option of choosing the language you wish to use. These include:

BASIC: This language is popular with those just starting out in programming, as the language is designed for beginners—in fact, the B in BASIC stands for beginner. It's also a favorite among those already familiar with a BASIC language for the PC, such as Visual Basic. BASIC is a more forgiving language than the others; for example, it's not case-sensitive. It doesn't care if you use different capitalizations to reference the same things.

C: The language of choice for programming professionals, C has more strict syntax rules, so it's often considered harder to learn. (In a spoken language, *syntax* is how the parts of speech are strung together to form a coherent statement. It's the same in a programming language.) For all the bad rep C gets for being a stern schoolmaster, it's actually not that much more difficult to learn than BASIC, at least not when it comes to microcontrollers.

C++: When C is mixed with object-oriented programming you get C++ (pronounced see-plus-plus). Object-oriented programming is a paradigm where virtual objects embody a collection of attributes. In the world of robotics, these objects can represent real-world devices, such as motors and sensors. C++ is the primary language used with the Arduino.

Python: Like C++ Python is object-oriented. It was designed to be self-descriptive and easy-to-read, and is a favorite choice of many programmers. For microcontrollers a subset of

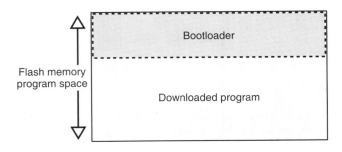

Flash memory
program space

Bootloader

Downloaded program

Figure 24-9 The bootloader program permanently resides in the Flash or other long-term storage memory of the microcontroller. Programs you write are added to the memory, and don't overwrite the bootloader.

the language, called MicroPython, is often used. It shares the main benefits of full Python, but is optimized for use in microcontrollers where memory may be more limited.

Pascal: Considered a blueprint for modern structured programming, Pascal is a somewhat lesser-used language when it comes to microcontrollers, but has the benefit of being easy to learn.

Graphical blocks: All of the programming languages above are textual, meaning you write code by typing text on the keyboard. With block programming, you create code by dragging icons and other graphic shapes into a script window. Blocks are said to be more intuitive for first-time learners.

There are several other programming languages used in microcontrollers, including Forth, Java, JavaScript, and C#. The choice of which language to use depends greatly on which one you're more comfortable with and which offers the feature set you want to exploit in your robot work.

Three Steps in Programming a Microcontroller

No matter what system or language you use, there are three basic steps to programming a microcontroller. They are:

1. Use your personal computer to write your program with a text editor or other application. Many commercial programming languages come with a fancy editor, called an *IDE*, for integrated development environment. This single application combines the programming step with the other two that follow.

2. You then compile your program into a form of data that the MCU will understand. The microcontroller doesn't know an "If" statement from a hole in the ground, and the job of the compiler is to convert the human-readable code you just wrote to the machine language the microcontroller understands.

3. After being compiled, the translated program is uploaded to the microcontroller. This is most often done using a USB or serial cable. Once uploaded, the program is immediately ready to be run inside the microcontroller. In fact, it will usually start running the moment the uploading process is complete.

What's in a word? Throughout this book I refer to the process of sending a program from a PC to a microcontroller as *uploading* in keeping with the terminology used by the Arduino and many other devices. Some sources use the term "downloading," but either way the idea is the same.

Microcontroller Shapes and Sizes

All microcontrollers are ICs, with anywhere from 8 to over 128 connection pins, depending on complexity. But you don't always work with a microcontroller as a separate IC. Popular form factors of microcontrollers include (see Figure 24-10):

Chip-only: You work with the microcontroller at the bare IC level. The chip may or may not need any external components to operate. At a minimum, some need a voltage regulator and an oscillator or resonator as a clock source.

Figure 24-10 Three typical microcontroller form factors: bare integrated circuit (chip-only), carrier board, and development board.

Carrier: In this variation the microcontroller IC is mounted on a carrier, which contains additional electronics. The carrier is itself the size and shape of a wide-profile IC: a 24-pin "double-wide" IC is common. This is the form factor of the BASIC Stamp and many of the small Arduino controllers. The carrier can be plugged into solderless breadboards for further experiments. Not all carriers are shaped like ICs. Some look more like long sticks, with one or two rows of connection pins.

Integrated development board: Physically the largest of the bunch, integrated (all-in-one) development boards contain the microcontroller and support electronics, plus extras like LEDs, switches, and header connections for experimenting. Using jumper wires, you can connect the development board to a solderless breadboard, where you might attach sensors, servo motors, and other external components.

Separate development boards are commonly available for chip- and carrier-style microcontrollers. These offer the best of both worlds; you get a compact controller when you must conserve space, but, when needed, you can transplant the MCU to the board.

Under the Hood of the Typical Microcontroller Chip

A key feature of microcontrollers is that they combine a microprocessor with various inputs and outputs (called *I/O*) that are needed to interface with the real world. For example, the Atmel ATmega328 28-pin microcontroller (Figure 24-11) sports the following features, many of which are fairly standard among microcontrollers.

● *Central processing unit (CPU).* This is the core of the microcontroller and performs all of the logic and arithmetic computations. The CPU takes your programming instructions and evaluates each one in turn.
● *Input/output pins.* The I/O lines are the gateway of information in and out of the microcontroller. An I/O line can be set to act as an input, in which case it can accept data from

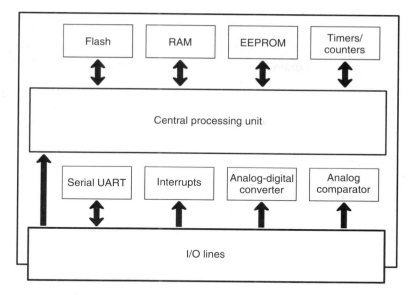

Figure 24-11 Basic block diagram of the Atmel AVR ATmega328 microcontroller, as the brains behind the Arduino Uno and others. The central processing unit forms the core of the controller. Additional built-in hardware provides special functions, such as timers, an analog comparator, and an analog-to-digital converter.

the outside, or it can act as an output, where it can control some external component. The ATmega328 has 28 pins total, of which 23 may be used for input/output chores—it's said to have 23 I/O lines or I/O pins. The remaining five pins are used for power supply connections.

- *Built-in analog converter.* Six of the chip's input/output pins are connected to an internal analog-to-digital converter (ADC), which translates analog voltages to digital binary values.
- *Built-in analog comparator.* The internal analog comparator can be used for basic go/no-go evaluation of analog voltage levels.
- *Flash program storage.* Programs you write and transfer to the microcontroller are stored in 32K bytes of rewritable flash memory. This allows the controller to be reprogrammed over and over again. Note that some MCUs are program-once only. Read more about these later in this chapter.
- *RAM and EEPROM data storage.* Small amounts of RAM (2K bytes) and EEPROM (1K bytes) storage are provided for the data that the controller uses during operation. Data in *RAM* (random-access memory) is lost when the controller loses power. Data in *EEPROM* (electrically erasable programmable read-only memory) retains its value even when the microcontroller is no longer powered.
- *Hardware interrupts.* Microcontrollers are designed to interact with the outside world, and hardware interrupts allow the chip to be literally "interrupted" by some external stimulus. It's not unlike asking someone to pinch you to wake you up. Interrupts make many common robotic programming tasks easier and more elegant. The ATmega328 has two primary hardware interrupts that are connected to two of its pins.

- *Built-in timers and counters.* These all-purpose accessories of microcontrollers operate separately from the CPU and provide a wide variety of useful services. For example, under program control, you can command a timer to generate a pulse every second. Counters are likewise special accessories separate from the CPU and literally count individual signal events, like the number of times a robot's wheels have rotated.
- *Programmable full-duplex serial port.* Serial data is a common means of communication with a microcontroller, not only for programming it, but also for the MCU to communicate with other devices.

PIN FUNCTIONS

Being ICs, the pins (connecting points) on microcontrollers serve as the means to power the chip, as well as get signals in and out of the device. On may microcontrollers each pin—other than power and ground—may support specialized functions.

For example, a microcontroller might have eight total input/output (I/O) pins for connecting to outside things, like motors or sensors. These are commonly referred to as general purpose input output (or GPIO) pins.

Some or all of these pins may serve special-duty uses, such as allowing for analog input or producing pulse width modulated signals (these topics were introduced above under the sections "Analog and Digital Conversion" and "Pulse and Signal Detection"). In order to use these functions you must connect your external devices to these pins. Refer to the pinout diagram for your microcontroller for specific details on which pins do what.

3.3V OR 5V OPERATION

Once upon a time most digital circuits were powered by 5 volts, making it easier to connect things together. Now, to make the electronics operate at higher speeds and at lower current consumption it's not uncommon for circuits to require 3.3 volts. Many microcontrollers now run off 3.3V; these include the Raspberry Pi, BBC micro:bit, and Parallax Propeller, as well as some versions of the Arduino, for example.

The operating voltage of the microcontroller directly determines the voltage requirements of its input/output pins. This effects any other circuit, such as sensors and motor drivers, that is connected to the controller. Read more about this very important topic in Chapter 29, "Interfacing Hardware with Your Microcontroller."

Providing Power

Power for your microcontroller has to come from somewhere. Common sources are wall adapters and batteries. Power you provide must both be at an appropriate voltage for your microcontroller, but it must have sufficient current to operate the circuitry of your microcontroller and everything else you've added.

Many microcontroller development boards (Arduino, Raspberry Pi, etc.) have their own voltage regulators on board. Separate regulators may be provided for both 3.3V and 5V; if you tap into this regulated power for your external circuitry make sure to use the correct voltages.

Important! Take care that your circuit doesn't draw more current than the regulator can provide. In the Arduino Uno the 3.3V provides only 50 mA (milliamps). That's not very much, so if your circuit requires more than that, you'll need to provide a separate regulated power supply. See Chapter 12, "Batteries and Power," for methods you can use.

MORE ON PROGRAM AND DATA STORAGE

At first glance, it looks like microcontrollers have somewhat limited memory space for programs. The typical low-cost microcontroller may have only a few thousand bytes (yes, *bytes,* not megabytes) of program storage. While this may seem terribly confining, in reality most microcontrollers are programmed to do single or well-defined jobs. These jobs may not require more than a few dozen lines of program code.

More elaborate microcontrollers may handle more program storage. Keep the program storage limits in mind when you're planning which brain to get for your robot. A single type of microcontroller is often available in slightly different versions; each version may accommodate a different amount of program data.

Data storage may seem even more restricted. The typical microcontroller may only provide 1K or 2K of RAM for data (if provided EEPROM space, a type of semipermanent memory, is even less). But again, most microcontroller applications make efficient use of data space. As long as you follow good coding practice, you should seldom run out of data storage space.

SPECIALITY MICROCONTROLLERS: ONE-TIME PROGRAMMABLE

So far I've talked about microcontrollers with program storage memory that can be repeatedly updated. Less costly microcontrollers are made to be programmed only once, and are intended for permanent installations. These *one-time programmable* (OTP) microcontrollers are popular in consumer goods and automotive applications.

For robotics applications, the OTP is useful for dedicated processes, such as controlling servos or triggering and detecting a sonar ping from an ultrasonic distance measurement system. You'll find a number of the ready-made hobby robotic solutions on the market today that have, at their heart, an OTP microcontroller. The microcontroller takes the place of more complex circuitry that uses individual ICs.

All About Microcontroller Speed

If you have a personal computer, you probably know that its microprocessor runs at a certain speed. Older PCs were rated in megahertz (millions of cycles per second); the latest models operate in the gigahertz (billions of cycles per second) range.

Likewise, microcontrollers operate at set speeds. These speeds are rather low for a modern computational device—most MCUs operate at 4 to 60 MHz. But again, the nature of microcontrollers doesn't require superspeeds. For one thing, many microcontrollers are more efficient in how they execute their program code.

Some controllers for robotics are speed demons. Though not a true microcontroller, as of this writing the latest Raspberry Pi models sport processor speeds of 1.4 GHz (not megahertz, but gigahertz). These speeds are not unusual for modern single-board computers. It's for this reason SBCs like the Raspberry Pi are great for offloading processes such as full-motion graphics.

On the Web: Coding 101

In order to use a microcontrollers you need to program it with code. If you're just starting out be sure to read the free *Coding 101* tutorial on the RBB Support Site. It details the most important programming concepts you'll encounter as you develop your robots.

CHAPTER 25

Using the Arduino

Microcontrollers are now so commonplace that you have your pick of hundreds of makes and models, from the supersimple to the confoundedly complex. Somewhere in the middle is the Arduino, a small and affordable microcontroller development board that's fast becoming something of a superstar.

What's made the Arduino a darling of geeks the world over is this: both its hardware design and its software are open source. That means others are able to take the best ideas and improve on them, all without paying licensing fees. This has created something of a cottage industry of fans and third-party support.

In this chapter you'll read about what the Arduino is and how to apply it to your robotics chores. Be sure to see the chapters in Parts 7 and 8 for numerous working examples of using the Arduino in real-world applications, and also check out the bonus programming examples on the RBB Support Site (see Appendix).

Arduino Under the Hood

First introduced in 2005, the Arduino has gone through numerous iterations, revisions, and improvements. Figure 25-1 shows what might be called the main or core Arduino board design, the Arduino Uno. It's a printed circuit board that measures 2-1/8" by 2-3/4", containing an Atmel ATmega microcontroller chip running at 16 MHz, a power jack for a 2.1mm (center positive) barrel connector, and a USB Type B jack for hooking up to a host computer.

A series of 32 female pin headers allow connection of external devices to the Arduino. The headers are largely separated into three groups: power, analog input, and digital input/output.

Of the 32 pins, 20 are devoted to input and output. There are six analog input pins, which can also serve as general-purpose digital I/O. The 14 digital input/output pins include 6 that can be used to generate PWM (pulse width modulated) signals, useful for such things as controlling the speed of motors. All I/O pins can be used as digital outputs and can sink or source up to 40 milliamps, well enough to light LEDs.

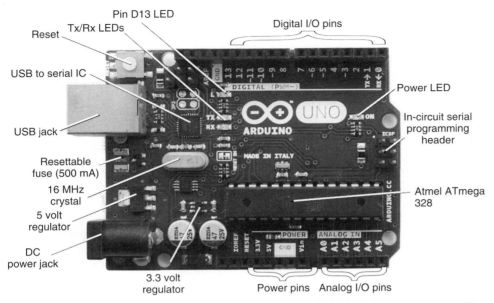

Figure 25-1   Points of interest on an Arduino Uno board include the barrel power connect, USB jack, function and power LEDs, and power and logic signal pins.

 The Arduino Uno has gone through a number of iterations since its first release. This book details version R3, the latest as of this writing. Earlier versions of the Uno are nearly identical except for some cosmetic differences.

At the heart of the Arduino Uno is an Atmel ATmega328 microcontroller chip. Many other Arduino models, which include the Mega and Leonardo, use different microcontrollers with differing capabilities. As the Uno is by far the most commonly used of the family, I'll be concentrating on it throughout the book.

 Let me pause here to point out that the microcontroller chips used on the Arduino are not "empty." They, in fact, come preloaded with a small *bootloader* program for use with the Arduino development editor, described later in this chapter. The bootloader assists in the upload process.

Ready Expansion via Shields

The Arduino itself has no breadboard area, but it's easy enough to connect any of the inputs or outputs to a small breadboard via wires. For an application like robotics, you'll want to expand the Arduino I/O headers to make it easier to plug in things like motors, switches, and ultrasonic or infrared sensors.

Figure 25-2 Expansion shield for the Arduino, offering a small solderless breadboard for experimenting. Use short jumper wires to connect the Arduino pins (top and bottom) to the contact points in the breadboard. (Photo courtesy Adafruit Industries.)

One method is to use an add-on expansion board known as a *shield*. Shields stick directly on top of the core board and Mega designs. Pins on the underside of the shield insert directly into the Arduino's I/O headers. Two popular expansion shields are the solderless breadboard shield (see Figure 25-2) and the proto shield; both provide prototyping areas for expanding your circuit designs.

Of course, you don't absolutely need a shield to expand the Arduino. You can place a breadboard—solderless or otherwise—beside the Arduino and use ribbon cables or hookup wire to connect the two together.

Variations on a Theme

The Arduino is available in many *form factors*. The most popular and commonly used form factor is the Arduino Uno, a mid-sized board that can fit in a shirt pocket. Unless otherwise noted all the Arduino projects in this book use the Uno.

Other form factors include (see Figure 25-3):

1. **Mega2560,** a physically larger board that also uses a larger Atmel chip that sports over three times the number of analog and digital I/O lines. Memory and program space are larger, too. The Mega2560 can use the same shields as the Uno.
2. **Mini,** a slimmed down stamp-sized version that is not shield-compatible, is ideal for very small bots with limited space. The Mini lacks its own USB jack and requires the use of a USB adapter so it can connect to a host PC for programming. Akin to the Mini is the Nano, another small-format Arduino that (in most versions) comes with the USB jack installed.

Other variations of the Arduino depart from the standard form factor and may not be able to use expansion shields. An example is the Adafruit FLORA, a special Arduino engineered for making wearable microcontroller projects. Think Borg implants, only more friendly looking. The flower-shaped FLORA has a flat profile and can be sewn into fabric. It has connection points on the ends of its 22 petals.

Figure 25-3 A trio of Arduinos: clockwise from top is the Mega2560, Mini, and Uno.

USB Connection and Power

To allow the easiest possible means of programming, all the core Arduino boards support USB. You need merely to connect a suitable USB cable between the Arduino and your computer. The necessary USB drivers are provided with the Arduino software. In most cases, installation of the drivers is not fully automatic, but the steps are straightforward, and the Arduino support pages provide a walk-through example.

The USB jack provides communications, both for uploading programs from your PC (discussed later), and for serial communications back to the PC. The USB link includes a 500 mA resettable fuse to guard against possible damage caused by a wayward Arduino to the USB ports on your PC. When plugged into a USB port, the Arduino takes its power from it. With USB 2.0, drive current is limited to 500 mA, depending on the port design.

Operating voltage of the Arduino circuitry is 5 volts, which is supplied either by the USB cable when it's plugged into a USB port on your computer or by a built-in low-dropout linear voltage regulator when the board is powered externally. The regulator is intended to be powered by 7 to 12 vdc; a 9-volt battery is ideal for most tasks. Anything higher than 12 volts is not recommended, as it could cause the regulator to overheat.

The Arduino is also equipped with a 3.3-volt low-dropout voltage regulator. Depending on the version of the Arduino board, the regulator is either built into the USB-to-serial chip or separate. In either case, maximum current output is rather low, on the order of 50 mA. The 3.3-volt regulator is good for powering small electronics that require the lower voltage. These include certain types of accelerometers and gyroscopes. Just be sure the circuit consumes less than about 10 mA.

Figure 25-4   Ready-made cable for connecting a 9-volt battery to the Arduino 2.1mm power jack.

For robotics I think it's best to power the Arduino from its own battery, and use different batteries for the motors. You can make your own 9-volt battery to 2.1mm barrel connector jumper cable or purchase one ready-made (see Figure 25-4).

Indicator LEDs are provided on the Arduino for testing and verification. One LED shows power; two other LEDs show serial transmit and receive activity, and should flash when the board is being programmed from your computer. A fourth LED is connected in parallel with digital I/O line 13 and serves as a simple way to test the Arduino and make sure it is working properly.

Arduino Pins

The Arduino uses its own nomenclature for its I/O pins, and the names and numbers of its pins don't correlate to those on the ATmega microcontroller. This can cause some confusion if you're already familiar with working with the bare ATmega chips.

Figure 25-5 shows the pinout diagram of the 28-pin ATmega328 microcontroller chip. The labels on the inside of the chip are the primary function names for each of the pins. The labels outside in parentheses are alternative uses, if any, for the pins.

Also shown in Figure 25-5 is pin mapping between the Arduino and the ATmega328. It's important to remember that the pin numbers are not the same between the two: pin 12 on the ATmega328 is actually mapped to digital pin D6 on the Arduino, for example. Pin mapping is not something you need to worry about in typical Arduino programming, but it's nice to know what leads to where.

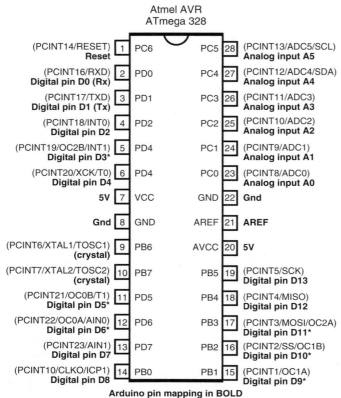

Atmel AVR
ATmega 328

(PCINT14/RESET) **Reset**	1	PC6	PC5	28	(PCINT13/ADC5/SCL) **Analog input A5**

Arduino pin mapping in BOLD

*denotes capable of
PWM output

Figure 25-5   Pinout diagram of the
Atmel ATmega328 chip, with the pin
mapping to the Arduino I/O lines.

Programming the Arduino

Microcontrollers depend on a host computer for developing and compiling programs. The software used on the host computer is known as an *integrated development environment*, or *IDE*. The "Arduino language" is essentially C++, which is itself a variation of good old-fashioned C.

If you are unfamiliar with this language, don't worry; it's not hard to learn, and the Arduino IDE provides some feedback when you make mistakes in your programs.

If you're brand new to programming be sure to check out the RBB Support Site for an in-depth primer on programming, and links to useful online learning guides.

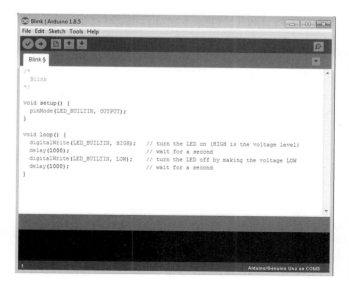

Figure 25-6 The Arduino integrated development environment (IDE) provides a centralized place to write, compile, and upload programs to the Arduino board.

To get started with programming the Arduino using its IDE, first go to *www.arduino.cc,* and then click on the Download tab. Find the platform link (PC, Mac, Linux) for your computer, and download the installation file. Step-by-step instructions are provided in the Getting Started section of the Arduino Web site. Be sure to read through the entire set of instructions.

Once installation is complete, you're ready to try out your Arduino. Start by connecting the board to your PC via a USB cable. If this is the first time you've used an Arduino on your PC, you must install the USB communications drivers, as detailed in the Getting Started guide.

First-time use of the environment requires you to specify the Arduino board you are using and, as necessary, the serial port that is connected to the board (the Arduino's USB connection looks like a serial port to your computer). You may then open an existing example program and upload it to your board. Or you may write your own program in the IDE editor. Figure 25-6 shows the Arduino IDE with a short sketch in the main window.

The next step is to compile the program—called "sketches" in Arduino parlance. This prepares the code for uploading to the Arduino. At the bottom of the text editor is a status window, which shows you the progress of compiling. If the program was successfully compiled, it can then be transferred to the Arduino, where it will automatically run once the upload is complete.

Programming for Robots

The Arduino supports the notion of *libraries,* code repositories that extend core programming functionality. Libraries let you reuse code without having to physically copy and paste it into all your programs. The standard Arduino software installation comes with several libraries you may use, and you can grab others from the Arduino support pages, and from third-party Web sites that publish Arduino library code.

A good example of a library you'll use with most any robot is *Servo*. This library allows you to connect one or more hobby R/C servos to the Arduino's digital I/O pins. The Servo library comes with the standard Arduino installation package, so adding it to your sketch is as simple as choosing Sketch->Import Library->Servo.

Structurally, Arduino sketches are very straightforward and are pretty easy to read and understand. All Arduino sketches have at least two parts, named *setup()* and *loop()*. These are called *functions,* and they appear in the sketch like this:

```
void setup() {
}

void loop() {
}
```

- The (and) parentheses are for any optional arguments (data to be used by the function) for use in the function. In the case of setup() and loop(), there are no arguments, but the parentheses have to be there just the same.
- The { and } braces define the function itself. Code between the braces is construed as belonging to that function—the braces form what's referred to as a *code block*. There's no code shown here, so the braces are empty, but they have to be there.
- The *void* in front of both function names tells the compiler that the function doesn't return a value when it's finished processing. Other functions you might use, or create yourself, may return a value when they are done. The value can be used in another part of the sketch.
- The setup() and loop() functions are required. Your program must have them, or the IDE will report an error when you compile the sketch. These are programming functions that do what their names suggest: setup() sets up the Arduino hardware, such as specifying which I/O lines you plan to use. The loop() function is repeated endlessly when the Arduino is operating.

Many Arduino sketches also have a *global declaration* section at the top. Among other things, the declaration is where you put variables for use by the whole program—see the example in the next section. It's also a commonplace to tell the IDE that you wish to use an external library, like Servo, to extend the base functionality of the Arduino.

USING VARIABLES

Arduino uses *variables* to store information while your sketch is running. The platform supports a number of *data types* for holding variables of different types and sizes. Among the most common are

- *int*—holds a signed (can be positive or negative) whole number (integer), where the value can be from −32,768 to 32,767.
- *unsigned int*—holds an unsigned (positive only) number from 0 to 65,535.
- *byte*—holds an unsigned number from 0 to 255.
- *boolean*—holds a true or false value.
- *float*—holds a floating-point value, where the digits to the right of the decimal can have up to 15 places.
- *String*—holds text, usually meant for display on an LCD panel or transmission back to the PC as a message.

To use a variable, you must first declare it. This may be done at the top of the sketch or anywhere within it. Where you declare a variable determines its *scope*: variables declared at the top of the sketch are global; that is, they can be used anywhere within the sketch. A variable declared inside a function can be used only within that function.

Declaring (or *defining*) a variable requires you to first specify its type. You then indicate its name, followed by an optional step of assigning a value to the variable just declared:

```
int myInt = 30;
```

declares an integer variable named *myInt*, plus it assigns a value of 30 to it. This one line is equivalent to:

```
int myInt;
myInt = 30;
```

Names for variables must contain only letters and numbers or the underscore (_) character. The name can't start with a number, and it can't contain a space. Capitalization matters, so *myInt* is distinctly different from *myint*.

Many programmers prefer a consistent naming convention for their variables. This includes using consistent capitalization. The common practice today is called *camelCase*—low on the ends, high in the middle.

An exception to this is when using constants, variables whose contents are defined once and never changed. The common practice here is to use all uppercase characters, as in

```
const int POT;
```

for a constant variable named *POT*.

USING ARDUINO PINS

The input/output pins of the Arduino are referenced by number. There are two numbering sequences: one for the analog pins and another for the digital pins.

- Analog pins are referenced as A*x*, where *x* is a number. For example, to reference analog pin number 0, you'd use A0.
- Digital pins are referenced in sketches just by their numbers. (Additionally, within this book the digital pins are described as D*x*—example: D13 or D9—to avoid any confusion about which pins to connect things to.)

In actual use, most of the Arduino programming statements are self-aware of the proper pins—analog or digital—to use. For example, when using the *analogRead* programming statement, which reads a voltage level at a pin, the compiler knows you're talking about an analog pin, as the digital pins don't support this feature.

By default, the digital pins are automatically considered as inputs, meaning they are set up to read a value, rather than set a value. At any point in a sketch you can inform the Arduino that you wish to use a pin as an output. The process is simple:

```
pinMode(PinNumber, Direction);
```

where *PinNumber* is the number of the pin you'd like to use, and direction is either INPUT or OUTPUT (these are predefined constants, by the way—that's why they're in uppercase). For example,

```
pinMode(LED_BUILTIN, OUTPUT);
```

makes pin D13 an output. Once made an output, it can do output-like things, such as light up an LED. Such an example follows later in this chapter.

Digital pins may be on or off, defined as 0 or LOW (off), or 1 or HIGH (on).

- Use *digitalWrite* when setting the value of a digital pin.
- Use *digitalRead* when testing the value of a digital pin. You typically use a variable to store the value of the pin.

For example,

```
digitalWrite(LED_BUILTIN, HIGH);
```

turns pin D13 on (sets it HIGH).

```
myVar = digitalRead(LED_BUILTIN);
```

reads pin D13, and assigns its value (either LOW or HIGH) to the *myVar* variable.

In the preceding examples I've used the *LED_BUILTIN* constant name for pin D13 on the Arduino Uno. This constant is built into the Arduino IDE, and for the Uno is defined internally as the number 13. Other Arduino boards could have their LED connected to a different pin. In that case the constant has a different meaning when that board is selected inside the IDE.

EXPERIMENT BY DOING

Program *arduino_test.ino,* shown below, demonstrates a few fundamental Arduino concepts useful in any robotics development, reading an analog sensor and providing visual feedback.

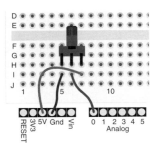

I've taken one of the examples that come with the Arduino IDE and modified it slightly to conform to the style used throughout this book. It uses a 10 kW potentiometer to alter how fast Arduino's built-in LED flashes.

Check out Figure 25-7 for a schematic of the circuit used for the program listing. The potentiometer is connected to the board as a common voltage divider. That way, the Arduino detects the value of the pot as a variable voltage from 0 volts (ground) to 5 volts.

Be careful with how you wire the potentiometer. Be sure the wiper is connected to the Arduino, and the legs of the pot are connected to 5V and ground. If you accidentally connect the wiper to 5V or ground, you'll create a dead short across the pot as you rotate it. The pot may burn out, and the Arduino will go into "fail-safe" protection mode by shutting down.

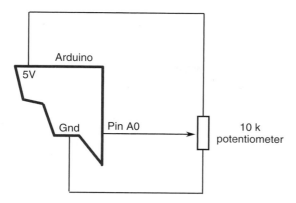

Figure 25-7 Schematic diagram for the arduino_test.ino sketch. The 10 kΩ pot is connected to the 5V and Gnd pins of the Arduino, as well as analog pin A0.

arduino_test.ino

```
#define LED 13              // LED to digital pin D13
#define POT 0               // pot to analog pin A0
int potValue = 0;           // variable for pot value

void setup() {
  // initialize D13 as an output
  pinMode(LED, OUTPUT);
}

void loop() {
  potValue = analogRead(POT);   // read pot value
  digitalWrite(LED, HIGH);      // turn LED on
  delay(potValue);              // wait for value (milliseconds)
  digitalWrite(LED, LOW);       // turn LED off
  delay(potValue);              // wait for value (milliseconds)
}
```

Here's how the program works:

The first two *#define* lines are known as compiler definitions. They work a little like constants, except they don't take up any memory space in the Arduino when the sketched is loaded.

The names *LED* and *POT* are defined as their associated pin numbers. When the sketch is compiled, these definitions are substituted for their numeric equivalents. It's merely a convenience, but a handy one! The built-in LED is connected to pin D13, and the potentiometer—named POT in the program—is connected to analog pin A0.

 Note: when using a *#define* leave out the = sign and semi-colon, or the IDE will signal a syntax error. This is just how C/C++ works.

A variable is defined to hold the current value of the potentiometer, which will be a number from 0 to 1023. This number is derived from the Arduino's integrated 10-bit analog-to-digital (ADC) converter, and it represents a voltage level from 0 volts to 5 volts.

The setup() section gets the Arduino hardware ready for the rest of the program. When first powered on, all the I/O lines are automatically defined as inputs. But the LED pin needs to be an output, so that distinction is defined here.

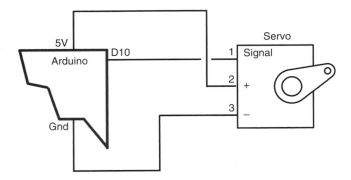

Figure 25-8 Connection diagram for testing servo functionality with the Arduino. Be absolutely sure of observing the correct polarity of the power and Ground connections to the servo.

The loop() section is automatically started the moment the program has been uploaded to the Arduino. Looping continues until either the board is unplugged, the Reset button on the Arduino is pushed, or a new program is loaded into memory. The loop begins by reading the voltage on analog pin A0. The program then turns the LED on and waits for a period of time defined by the current position of the potentiometer before turning the LED off again.

The waiting period is in milliseconds (thousandths of a second), from 0 to 1023, the range of values from the Arduino's ADC. Very fast delays of about 100 milliseconds or less will appear as a steady light.

Using Servos

The following program swings the servo motor in one direction, then the other, briefly pausing in between. You can use either an unmodified or a modified (continuous rotation) servo to see the code in action. Refer to Figure 25-8 for a diagram on hooking up the servo. Use a standard-size (or smaller) analog servo; stay away from larger or digital servos, as they may draw too much current for the USB port on your computer to handle.

 In the following example, text after the double slash // characters means a comment. It's for us humans. During the compiling phase, comments are ignored, as they are not part of the functionality of the sketch.

basic_servo_test.ino

```
#include <Servo.h>

Servo myServo;          // Create Servo object to control the servo
int delayTime = 2000;   // Delay period, in milliseconds

void setup()
{
  myServo.attach(9);    // Servo is connected to pin D9
}
```

```
void loop()
{
  myServo.write(0);    // Rotate servo to position 0
  delay(delayTime);    // Wait delay
  myServo.write(180);  // Rotate servo to position 180
  delay(delayTime);    // Wait again
}
```

The first line, *#include <Servo.h>*, tells the IDE that you want to use the Servo library, which is a standard part of the Arduino IDE installation. (Other libraries may require a separate installation, but they are used in the same way.) The name of the main Servo library file is *Servo.h,* so that is what's provided here.

Servo is actually a name of a *class*; that's how Arduino uses its libraries. With a class you can create multiple instances (copies) of an object without having to duplicate lots of code. Note that Servo is the name of the class to use, and myServo is the name I've given to the object just created.

The setup() function contains one statement, *myServo.attach(9)*. Here's what it all means:

- *myServo* is the name of the servo object that was defined earlier.
- *attach* is what's known as a *method*. Methods are actions that you use to control the behavior of objects. In this case, *attach* tells the Arduino that you have physically connected the servo to digital pin D9, and you want to activate it. A period separates the object name and method—*myServo.attach.*

Notice the ; (semicolon) at the end of the statement. It's a statement terminator. This tells the compiler that the statement is complete and to go on to the next line.

The loop() function contains the part of the sketch that is repeated over and over again, until you download a new program or remove power from the Arduino.

- *myServo.write(0)* is another method using the myServo object. The *write* method instructs the servo to move all the way in one direction. When using a modified servo, this statement causes the motor to continually rotate in one direction.
- *delay(delayTime)* tells the Arduino to wait the period specified earlier in the delayTime variable, which is 2000 milliseconds (2 seconds).
- The two statements are repeated again, this time with *myServo.write(180)* to make the servo go in the other direction.

Here's an important note about capitalization of variables, objects, and statement names. Like all languages based on C, these names are case sensitive. This means that myServo is distinctly different from myservo, MYSERVO, and other variations. If you try to use

```
myServo.attach(9);
```

(note the lowercase "s") when you've defined the object as *myServo*, the Arduino IDE will report an error—"myservo not declared in this scope." If you get this error, double-check your capitals.

Creating Your Own Functions

The flexibility of any programming language, Arduino included, comes in the ways you can develop reusable code, such as creating user-defined functions. To create a user-defined function, you give it a unique name and place the code you want inside a pair of brace characters, like so:

```
void forward() {
  myServo.write(0);
  delay(delayTime);
}
```

All user-defined functions must indicate the kind of data they return (for use elsewhere in the sketch). If the function doesn't return any data, you use void instead. You must also include parentheses to enclose any parameters that may be provided for use in the function. In the case of the forward user-defined function, there are no parameters, but remember that you need the (and) characters just the same.

That defines the function; you need only call it from elsewhere in your sketch to use it. Just type its name, followed by a semicolon to mark the end of the statement line:

```
forward();
```

See *arduino_servo.ino* for a full demonstration of an Arduino sketch that runs a servo forward and backward, then briefly stops it, using the detach method. The effect of the sketch is most easily seen when using a servo modified for continuous rotation.

arduino_servo.ino

```
#include <Servo.h>

Servo myServo;              // Create Servo object
#define servoPin 10         // Use pin D10 for the servo

int delayTime = 2000;       // Standard delay period (2 secs)

void setup() {              // Empty setup
}

void loop() {               // Repeat these steps
  forward();                // Call forward, reverse, servoStop
  reverse();                //   user-defined functions
  servoStop();
  delay(3000);
}

void forward() {            // Attach servo, go forward
  myServo.attach(servoPin); //   for delay period
  myServo.write(0);
  delay(delayTime);
  myServo.detach();         // Detach servo when done
}
```

```
void reverse() {              // Do same for other direction
  myServo.attach(servoPin);
  myServo.write(180);
  delay(delayTime);
  myServo.detach();
}

void servoStop() {            // Stop the servo by detaching
  myServo.detach();
}
```

Running Two Servos

Most robots have two motors. You can easily operate two servos from an Arduino, but you need to be mindful of connecting the servos to an external battery pack. The Arduino's built-in 5V voltage regulator can handle one standard size servo, but with two you begin to push your luck.

See Figure 28-9 for a schematic view of how to rig two servo motors to a separate battery. Figure 28-10 shows the same circuit in breadboard view. For servo power use a 4AAA or 4AA battery holder. You can use rechargeable or alkaline batteries.

 It is *very* (as in *VERY*) important that you do not reverse the power connections to the servos, or else they may be instantly and permanently fried! Connect the red power wire to the positive terminal of the battery pack. The black (or brown) wire connects to the negative terminal.

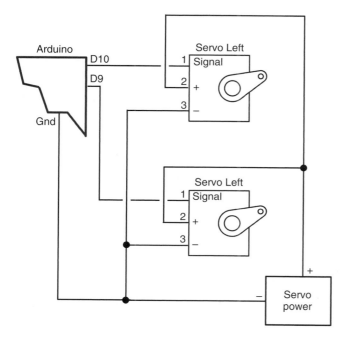

Figure 25-9 Connection diagram for using two servos. For most consistent results, the servos get their power from a separate battery pack.

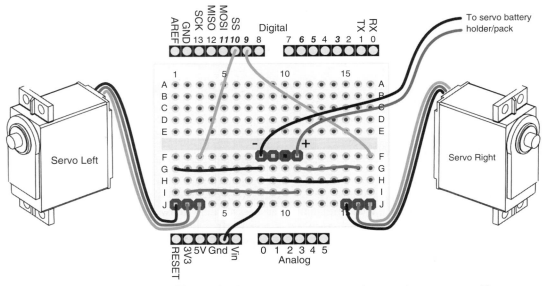

Figure 25-10 Breadboard view of how to hook up two servos to an Arduino and separate set of batteries.

Check out *arduino_2servos.ino* for a full demonstration of an Arduino sketch that runs two servos forward and backward. This sketch is intended for use with continuous rotation servos (see Chapter 15), mounted on either side of the robot.

arduino_2servos.ino

```
#include <Servo.h>

Servo servoLeft;            // Define left servo
Servo servoRight;           // Define right servo

void setup() {
  servoLeft.attach(10);     // Set left servo to digital pin 10
  servoRight.attach(9);     // Set right servo to digital pin 9
}

void loop() {               // Loop through motion tests
  forward();                // Example: move forward
  delay(2000);              // Wait 2000 milliseconds (2 seconds)
  reverse();
  delay(2000);
  turnRight();
  delay(2000);
  turnLeft();
  delay(2000);
  stopRobot();
  delay(2000);
}
```

```
// Motion routines for forward, reverse, turns, and stop
void forward() {
  servoLeft.write(0);
  servoRight.write(180);
}

void reverse() {
  servoLeft.write(180);
  servoRight.write(0);
}

void turnRight() {
  servoLeft.write(180);
  servoRight.write(180);
}
void turnLeft() {
  servoLeft.write(0);
  servoRight.write(0);
}

void stopRobot() {
  servoLeft.write(90);
  servoRight.write(90);
}
```

Flow Control Structures

Flow control structures tell the sketch what conditions must be met to perform a given task. By far the most common flow control structure is the *if* command. It tests if a condition exists or doesn't exist, then branches off execution of the program accordingly. The if structure looks like this:

```
if(condition) {
  // True: condition is met
} else {
  // False: condition is not met
}
```

Condition is an expression, usually something that determines if A equals B, as in:

```
if(digitalRead(9) == HIGH)
```

This tests if digital pin D9 is HIGH. If it is, then the program performs the *True* code, because the condition is met. Otherwise, it performs the *False* code. A complete example goes like this, where the value of digital pin D9 makes the LED on pin D13 briefly flash:

```
pinMode(13, OUTPUT);
if(digitalRead(9) == HIGH) {  // Remember: no semicolon here)
  digitalWrite(13, HIGH);
  delay (1000);
  digitalWrite(13, LOW);
}
```

Be mindful of the curly braces! You need a set of curly braces to enclose the code for any True condition.

```
if(blah-blah) {
   // True code here
}
```

And if you want separate code for when the condition is False, you must add the else command and its set of braces:

```
else {
   // False code here
}
```

Note that in this case there is no code for when the condition is False. This is acceptable, and quite common. Because there is no code for when the condition is False, the *else* command is omitted, as are its corresponding brace characters.

Conditions for the if test typically test that something is equal to something else, but there are other possible comparisons. Here are the most common:

==	equal to
<>	not equal to
>	greater than
<	less than
>=	greater than or equal to
<=	less than or equal to

Take special note of double equals signs for *equal to*. Using = will not give you what you want. You need to use ==.

A practical example is testing if the voltage on an analog pin is above or below the halfway point—the halfway point being 511 (possible values are 0 to 1023). The following short example tests if the value at analog pin A0 is above the halfway point, or 512 and over. If it is, the LED on digital pin D13 is turned on.

```
if(analogRead(A0) > 512) {
   digitalWrite(13, HIGH);
}
```

Using the Serial Monitor Window

The Arduino provides a quick and easy way to get feedback about any running program when it's tethered to its host PC via the USB cable. Built into the Arduino is the ability to talk to the PC via serial communications. You merely set up the serial link with

```
Serial.begin(9600);
```

and then send values or text to the PC with

```
Serial.println(value);
```

The result is shown in the Serial Monitor window, which you display by choosing Tools->Serial Monitor.

For example, suppose you want to see the digital value of a voltage applied to analog input pin A0. Set up the communications link in the setup() function, then repeatedly read the analog pin and return its value into the Serial Monitor window.

```
void setup() {
  Serial.begin(9600);
}

void loop() {
  int sensorValue = analogRead(A0);
  Serial.println(sensorValue, DEC);
  delay(100);
}
```

Note the optional data format in the *Serial.println* statement. The *DEC* tells the Arduino to return the value as a DECimal number. Also notice the delay statement. It's usually a good idea to include a slight delay to slow down processing when communicating back with your PC. The Arduino can process these commands faster than the serial link can accommodate, and there's a risk of losing data if the data is shuttled too quickly.

Some Common Robotic Functions

Arduino supports a number of built-in programming statements specifically useful for robotics. I'll briefly review just a few of them here, but you'll want to review these and others more fully in the Arduino language reference, available on the *www.arduino.cc* Web site.

- *tone.* Outputs a frequency on an I/O pin that, when connected to a speaker or piezo buzzer element, generates a tone. The tone to play is specified by its frequency, with a value of 440 being concert pitch A.
- *shiftOut.* Converts a byte (or more) of data to a stream of bits, which can then be sent one at a time to some external device, typically a *shift register* IC. The shift register receives the serial data and reconstructs it as parallel data outputs. The *shiftOut* statement is useful for minimizing the number of I/O pins used with external devices.
- *pulseIn.* Measures the width of a single pulse, from 10 milliseconds to 3 minutes. You can specify which I/O pin to use, whether you're looking for a LOW-to-HIGH or HIGH-to-LOW transition.
- *analogWrite.* Outputs a series of pulses whose duration, or *duty cycle,* is controlled via software—so-called pulse width modulation, or PWM. It's very commonly used to control the speed of a DC motor. The *analogWrite* statement may be used only with specifically marked digital pins. On Arduino Uno these pins are 3, 5, 6, 9, 10, and 11.

See Chapter 14, "Using DC Motors," for more information on PWM, especially as it relates to controlling the speed of motors.

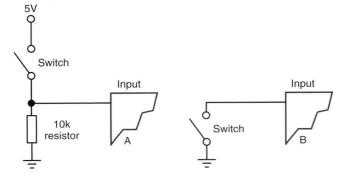

Figure 25-11 Standard connections for reading the value of a simple switch. In A, the switch is used with a 10 kΩ pull-down resistor. This keeps the value of the pin LOW until the switch is closed. In B, the switch relies on the Arduino's built-in pull-up resistor to keep the pin HIGH until the switch is closed.

USING SWITCHES AND OTHER DIGITAL INPUTS

Reading the value of a switch or other digital input is a straightforward affair with the Arduino. Simply use the digitalRead statement and indicate the digital pin you wish to use:

```
if(digitalRead(10) == HIGH) {
  Serial.writeln("Pin 10 is high.");
}
```

Figure 25-11A shows the traditional method of connecting a switch to a digital input pin. The 10 kΩ pull-down resistor ensures that the pin stays LOW as long as the switch is open. When the switch closes, the pin goes HIGH.

While resistors are the most common approach for wiring a switch to an Arduino, they are not strictly required, as the Arduino has its own pull-up resistors that can be turned on and off. When the pull-up resistors are engaged, the pins stay at a HIGH level, unless taken LOW by a switch or other input. To set the pull-up resistor, specify that the pin is an INPUT, then set its value to HIGH with *digitalWrite.*

```
pinMode(10, INPUT);        // set pin to input
digitalWrite(10, HIGH);    // turn on pull-up resistor
```

In this case, you can do away with the resistor, as shown in Figure 25-11B. Note that the built-in resistors are pull-ups, which means the switch goes LOW when activated. Your program code needs to be changed accordingly:

```
if(digitalRead(10) == HIGH)
```

INTERFACING TO DC MOTORS

You've already seen how to use the Arduino with servo motors. You can also use the microcontroller with DC motors. The I/O pins on the Arduino can provide only about 40 milliamps of current, not enough to directly power the typical DC motor. But as described in Chapter 29, "Interfacing Hardware with Your Microcontroller," you can use a motor H-bridge or other driver to run most any size of DC motor with your Arduino.

- To turn the motor on, apply HIGH to the Enable/PWM line.
- To control the direction of the motor, apply HIGH or LOW to the Direction line of the H-bridge.

- To control the speed of the motor, use the *analogWrite* statement to send a PWM signal to the Enable/PWM line. (Remember: To do this, you must connect the line to one of the digital I/O pins on the Arduino that supports PWM.)

In the following sketch, a motor bridge module is connected to digital pins D11 (for Enable/PWM) and D12 (for Direction). After setting up the two pins as outputs, the sketch repeats the steps of turning the motor on fully, reversing its direction, then slowing it down to 50 percent.

arduino_motor_control.ino

```
#define motDirection 12              // Direction line to pin D12
#define motEnable 11                 // Enable/PWM to pin D11

void setup() {
  // Pins 11 and 12 as outputs
  pinMode(motDirection, OUTPUT);
  pinMode(motEnable, OUTPUT);
}

void loop() {
  digitalWrite(motDirection, LOW);    // Set direction
  digitalWrite(motEnable, HIGH);      // Turn motor on
  delay(2000);                        // Wait 2 seconds
  digitalWrite(motDirection, HIGH);   // Reverse direction
  delay(2000);                        // Wait 2 seconds
  digitalWrite(motEnable, LOW);       // Turn motor off
  digitalWrite(motDirection, LOW);    // Set direction
  analogWrite(motEnable, 128);        // Set motor to half speed
  delay(2000);                        // Wait 2 seconds
}
```

The PWM pulsed output is disabled when you use *digitalWrite* on the same pin. In the example, *digitalWrite* is used to enable the motor at full speed, but you can also use *analogWrite* and set the PWM duty cycle to 255. However, *digitalWrite* is a more efficient use of the microcontroller when controlling the speed of the motor is not needed.

If your motor doesn't turn when its PWM duty cycle is set to 128 (50 percent), try a higher value. Not all motors will run at duty cycles below 50 percent.

Besides fashioning your own H-bridge circuit or board, you can use any of a number of third-party expansion shields and boards that contain their own H-bridge modules. Read more about one such board in Chapter 38, "Make R/C Toys into Robots."

CHAPTER 26

Using the BBC Micro:bit

Looks can be deceiving, and that also applies to microcontrollers. On the outside, the BBC micro:bit looks unassuming: measuring just 2″ by 1-3/4″, it has only a few components soldered to it, and even lacks the usual pin headers common in other microcontrollers. Its maker even promotes the micro:bit as a learning tool for children—so it couldn't be very powerful, right?

Think again. This little wonder has everything kids *or* adults need to create their own robust inventions. It easily has the horsepower to run the typical desktop robot, and comes with numerous sensors, such as an accelerometer and magnetic compass, already built-in.

The micro:bit is championed by the UK's British Broadcasting Corporation (BBC). In addition to producing such iconic television shows as *Monty Python, Doctor Who,* and *EastEnders,* the BBC has long supported computer literacy, especially for young learners. The micro:bit could be said to be a natural successor to the BBC Micro, a small home computer first introduced in the United Kingdom in 1981. As an educational product, the micro:bit enjoys substantial corporate support, including from Microsoft, Cisco, and Samsung.

Read now about the micro:bit, and how it can be used for all kinds of robotic endeavors. Additional hands-on projects using the micro:bit appear throughout this book, including a flashlight follower in Chapter 37, "Make Light-Seeking Robots."

A Closer Look at the BBC Micro:bit

Figure 26-1 shows the front and back of the BBC micro:bit.

Front side:

Miniature LEDs: A set of 25 LEDs are lined up in a 5 × 5 grid. There are enough pinpoint lights to display any number or Roman letter. And using side-scrolling, the LEDs can write out complete messages.

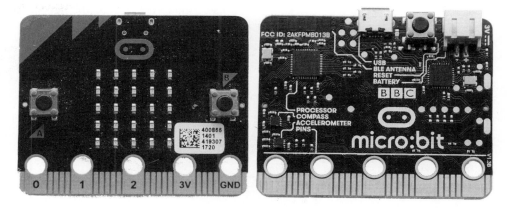

Figure 26-1   Front and back of the BBC micro:bit.

Push buttons: Directly interact with the micro:bit with two tactile switches. They're programmable: use them to provide any kind of user input.

Edge connector: To interface with the outside world the micro:bit uses a set of "fingers" on an edge connector. Five of the fingers have large holes for easy connection to jumpers with alligator clips or banana plugs. But you're not limited to just jumpers. With an appropriate adapter you can plug the micro:bit into a solderless breadboard and access all of its available input/output lines.

Back side:

USP jack: A USB cable is used to transfer programs you write on your PC into the micro:bit. (As an option you can also connect to the micro:bit wirelessly, and upload programs from a mobile device such as a tablet.)

Power plug: The micro:bit is powered by 3 volts. Plug in a set of two 1.5V alkaline batteries (rechargeable batteries don't provide enough voltage). Micro:bit starter kits include a suitable 2xAA battery holder and USB cable.

Reset button: Push the button to restart your micro:bit programs.

Labels for onboard accessories: The micro:bit knows you're curious, so it includes handy labels that identify the main points of interest, such as the main processor, compass, and accelerometer.

BUILT-IN FEATURES

At the heart of the micro:bit is a 32-bit ARM processor that runs at 16 MHz. There's 256K of Flash memory for program storage, and 16K of RAM for running programs. Purely for comparison purposes, an Arduino Uno has an 8-bit processor operating at 16 MHz, 32K of Flash program space, and 2K of RAM for running programs.

Figure 26-2 The edge connector on the micro:bit allows you to make quick external connections using jumpers.

Other features of the micro:bit worthy of note:

- You've already read about the 5 × 5 grid of LEDs, useful for basic signaling and text displays, as well as the two tactile buttons for basic interaction with the micro:bit.
- 17 general purpose I/O pins provide extended input and output. Three of the pins—labeled P0, P1, and P2—sport extra large connection points for using simple jumpers with alligator clips or banana plugs (see Figure 26-2).
- An integrated Low-Energy Bluetooth transceiver allows you to communicate with a compatible device, including other micro:bit boards, or a mobile device (e.g., a tablet). Tablet connection provides a way to wireless program the micro:bit; no USB cable needed, and you can do it from an Android or iOS app.
- A built-in 3-axis compass and 3-axis accelerometer helps you to place the micro:bit in world space. These features are useful for building robots that can self-navigate.
- The micro:bit supports both SPI and I2C high-speed serial communications. Ports for both are provided on specific I/O pins; see the section "Edge Connections" for more details.
- A 3.3V regulator provides proper voltage to the main microprocessor chip and all circuits on the micro:bit board. The regulator is active when you've plugged the micro:bit into a PC via the USB cable (not used when 3V battery is directly connected to power jack).

EDGE CONNECTIONS

The method of attaching the micro:bit to external devices using alligator clips is fine for getting started, but it has its limitations. The most critical is that you can connect to only three I/O ports. That's just enough for two servo motors and one external bumper switch—it's enough for a basic robot, but not much else.

Fortunately, you can get breadboard-compatible edge connector adapters that let you access all of the micro:bit's I/O pins. Just plug the board into the adapter, and use the adapter with a standard solderless breadboard. There are also special purpose expansion boards for the micro:bit that come with their own edge connector. I use one of these for the light-seeking robot in Chapter 37.

For your reference, Figure 26-3 shows the edge connector pinouts of the micro:bit. The pins are in two basic groups: GPIO (labeled 0 through 19) and power. Pin numbers are marked; where there is no number, that pin is either used for power, or is otherwise reserved.

In documentation and programming examples, pins are notated using the letter P or p, and their corresponding number.

```
    Pin0
or
    pin0
```

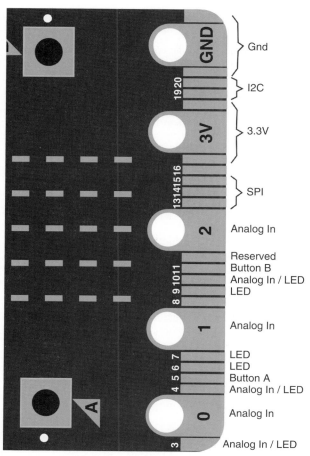

Pins without labels have no
defined alternate function

Figure 26-3 Pinout diagram of the edge connector on the micro:bit. Use an edge card adapter to connect to the pins beyond P2.

ALTERNATIVE PIN USES

Some of the I/O pins are shared with onboard components, and have alternative uses, as shown in Figure 26-3. Some pins are shared with internal hardware, including the LED grid and I2C/SPI communications ports. Avoid using these I/O pins if you're also accessing the internal hardware connected to them.

I2C Pinout

The I2C interface is used by the built-in compass and accelerometer. You should ideally use these pins only for connecting external I2C-based components to your micro:bit. If you use these pins as general I/O the accelerometer and compass functions will not work.

Pin	Function
P19	SCL
P20	SDA

SPI Pinout

While the micro:bit supports SPI serial communications none of its onboard components currently make use of it. You are free to use the pins as general I/O, or for connecting SPI-based components.

Note that the SPI protocol allows you to share many devices on the same three connection pins. You will need to allocate an I/O pin for each device to serve as a Chip Select function.

Pin	Function
P13	SCK
P14	MISO
P15	MOSI

LED Pins

The 25 LEDs are in a grid of five columns by five rows. Six of the available GPIO pins are used to address the LEDs (the micro:bit documentation refers only to columns). If you are using the LED display the following pins are not available to you:

P3, P4, P6, P7, P9, P10

You may access these pins by disabling the LED display, as detailed in Chapter 36 "Visual Feedback from Your Robot." When disabled, the I/O pins may be freely accessed by outside hardware.

Analog Input Pins

All GPIO pins on the micro:bit are capable of digital input and output. Five pins, marked in Figure 26-3, are also capable of analog input. The analog-to-digital converter built into the micro:bit accepts a variable voltage between 0V and 3.3V, and outputs a 10-bit value of between 0 and 1023.

PWM

In addition to some pins capable of analog input, all pins on the micro:bit can output a pulse width modulation (PWM) signal. The duty cycle (on versus off) time of the PWM signal varies from 0 to 100 percent by specifying a value from 0 to 1023.

Figure 26-4   Access all of the I/O pins on the micro:bit with an edge connector adapter. This adapter plugs into a solderless breadboard.

A typical use of PWM is controlling the speed of DC motors (read more about this technique in Chapter 14 "Using DC Motors"). You can also use it to vary the brightness of LEDs, change the volume of piezo sound making elements, and more.

USING AN EDGE CONNECTOR ADAPTER

The spacing between the connector lands of the micro:bit are not at the 0.100″ you find in most microcontroller boards aimed at the education and amateur market. This means traditional interconnection schemes won't work—you can't just plug in stuff to a micro:bit and expect it to work.

But you can use adapter boards, like the one in Figure 26-4. These provide a means to firmly attach to the micro:bit's edge pins and feed those to a standard breadboard. This allows you to use any standard 0.100″ connector to make circuits. I like the adapters that provide access to all pins on the micro:bit edge connector.

Choice of Programming Languages

The micro:bit is supported by a number of programming languages, though not all are considered to be in active development. A few are considered "legacy" so they aren't covered here.

The active programming platforms for micro:bit include:

MakeCode JavaScript Blocks Editor: Visual block programming has become a popular way to learn programming. The micro:bit block editor lets you program the controller by dragging colored blocks around the screen, connecting functions, and setting runtime conditions. Figure 26-5 shows an example of the blocks editor with a simple micro:bit

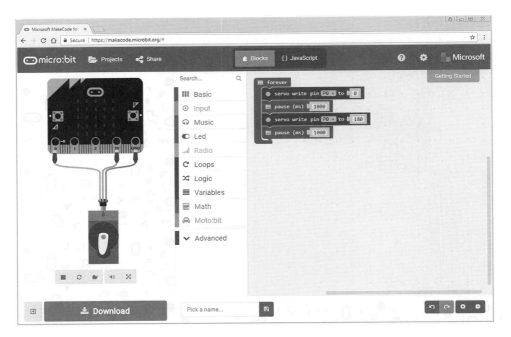

Figure 26-5   Program the micro:bit using graphics blocks in the MakeCode JavaScript Blocks editor.

program that controls a servo. The editor is available online, and can be used by desktop PCs and mobile devices.

JavaScript: Underlying the MakeCode Blocks Editor is JavaScript, a text-based scripting language commonly used to add interactive features to Web pages. As you drag and connect blocks, corresponding JavaScript code is automatically generated for you. You can switch between blocks and JavaScript in the MakeCode editor.

MicroPython: MicroPython is a variation of the Python programming language designed for use in microcontrollers and other devices with less memory than traditional computers. It is one of the favorite programming choices among more seasoned micro:bit users.

Though an online browser-based version of MicroPython is available, if you are using a PC to develop for the micro:bit I strongly recommend that you install the MicroPython environment and *mu* code editor on your computer.

The steps are not difficult, and automatic installers are provided for Windows, OSX, and Linux on the micro:bit site. See also the RBB Support Site (see Appendix) for helpful tips.

Which to choose? In this book most of the examples for the micro:bit are in MicroPython (some JavaScript), if for no other reason they are easier to present in printed form.

Extending Micro:bit via Packages

Regardless of programming editor you choose, you can extend the functionality of the micro:bit by adding *modules* or *packages* to the core platform. In the MakeCode editor, packages are added using the Add Package option, under the Advanced pull-down list. The packages include the visual blocks needed to display the programming options, plus the underlying JavaScript code to make it all work.

When you choose this option a few popular packages are listed, and you can search for others that third-party developers have created. Packages include helper libraries for Adafruit Neopixel (their brand of addressable red-green-blue LEDs), sensors, and motor drivers.

See Chapter 37, "Make Light-Seeking Robots," for a micro:bit-based robot project that uses SparkFun *moto:bit* motor driver board. The board uses a package provided by SparkFun to simplify programming.

MicroPython likewise has library code packages, called *modules*. The core functionality of the micro:bit is contained, appropriately enough, in a module named *microbit*. To use a module you merely reference it at the top of your Python code, similarly to how you use the #include statement in an Arduino sketch.

```
from microbit import *
```

Or more simply:

```
import microbit
```

Other modules are provided with the basic micro:bit MicroPython distribution for such tasks as manipulating the LED grid, taking over the compass and accelerometer, using the I2C and SPI serial interface for external devices, and more.

You only need to import modules you need. This keeps memory usage to a minimum. For example, if you're creating a program that adds functionality for the micro:bit's Bluetooth transceiver and random number generator you'd add these modules at the top of your program:

```
import radio
import random
```

You can even specify certain code libraries contained in a module, which further streamlines memory use. For instance, this code imports only the display, Image, and sleep functions from the *microbit* package:

```
from microbit import display, Image, sleep
```

 In Python capitalization matters. If a library is defined as Image, you need to use that capitalization or an error will result. If the editor complains of a NameError, look for capitalization mistakes in your code.

When using the recommended mu code editor you can add third-party micro:bit MicroPython modules. Once you're familiar with MicroPython on the micro:bit, check out repositories like GitHub to see what others have coded and shared. You can even write your own and add them as modules for your future work.

Uploading Programs to the Micro:bit

Among microcontrollers the micro:bit uses a fairly unique method of uploading programs to it. All of the program editors generate what's commonly referred to as a *hex file,* a text-based file that contains nothing but hexadecimal (base-16) characters in it. The hex file is transferred to the micro:bit in either of two principle ways:

- Transfer via Bluetooth, if using a mobile device that's been paired to the micro:bit:
- If using a traditional wired connection, save the hex file to your PC, then transfer it via USB cable.

The micro:bit doesn't use the typical communications port method of linking to PC. Instead, your PC "sees" the micro:bit merely as a thumb drive. Simply plug the micro:bit into a USB port, and it'll appear as a drive. Drag the saved hex file into the drive. The file is automatically uploaded into the micro:bit's internally program memory.

An uploaded program is retained in memory until you replace it with something else.

Useful Robotics Tasks

The micro:bit is a preeminent microcontroller for robotics. Using its standard programming tools and built-in functionality, it can control LEDs, watch for button presses, drive servo motors, monitor direction of travel, detect changes in tilt and acceleration, and more. Most of the examples that follow are written for MicroPython.

LIGHT AN LED

You already know that the micro:bit has 25 built-in LEDs. You can control each of the 25 individually or in groups. See Chapter 36, "Visual Feedback from Your Robot" for details, but here's a quick overview:

Controlling One Built-in LED

Turn on any single LED in the 5 × 5 grid with

```
from microbit import *
display.set_pixel(0, 0, 9)
```

This activates the LED at row 0, column 0 (upper-left corner). The 9 means full brightness. You can use values from 0 (off) to 9 (fully bright). Values in between act to dim the brightness of the LED.

Displaying a Predefined Image

The micro:bit contains several dozen predefined icon images. The following displays the "happy face" icon:

```
from microbit import *
display.show(Image.HAPPY)
```

Controlling an External LED

You can just as easily control an LED connected externally to the micro:bit. See Figure 26-6 for a sample connection. Be sure to include the 330 Ω current-limiting resistor, or you'll burn out the LED.

```
from microbit import *
pin0.write_digital(1)
```

This code tells the micro:bit to output a digital value of 1 (logical HIGH) to pin0. This causes the LED to illuminate. To turn it back off you'd use:

```
pin0.write_digital(0)
```

Note the capitalization of *pin0*. Don't use Pin0, or the code won't work.

REGISTER A BUTTON PRESS

The micro:bit has ready code for using its two push buttons, marked A and B, so let's cover these first. The following code tests if the A button is depressed, and if it is, turns the upper-left LED on.

```
from microbit import *
while True:
  if button_a.is_pressed:
    display.set_pixel(0, 0, 9)
  else
    display.set_pixel(0, 0, 0)
```

There are some other interesting things about this code.

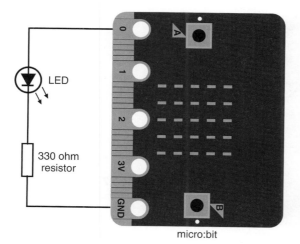

Figure 26-6 How to connect an external LED to the micro:bit. Don't forget the 330 Ω resistor inline with the LED.

- The if test checks to see if Button A (defined as *button_a*) is pressed. If it is, the LED is turned on (brightness=9). If the button is not pressed, the LED is turned off (brightness=0).
- The while True loop causes the micro:bit to repeat the code after it forever—or at least until you turn it off!
- Python uses indenting to tell it the hierarchy of code. Because all the code after the while True statement is indented, it is considered to be part of the while loop structure.

Using slightly different code you can determine if an externally connected switch is depressed. See Figure 26-7 for the external switch connection diagram. The example MicroPython program reads the value of pin0 and turns the upper-left LED on (switch is depressed) or off (switch is not depressed).

```
from microbit import *
while True:
  if pin0.read_digital():
    display.set_pixel(0, 0, 9)    # True: HIGH or logical 1
  else
    display.set_pixel(0, 0, 0)    # False: LOW or logical 0
```

DRIVE A SERVO MOTOR

The micro:bit is adept at driving one or more servo motors, but as of this writing, there's no built-in support for servos in MicroPython. Never fear: switching to the MakeCode blocks editor and JavaScript cures that in no time!

As noted above, the MakeCode JavaScript editor is available online from the *microbit.org* Web site. Starting the editor displays a new, empty project. From here, one can select blocks from the block pull-down lists, and drag them into place in the editor pane. There's no better teacher than experimenting. Use the graphical simulator to see what effect your blocks have.

Figure 26-8 shows a view of the editor with a rudimentary servo "exerciser" script, both in blocks and JavaScript. Four code blocks are wrapped inside a *forever* block so that the code repeated indefinitely. This example spins a standard servo, which is connected to pin0 of the micro:bit, from one extent to the other.

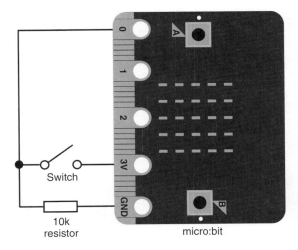

Figure 26-7 Basic connection diagram for using an external switch with the micro:bit. The resistor assures a consistent digital value when the switch is not depressed.

```
forever
    servo write pin P0 ▾ to  0
    servo write pin P1 ▾ to  180
    pause (ms)  1000
    servo write pin P0 ▾ to  180
    servo write pin P1 ▾ to  0
    pause (ms)  1000
```

```
basic.forever(() => {
    pins.servoWritePin(AnalogPin.P0, 0)
    pins.servoWritePin(AnalogPin.P1, 180)
    basic.pause(1000)
    pins.servoWritePin(AnaloPin.P0, 180)
    pins.servoWritePin(AnaloPin.P1, 0)
    basic.pause(1000)
})
```

Blocks   **JavaScript**

Figure 26-8   Block code for "exercising" one servo connected to pin0, alongside the JavaScript code equivalent.

You don't *have* to drag blocks around to create programs in the MakeCode JavaScript editor. In fact, once you get the hang of coding the micro:bit using visual blocks you can—and probably will want to—switch to the JavaScript view. In this view, it's the same program, but shown using JavaScript code.

Recreate the code below by first clicking the Projects button along the top of the editor, and then selecting New Project. Switch to the JavaScript view by clicking its button at the top, and then enter the following:

```
basic.forever(() => {
  pins.servoWritePin(AnalogPin.P0, 180)
  basic.pause(1000)
  pins.servoWritePin(AnalogPin.P0, 0)
  basic.pause(1000)
})
```

Once all the code is in place the simulator will start, with an animation of a servo motor spinning back and forth.

Adjust the rotation of the servo motors by altering the second parameter of the servoWrite-Pin statement line. Try changing the 180 to 90; now the servo only moves to its midpoint.

```
pins.servoWritePin(AnalogPin.P0, 90)
```

Or vary the time between servo moves by changing the *pause* statement. Changing from 1000 to 2000 extends the delay between servo movements from 1 to 2 seconds.

```
basic.pause(2000)
```

Note that the version of JavaScript used in the MakeCode editor doesn't use semicolons at the ends of statement lines. If you're an experienced JavaScript programmer you might be tempted to add them, but they will be lost when you save your work or switch back to Blocks view.

In similar fashion you can control two servo motors, simply by adding additional servoWritePin statements controlling other ports. Assuming a traditional two-servo robot, this code drives the bot forward and backward in 2-second intervals

```
basic.forever(() => {
    pins.servoWritePin(AnalogPin.P0, 0)
    pins.servoWritePin(AnalogPin.P1, 180)
    basic.pause(2000)
    pins.servoWritePin(AnalogPin.P0, 180)
    pins.servoWritePin(AnalogPin.P1, 0)
    basic.pause(2000)
})
```

Be sure to check out Chapter 39, "Make Line-Following Robots," for an example of using servo motors to run a micro:bit-based robot.

CHAPTER 27

Using the Raspberry Pi

On a purely semantic level, the Raspberry Pi is not a microcontroller, but a single-board computer. It lacks many features commonly associated with the typical microcontroller, though these can often be replicated in software (or as necessary by adding external hardware).

Because it is essentially a full Linux-based computer on a small circuit board, it's also considered harder to use than the archetypal microcontroller, such as the Arduino or BBC micro:bit. The learning curve is a little less steep if you already know how to get around a Linux computer.

What makes the Raspberry Pi special is its horsepower. On one credit card sized board you get a 32- or 64-bit quad core ARM processor running at 1.4 GHz. That's considerably more powerful than something like an Arduino Uno, which is a single core 8-bit processor running at a "paltry" 16 MHz.

But this compares apples to oranges. Using a Raspberry Pi on your average two-servo bump-and-go robot may be a bit of overkill. Save the simple jobs for the simpler microcontrollers. Where the Pi shines is extending the features of your bot, like giving it machine vision, or a full-color touch screen front full motion interface. Let's explore the Raspberry Pi and see what it can do.

Inside the Raspberry Pi

At the heart of the Raspberry Pi 3, the latest model as of this writing, is a 1.4 GHz 64/32-bit quad-core ARM Cortex-A53 microprocessor. This is an industrial strength chip used in some commercial products, such as a number of the Roku media players.

Additional standard features of the Raspberry Pi 3 include:

- 1Gb RAM, running at 900 MHz
- Broadcom VideoCore IV video, supporting OpenGL and OpenVG 2D and 3D graphics
- General purpose I/O (GPIO) pins provided on a double-row header
- SD slot for operating system firmware card

- HDMI video with support for SD and HD resolution
- Sound output via 3.5mm audio jack
- Wired Ethernet on many versions, and wireless on specific versions (example: Zero W and Pi 3)
- USB ports for attaching a keyboard or mouse
- Direct interface to a color camera
- Support for multiple flavors of Linux, including the quasi-official Raspbian OS, provided by the Raspberry Pi Foundation.

Figure 27-1 shows two popular versions of the Raspberry Pi, discussed in more detail in the next section.

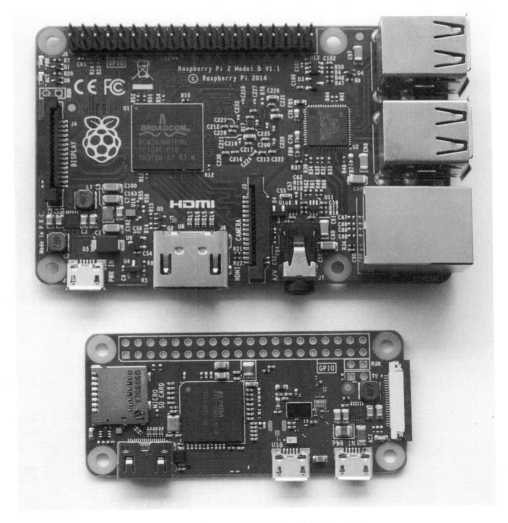

Figure 27-1 Two versions of the Raspberry Pi; the Pi 3 and the Zero W.

Variations in Raspberry Pi Board Designs

Since its inception in 2013, the Raspberry Pi has undergone more than a dozen variations. In many instances differences between the boards are subtle, but in others—notably the amount of RAM, number of pins on its GPIO header, or the type of microcontroller used—the variations are notable.

Rather than go through all the models here, I direct you to the main Raspberry Pi Web site at *raspberrypi.org,* and just highlight the main variations here.

1. **Models A and B.** Model A comprises the original two versions of Raspberry Pi, denoted as A and A+. Model B comprises later generations, with (as of this writing) the Raspberry Pi 3 Model B+ being the latest model. Figure 27-2 shows the general layout of the Pi 3 board.
2. **The Zero model** comes in a smaller form factor than the A and B models. Many of the connectors are miniature versions. The Zero W version allows you to connect to the Internet wirelessly. Figure 27-3 shows the general layout of the Zero board.
3. **The Compute Module** series is designed for embedded industrial applications. They share many of the same internal specifications, but use edgecard connectors for plugging into a motherboard, similar to inserting RAM sticks in a PC. They are not ideal for education or hobby use.

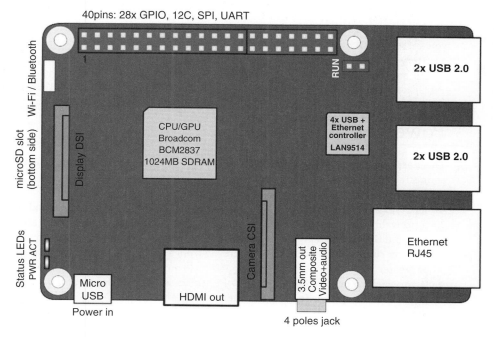

Figure 27-2   Layout of the Raspberry Pi 3 Model B+. (*Art courtesy Efa.*)

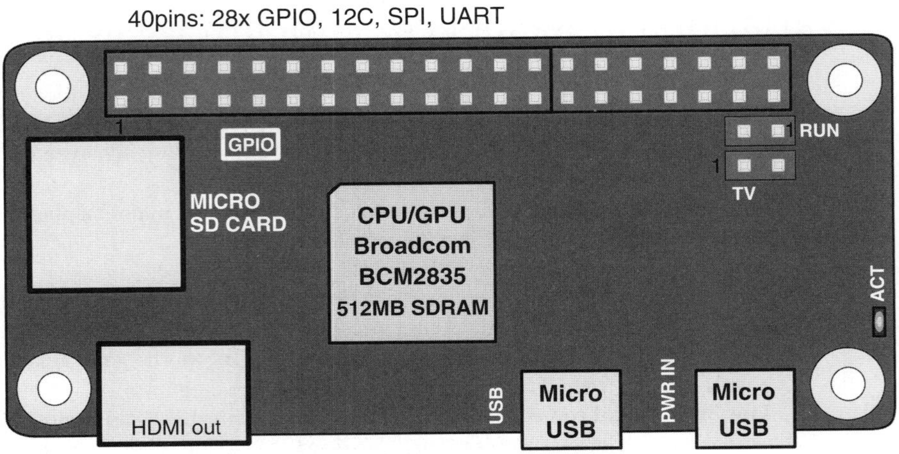

40pins: 28x GPIO, 12C, SPI, UART

GPIO

1

MICRO
SD CARD

CPU/GPU
Broadcom
BCM2835

512MB SDRAM

RUN

1

TV

ACT

HDMI out

USB

Micro
USB

PWR IN

Micro
USB

Figure 27-3   Layout of the Raspberry Zero. (*Art courtesy Efa.*)

Unless you already have a Raspberry Pi kicking around your closet, you're best off getting the latest version, so you'll be as up-to-date as possible. The recent hardware is more capable, with more projects examples written for them.

Which one to get, the standard model or Zero form factor? Either is perfectly suitable unless you are restricted in size. In that case, the Zero is the better option. It also uses less current, so a battery-based project will have better life between charges.

But be aware that the Zero has a somewhat slower processor (1 GHz, single-core CPU versus 1.4 GHz four-core in the Pi 3 Model B), and includes 512 KB of RAM, rather than 1 GB.

Powering the Pi

The Raspberry Pi may be powered via its micro USB connector, or by *regulated* 5V power connected to its 5V input.

- The typical USB 2.0 port on a PC is rated at 500 mA. The Raspberry A and B models can conceivably consume more current than that, depending on the peripherals attached to it. A powered USB hub can provide additional current for devices connected to your PC.
- A better choice is to power the Raspberry Pi from a regulated 5V adapter capable of 2.5 amps or greater.
- If you wish to power your Raspberry Pi project with batteries, a USB battery pack/charger is a good option. It provides the rechargeable battery, USB cable, and internal 5V regulator.

Note that the USB connector on the Raspberry Pi is the micro type, the same commonly used in smartphones and mobile devices. Micro and mini USB plugs are similar in overall size, but they are not compatible with one another. Trying to fit a mini plug into a micro jack may damage both components.

Choice of Operating Systems

For newcomers one of the most perplexing aspects about the Raspberry Pi is its wide range of available operating systems. Actually, most of these choices are really the same basic *Unix-like OS*—that is they borrow in one measure or another elements of the Unix operating system.

One of the most popular Unix-like operating systems in Linux. Several of the Raspberry Pi operating systems are likewise based on Linux. These are often referred to as *Linux distributions*. CentOS and Fedora are examples of Linux distributions available for the Raspberry Pi.

If you're a dyed-in-the-wool Unix/Linux user, you'll have your favorite version already. But if you're just starting out, one good choice is NOOBS, available on an already-formatted SD card. You need only insert the card into Raspberry Pi's reader, and you're set to go. NOOBS (for "newbies") is available from a number of online sources, and is often included in Raspberry Pi starter packages.

NOOBS is basically an installer for Raspian, which is the de facto standard OS for the Raspberry Pi. You can alternatively provide your own SD card, format it, then download the Raspian *image* from the Raspberry Pi Web site.

Accessing the Raspberry Pi

With most microcontrollers you program them from a PC, and then upload the prepared program to the device. Once on the device the program executes. Because the Raspberry Pi is really a computer, everything is done within the device itself. To program it you access it as you would any desktop computer.

There are several methods of accessing a Raspberry Pi. Among the more common are:

- **Direct connection.** Use an HDMI cable to hook up to a monitor, and a keyboard from which you can type commands. If your Raspberry Pi starts (boots) with a window-style graphical interface, you may also need a mouse to navigate through the menus and icons.
- **Through Secure Shell (SSH) and an Ethernet connection.** With SSH you remotely log into your Raspberry Pi and operate it from another computer. You need an SSH program on your computer; they're available for all operating systems. For Windows a popular choice is PuTTY.
- **Through a USB serial console.** A USB-to-serial adapter can provide a link between your computer and Raspberry Pi. The adapter connects to the power and serial port pins on your Pi (see Figure 27-4). You need a terminal program on your computer (like PuTTY for Windows) where you can type commands and see the results.

Figure 27-4   A USB serial adapter lets you access the Raspberry Pi from a console (terminal) program running on a separate computer.

The method you use depends on your experience and preference. If you're just starting out, the direct connection provides the most flexibility, and takes only a little more effort to sort out a keyboard, mouse, and monitor (and the mouse may not even be needed). You can set up your Pi just the way you want it. But it can also be more time consuming as you connect and disconnect monitors and keyboards.

The Raspberry Pi is an Internet-centric computer. All but the very early Raspberry Pi's have built-in Ethernet support: for a wired connection just plug in a network cable; for wire-free, log into your wireless router. Setting up Internet access is one of the first things you must do when you begin work with a Raspberry Pi.

While you could conceivably use a Pi without access to the Internet, you'd be quite stymied in the process. A common task is to add or update programming packages—modules that extend the functionality of the Raspberry Pi. The process is easy when you're connected to the Internet, but aggravatingly time consuming when you're not.

Establishing a connection to the Internet depends on the model of Raspberry Pi, its operating system, and the method of connection. You can find specific steps on the Raspberry Pi Web page at *raspberrypi.org*.

Figure 27-5 A servo HAT allows you to connect and control multiple R/C servos via a serial communications link.

Hardware Expansion

As capable as the Raspberry Pi is, some jobs call out for more hardware. Taking a cue from the popular shields for the Arduino, the Pi supports a similar concept, called a HAT—for *Hardware Attached on Top*. HATs plug into the GPIO header on the top of the board, either directly or with a ribbon cable.

HATs are often used to overcome a hardware limitation in the Raspberry Pi. For example:

1. **Servo HAT.** Controlling more than two or three servos on a Raspberry Pi can get complicated. A servo HAT (see Figure 27-5) provides not only the 3-pin headers to connect to servos, but a coprocessor that takes over managing all the servo. You use an I2C or SPI serial link to send commands to the HAT, which in turn moves the servos.

2. **Analog-to-Digital Converter HAT.** The Raspberry Pi doesn't have any pins with built-in analog-to-digital (ADC) conversion. If you want to use an analog sensor with a PI you need to provide the converter yourself. An ADC function is a common feature of general-purpose prototyping HATs, which may provide analog inputs and high-current outputs (for motors and relays).

Some types of sensors, such as a light-dependent resistor (CdS cell), phototransistor, or flex sensor, can use a different technique to convert analog-to-digital voltage. Using an RC circuits, which combines a resistor and a capacitor, you can simulate the voltage at the sensor by timing how long it takes for the capacitor to charge. Read more about RC circuits in Chapter 29, "Interfacing Hardware with Your Microcontroller."

Breakout boards and adapters provide additional methods to expand a Raspberry Pi. For example, rather than connect jumpers directly to the 40-pin GPIO header, you might prefer a breadboard adapter, which permits easier prototyping. Plug the adapter into a breadboard, and connect it to the Raspberry Pi using a ribbon cable. Depending on the design of the adapter you have access to some or all of the 40 pins.

Understanding GPIO Pins

Except for the oldest models, the Raspberry Pi sports a set of double row of header pins for power and I/O. There are 40 pins total; 12 are for power (5V, 3.3V, and ground), and the remainder for various input/output tasks. The pinouts of the 40-pin GPIO header is shown here.

GPIO #	Alt function*	Pin #		Pin #	Alt function*	GPIO #
—	+3.3 V	1		2	+5 V	—
2	SDA1 (I2C)	3		4	+5 V	—
3	SCL1 (I2C)	5		6	GND	—
4	GCLK	7		8	TXD0 (UART)	14
—	GND	9		10	RXD0 (UART)	15
17	GEN0	11		12	GEN1	18
27	GEN2	13		14	GND	—
22	GEN3	15		16	GEN4	23
—	+3.3 V	17		18	GEN5	24
10	MOSI (SPI)	19		20	GND	—
9	MISO (SPI)	21		22	GEN6	25
11	SCLK (SPI)	23		24	CE0_N (SPI)	8
—	GND	25		26	CE1_N (SPI)	7
EEPROM	ID_SD	27		28	ID_SC	EEPROM
5	N/A	29		30	GND	—
6	N/A	31		32		12
13	N/A	33		34	GND	—
19	N/A	35		36		16
26	N/A	37		38	Digital IN	20
—	GND	39		40	Digital OUT	21

* For non-power pins, *Alt function* is the alternative function of the pin beyond generic I/O.

When programming the Raspberry Pi you can opt to refer to I/O by their physical pin (or board) number, or by their GPIO ID. For example, Board pin 12 is GPIO pin 18. This difference is a common source of confusion to those new to programming on a Raspberry Pi. You can use either method to reference pins, but you should be consistent to avoid errors.

Programming Options

In the world of Unix and Linux, why settle on just one way of doing things when you can have six?! For every task you'll do with the Raspberry Pi, odds are you'll encounter at least a handful of other approaches, many of them equally capable in their goal, but differing in their technique. This certainly applies to coding options for the Raspberry Pi. Virtually every programming language ever developed is available to you, and several are provided with the Raspberry Pi, including C, Java, JavaScript, Scratch, and Perl.

So many choices can be daunting; fortunately one language stands out as a favorite: Python. It's easy to learn and use, and can be expanded with new functionality by using third-party modules. You'll find large libraries of ready-made code for robotics tasks in Python.

Most of the time you'll save your Python scripts into files stored on the Pi's Flash memory. To execute (run) the script, you provide a command at the command prompt, like:

```
sudo python myscript.py
```

- *sudo* tells the Pi that you want to run the command as a superuser, a user of the operating system who can run any program and access any file. Your superuser status lasts only as long as the one command.
- *python* is the name of the main Python language interpreter
- *myscript.py* is the name of the Python script you want to run

For this to work the script—and possibly the Python interpreter itself—needs to be in the current directory that you are executing the command from. Otherwise you may need to provide the full path for each in your command. Check out the Raspberry Pi documentation for additional details on the various methods of running commands at the command prompt.

You can create and store as many scripts as you'd like, and experiment with each. For convenience store your Python scripts in one directory so you have access to them. Start it at the command prompt, then refine it as needed. You create and edit program scripts using a text editor. Many editors are provided in Raspian and other Raspberry Pi Oss; among the most popular are *nano* and *IDLE*. Try them to see which one you like best.

Some Common Robot Functions

The Raspberry Pi is adept at any task that requires the control of a digital input or output. This makes common robotic functions like testing for bumper switch contact and other digital input a snap.

As noted above, the Pi is less capable of reading analog signals because it lacks an onboard ADC; you must add your own if you wish to use analog sensors. The easiest method is to fit the Raspberry Pi with a HAT that incorporates the requisite ADC circuitry on it.

Let's take a look at both controlling a digital output, and reading a digital input to determine its state. The example code assumes you are using a current version of Raspian, which includes all the required Python modules for accessing the GPIO pins.

LIGHTING AN LED

Figure 27-6 shows the familiar hookup diagram for lighting an LED from a digital pin. The LED is connected to physical pin 7, which can be referred to in code either by its *board number,* or its *BCM number.* More about this in a bit. Don't forget the 680 Ω current limiting resistor between the LED and the IO pin.

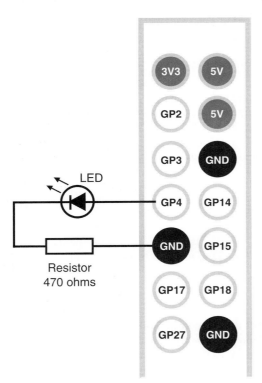

Figure 27-6 Connection diagram for lighting an LED from a Raspberry Pi.

The code is straightforward

```
import RPi.GPIO as GPIO
from time import sleep

GPIO.setmode(GPIO.BCM)
GPIO.setup(4, GPIO.OUT)

i = 1
while i <= 5
  GPIO.output (4, True)
  sleep (1)
  GPIO.output(4, False)
  sleep (1)
  i += 1
GPIO.cleanup()
```

The lines

```
import RPi.GPIO as GPIO
from time import sleep
```

add code libraries from standard modules that are a part of the Python installation. The *RPi. GPIO* module provides useful coding shortcuts for using the GPIO pins, and the *time* module gives you ready access to handy clock-related tools, such as providing a momentary delay with the *sleep* statement.

The lines

```
GPIO.setmode(GPIO.BCM)
GPIO.setup(4, GPIO.OUT)
```

set up the Raspberry Pi's GPIO pins. The *setmode* method tells the Pi you want to refer to the I/O pins from the Broadcom channel numbers rather than the physical board numbers (Broadcom is the maker of the "system-on-chip" that forms the heart of the Raspberry Pi). You're using BCM Pin 4, which is the same as Board pin 7. The *setup* method indicates that the pin i an output.

A while loop repeats a set of code so that the LED flashes on and off a total of five times:

```
i = 1
while i <= 5
  GPIO.output (4, True)
  sleep (1)
  GPIO.output(4, False)
  sleep (1)
  i += 1
```

And the last line performs some housekeeping cleanup to reset the GPIO to its default. While the cleanup method isn't always required, it's a good idea while you' menting with the Raspberry Pi.

```
GPIO.cleanup()
```

You can use the Raspberry Pi console to display messages from your script. Sprinkle some *print* statements throughout your code to help you follow what it's doing:

```
while i <= 5
  GPIO.output (4, True)
  print ('Light on')
  sleep (1)
  GPIO.output (4, False)
  print ('Light off')
  sleep (1)
  i += 1
```

Each *print* statement sends the text message in quotes to the console during script execution. For this script you'll get something like:

```
Light on
Light off
Light on
Light off
```

. . . and so on

DETECTING A PUSH-BUTTON SWITCH

The momentary push-button switch is the world's simplest robotic sensor. Figure 27-7 shows the basic connection diagram of two switches and the LED from the previous example. One switch is used to light up the LED in response to the switch state, and the other switch is used to gracefully exit the program.

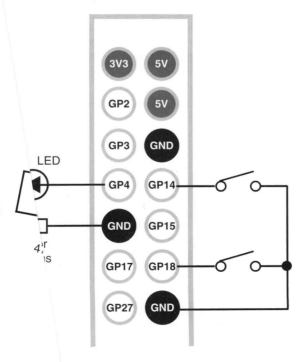

Figure 27-7 Hookup diagram for connecting a push-button switch to a Raspberry Pi.

```
import RPi.GPIO as GPIO
import time

GPIO.setmode(GPIO.BCM)
GPIO.setup(4, GPIO.OUT)
GPIO.setup(14, GPIO.IN, pull_up_down=GPIO.PUD_UP)
GPIO.setup(18, GPIO.IN, pull_up_down=GPIO.PUD_UP)

while True:
  input_state = GPIO.input(14)
  if input_state == False:
    GPIO.output (4, True)
  else
    GPIO.output (4, False)
  if GPIO.input(18) == False:
    break
GPIO.cleanup()
```

Now there are three *setup* methods, one for the LED connected to BCM pin 14, and two more for the switches attached to BCM pin 14 (physical pin 8) and BCM pin 18 (physical pin 12).

Note also a third parameter in the *setup* method for the switches: the code *pull_up_down=GPIO. PUD_UP* establishes an internal pull-up resistor, provided by the Raspberry Pi, for use with the switch. The pull up sets the default state of the switch as True (1 or HIGH) when it is not depressed.

A while statement creates an infinite loop. Inside the loop the program constantly checks if Switch #1 is depressed. When it is, its value goes to False (0 or LOW). The LED is lit accordingly.

```
while True:
  input_state = GPIO.input(14)
  if input_state == False:
    GPIO.output (4, True)
  else
    GPIO.output (4, False)
```

The code

```
if (GPIO.input(18) == False:
    break
```

literally "breaks" out of the infinite while loop. The program cleans up the GPIO and then terminates.

Note the inverted logic of the switches. Because you're using a pull-up resistor, the normal (non-depressed) condition of the switch is True (also notated as 1 or HIGH). When the switch closes, it creates a path to ground; that produces a value of False (0 or LOW).

With this in mind you can simplify the code by doing away with the first *if* test, and simply apply *not-logic* to the *input_state value*:

```
GPIO.output (not input_state)
```

The *not* acts to invert the value of *input_state*: False becomes True, and True becomes False. You can use these values to directly drive the LED status.

High-Level Functions with the Raspberry Pi

For me this is where the Raspberry Pi really shines. The average microcontroller is great for running motors and interfacing with sensors. They're cheap and easy to program, requiring little more than a USB cable. But it lacks the horsepower for really big tasks, like machine vision or full motion video display.

Why not combine the best of both worlds: use a simpler microcontroller for all the low-level *roboty* stuff, and augment it with a Raspberry Pi that performs some high-level function: How about vision to locate humans or map a room, or a colorful touch screen interface that lets you interact with your creation *a la* the fancy control panels in Star Trek.

On the surface this kind of project may sound like tremendously complex, and it is if you're trying to do it all from scratch. But you'd be surprised at the number of turnkey kits and products that use a Raspberry Pi to do the heavy-duty lifting. These expansion systems offload the very resource intensive requirements of tasks like full motion video and machine vision from your main microcontroller.

Among the preeminent ready-made projects using the Raspberry Pi are Google's *AIY* line of experimenter tools. As of this writing there are the AIY Voice and AIY Vision kits; the second generation of these kits come with a Raspberry Pi Zero W (the version with wireless Ethernet already inside), and software ready-to-go on a micro SD Flash card.

Both kits come with funky cardboard enclosures, but there's no reason you can't mount the electronics and mechanics atop your robot. With these and similar expansion kits you can use I2C or even a standard serial link to communicate between the Raspberry Pi and your robot's main microcontroller.

AIY VOICE KIT

With the Voice Kit you can add speaker-independent natural language voice commands to your robot.

- *Speaker-independent* means you don't need to "train" the system to respond to the uniqueness of your voice. That's usually a requirement of low-end voice recognition systems.
- *Natural language* means you can speak in normal phrases and sentences, and the system will *parse* out your speech, then piece together what you want out of the important bits.

From the outside this may sound a little boring, and not at all robot-like. But consider the growing interest in home robots that act as personal assistants. Devices like Apple's Siri, Amazon's Echo, and Google's Assistant are all examples of speaker-independent natural language voice systems.

The Voice Kit connects to the Google Assistant, which lives in the cloud. Push a "listen" button, say something, and your speech is processed on Google's servers. You get back keywords that you can then use as prompts or commands.

For example, suppose you have two motors connected to an Arduino (or maybe directly to an H-bridge) using GPIO BCM pins 14 and 15. In your code you look for certain keywords for commanding the robot, including "Robot stop" and "Robot go."

```
if 'robot stop' in text:
  aiy.audio.say('OK, stopping')
  GPIO.output(14, False)  # turn motors off
  GPIO.output(15, False)  # text after the hash is a comment
if 'robot go' in text:
  aiy.audio.say('OK, going')
  GPIO.output(14, True)   # turn motors on
  GPIO.output(15, True)
```

Two GPIO pins are connected to the I/O pins on another microcontroller, such as an Arduino. A sketch running on the Arduino checks its I/O pins for changes. The "low-level" microcontroller in turn controls DC motors, servo motors, ultrasonic sensors, analog light sensors—you name it.

 Of course you can always skip the intermediate microcontroller and operate everything from the Raspberry Pi. I like to unload higher level functions to dedicated hardware to make troubleshooting and debugging easier.

The Voice Kit requires you to register a developer account with Google, and may include proving payment information if you go over your free 60 minutes per month.

The main parts of the second generation Voice Kit are shown in Figure 27-8. The kit includes the Raspberry Pi Zero W with header pins already soldered on, plus a preformatted micro SD card with the Raspian OS and all software and demo files already included. The *Voice Bonnet* card is a HAT for connecting the large arcade button and speaker, and fits atop the Zero. It also includes two small microphones used to listening for command phrases.

 The arcade button grabs the attention of the voice recognizer: press it, and say your command. You can also clap your hand, say a programmed hotword, or use an external trigger like a digital infrared sensor.

AIY VISION KIT

It's easy enough to endow a robot with a basic kind of sight by using light or sound sensors. These detect an object—any object—that may be in front of it. These kind of sensors, as easy as they are to use, don't provide a world view for the robot. Is that a couch or a person standing in the way? Is there a piece of clothing on the floor ahead, or is a cat sleeping peacefully?

Machine vision is the scientific field of giving sight to a computer. The source image is often from a video camera taking rapid snapshots of the outside world. Software running on the computer analyzes each of these snapshots, attempting to recognize images and shapes from it. The visual scene is converted into a (usually textual) description that the computer can then process, catalog, and act upon.

Armed with a halfway decent color camera, your desktop computer has enough computing power for relatively sophisticated machine vision. Only your PC is too big and heavy to use on the average robot. Enter the Raspberry Pi, equipped with a color camera and open source vision analysis software.

Figure 27-8 Main parts from the second-generation Google AIY Voice Kit.

That's the idea of the Google AIY Vision Kit (see Figure 27-9), an affordable self-standing machine vision solution that includes a Raspberry Pi, color camera, and pre-loaded software. Like the Voice Kit above, the Visual Kit comes with a foldup cardboard enclosure which you can perch on the top of your robot. Or if you want something that looks more integrated, you can mount the individual pieces of the kit on or in your robot.

At the heart of the Vision Kit is the *Vision Bonnet,* a custom HAT, created by Google for the Raspberry Pi Zero using an Intel Movidius vision processing chip. Unlike the Voice Kit, which requires access to the cloud, everything you need to detect people, faces, objects, shapes, and colors is canned inside the *Vision Bonnet.*

Augmenting the hardware is a set of software modules and demos that provide useful examples for further exploration. Perhaps the most useful for robotics is the object classification demo, where the kit will identify some 1000 common objects, plus over 2000 predefined food dishes, including leek soup and onion rings.

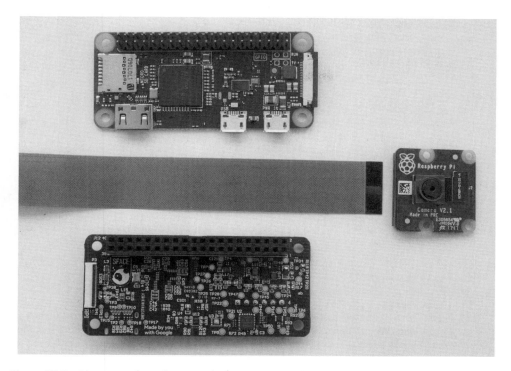

Figure 27-9   Main parts from the second-generation Google AIY Vision Kit.

The main object analysis software is TensorFlow, open source "deep learning" neural network software developed by Google and used by them in various commercial and experimental projects. TensorFlow can scan an image and identify objects, noting their size and location on the screen (see Figure 27-10) along with a percentage of certainty that the object is what it thinks it is. You can add your own images to the database.

Simpler vision analysis involves looking for outlines, basic shapes, and colors. For instance, a ball playing robot that looks for a red ball will first identify the object as a ball, move toward it, and kick it or grab it.

EXPANDING WITH THE VISION BONNET

If you're wanting to keep all your development on the Raspberry Pi, the *Vision Bonnet* has its own microcontroller that can off-load many of the more common robotic functions. The extra functionality of the Bonnet is supported by an additional software package named *gpiozero* that you must install; once installed the software modules containing code or the extra functionality is available for your projects.

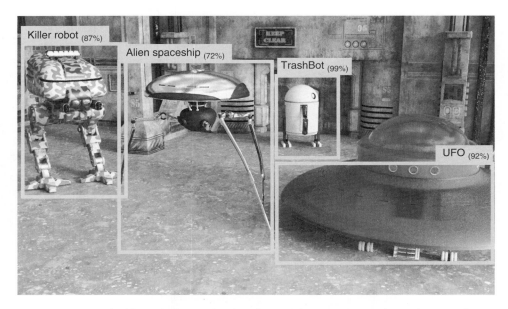

Figure 27-10   Backed by Google's artificial intelligence the Vision Kit analyzes objects in the picture and detects known objects, giving each a relative percentage of certainty that the object is what it thinks it is.

For example, if you want to use an HC-SR04 ultrasonic distance sensor, you can connect the sensor to the Bonnet through two voltage divider resistors (noted in the *gpiozero* documentation), and obtain the distance with a simple bit of code:

```
from gpiozero import DistanceSensor
from time import sleep

sensor = DistanceSensor(echo=18, trigger=17)
while True:
  print('Distance: ', sensor.distance * 100)
  sleep(1)
```

Or to read a CdS cell, the library provides a handy LightSensor library:

```
from gpiozero import LightSensor

ldr = LightSensor(18)
ldr.wait_for_light()
print("Light detected")
```

If the cell receives sufficient light to trigger the circuit the program will output "Light detected."

CHAPTER 28

More Microcontrollers for Robots

The world is awash with microcontrollers, and many of them make for handy robot brains. I can't describe all of the available microcontrollers here, but what follows is a short list of three popular controllers commonly used in small robotics.

- Revolution Robotics PICAXE
- Parallax BASIC Stamp
- Parallax Propeller

All are affordable with free programming tools and plenty of online examples.

Using the PICAXE

PICAXE is a family of low-cost single-chip microcontrollers that use the BASIC programming language. Versions are available with as few as 8 pins and 6 I/O lines, to 40 pins and over 30 I/O lines. You can plug a PICAXE into a solderless breadboard and, with just a few resistors, have a complete microcontroller system ready for your use.

Figure 28-1 shows the pinouts for the 08M2, the smallest of the PICAXE chips, and its larger cousin the 18M2. Both are among the more popular microcontrollers of the family. Other chips differ not only in the number of physical pins on the chip, but in the ordering of the functions for the inputs and outputs.

For example, the 18M2 chip supports additional programming commands not found in the 08M. The variations within the PICAXE chips—what features they support and what commands they'll run—can be a point of confusion. Always refer to the PICAXE documentation for a full understanding of the differences and capabilities within each chip in the family.

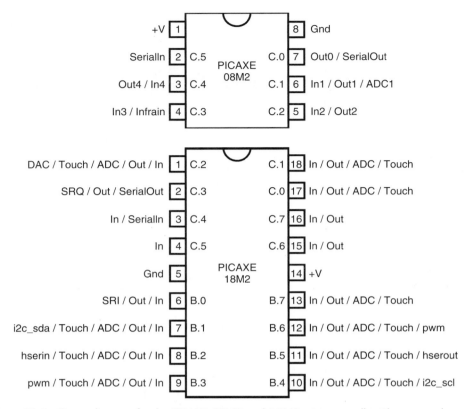

Figure 28-1 Pinout diagram for the PICAXE 08M2 and 18M2 microcontroller. There are other PICAXE chips available; refer to the PICAXE documentation for more details on the rest of the family.

For all PICAXE chips it's important to remember the difference between the physical pins of the IC and the names of the I/O lines. What's referred to as *C.4* is the name given to an I/O line, for example not the actual pin on the chip.

The reason for this is consistency in sharing programs across the different PICAXE chips. The functions available at the different physical pins (legs) of the chip can vary, meaning that physical pin 4 on one PICAXE controller may have a completely different function than physical pin 4 on another PICAXE.

BASICS OF ALL PICAXE CONTROLLERS

All PICAXE chips require connections to power and ground. Except for some special low-voltage varieties, the PICAXE runs on power from 3 to 5.5 volts. It's common to power the chip with a set of two or three AA or AAA batteries. This provides 4.5 volts when using nonrechargeable batteries. If using rechargeable batteries rated at 1.2 volts per cell, you're better off using four cells, which provide a total of 4.8 volts.

Don't power the PICAXE with more than 5.5 volts. Otherwise, the chip will likely be destroyed within seconds.

The PICAXE is programmed via a USB connection from your PC. You need a cable with a USB-to-serial adapter built into it, available from Revolution Education, the makers of the PICAXE. The cable has a USB Type-A connector on one end and a "stereo"-style 1/8″ mini plug on the other. To use the USB cable you'll need a USB driver for your operating system, also available from Revolution. Note that unlike the Arduino and some other microcontrollers, the USB cable does not provide power for the device. You need to connect to a battery separately.

PICAXE controllers require at least two external resistors to function. The minimal circuit is illustrated in Figure 28-2, showing how the two resistors are used for the program uploading process. Even when not uploading a program, the resistors must remain part of the circuit, or else the PICAXE will not behave correctly.

- The 22 kΩ series resistor limits current from the PC serial port to the chip.
- The 10 kΩ pull-down resistor prevents the *SerialIn* pin from floating, which can happen when the PICAXE is not actively connected to a serial port. If left floating, operation of the chip can become unstable.

Additionally: on PICAXE chips that have a Reset pin you need to connect a 4.7 kΩ resistor to the +V power; chips that have an external clock need a ceramic resonator.

In newer documentation, PICAXE input/output pins are referenced using a *port.letter* identifier, and a number. This method helps to provide consistency across devices when programming. For example on the 08M2, pin C.1 can be used as an output, digital input, and analog input, among other tasks. On the 18M2, pin C.1 is physically located on a different "leg" of the IC, and is capable of much of the same functionality.

Over the years the architecture of the PICAXE chips has changed, and the *port.letter* nomenclature is not used in older documentation or examples. For instance, you might see code that identifies the pin as 4, rather than *C.4*. You will need to manually translate these older pin assignments if you're using a newer chip.

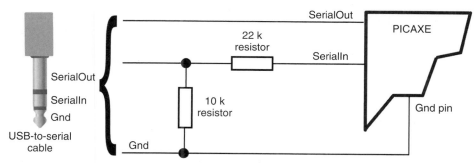

Figure 28-2   The minimal parts setup for all the PICAXE controllers consists of the microcontroller itself, plus two resistors used for uploading programs. The resistors must remain even when the cable is removed.

A CLOSER LOOK AT THE PICAXE 08M2

Let's now take a closer look at the 08M2, the latest generation 8-pin PICAXE controller.

- Two of its eight pins are dedicated for specific use—power and ground—leaving the remaining six for general input and/or output.
- Pin C.3 (physical pin 4), is an input only.
- Pin C.0 (physical pin 7), works as an output only.
- The remaining pins can serve as either inputs or outputs. Their respective basic and extra functions are listed.

Note in Figure 28-2 the wire from *SerialOut,* the link back to your PC's serial port. During programming this I/O line is occupied, but it can be used as another output if you disconnect the USB cable after upload is complete.

The 08M2 has somewhat limited memory available for programs, though much more than previous generations of the 08 family. The 08M2 contains 2K of memory area for program storage, and 128 bytes of RAM for program execution. While this may not seem like a lot of memory, for many single-function applications—a typical use for a low-cost 8-pin controller like the 08M2—it's more than adequate.

The 08M2 is particularly well suited as a coprocessor that offloads tasks from a main processor. For instance, a main feature of all the current PICAXE chips is its ability to listen for standard universal remote control signals. A simple program running in the chip can detect and decode IR signals, and send them (via simple serial) to your main controller. In this way, you can handle background tasks on one or more inexpensive PICAXE chips. This frees up your robot's main microcontroller, allowing it to be more nimble.

PROGRAMMING THE PICAXE

All PICAXE controllers may be programmed using editor software provided by Revolution Education. Check their Web site at *picaxe.com* for details. Editors are available for Microsoft Windows, Macintosh OS X, and Linux. You can select from among several programming styles, including textual and graphic blocks.

If you're just starting out with the PICAXE you might want to opt for a starter pack or project board. It comes with a PICAXE chip, USB cable, battery holder, and a PCB for prototyping. After use, be sure to remove at least one battery from the holder, or disconnect the holder from the board, to prevent the batteries from being slowly drained.

The program editor, shown in Figure 28-3, supports the PICAXE BASIC language. Here you can write, save, and print your programs. PICAXE programs are stored as standard ASCII files, so they can be opened by any text editor.

The program editor has some nifty features you'll want to try. These include:

- *Workspace settings* where you can specify the specific PICAXE microcontroller you're using, and the serial port on your computer that's connected to the chip.

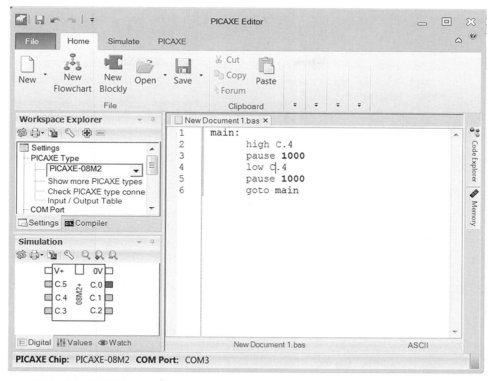

Figure 28-3   The PICAXE integrated programming editor. From the editor you can write, edit, simulate, compile, and upload your PICAXE programs.

- Mistakes in your program are flagged using the *Syntax checker*. Errors are flagged by line, where you can make necessary corrections.
- Numerous *wizards* help in constructing several kinds of programming code. Examples include a tune wizard, used to build a progression of musical notes, and a data logger wizard for use in storing information collected over long periods of time.
- The built-in *simulator* will run through your code and show you how it'll function inside the PICAXE chip. You see how memory is used, and what effect your code has on the I/O pins of the controller.

Your programs must be compiled before they can be sent to the PICAXE. This is done in the editor by clicking on the *Program* button. Syntax errors are flagged, and if they are found, compiling stops. When you have successfully compiled the program it is automatically uploaded to the PICAXE.

See the RBB Support Site (see Appendix) for more complete details on the PICAXE programming language.

PICAXE FUNCTIONS FOR ROBOTICS

Much of the power of the PICAXE comes in its set of built-in commands that reduce the complexity of programming. Most are designed to control some activity of the chip, like to make sound through one of the output pins, or to produce timed signals to control one or more R/C servo motors. I'll briefly review the special functions most useful for robotics, but you'll also want to look over the PICAXE manual for more.

button The *button* command momentarily checks the value of an input and then branches to another part of the program if the button is LOW (0) or HIGH (1). The *button* command lets you choose which input pin to examine, the "target state" you are looking for (either 0 or 1), and the delay and rate parameters that can be used for such things as switch debouncing. The *button* command doesn't stop program execution, which allows you to monitor a number of pins at once.

irin and **irout** These commands are used to receive and send (respectively) infrared remote control signals that use the *Sony SIRC* protocol. *Irin* is used in conjunction with an infrared receiver/modulator, which detects the signals from the remote, and converts them to on/off pulses. Conversely, *irout* is used in conjunction with an infrared LED to produce Sony-compatible IR signals. You might use these features to communicate with other robots via an infrared link, or to remotely control your TV and other consumer electronics gear that work with the Sony standard.

servo The *servo* command (and its lieutenant, *servopos*) allows you to control the operation of up to eight R/C servo motors (depends on the number of output pins available on the chip, of course). With the command you specify the output port to use, along with the position of the motor, in tenths of a microsecond. Valid values are from 75 to 225 (note: you must use whole numbers only). For example, 75 sets the speed to 750 microseconds (0.75 millisecond); 225 sets it to 2250 µs (2.25 ms).

sound The *sound* command is used to generate tones primarily intended for audio reproduction. You can set the I/O pin, note (frequency in hertz) and duration. You can string a series of notes (each with a separate duration) to produce simple monophonic music or sound effects.

pause The *pause* command is used to delay execution by a set amount of time. To use *pause* you specify the number of milliseconds (thousandths of a second) to wait. For example, *pause 500* pauses for half a second.

pulsin The *pulsin* command measures the width of a single pulse. You can specify which I/O pin to use, whether you're looking for a 0-to-1 or 1-to-0 transition, as well as the variable you want to store the result in. *Pulsin* is handy for measuring time delays in circuits, such as the return "ping" of an ultrasonic sonar.

pulsout The *pulsout* command is the inverse of *pulsin:* with *pulsout* you can create a finely measured pulse with a duration of between 10 and 65535 µs (the resolution is increased when running the PICAXE at higher speeds). The *pulsout* command is ideal when you need to provide accurate waveforms.

readadc The *readadc* command reads a linear voltage at any ADC (analog-to-digital converter) input pin. The voltage is then converted to an 8-bit (0 to 255) digital equivalent. A similar command, *readadc10,* provides a 10-bit (0 to 1023) resolution.

serin and **serout** The *serin* and *serout* commands are used to send and receive asynchronous serial communications using any output pin. They represent one method for communicating with other devices. Both commands require that you set the particulars of the serial

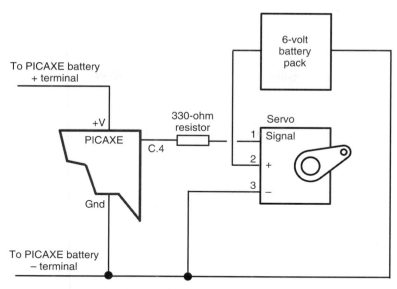

Figure 28-4 Connection diagram for attaching a standard R/C servo to a PICAXE. You must power the servo from its own 6-volt battery. Be sure to connect together the ground (Gnd) leads of the PICAXE battery and the servo battery.

communications, such as data (baud) rate, and the number of data bits for each received word. One application of *serout* is to interface a serial liquid crystal display to the PICAXE. You use *serout* to send commands and text to the LCD.

USING SERVOS WITH THE PICAXE

Let's take a deeper look at the *servo* command, to show how easy it is to control R/C servos with the PICAXE. Figure 28-4 shows a typical connection diagram between PICAXE and servo, with an optional 330 Ω resistor in series on the signal line.

This resistor is recommended in the PICAXE documentation, and limits the current draw from the servo, should it go haywire. Note the separate battery supply for the servo, and the connected ground wires between the PICAXE and the servo. Both are required for proper operation.

Code for the 08M2 is straightforward:

```
init:
   servo C.4,150      ' Set up servo

main:
   servopos C.4,100   ' Move servo to one end
   pause 2500         ' Pause 2.5 seconds
   servopos C.4,200   ' Move servo to other end
   pause 2500         ' Pause again
   goto main          ' Repeat
```

The servo is first initialized with a value of 150, which equates to 1500 microseconds, or 1.5 milliseconds. This centers the servo midway in its travel. The main body of the program then

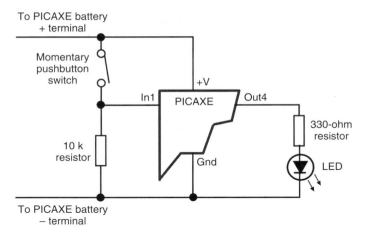

Figure 28-5 Connection diagram for demonstrating a pushbutton input and LED output. The pushbutton switch is a normally open momentary type.

repeats a loop where the servo is positioned clockwise and counterclockwise, pausing between each transit for two and a half seconds.

Note that the PICAXE can support more than one servo at a time, up to a maximum of eight. The more servos you add the more jitter may occur on the servos, due to timing instabilities during program execution. This is especially true if the PICAXE program must branch off to do other tasks, including using the *debug* command to return messages to you.

USING DIGITAL INPUT AND OUTPUT WITH THE PICAXE

A common robotics application is reading an input, such as a switch, and controlling an output, such as an LED or motor. The following example shows a simple method of reading the state of a switch, and then displaying the result on an LED. The switch is connected to pin In1; the LED to pin Out4. Figure 28-5 shows a connection diagram.

The code notes when the switch closes and produces a 1 (HIGH) on pin In1. The LED connected to Out4 lights up, and the program pauses for 1 second before continuing.

```
main:
  if C.1 = 1 then flash   ' 1 = switch closed
  low C.4                 ' Turn off C.4
  pause 10
  goto main               ' Repeat

flash:
  high C.4                ' Turn off C.$
  pause 1000              ' Wait 1 second
  goto main
```

Using the Parallax BASIC Stamp

Since its inception, the Parallax BASIC Stamp has provided the onboard brains for countless robot projects. This thumbprint-sized microcontroller uses BASIC-language commands for instructions and is popular among robot enthusiasts, and electronics and computer science instructors.

INSIDE THE BASIC STAMP

Embedded in the BASIC Stamp is a proprietary BASIC-like language interpreter called *PBasic*. The first order of business is to write a BASIC program on your PC. Once complete you upload these commands to the BASIC Stamp, which then executes them one after the other.

The Stamp is available in several versions:

- *BS2 Module.* The BASIC Stamp module (see Figure 28-6) contains the actual microcontroller chip as well as other support circuitry. All are mounted on a small printed circuit board that is the same general shape as a 24-pin IC. In fact, the BS2 is designed to plug into a 24-pin IC socket. The BS2 module contains the microcontroller that holds the PBasic interpreter, a 5-volt regulator, a resonator (required for the microcontroller), and a serial EEPROM chip.
- *BASIC Stamp Board of Education.* Typically sold without a BS2 module, the Board of Education, known also as BOE, offers you a convenient way to experiment with the controller. It contains connectors for four R/C servo motors, and a mini-size solderless breadboard. The BOE provides connectors for quick hookup to a PC to program the Stamp chip. It's available in serial and USB versions.
- *Boe-Bot.* The Boe-Bot is a small metal-fabricated mobile robot kit built around the BS2 chip and includes the BASIC Stamp Board of Education. If you already have a BS2 and BOE, you can purchase just the Bot chassis, which comes with motors, wheels, and all hardware. See the end of this chapter for a quick tour of the Boe-Bot, and discover how easy it is to build your own thinking automaton.

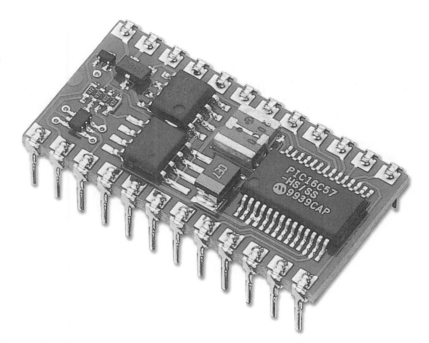

Figure 28-6   The BASIC Stamp 2 is a self-contained microcontroller, complete with the controller chip itself, plus voltage regulator and other parts. It comes on a small circuit board that has the same size as a 24-pin-wide integrated circuit. (*Photo Courtesy Parallax Inc.*)

HOOKING UP: CONNECTING THE BASIC STAMP TO A PC

The BASIC Stamp was engineered to make it easy to connect to a personal computer, using only a serial port or a USB port. These days many computers don't have a serial port; they use USB instead. If your computer lacks a serial port, you're best off getting a USB-to-serial adapter, or opt for a BASIC Stamp experimenter's board that already has a USB connector on it.

You shouldn't rely on power provided by the USB connection to operate your BASIC Stamp. Provide the BASIC Stamp with its own power source. If using the BASIC Stamp by itself, you can connect a 9-volt battery to its power terminals; the built-in voltage regulator will provide the proper volts for the chip.

Once connected to the PC (USB or serial), you need only install the software from the CD or downloaded from the Parallax site, and you're set to go.

 See the RBB Support Site (see Appendix) for more complete details on the BASIC Stamp programming language.

INTERFACING SWITCHES AND OTHER DIGITAL INPUTS

You can easily connect switches, either for control or for "bump" or other contact sensors, to the BASIC Stamp using either of the approaches shown in Figure 28-7. The simplest way to detect a switch closure is with the *Inx* statement (*x* is a number from 0 to 15 that denotes a pin). For example:

```
if In3 = 1 then
```

checks if input pin 3 is 1 (HIGH).

INTERFACING SERVO MOTORS TO THE BASIC STAMP

Servo motors are easily connected to and controlled with a BASIC Stamp. You may connect any I/O pin of the BASIC Stamp directly to the signal input of the servo. Keep in mind that the BASIC Stamp cannot provide operating power to the servo motor; you must use a separate battery or power supply for it.

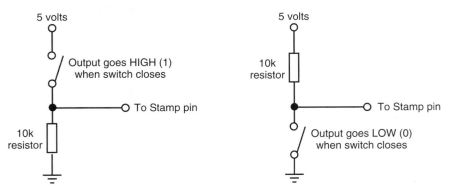

Figure 28-7   When using switch inputs, use a 10 kΩ resistor connected to ground or 5 volts, depending on whether you want the input pin to be LOW or HIGH when the switch is open.

Use the hookup diagram in Figure 28-4 for connecting an R/C servo to the BASIC Stamp by using a separate battery supply for the servo. Note that the ground connections of the power supplies for both the BASIC Stamp and the servo are connected.

To operate the servo, use the *pulsout* statement, which sends a pulse of a specific duration to an I/O pin. Servos need to be "refreshed" with a pulse about 50 times each second to maintain their position.

By adding a delay (using the *pause* statement) and a loop, your BASIC Stamp can move and maintain any position of the servo. The following examples show the basic program, using pin 0 as the control signal line to the servo. It sets the servo in its approximate center position:

```
low 0
Loop:
  pulsout 0,750
  pause 20
  goto Loop
```

To change the angular position of the servo motor, merely alter the timing of the *pulsout* statement:

```
pulsout 0,1000      ' 2000 usec pulse, approx. 180
                    '    degrees position
pulsout 0,500       ' 1000 usec pulse, approx. 0
                    '    degrees position
```

These values assume a strict 1000- to 2000-μs operating range for a full 0 to 180° rotation. This is actually *not* typical. You will likely find that your servo will have a full 180° rotation with higher and lower values than the nominal 1000 to 2000 μs. You can determine this only through experimentation. See Chapter 15, "Using Servo Motors," for more details on this topic.

Using the Parallax Propeller

On the surface, the Parallax Propeller is like many other programmable microcontrollers that can be used to interface with external devices. Common applications include running motors, reading the status of switches, checking temperature probes, and similar tasks common in robotics.

But what sets the Propeller apart from most other microcontrollers is that it's designed from the ground up to be *multi-tasking*. Inside one Propeller are actually eight individual 32-bit microcontrollers, each operating independently but able to cooperatively share resources, like memory and I/O pins. You don't have to use the multi-tasking feature of the Propeller, but it's nice to know the ability is there should you need it.

What follows is information on the Propeller 1. At this time of this writing Parallax was developing the Propeller 2, a next generation multi-tasking microcontroller with additional features.

INSIDE THE PROPELLER

The individual processors in the Propeller are commonly referred to as *cores* or *cogs*. Rather than rely on a specific arrangement of internal hardware like PWM generators, serial ports, and interrupts, each cog can synthesize any of these through software control.

For example, if you just happen to need four high-speed serial ports, you can define multiple cogs to handle all the communication. On the other hand, if you don't need that many serial ports, there's no wasted hardware. The cogs are free for use as something else.

The Propeller is just an IC, available in both surface mount and 40-pin DIP. While you can construct circuits with a bare Propeller chip, more often you'll want to use a premade development or prototyping board, which provides a voltage regulator to supply the required 3.3 volts to the Propeller; USB connection for programming from a computer; plug-in headers for wiring components to the Propeller; and various other support electronics.

There are a number of Propeller-based development and prototyping boards available. Two of the most flexible are

- Propeller Board of Education (*PropBOE*). This full-featured development board contains Propeller chip, small solderless breadboard, headers for connection to sensors and servos, analog-to-digital converter, and built-in hardware for video and audio.
- Propeller Activity Board. The *PAB* (see Figure 28-8) is a lower cost option with similar features to the PropBOE, but without the built-in hardware connectors for video output.
- FLiP. The FLiP is a chip-on-board design made to plug into a 400 point or larger solderless breadboard.

Figure 28-8   The Parallax Propeller Activity Board contains a multi-tasking Propeller chip, USB, small prototyping board, 3-pin headers for servos and sensors, micro-SD card reader, and audio jack on one board. (*Photo courtesy Parallax Inc.*)

The PopBOE and PAB provide many built-in features of particular use for robot builders. These include:

Integrated USB adapter, connects the Propeller to your Windows-based PC for programming using a USB cable. Before first use, be sure to download the Propeller software package, which includes the necessary USB driver to connect your computer to the Propeller.

Solderless breadboard area, with connection points to power, as well as the first 16 I/O pins of the Propeller.

Headers for directly connecting up to six R/C servos, or combination of servos and sensors.

Integrated micro-SD data card, for reading and storing data in permanent and replaceable memory.

Plug-in headers for wireless modules, including standard XBee transceivers.

Audio output jack, for quick connection to a battery-operated amplified speaker.

Built-in analog-to-digital converter, for sampling the voltage level of several analog sensors at once.

THREE EDITORS, MANY LANGUAGES

The Propeller is supported by three official program editors from Parallax.

PropellerIDE Supports three languages, PASM (Parallax assembly language, typically used by developers), PropBasic, and Spin. PropBasic is a variant of the BASIC language for the Propeller. Spin is a general-purpose but proprietary language originally developed to support the Propeller's unique mix of hardware.

SimpleIDE Supports Propeller C, a variant of the C language with an extensive library of supporting functions, such as servo motors.

BlocklyProp Supports a version of Google's Blockly graphic block programming platform.

IDE standard for integrated programming environment.

Of these programming platforms and languages, the SimpleIDE/C and BlocklyProp are among the most commonly used today by Propeller newcomers. Spin (and to a lesser extent PASM) remains a favorite of experienced Prop developers.

SimpleIDE and BlocklyProp are supported in multiple operating systems including Windows, Macintosh, and Linux.

INSTALLING THE PROGRAMMING SOFTWARE

The Prop's programming software is fairly easy to install, with just a few qualifications and special steps. Excellent step-by-step documentation is provided on the Parallax site so I won't duplicate it here.

- PropellerIDE and SimpleIDE are desktop applications. Download the installers for these programs and run them with Administrator/Owner privileges.

- BlocklyProp visual editor is an online tool. Use Google Chrome or other supported browsers to open the BlocklyProp page; see the Parallax support pages for the exact URL, as it might change from time to time.

Don't plug Propeller board into your computer until the software and USB drivers have been installed. During the installation process you'll be prompted if you wish to install the USB drivers.

SPECIAL BLOCKLYPROP SETUP

The BlocklyProp program editor is browser-based, and requires an extra program component, the BlocklyProp Client, if you want to upload your finished programs to your Propeller. You'll be prompted to download and install this tool the first time you attempt to upload a program. The tool acts as a trusted bridge between the BlocklyProp programs in the Web browser and the Propeller hardware. As with the Propeller IDE software programs are uploaded via a USB cable.

Alternatively, you can save your BlocklyProp programs as a C file, and then use SimpleIDE to review, modify, and upload to the Propeller. When saving the BlocklyProp visual editor converts the blocks to their C-code equivalents—handy if you want to learn how your code translates to C.

Two files are saved: a main program file with a .c file extension, and a project file with a *.side* extension. Save them both to the same folder. In SimpleIDE, choose Project->Open, navigate to the .side project file you want, and select it.

USING THE SIMPLEIDE SOFTWARE

For the purposes of conciseness I'll concentrate from here on out with SimpleIDE and its C-based programs and paired to a PAB.

As with all well-written programming suites, SimpleIDE comes with a "Hello world" example program for testing that everything is connected and working properly. The first time Simple IDE is run, a verification program, *Welcome.c*, is automatically loaded for you. See Figure 28-9.

1. Be sure the PAB is switched on, with its power switch set to Position 1. This provides juice to all of the electronics, but not to the servo headers.
2. Double check that the port setting in the SimpleIDE window matches the port for the USB connection to your PAB. This number will vary, depending on other USB-based drivers you've installed. For me, running Windows, the PAB installed as COM7. SimpleIDE will make an educated guess as to the correct port if your PAB is connected and turned on.
3. Find the *Run with Terminal* button on the toolbar. This action (1) complies the program; (2) uploads the compiled code to the PAB; and (3) displays a special Terminal window so that you can view messages sent back from the board. Upon successful completion of the program the terminal will show "Hello!"

See the RBB Support Site for more complete details on the Propeller programming languages.

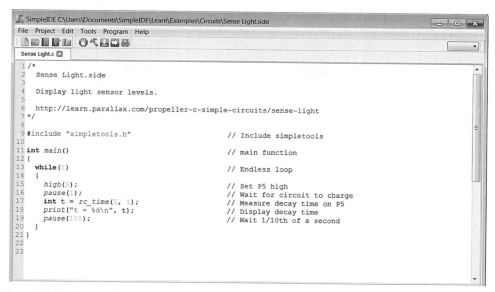

Figure 28-9   The SimpleIDE software allows you to program a Propeller using the C language. Parallax provides a large array of coding examples and support libraries for SimpleIDE.

USING SWITCHES WITH THE PROPELLER

To avoid having robots that roam around like mindless zombies, we attach switches and other sensors to them, then write programming code to sense when those sensors are activated. Use Figure 28-10 as a guide for connecting a switch to the Propeller.

Now for code to see what happens:

```
#include "simpletools.h"

int main {
  while(1) {
    int mySwitch = input(5);
    print (mySwitch);
    pause (100);
  }
}
```

The switch is connected to digital pin 5, which is labeled P5 on the PAB. This pin is referenced in the code as

```
int mySwitch = input(5);
```

which creates a variable named *mySwitch* to hold the instantaneous value of digital pin 5.

The *int* at the beginning indicates the type of variable to create—in this case an integer type, which is simply a whole number. This is followed by the name of the variable, and then finally the actual assignment to the variable of the current state of the switch.

To see the code in action use the Run with terminal button to upload the program and display the terminal window.

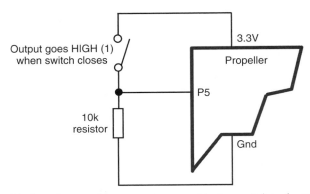

Figure 28-10   Hookup diagram for connecting a pushbutton switch to the Propeller.

Things to notice:

- An *#include statement* references a standard library that has common functions built in. These functions provide convenient shortcuts for such tasks as printing text to the terminal window and pausing for a given period of time.
- The principle body of the program is contained in a function named main.
- Code is repeated forever (as long as the Propeller is powered) using *while(1)*. This syntax produces a simple infinite loop. If you're familiar with the Arduino, it's the same as putting your code in the *loop()* function.

USING SERVOS WITH THE PROPELLER

The Propeller readily supports multiple servos. Because both the PropBOE and PAB boards already have 3-pin headers for connecting servos, you merely need to plug them in. Well ... almost.

The V+ pin of the 3-pin headers is switchable between two power sources: regulated 5V and the input source providing power to the board. As needed change the jumper for the header you are using to its VIN position, and be sure you're powering the board from a 6V battery pack.

For this demonstration plug a servo into the P16 3-pin header along the top of the board, as shown in Figure 28-11.

```
/*
 Test one servo, one time through
 Direction of servo (clockwise, counter-clockwise
    depends on make and model of servo
*/

#include "simpletools.h"          // Include simpletools library
#include "servo.h"                // Include servo library

int main() {                      // main function
    servo_angle(16, 0);           // P16 (right) servo turn clockwise
    pause(3000);                  // Wait 3 seconds
    servo_angle(16, 1800);        // Turn counter-clockwise
    pause(3000);                  // Wait 3 seconds again
```

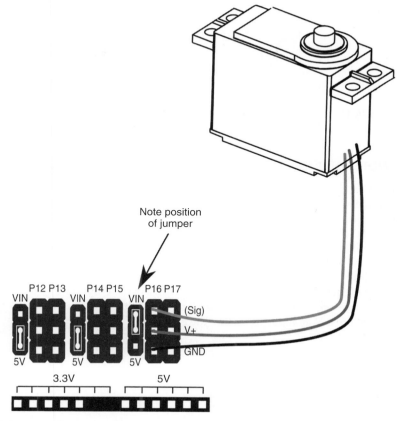

Figure 28-11 Directly connect the R/C servo to the P16 3-pin header. Be sure to change the power jumper to VIN.

```
        servo_angle(16, 0);              // Repeat action
        pause(3000);
        servo_angle(16, 1800);
        pause(3000);
        servo_stop();                    // Stop servo
}
```

1. Connect your PAB to your computer via a USB cable and start the SimpleIDE software.
2. Turn on the power to the PAB by sliding its power switch to Position 2. In Position 2, the main board AND the servo headers receive power.
3. Upload the program to your Activity Board by clicking on the *Load EEPROM & Run* button in the toolbar.

After uploading, the program should start after a few seconds, and cause the servo motor to run.

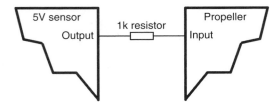

Figure 28-12 The Propeller is a 5V tolerant device, but you still need to use a current limiting resistor when a 5V output is connected to a Prop I/O pin.

A NOTE ON USING 5-VOLT SENSORS ON THE 3.3-VOLT PROPELLER

You can add many types of sensors to a Propeller, but remember that the chip is intended to be operated at 3.3 volts. This means the sensors should either also operate at 3.3 volts, or use level shifting circuitry to avoid harming the Propeller. See Chapter 29, "Interfacing Hardware with Your Microcontroller" for more information.

Another method is to use a current limiting resistor in series between the Propeller and sensor, as shown in Figure 28-12. This works because the I/O pins on the Propeller are 5V tolerant. The resistor—1 kΩ is commonly used—acts to limit current from the sensor to a save level for the Prop. This method ONLY works for a device like the Propeller that is 5V tolerant.

The series resistor is not required when using a 3.3V sensor, or a device (like a servo) that does not provide an input signal into the Propeller.

Interfacing Hardware with Your Microcontroller

The brains of a robot don't operate in a vacuum, even if you've built a vacuum-cleaning robot. They need to be attached to motors to make the robot move and to sensors to make the robot perceive its surroundings. In most cases, these outside devices cannot be directly connected to the computer or microcontroller of a robot. Instead, it is usually necessary to condition these inputs and outputs so they can be used by the robot's brain.

In this chapter I'll show you the most common and practical methods for interfacing real-world devices to computers and microcontrollers. For your convenience, some of the material presented in this chapter is replicated, in context, in other chapters of the book. It's also here to bring everything into focus.

Sensors as Inputs

By far the most common use for inputs in robotics is sensors. There are a variety of sensors, from the supersimple to the amazingly advanced. All share a single goal: providing the robot with data it can use to make intelligent decisions. For example, using a low-cost temperature sensor, an "energy watch" robot might record the temperature as it strolls throughout the house, looking for locations where the temperature varies widely, indicating a possible energy leak.

TYPES OF SENSORS

Broadly speaking, there are two types of sensors: analog and digital, so-called because of the output signal they produce:

- Basic *digital sensors* provide on/off results. A switch is a good example of a digital sensor: either the switch is closed or it's open. Closed and open are analogous to on and off, True and False, HIGH and LOW (these are called *logic levels*). More complex digital sensors can provide

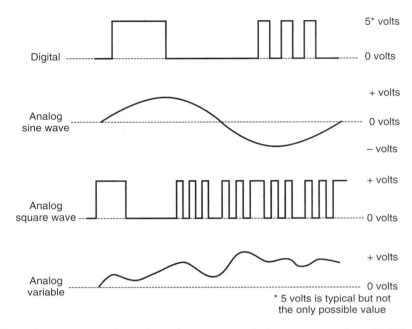

Figure 29-1   Common types of signals used in sensors and other circuits in robots. Digital signals contain information in the sequence of LOW and HIGH pulses. Analog signals contain information either as the instantaneous voltage level or the number of changes in the voltage over a certain period of time, such as 1 second.

a stream of HIGH/LOW pulses that represent a value other than on or off—for example, they can represent a temperature or an angle. These sensors must connect to your robot's brain in very specific ways.

- *Analog sensors* provide a range of values, usually a voltage. In many cases, the sensor itself provides a varying resistance or current, which is then converted into a voltage by an external circuit. For example, when exposed to light, the resistance of a CdS (cadmium sulfide) cell changes dramatically. In a simple circuit with a second fixed resistor, the resistance of the CdS cell is used to provide a voltage.

As shown in Figure 29-1, in both digital and analog sensors, the result is a voltage level that can be fed to a computer, microcontroller, or other electronic device. In the case of a simple digital sensor like a switch, the robot electronics are only interested in whether the voltage is a logical LOW (usually 0 volts) or a logical HIGH (usually 5 volts). As such, these simple digital sensors can often be directly connected to a robot control computer without any additional interfacing electronics.

In the case of an analog sensor, you need additional robot electronics to convert the varying voltage levels into a form that a control computer can use. This typically involves an analog-to-digital converter or RC (resistor capacitor) circuit; both are discussed later in this chapter.

And in the case of a digital sensor that provides a stream of HIGH/LOW pulses to represent more complex data, this stream must be captured and analyzed by the robot's control computer. Exactly how this is done depends on the type of digital signal the sensor produces. In many cases, the value provided by the sensor can be interpreted using software running on the control computer. Sophisticated interfacing electronics are not required.

There are many kinds of analog signals. Shown in Figure 29-1 are three common types, analog sine wave, analog square wave, and analog variable voltage.

The actual voltages involved in these signals can vary depending on the circuitry, whereas with digital signals it's usually 0 and 5 volts (0 and 3.3 volts is common too). Note that analog square waves may look like a digital signal, but it's the frequency (number of waves per second) that is of most interest.

EXAMPLES OF SENSORS

One of the joys of building robots is figuring out new ways of having them react to things. This is readily done with the wide variety of affordable sensors now available. New sensors are constantly being introduced, and it pays to stay abreast of the latest developments. Not all new sensors are affordable for the hobby robot builder, of course—you'll just have to dream about getting that $5000 vision system. But there are plenty of other sensors that cost much less. Many are just a few dollars.

Part 8 of this book discusses many different types of sensors commonly available today that are suitable for amateur robotic work. Here is just a short laundry list to whet your appetite:

- *Contact switches.* Used as "touch sensors," when activated these switches indicate that the robot has made contact with some object.
- *Sonar range finder.* Reflected sound waves are used to judge distances.
- *Infrared range finder or proximity.* Reflected infrared light is used to determine distance and proximity.
- *Pyroelectric infrared.* An infrared element detects changes in heat patterns and is often used in motion detectors.
- *Speech input or recognition.* Your own voice and speech patterns can be used to command the robot.
- *Sound.* Sound sources can be detected by the robot. You can tune the robot to listen to only sounds above a certain volume level.
- *Accelerometer.* Used to detect changes in speed and/or the pull of the Earth's gravity, accelerometers can be used to determine how fast a robot is moving, whether it's tilted dangerously from center, or even when it's abruptly stopped.
- *Magnetometer.* Acts as a compass for the Earth's magnetic poles, or as a sensor to any strong magnet field. Can be used to determine the robot's relative position in a room.
- *Gyroscope.* Detects the rate of turning, or angular velocity of the robot. As with a magnetometer, it can be used to determine the robot's relative position in a room.
- *Gas or smoke.* Gas and smoke sensors detect dangerous levels of noxious or toxic fumes and smoke.
- *Temperature.* A temperature sensor can detect ambient or applied heat. Ambient heat is the heat of the room or air; applied heat is some heat (or cold) source directly applied to the sensor.
- *Resistive.* A resistive sensor detects touch, force, pressure, or strain. One inexpensive form of resistive sensor is the round or square touch pad. It provides a varying resistance depending on the amount of pressure against it.

- *Light sensors.* Various light sensors detect the presence or absence of light. Light sensors can detect patterns when used in groups (called "arrays"). Though not a camera by any means, the greater the number of sensors, the more detail the robot can discern.
- *Vision.* A sensor with an array of thousands of light-sensitive elements is essentially a video camera, which can also be used to construct the eyes of a robot.

Motors and Other Actuators

A robot uses *actuators* to take some physical action. Most often, one or more motors are attached to the robot's brain to allow the machine to move. On the typical mobile robot, the motors serve to drive wheels or legs, which scoot the bot around the floor. On a stationary robot, the motors are attached to arm and gripper mechanisms, allowing the robot to grasp and manipulate objects.

Motors aren't the only ways to provide motility to a robot, though they are the most common. Your robot may use solenoids to "hop" around a table, or pumps and valves to power pneumatic or hydraulic pressure systems.

No matter what system the robot uses, the basic concepts are the same: the robot's control circuitry (i.e., a computer or microcontroller) provides a voltage to the output, which turns on the motor, solenoid, or pump. When voltage is removed, the motor (or whatever) stops.

OTHER COMMON TYPES OF OUTPUTS

Other types of outputs are used for things like:

- *Sound.* The robot may use sound to warn of some impending danger ("There'll be no escape for the princess this time!") or to scare away intruders. If you've built a BB-8 style robot (again this *Star Wars* thing), your robot might use chirps and bleeps to communicate with you. Hopefully, you'll know what "bebop, pureeep!" means.
- *Voice.* Either synthesized or recorded, a voice lets your robot communicate in more human terms.
- *Visual indication.* Using light-emitting diodes (LEDs), numeric displays, or liquid-crystal displays (LCDs), visual indicators help the robot communicate with you in direct ways.

CONSIDERING POWER-HANDLING REQUIREMENTS

Outputs typically drive heavy loads: motors, solenoids, pumps, and even high-volume sound demand lots of current. The typical robotic control computer cannot provide more than 15 to 40 mA (milliamps) of current on any output. That's enough to power one or two LEDs, but not much else.

To use an output to drive a load, you need to add a power element that provides adequate current. This can be as simple as one transistor or it can be a ready-made power driver circuit capable of running large, multihorsepower motors. One common power driver is the H-bridge, so-called because the transistors used inside it are in an "H" pattern around the motor; see Chapter 14, "Using DC Motors," for more information on H-bridges. The H-bridge can connect directly to the control computer of the robot and provides adequate voltage and current to the load.

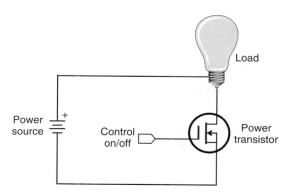

Figure 29-2   Basic electronic control of a "load," in this case a lightbulb. Depending on how the load is wired, it can draw current either into or out of the circuit.

CURRENT SOURCE OR CURRENT SINK

An electric circuit is said to "source" or "sink" current. The terms are relative to the load, that is, the part of the circuit that is drawing and consuming the current. A good example of a load is a lightbulb, shown in the very simplified circuit in Figure 29-2. The lightbulb is turned on and off using a transistor.

- When *sourcing* current, the circuit supplies current to the load. In the typical wiring, current is sourced when the output goes HIGH in order to turn the load on.
- When *sinking* current, the load draws current from the circuit. The wiring is the reverse of that above: current is sinked when the output goes LOW in order to turn the load on.

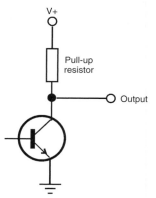

The difference between sourcing and sinking is not irrelevant, though it may seem so at first glance. It's important because most circuits can sink more current than they can source. This is why on many chips you'll see different current-handling capacities for source and sink. The source is often less—for example, source 40 milliamps (mA) current and sink 50 mA.

Depending on how they are wired, certain types of semiconductors, such as NPN transistors, cannot source current. To provide proper operation of the transistor, a *pull-up resistor* is usually added between the collector pin of the transistor and the V+ power connection. (For more on NPN and other types of transistors, see Chapters 22, "Common Electronic Components for Robotics"). A pull-up resistor is just a normal resistor. It's used to literally "pull up" the output of the transistor to the level of V+ when the transistor is not conducting current.

The output of the LM339 comparator IC is what's known as *open collector*—it's just the collector pin of an NPN transistor, without anything else added. This makes the chip more flexible in different kinds of circuits, but it also means you need to add that pull-up resistor to the comparator output, or else the chip won't work properly.

The value of the pull-up resistor depends on the application and the characteristics of the transistor. Without getting into the nuances of circuit design, so-called *weak pull-up* resistors have values of 20 kΩ and above. They're often used to reduce power usage of the transistor, chip, or other circuit that uses them. Many microcontrollers, such as the Arduino, have weak pull-up resistors built inside them; a software setting can disable the resistors.

Strong pull-up resistors have values of 2 to 10 kΩ. They're preferred in applications where very fast signal changes are required or when noise in the circuit may cause problems for other components.

Interfacing Digital Outputs

As mentioned previously, most output circuits require more voltage or current than the control electronics (computer, microprocessor, microcontroller) of your robot can provide. You need some way to boost the current needed for proper operation. Techniques include:

Direct connection: Some types of output devices can be driven directly by your robot electronics because their current consumption is low. These typically include LEDs and small piezo buzzers.

Bipolar and MOSFET transistors: Transistors are used to amplify current. With a transistor on the output of a microcontroller it's the transistor that provides the current to the load, and not the microcontroller.

Motor bridge module: A motor bridge module is a special type of integrated circuit (IC) that's intended to be used to interface to a DC motor. While motor drive is the most common application of these ICs, in fact they can be used to operate just about any high-current application.

Relay: A low-tech but still usable output interface is the mechanical relay. For very small relays you can sometimes connect it directly to your robot's microcontroller or other electronics, but in most cases you need a resistor and transistor to boost the current.

Interfacing Digital I/O

Let's talk now about common ways to connect digital inputs and outputs to the control electronics (microprocessor, computer, or microcontroller) of your robot.

BASIC INTERFACE CONCEPTS

Switches and other strictly digital (on/off) sensors can be readily connected to your robot. The most common methods are:

Direct connection: The most basic interface is simply a wire between devices. This is an acceptable approach when there are no voltage or current issues involved. This is the most common method when connecting switches as inputs to a microcontroller, for use as bump detectors (Figure 29-3), or when connecting a piezo element to produce sound.

Switch debounce: Mechanical switches give dozens to hundreds of false triggers each time they change from opened to closed state. Call these "glitches" or "transients" or *bounces,* the net effect is the same: your microcontroller may react to each of these false triggers,

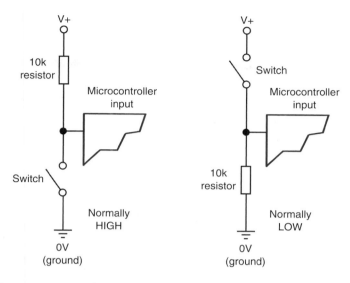

Figure 29-3   Direct connection of an input, in this case a simple switch, to a microcontroller or other circuit.

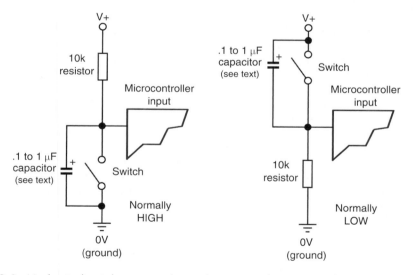

Figure 29-4   Mechanical switches create electrical noise—or "bounces"—when opened or closed. A circuit like this one removes the extra noise, helping the control electronics better determine exactly when the switch is opened or closed.

rerunning the same code multiple times, and causing potential behavioral problems in your robot (if there's such a thing as attention deficit disorder in a robot, this is it!). A (very) basic switch debounce circuit like that in Figure 29-4 provides a clean transition when the switch changes state. The value of the capacitor can be selected empirically, a fancy word for trial and error—I like to think of it as *learning through experimentation!*

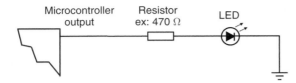

Figure 29-5 Some electronic devices, such as light-emitting diodes, require current limiting, or else damage to the circuit could result. The resistor reduces current to the LED.

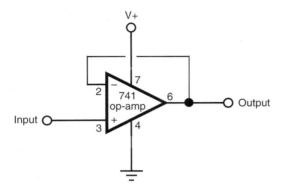

Figure 29-6 A buffer is used to isolate one circuit from another, but it may also be used to provide additional drive current, to invert the digital signal from LOW to HIGH (or vice versa), and for many other applications.

On a switch with lots of glitches you'll need to select a higher value, but the higher the value, the more sluggish the switch reaction gets. Note the polarity of the capacitor if using a tantalum or aluminum electrolytic components.

Limit current: Too much current can kill a microcontroller, and some hardware interfacing may require ways to limit current going into, or out of, the controller. An ordinary resistor (see Figure 29-5) is typically used to limit current. One common use: a resistor between microcontroller and LED. The resistor prevents the LED from drawing too much current, which will definitely destroy the LED and could also damage the microcontroller. Other instances of current limiting include inserting a resistor inline with servo motors and from switches that could carry an electrostatic discharge into the microcontroller. For 5-volt sources, the typical resistor value is between 330 and 680 Ω.

Buffer input: A buffer is any of a number of active electronics between input/output and the controller. There are lots of kinds and uses of buffers; a common form is a transistor between a microcontroller and a relay that operates a motor. The transistor not only helps to isolate the I/O line of the controller, but also boosts current needed to drive the relay. Other forms of buffers include an op-amp and specialty ICs that contain several independent buffered inputs and outputs (Figure 29-6). Note that with the hookup shown in the diagram, the output is inverted from its input.

USING OPTO-ISOLATORS

Sometimes you might wish to keep the power supplies of the inputs and control electronics totally separate, in order to provide the maximum of protection between input and output. This is most easily done using opto-isolators, which are readily available in IC-like packages. Figure 29-7 shows the basic concept of the opto-isolator: the source controls an LED. The input of the control electronics is connected to a photodetector of the opto-isolator.

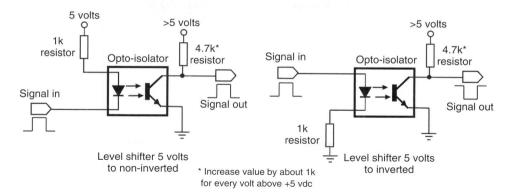

Figure 29-7   An opto-isolator keeps two circuits separate from one another, yet provides a way for one circuit to influence another.

Since each "side" of the opto-isolator is governed by its own power supply, you can also use these devices for simple level shifting, for example, changing a +5 vdc signal to +12 vdc, or vice versa. Figure 29-7 shows going from 5 volts to more than 5 volts, but other voltages are possible, too (the practical minimum voltage is 3.3 volts). You may need to adjust the values of resistors to compensate:

● Adjust the value of the current-limiting resistor to the LED up or down to match the input voltage. A lower voltage may need a lower value resistor. See the Ohm's Law section in Chapter 22 for details.
● Adjust the value of the pull-up resistor on the phototransistor up or down to match the output voltage. As noted try values ±1k for each voltage difference from that shown.

USING BIDIRECTIONAL LEVEL SHIFTING

While opt-isolators can help you to bridge a lower-voltage to a higher one—useful when mixing a 3.3-volt microcontroller with a 5-volt sensor—it can only do it in one direction. Enter the bidirectional level shifter. These use various tried-and-true techniques to provide reliable shifting between low- and high-voltage inputs.

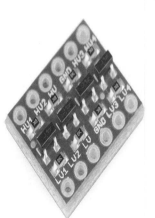

Two-way level shifting is handy, among other things, when using a bidirectional communications methods such as I2C. With these methods a single wire handles talking both ways, which may involve interfacing a 3.3-volt component with a 5-volt component.

Figure 29-8 shows one popular method of bidirectional level shifting, which employs an N-channel MOSFET transistor that has a particularly low gate voltage threshold. Unfortunately, there aren't a huge variety of such MOSFET transistors available in the convenient TO-92 thru-hole package—this sort of transistor is more the realm of surface mount components, which require soldering and a small circuit board.

Fortunately, a number of online electronics retailer offer low-cost level shifter boards that combine four, eight, and even more of these MOSFET circuits (or they use a different technique, but the end result is the same). Figure 29-9 shows the basic hookup. Check the documentation that comes with your board for particulars.

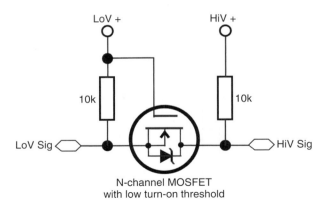

Figure 29-8 Concept of using a MOSFET transistor to translate a digital signal between two voltages.

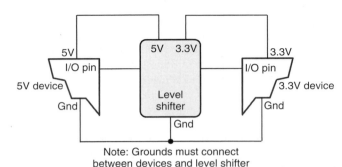

Note: Grounds must connect between devices and level shifter

Figure 29-9 Basic hookup for the typical level shifting breakout board.

USING VOLTAGE DIVIDER LEVEL SHIFTING

When you don't need bidirectional level shifting you can often employ a simpler voltage diver circuit. It consists of just two 1/8 or 1/4 watt resistors, wired in series. The output is the tap between the resistors; the amount of drop is determined by the ratio of the resistor values.

Combining a 1 and 2 kΩ resistor about does the trick of reducing a 5-volt signal to approximately 3.3 volts. Figure 29-10 shows the circuit. In case you're curious, here's the equation for calculating the resistances.

$$V_{out} = V_{in} \, [[mult]] \, \frac{R_2}{R_1 + R_2}$$

R1 and R2 are the values of the two resistors in ohms; V*in* is the input voltage, and V*out* is the output voltage. If you prefer you can use one of the many online Ohm's Law calculators to check your math.

WORKING WITH 5-VOLT TOLERANT COMPONENTS

Just because you're connecting a 5-volt component to a 3.3-volt microcontroller (or vice versa) doesn't always mean you need to use level shifting. Some devices, such as the Parallax Propeller,

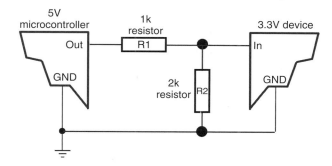

Figure 29-10   Use two resistors with about a 2:1 value ratio to drop 5 volts to approximately 3.3 volts.

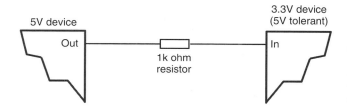

Figure 29-11   Add a series resistor to limit current when using a 5-volt tolerant 3.3-volt device. Check the datasheet for the 3.3-volt device to verify the value of the resistor.

are designed to be 5-volt tolerant—meaning the device's I/O pins can handle 5 volts, even if the device itself only operates at 3.3 volts.

When connecting 3.3 volt to 5 volt components that are voltage tolerant it's usually necessary to add a 1 kΩ series resistor between the two devices. The resistor helps to limit the current that passes from the 5-volt output. Figure 29-11 shows an example connection.

How do you know if your 3.3-volt component is 5-volt tolerant, and that the 1 kΩ resistor is the correct interfacing method? Read the documentation for all the devices you use. Don't assume; look to see if the datasheet says the device is 5-volt tolerant, and check out any circuit examples.

ZENER DIODE INPUT PROTECTION

If a signal source may exceed the operating voltage level of the control electronics, you can use a zener diode to "clamp" the voltage to the input. Zener diodes act like valves that turn on only when a certain voltage level is applied to them. By putting a zener diode across the V+ and ground of an input, you can basically divert any excess voltage and prevent it from reaching the control electronics.

Zener diodes are available in different voltages; 4.7- or 5.1-volt zeners are ideal for interfacing to inputs in most robot electronics. You need to use a resistor to limit the current through the zener. Using zeners, and selecting the proper resistor value, requires some simple math, which is detailed in Chapter 12, "Batteries and Power."

Interfacing Analog Input

In most cases, the varying nature of analog inputs means they can't be directly connected to the control circuitry of your robot. If you want to *quantify* the values from the input, you need to use

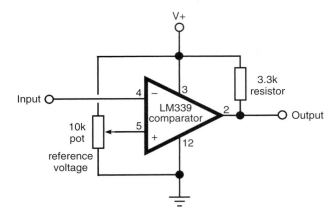

Figure 29-12 In a voltage comparator, an input voltage is compared against a reference voltage. A common scheme is shown here, where the reference voltage is provided by a potentiometer.

some form of analog-to-digital conversion, detailed later in this chapter. (Relax: most modern microcontrollers have analog-to-digital conversion built in.)

Additionally, you may need to condition the analog input so its value can be reliably measured. This may include amplifying, as detailed in this section.

VOLTAGE COMPARATOR

A voltage comparator takes an analog voltage and outputs a simple off/on (LOW/HIGH) signal to the control electronics of your robot. The comparator is handy when you're not interested in knowing the many possible levels of the input, but you want to know when the level goes above or below a certain threshold.

Figure 29-12 shows a common voltage comparator circuit. The potentiometer is used to determine the "trip point," or threshold, of the comparator. To set the potentiometer, apply the voltage level you want to use as the trip point to the input of the comparator. Adjust the potentiometer so the output of the comparator just changes state.

Note that a pull-up resistor is used on the output of the comparator chip (LM339) used in the circuit. The LM339 uses an open collector output, which means that it can pull the output LOW, but it cannot make it HIGH. The pull-up resistor allows the output of the LM339 to go HIGH when it needs to.

The LM339 integrated circuit actually has four independent voltage comparators in it. This is handy in many robotics experiments, where you want to provide multiple inputs for sensors for the left and right of the robot, or for the front and back.

You might have noticed that the voltage comparator has two inputs, one marked + and the other marked −. These are more commonly referred to as the *noninverting* and *inverting* inputs, respectively. You can alter the operation of the comparator by switching the roles of the inputs. That is, instead of having the output of the comparator switch *off* when the threshold voltage is reached, by flipping the inputs the comparator will turn *on*.

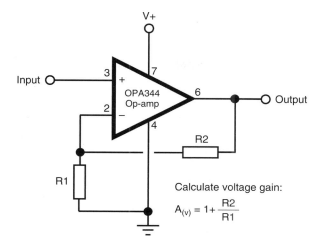

Figure 29-13 The op-amp is a highly versatile circuit, able to amplify and condition signals in hundreds of ways. This is the basic connection scheme for an op-amp designed to amplify a signal.

Calculate voltage gain:

$$A_{(v)} = 1 + \frac{R2}{R1}$$

SIGNAL AMPLIFICATION

Some types of analog sensors don't provide a signal that is strong enough to be directly used by the rest of your robot's circuitry. In these instances you must amplify the signal, which can be done by using a transistor or an operational amplifier.

The op-amp method is the easiest in most cases. While the LM741 op-amp is perhaps the most famous, it's not always the best choice, depending on the application. So I've specified an OPA344, which is a low-cost op-amp available from a number of online sources. It provides two benefits over the LM741: its output is rail-to-rail, meaning that assuming a 5-volt supply, the full voltage swing is 0 to 5 volts (or very nearly so). Second, it is made to operate from a single-ended power supply. No need for a split (+ and –) power supply. The chip will work at supply voltages of 2.7 to 5.5 volts.

If you can't locate an OPA344 chip, you can substitute most any other op-amp that is both rail-to-rail and single-supply.

Figure 29-13 shows the basic op-amp as an amplifier. The resistors marked *R1* and *R2* set the *gain,* or amplification, of the circuit. The basic formula to use to calculate approximate voltage gain is provided in the illustration.

OTHER SIGNAL TECHNIQUES FOR OP-AMPS

There are many other ways to use op-amps for input signal conditioning, and they are too numerous to mention here. A good source for simple, understandable circuits is the *Forrest Mims Engineer's Notebook,* by Forrest M. Mims III, available at most online bookstores. No robotics lab should be without Forrest's books.

COMMON ANALOG INPUT INTERFACES

Many types of analog devices can be connected to robot electronics through simple interfaces. Most are engineered to provide a varying voltage, which can then be applied to the

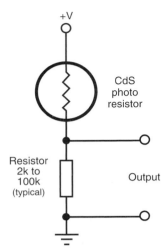

Figure 29-14 A cadmium-sulfide photo cell is a variable resistor. Its resistance changes depending on the amount of light falling on the sensor. When connected with another resistor, the output of the cell can be read as a varying voltage.

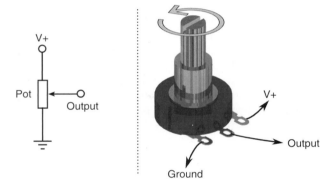

Figure 29-15 A potentiometer provides a convenient way to detect a varying voltage. The shaft of the potentiometer can be connected to a moving part of your robot (like an arm), and the voltage may be used to indicate position.

analog-to-digital converter of a microcontroller (see later in this chapter), or a voltage comparator, to determine if the voltage exceeds a certain threshold. Among the most common interfaces for robotics are:

- *CdS* cells are, in essence, variable resistors that are sensitive to light. By putting a CdS cell in series with another resistor between the V+ and ground of the circuit (Figure 29-14), a varying voltage is provided that can be read directly into an analog-to-digital converter or voltage comparator. No amplification is typically necessary.
- A *potentiometer* forms a *voltage divider* when connected as shown in Figure 29-15. The voltage varies from 0 volts (ground) to V+. No amplification is necessary.
- The output of a *phototransistor*, or light-sensitive transistor, is a varying current that can be converted to a voltage by using a resistor (see Figure 29-16). The higher the resistance, the higher the sensitivity of the device. The output of a phototransistor is typically from 0 volts (ground) to close to V+, and therefore no further amplification is necessary.
- Like a phototransistor, the output of a *photodiode* is a varying current. This output can also be converted into a voltage by using a resistor. The output of a photodiode tends to be fairly low. That means amplification is usually required.

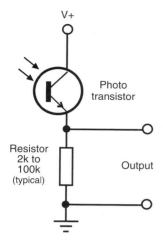

V+

Photo
transistor

Resistor
2k to
100k
(typical)

Output

Figure 29-16   A phototransistor is a transistor that is sensitive to light. As light falls on the device, it turns on (conducts current). The more the light, the more the transistor turns on. When used with a resistor, the output of the transistor is a variable voltage.

Using Analog-to-Digital Conversion

Computers are binary devices: their *digital* data is composed of 0s and 1s, strung together to construct meaningful information. But the real world is analog, where data can be most any value, with literally millions of values between "none" and "lots"!

   Analog-to-digital conversion is a system that takes analog information and translates it into a digital, or, more precisely, *binary*, format suitable for your robot. Many of the sensors you will connect to the robot are analog in nature. These include temperature sensors, microphones and other audio transducers, variable output tactile feedback (touch) sensors, position potentiometers (the angle of an elbow joint, for example), light detectors, and more. With analog-to-digital conversion you can connect any of these to your robot.

HOW ANALOG-TO-DIGITAL CONVERSION WORKS

Analog-to-digital conversion (ADC) works by converting analog values into their binary equivalents. Low analog values (like a weak light striking a photodetector) might have a low binary equivalent, such as "1" or "2." But a very high analog value might have a high binary equivalent, such as "1023."

   The *smaller* the change in the analog signal required to produce a different binary number, the *higher* the "resolution" of the ADC circuit. The resolution of the conversion depends on both the voltage span (0-to-5 or 0-to-3.3 volts are most common) and the number of bits used for the binary value to represent the analog voltage.

   Suppose the signal spans 5 volts, and 10 bits are used to represent the levels of this voltage. There are 1024 possible combinations of 10 bits, which means the span of 5 volts will be represented by 10 discrete values or steps—from 0 to 1023. Figure 29-17 shows a changing analog signal of 1 to 5 volts, with equivalent digital values showing at 0, 2.5, and 5 volts.

- 0 volts is binary 0.
- 2.5 volts is binary 511.
- 5 volts is binary 1023.

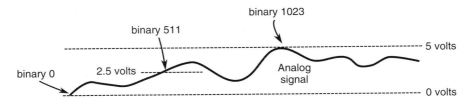

Figure 29-17 How an analog voltage (in this case 0 to 5 volts) is converted to digital values 0 to 1023, or 1024 total steps.

Given 5 volts and 10 bits of conversion, the ADC system will have a resolution of 0.0048 volts (4.8 millivolts) per discrete step. Obviously, the resolution of the conversion will be finer the smaller the span or the higher the number of bits. With an 8-bit conversion, for instance, there are only 256 possible combinations of bits, or roughly 0.019 volts (20 millivolts) per step.

ANALOG-TO-DIGITAL CONVERSION ICs

You can construct analog-to-digital converter circuits using discrete logic chips—basically a series of voltage comparators strung together. One *much* easier approach is a special-purpose ADC integrated circuit. While somewhat "old skool" these days, ADC chips are still widely available, are fairly cheap, and come in a variety of forms. Here are some:

- *Single or multiplexed input.* Single-input ADC chips, such as the ADC0804, can accept only one analog input. Multiplexed-input ADC chips, like the ADC0809 or the ADC0817, can accept more than one analog input (usually 4, 8, or 16). The control circuitry on the ADC chip allows you to select the input you wish to convert.
- *Bit resolution.* The basic ADC chip has an 8-bit resolution. Finer resolution can be achieved with 10- and 12-bit chips. For example, the LTC1298 is a 12-bit ADC chip that can transform an input voltage (usually 0 to 5 volts) into 4096 steps.
- *Parallel or serial output.* ADCs with parallel outputs provide separate data lines for each bit. Serial ADCs have a single output, and the data is sent 1 bit at a time. The LTC1298 is an example of an ADC that uses a serial interface.

INTEGRATED MICROCONTROLLER ADCs

Many microcontrollers and single-board computers come equipped with one or more analog-to-digital converters built in. This saves you the time, trouble, and expense of connecting a stand-alone ADC chip to your robot. You just tell the system to fetch an analog input, and it tells you the resulting digital value.

USING RC CIRCUITS TO APPROXIMATE ANALOG VOLTAGE

When a microcontroller lacks an ADC input, and an external converter is overkill, you may be able to fall back to a time-and-tested method using a resistor and capacitor to approximate analog voltages. This type of connection is often referred to as an RC circuit or network.

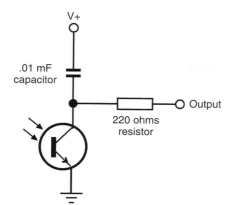

Figure 29-18 A common RC (resistor capacitor) circuit for translating an analog signal—in this case a phototransistor—into a digital output.

Figure 29-18 shows one of the more common variations. In use the microcontroller first discharges the capacitor, then times how long it takes for the capacitor to charge. This reveals the approximate voltage at the output terminal.

The time it takes to charge the capacitor is specified using some simple math and this well-known formula:

$$t = RC$$

- t is time
- R is resistance, in ohms
- C is capacitance, in farads

The result is the time it takes to charge or discharge the capacitor to about 63.2 percent of the difference between old and new value.

Since it takes longer to charge a capacitor if it has a large value, the capacitor is typically 1.0 μF or less. The value of the resistor is calculated so that it provides a reasonable time measure for the microcontroller, usually 50 to 500 microseconds. For example, given a voltage of 5 volts, a 0.1 μF capacitor (0.0000001 farads) and a 4.7 kΩ resistor, the charge or discharge time is 470 milliseconds.

Most microcontrollers provide a function for measuring these kinds of time delays. For example, the Parallax BASIC Stamp—which lacks a built-in ADC—provides the *rctime* statement for this very purpose.

Using Digital-to-Analog Conversion

Digital-to-analog conversion (DAC) is the inverse of analog-to-digital conversion. With a DAC, a digital signal is converted to a varying analog voltage. DACs are common in some types of products, such as audio compact discs.

In the field of robotics, DAC is typically performed indirectly using an approach referred to as pulse width modulation (*PWM*). It's most common in controlling the speed of motors. In operation, a circuit applies a continuous train of pulses to the motor. The longer the pulses are "on," the faster the motor will go. This works because motors tend to "integrate" the pulses to an average voltage level; no separate DAC is required. See Chapter 14 for additional information on PWM with DC motors.

When needed, you can accomplish DAC using ICs specially designed for the task. The DAC08, for example, is a time-honored 8-bit digital-to-analog converter IC. It's inexpensive (a couple of dollars) and is easy to use.

Multiple Signal Input and Output Architectures

As we've already seen, the basic input and output of a computer or microcontroller are a two-state binary voltage level (off and on), usually 0 and 5 volts. Two basic types of interfaces are used to transfer these HIGH/LOW digital signals to the robot's control computer. They are parallel and serial.

PARALLEL INTERFACING

In a *parallel interface,* multiple bits of data are transferred at one time using separate wires. Parallel interfaces enjoy high speed because multiple data bits are transferred at the same time. In a typical parallel interface with an 8-bit-wide *port,* there are eight data lines, and 8 bits are transferred simultaneously. Circuits connected to the parallel interface know the data is read when a clock (or strobe) bit is toggled from state to another (see Figure 29-19).

One of the most common parallel interfaces you'll encounter in robotics work is when using the HD7740 controller with an LCD. The HD7740 is the de facto standard controller for all-text LCDs, and it supports a 4-bit-wide port. You communicate with it using four data lines, plus three additional data lines for control. See a working example in Chapter 36, "Visual Feedback from Your Robot."

SERIAL INTERFACING

The downside to parallel interfaces is that they consume many input and output (*I/O*) lines on the robot computer or microcontroller. There are only a limited number of I/O lines on the control computer—typically 12 to 16, sometimes fewer. If the robot uses even one 8-bit parallel port, that leaves precious few I/O lines for anything else.

Serial interfaces, on the other hand, conserve I/O lines because they send data on just one or two wires. They do this by separating a byte of information into its constituent 8 bits, then sending each bit down the wire at a time, in single-file fashion (Figure 29-20). Most serial communications schemes use two I/O lines, but there are some that use just one, and others that use three or four (see the next sections). The additional I/O lines are used for such things as timing and coordinating between the data sender and the data recipient.

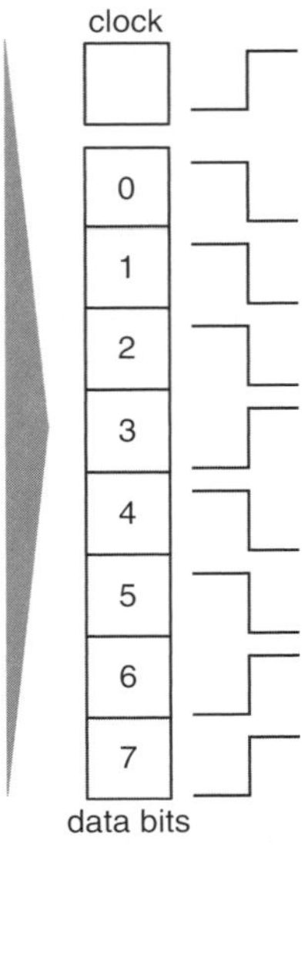

Figure 29-19   In a parallel interface, individual wires are used to carry all the data bits of the communication. A separate clock wire is used to synchronize when the data bits should be read.

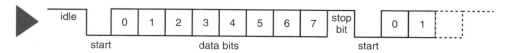

Figure 29-20 Asynchronous serial communication uses one wire to carry data bits. Synchronizing the communications—to know when data bits can be sent and received—is handled by so-called start and stop bits embedded in the data.

A number of the sensors you may use with your robot have serial interfaces, and on the surface it may appear they are a tad harder to interface than parallel connections. But, in fact, this is not the case if you use the right combination of hardware and software.

Before you can use the serial data from the sensor, you have to "clock out" all of the bits and assemble them into 8-bit data, which is used to represent some meaningful value—such as distance between the sensor and some object. The task of reconstructing serial data is made easier when you use a microcontroller with built-in serial communications commands, which are supported by microcontrollers including the Arduino, Propeller, and others.

Serial communications can be broadly classified into two groups: asynchronous and synchronous.

In *asynchronous serial*, a single wire is used to convey the bits of data, and to provide necessary timing signals so the listener can follow the conversion. In Figure 29-20, note the idle, start, and stop signals that are part of the 8 bits of data. When *idle*, the listener knows the talker isn't sending any data. But the moment a *start* bit is encountered, the listener knows data are to follow.

In *synchronous serial*, at least two wires are used: one contains just the data bits, and the other contains a control clock; at each pulse of the clock the listener accepts 1 bit of data. The I2C method, detailed next, is an example of one kind of synchronous serial communication.

INTERFACING WITH I2C AND SPI

Because most microcontrollers readily support serial communications via simple-to-use commands, a rising trend in robotics is using serial-based hardware. LCD modules with serial, rather than the typical parallel, interface have long been popular. The same concept is now found on a growing number of other kinds of hardware, such as motor control modules and ultrasonic sensors.

As noted, synchronous serial communication uses two I/O lines: one for data and another for a clock. With each pulse of the clock line, another bit of data is transferred from sender to receiver. One problem with this approach is that you need two I/O lines for every serial link in your system. If you have four serial links, you need eight I/O lines. You need even more I/O lines if you want bidirectional (two-way) communications between the devices.

Several alternative *serial protocols* are commonly used in microcontrollers that make serial communications more resistant to data errors and reduce the overall pin count when working with multiple serial devices. The two most common are I2C and SPI.

I2C

I2C uses two bidirectional data lines (hence, it's often referred to as a two-wire interface, shown in Figure 29-21) to connect one or more *slave nodes* to a single *master node*. Communication is bidirectional: master and slave can talk to one another.

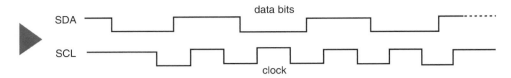

Figure 29-21   Synchronous serial communication uses two (sometimes more) wires; one wire carries the data bits themselves, and the other wire, a clock signal to provide synchronization.

The two lines of an I2C connection are referred to as SDA and SCL. *SDA* stands for Serial Data Line, and it contains the data bits themselves; *SCL* stands for Serial Clock Line, and it contains the clock pulses used to marshal the data.

Technically speaking, I2C is I²C, but the superscript isn't always shown in discussions or documentation, and is a source of confusion for some. For simplicity, we show it as I2C, and, in fact, many people now refer to it as "eye two see" rather than "eye squared see." Use whatever form you're comfortable with.

Common uses of I2C include communicating with sensors and other devices that support it, memory expansion (additional RAM and more EEPROM), and allowing two (or more) microcontrollers to talk to one another. This latter example has numerous applications when using so-called subcontrollers: a main controller that does most of the work, but additional microcontrollers for specific tasks, such as operating motors.

Implementing I2C is fairly simple, when given the right supporting software. For instance, two of the I/O lines on the Arduino microcontroller are designed with I2C in mind. The controller comes with a function library that makes setting up and using I2C relatively painless. In Part 8 there are several examples of using I2C with sensors and the Arduino.

SPI

SPI, or serial peripheral interface, uses three I/O lines to communicate between master and slave devices, with a fourth I/O line reserved as a kind of "raise your hand" signal to keep everyone from talking at the same time. This fourth I/O is referred to as the *Slave Select* (SS) line. The functions of all the I/O lines in SPI are aptly named and indicate their roles:

- *SCLK*. Serial Clock (provided by the master).
- *MOSI*. Master Output, Slave Input.
- *MISO*. Master Input, Slave Output.

To connect two slaves to one master, for example, you need the three main I/O lines wired to each slave. And then you need two additional lines, each one separately wired from the master to the slave. See Figure 29-22 for an example.

Connecting with USB

USB stands for universal serial bus, and it's now the most common way for a computer to communicate with devices connected to it. You may frequently use a USB to download programs that

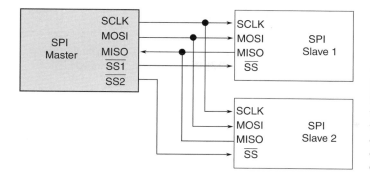

Figure 29-22 Serial peripheral interface (SPI) uses a minimum of four wires for bidirectional communications. But you can set up additional communications links with other devices by adding only one wire for each one.

you develop on your computer to your microcontroller. In order for your PC and microcontroller to talk via USB, the controller must have a USB connector on it. Microcontrollers that are built on a larger circuit board, such as the Arduino, have integrated USB; when not built-in, USB is available as an option.

Other methods of communicating between PC and microcontroller include using the older-style serial ports and even the now-almost-forgotten parallel port. Parallel port connections are not as common these days, but many single-chip controllers (PICMicro, AVR, PICAXE) are programmed via the PC serial port. If your computer lacks a serial port, you can use a USB-to-serial converter. These are available from most any local or online store that specializes in personal computers.

Some robo-experimenters prefer to use self-powered USB hubs. Not only do the hubs provide isolation between your project and your PC (protecting your PC in case of a bad short), but because they use their own power supply, they are better able to deliver the requisite current to each port (500 mA for USB 2.0, 900 mA for USB 3.0).

On the Web: Fiddling with Bits

It always seems your robot needs one more I/O pin than what's available. You could: drop a feature or two from the robot, add a second microcontroller, or select a controller that has more pins. But these are all fairly drastic—not to mention unnecessary—measures; a better approach is to use various methods of *multiplexing, demultiplexing,* and port expanding.

See the RBB Support Site for ways to virtually increase the pins on your controllers. Learn how to read a dozen or more switches, or light a panel of LEDs, with just a few pins on your microcontroller. You'll also find useful tips on how to convert serial data to parallel, and vice versa.

Plus, there are several hands-on projects that use mux/demux and serial-parallel conversion techniques in Parts 7 and 8 of this book, so be sure to check these out as well!

Following Good Design Principles

While building circuits for your robots, observe the good design principles described in the following sections, even if the schematic diagrams you are working from don't include them.

 Some of these concepts assume you already know what a resistor and capacitor are. If you don't, then no worries; these topics are covered in Chapter 22, "Common Electronic Components for Robotics." This is just a review of ways to improve the functionality of your robot circuits.

USE PULL-UP/PULL-DOWN RESISTORS

This topic has been covered a few times, but it doesn't hurt to mention it again in the context of good design. When something is unplugged in your robot, the input voltage might waver back and forth. This can influence the proper functioning of your robot. Use pull-up or pull-down resistors on any circuit inputs where this could be a problem. A common value is 10 kΩ. In this way, the input always has a "default" state, even if nothing is connected to it.

A pull-up resistor is connected between the input and the + (positive) power supply of the circuit; a pull-down resistor is connected between the input and – (negative or ground), as shown in Figure 29-23.

TIE UNUSED INPUTS LOW

Unless the instructions for a component say otherwise, tie unused inputs to ground to keep them from "floating"—*floating* means an indeterminate voltage state. A floating input can cause the circuit to go into oscillation, rendering it practically unusable.

USE DECOUPLING CAPACITORS

Some electronic components, especially fast-acting logic chips and the venerable LM555 timer IC, generate a lot of electrical noise that can spread through the power supply connections. You can reduce or eliminate this noise by using *decoupling* (also called *bypass*) capacitors.

These aren't specific types of capacitors; rather, "decoupling" refers to the job they perform. The value of the capacitor isn't supercritical. I like to use 1- to 10-µF (1 to 10 microfarad) tantalum electrolytic capacitors positioned between the positive and ground terminals of the noisy

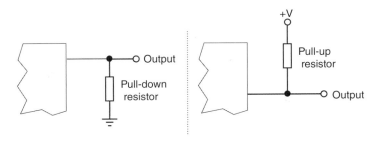

Figure 29-23 Concept of the pull-down and pull-up resistor. A common value of the resistor is 10 kΩ.

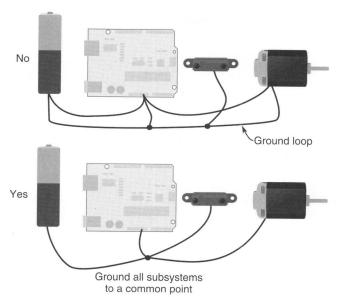

Figure 29-24 A ground loop is when there is more than one path for the ground connection. Ground loops can cause erratic behavior in circuits. Always connect the ground leads of components to one central point.

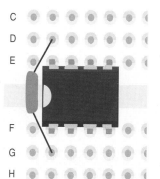

component. Some designers like to use a decoupling capacitor on every IC, while others place them beside every third or fourth IC on the board.

It's also a good idea to put decoupling capacitors between the positive and ground connections of any circuit at the point of entry of the power supply wires. Many engineering texts suggest the use of 1- to 100-μF tantalum capacitors for this job. Remember that tantalum capacitors are polarized—they have a + and a – side. Be sure to properly orient the component in the circuit, or the capacitor (and maybe some other parts) will be ruined.

KEEP LEAD LENGTHS SHORT

Long wire leads on components can introduce electrical noise in other parts of a circuit. The long leads also act as a virtual antenna, picking up stray signals from the circuit, from overhead lighting, and even from your own body. When designing and building circuits, try to keep lead lengths as short as you can for everything. When soldering, this means soldering the components close to the board and clipping off any excess lead length.

AVOID GROUND LOOPS

A ground loop is when the ground wire of a circuit comes back and meets itself. The positive and ground connections of your circuits should always have "dead ends" to them. Ground loops can cause erratic behavior and excessive noise in the circuit. See Figure 29-24 for a visual depiction of a nasty ground loop that almost guarantees problems.

PART 7

Robot Sensors

CHAPTER 30

Touch

A sure way to detect objects is to make physical contact with them. Contact is the most common form of object detection, and it's also the cheapest to implement—often with just an inexpensive switch. Touch sensing can be used on the base of a robot for when it's crisscrossing the living room carpet, or on arms and grippers, to detect when objects have been grasped or even when they're about to be crushed to smithereens.

Here you'll learn about touch-sensing systems, whether it's a simple switch on the back of the bot so it knows when it's bumped into something, or artificial "skin" that detects the amount of pressure applied to it.

This chapter deals only with "touch," your robot making actual contact with something. Other principal forms of robotic sensing are proximity detection and distance measurement. These are used to detect objects before your robot bounces into them. Read more about proximity and distance sensing in Chapter 31.

Source code for all software examples may be found at the RBB Support Site. See the Appendix for more details. To save page space, the lengthier programs are not printed here.

Understanding Touch

Touch, also called *tactile feedback,* is a reactive event. The robot determines its environment by making physical contact; this contact is registered through a variety of touch sensors. Most often, a collision with an object is a cause for alarm, so the reaction of the robot is to stop what it's doing and back away from the condition.

But in other cases, contact can mean the robot has found its home base, or that it's located an enemy bot that is about to pound the living batteries out of it. For the typical robot, touch is reduced to sensing only mechanical pressure.

Because robotic touch is based on pressure, the amount of pressure dictates how sensitive the robot is. A large mechanical switch that requires a great deal of force to actuate will be insensitive to routine contact. In order for the switch to activate, it may require a forceful collision with an object.

Conversely, a lightweight wire that is actuated by a slight sideways pressure may be activated by a gentle nudge. You can tailor your robot's contact sensitivity by altering the type and size of its touch detectors.

The Mechanical Switch

The lowly mechanical switch is the most common, and simplest, form of tactile (touch) feedback. Just about any momentary, spring-loaded switch will do (see Figure 30-1). When the robot makes contact, the switch closes, completing a circuit—or in some cases, the switch opens, breaking the circuit. Either way works.

The switch may be directly connected to a motor, or, more commonly, it may be connected to a microcontroller or other electronic circuit. A typical wiring diagram for the switch is shown in Figure 30-1. The pull-down resistor is there to provide a consistent LOW output for the switch when there is no contact. When contact is made, the switch closes, and the output of the switch goes HIGH.

When using a microcontroller, you can determine how the robot reacts to the physical collision by altering its programming. Typically, for a switch used for a touch sensor, the programming instructs the robot to stop, back up, and head in a new direction.

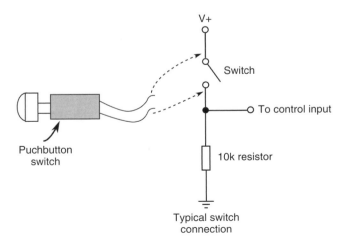

Figure 30-1 Mechanical and electrical connection of a momentary (spring-loaded) pushbutton switch. The 10 kΩ resistor ensures that the output is LOW until the switch closes.

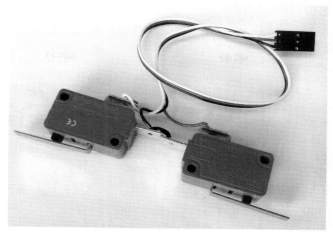

Figure 30-2   Leaf or lever switches make for ideal bumpers for robotics. You can choose from among switches with a long or a short leaf.

PHYSICAL CONTACT BUMPER SWITCH

You can choose from a wide variety of switch styles when designing contact switches for tactile feedback. A *leaf* or *lever* switch (sometimes referred to as a Microswitch, after a popular brand name) comes with plastic or metal strips of different lengths that enhance the sensitivity of the switch. See Figure 30-2 for an example of a lever switch ideal for use as a contact bumper for a small robot.

Leaf switches require only a small touch before they trigger. The plunger in a leaf switch is extra small and travels only a few fractions of an inch before its contacts close. As the leaf is really a mechanical lever, lengthening it increases sensitivity. But it also increases the distance (called *throw*) that the end of the lever must travel before the switch makes contact.

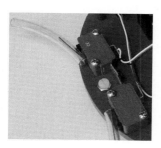

ENLARGING THE CONTACT AREA OF THE SWITCH

The surface area of most switches is pretty small. You can enlarge the contact area by attaching rubber, metal, or plastic (or a length of wire) to the switch. For leave switches a piece of aquarium tubing works well if you'd like a compliant bumper; if you need something more rigid, 1/16"-thick plastic or aluminum tubing is suitable. For tubing just slip over the switch or glue it into place.

For any kind of switch you can mount a plastic (preferred for weight) or metal plate to the plunger to increase surface area.

 Low-cost pushbutton switches are not known for their sensitivity. The robot may have to crash into an object with a fair amount of force before the switch makes positive contact, and for most applications that's obviously not desirable. Spend a little more for higher-quality switches.

EXTENSION WHISKER

The whiskers of a cat help its brain form a 3D topographical map of the animal's surroundings. At its most simplistic level, the whiskers can be used to measure space. We can apply a similar technique to our robot designs—whether or not kitty whiskers are actually used for this purpose.

By bending the whiskers, you can extend their usefulness and application. The whiskers in Figure 30-3 make contact with a small (and sensitive) leaf switch. You can cement or tape the wire to the switch, and bend the wire to make a whisker of the size and shape you want. When the switch and whiskers are positioned so they detect vertical motion, they can determine changes in topography to watch for such things as the edge of a table or the corner of a rug.

You can use thin 20- to 25-gauge piano ("music") wire for the whiskers of the robot. Attach the wires to the ends of switches, or mount them in a receptacle so the wires are supported by a small rubber grommet.

You can extend the cat's whiskers concept by pairing two leaf switches, with different lengths of wire. Make one wire short to catch objects that are in very close proximity to the robot, and a longer wire to detect objects in the general vicinity. Position the whisker pairs on either side of the robot to detect objects to the left and right, as shown in Figure 30-4.

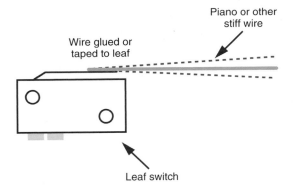

Figure 30-3 Use piano (music) wire or small rigid tubing to extend the length of a leaf switch. Use glue or tape to hold the wire in place.

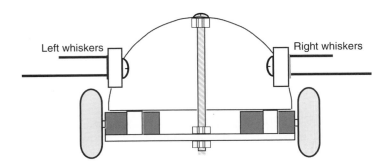

Figure 30-4 Use pairs of whisker switches on each side of the robot to detect both close and very close objects.

Avoid sharp wires sticking out of your robot. Not only can they snag on objects, they can poke at skin, eyeballs, and other sensitive body parts of humans and animals. Bend the ends of the whiskers or other wires to form a small blunted loop. Or if the wires are heavy enough, insert a rubber stopper, like that used for knitting needles.

HOMEBREW CONTACT WHISKER

With some stiff music wire, you can build your own contact whiskers and form them into any shape you like. Figure 30-5 shows the basic idea of a homebrew wire whisker for a robot. Only one switch is shown; duplicate for the other side of the board. The design works best when made directly on an experimenter's solderboard (don't use a solderless breadboard).

How it works: the wire goes between a pair of male header pins, where the middle pin of the three has been cut off. The whisker makes contact with any side-to-side movement. If the fit between the pins is too small for the board you have, use instead a set of four male header pins and cut off the middle two.

Anchor the end of the wire using glue or solder or some other fastener. If you solder the wire should be copper or brass; metals like stainless steel are harder to solder to. You should apply a soldering iron or pencil with a higher wattage than you'd normally use for electronics. Solder the wire connections first, before adding the other components.

Note the 10 kΩ resistors. One resistor is used as a *pull down*; it keeps the output of the switch from "floating" when there is no contact. The other resistor is used to limit any stray currents that may come through the bare wire. The *control input* is a microcontroller I/O line, a connection to a motor, and so forth.

 Avoid damage or injury by adding a small loop at the outside end of the whisker. This prevents the wire from poking into ankles, eyeballs, and antique furniture legs.

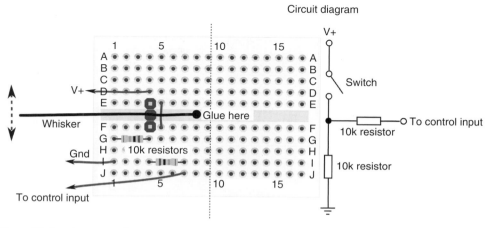

Figure 30-5   Plan view of two ways to implement the electronics of homebrew whisker switches. The one on the left is most sensitive to side-to-side motion; the one on the right, back-and-forth motion. See the text for how to use the male header pins as whisker contacts.

MULTIPLE BUMPER SWITCHES

What happens when you have many switches scattered around the periphery of your robot? You could connect the output of each switch to your microcontroller, but that's a waste of I/O pins. A better way to do it is to use a priority encoder or I/O expander. These schemes allow you to connect several switches to a common control circuit.

Using a Priority Encoder

The circuit in Figure 30-6 uses a 74148 family IC (such as the 74HC148) priority encoder IC. Switches are shown at the inputs of the chip. When a switch is closed, its binary equivalent appears at the A-B-C output pins. With a priority encoder, only the switch that represents the most significant bit is indicated at the output (pin 4 is the most significant bit; pin 10 the least significant bit).

In other words, if switches connected to pins 4 and 1 are both closed, the output will reflect only the closure of switch 4, as 4 has a higher priority. Sometimes this is useful information (some bumper switches are more important than others); sometimes it's not. When it's not, use one of the other approaches discussed next.

When using a 74148 chip with fewer than eight switches, be sure to tie any unused inputs HIGH. Connect the switches in descending significant order: 4, 3, 2, 1, 13, 12, 11, 10.

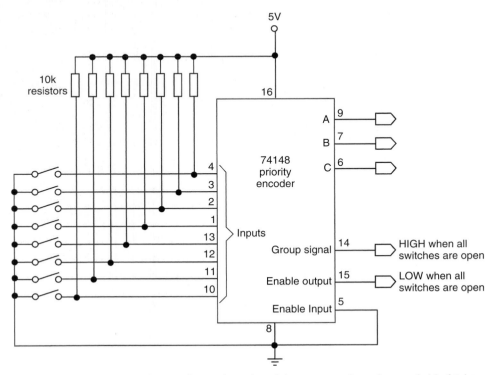

Figure 30-6   A priority encoder IC tells you the value of the most significant bit in a field of 8 bits. The 10 kΩ resistors are used as pull-ups for the switches.

Back up the truck a bit before going on. The 74148 has an interesting feature you might not want to miss. Notice pins 14 and 15. These are "group" outputs, meaning their state changes when *any* of the switches are closed. You can use this feature with a microcontroller that has a hardware interrupt pin (like the Arduino).

If any of the eight switches close, the group output can signal the interrupt. The microcontroller can then check the A-B-C output lines to see which switch is closed (if more than one switch is closed, the most significant switch is indicated). Use this feature when your microcontroller has only a few hardware interrupt pins and your robot has a lot of switches.

There are methods of cascading (daisy-chaining) priority encoders, so you can check 16 or more switches. See the datasheet for the 74148 chip for ideas.

Using a Port Expander

The 74148 IC is admittedly old school. While it works well and is simple to use, a more modern solution is to use an I/O expander, a specialty IC that uses serial communications to amplify the number of pins available to read switch values (they can also be used for output tasks, like lighting LEDs).

Let's take the SX1509 I/O expander as an example. Available from SparkFun and other online sources as a handy breakout board, it's one of a number of expanders that can be used with the Arduino, BBC micro:bit, and other microcontrollers that support the I2C serial interface.

Connection is simple as shown in Figure 30-7: just two wires to the microcontroller, and up to 16 switches connected to the I/O pins. These devices use I2C, so you'll either need to dig into the datasheet to discover how to communicate with the thing, or find a software library designed for your microcontroller.

The SX1509 has a number of handy features you'll appreciate. All of its pins have built-in pull-up resistors you can set or unset, and all pins support PWM and an LED "blink" function. These are set-and-forget behaviors that can free up your microcontroller for other tasks.

More Pin Expansion on the Web

Other ICs are available for I/O expansion, including multiplexers, serial-to-parallel converters, and parallel-to-serial converters. See the RBB Support Site for handy how-to articles on using these other techniques.

Using a Button Debounce Circuit

Bounce is what happens when the contacts in a switch open or close. The contacts don't just immediately open or close once when the switch is pushed. There may be dozens of "tentative" opens and closes each time the switch changes state. The bounces are a kind of electrical noise that can influence the operation of your circuits and programming.

There are numerous ways to remove the extra bounces when a switch opens or closes. They all operate on the principle of stretching the duration of the first switch event that occurs. Most bounces are less than 10 or 20 milliseconds (often much shorter than that, depending on the switch); by stretching out the switch change, all the other bounces that come after are simply missed.

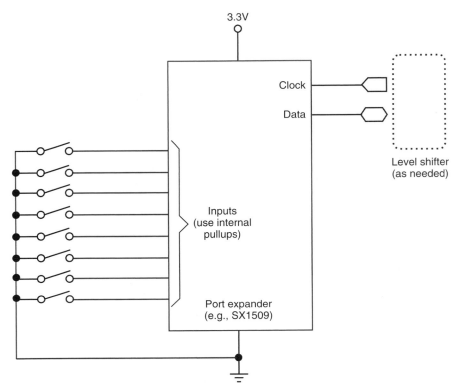

Figure 30-7 Hookup to an I/O expander is simple, but you need a software library to access its features, or else discover these yourself by reading the datasheet for the device.

Figure 30-8 shows a common approach to hardware button debouncing. It uses one-sixth of a 74HC14 Schmitt trigger inverter IC, along with a resistor and capacitor to form an *RC timing* network. Note that the 74HC14 contains inverting buffers, meaning that the polarity of the input signal is reversed on the output—LOW becomes HIGH, and HIGH becomes LOW. Remember this when you connect your circuit to your microcontroller. (If you really, really hate this aspect, and you have inverters to spare, connect two in daisy-chain fashion. This method inverts that which was inverted, and you're back to life as usual.)

Schmitt triggers do their magic because their output is either LOW or HIGH. There are never any in-between voltages, which can occur because of the introduction of the capacitor in the switch network. The capacitor changes the switch transitions from sharp cliffs to gradual slopes

You can implement switch debouncing in many different ways. Check out the RBB Support Site for several additional methods, including using an LM555 timer IC to literally "stretch out" the pulse. This circuit is useful for debouncing, and to make even the most fleeting switch contact last longer. That can be handy when programming, to keep your microcontroller or other circuit from missing the switch action.

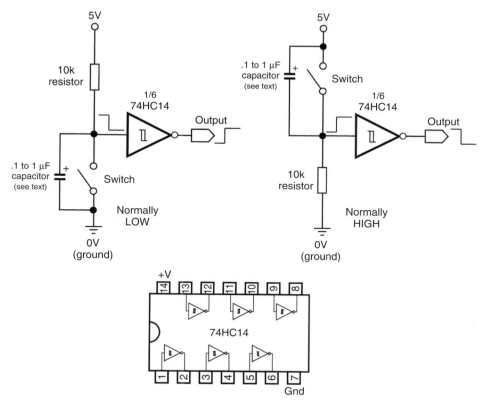

Figure 30-8   Switch debouncer built around a Schmitt inverter buffer. The 7414 IC has six Schmitt inverters in it. Vary the value of the capacitor (from about .1 to 1 µF) for a longer or shorter pulse output. Higher values produce longer delays and are useful with really dirty (electrically speaking) switches.

(doing so masks the bounces), and the gradual slopes can cause your microcontroller to confuse what is and isn't a proper LOW or HIGH. The Schmitt trigger eliminates this problem.

Debouncing Switches in Software

Most or all of the functionality of a debouncer circuit can be built in software—good when you're already using a microcontroller or computer for your robot's brain. Like the capacitor, the software technique uses delays to momentarily slow down the program when the first switch transition occurs.

Most microcontrollers have programming statements that provide debounce delays, saving you the hassle of adding the code yourself. The Arduino has available a *Bounce* library that you can download and add to your sketches. The Parallax Propeller, Raspberry Pi, and many other popular microcontrollers have their own debounce libraries or examples. Search the repositories for your controller of choice and use these libraries whenever you find it necessary to debounce a switch.

Programming for Bumper Contacts

Bumper switches and other forms of touch produce what's known as transitory events: they may not occur for long periods of time, and when they do, they may not last long. When using a microcontroller, you need to program it to watch for these events, so that your robot can take the appropriate action when contact is made.

There are three general approaches for programming a microcontroller for bumper contacts and other events of short duration: preemptive wait, polled, and interrupt.

- *Preemptive waiting* involves using a command of your microcontroller to continuously check the state of a single switch or other transitory input. Because of how preemptive waiting works, the microcontroller is not able to perform other parts of its regular program.
- *Polling* involves periodically checking the state of any switches or other transient event sensors in your robot, while allowing the rest of your program to run. Because most microcontrollers are quite fast compared to the various functions of your robots, this is often a perfectly acceptable approach. As part of the main program code, your robot checks each switch in turn, testing whether the switch has been activated.
- *Interrupts* are handled internally by the microcontroller, and trigger by themselves. Your software need only ready the interrupt, telling the microcontroller what you want to have happen when it's triggered. When an event occurs, the interrupt momentarily stops the regular program and performs whatever special task you've set up.

Program arduino-*button-press.ino* shows how polling and interrupts are handled on the Arduino. To simplify things, only two switches are used; bumperA is polled, and bumper is set up for an interrupt. Two of the digital input pins on the Arduino may be used as hardware interrupts. Depending on the exact model of the Arduino hardware you are using, this is usually pins D2 and D3. To try this program, connect one switch to pin D12 and another to D2.

The program sets up pin D13, which has an integrated LED already on it, as an output. It also attaches an interrupt to watch for any change on pin D2 (known as interrupt 0). Ordinarily, the LED shows the value of the switch on D12. It's off if the switch is open, on if the switch is closed. This is handled by the *poll* routine.

When the switch connected to D2 opens or closes, the microcontroller immediately branches to the *handle_interrupt* routine, which blinks the LED on and off for 1 second. After the *handle_interrupt* routine is finished, the regular program resumes.

This example also demonstrates that when the Arduino is "servicing" the interrupt, the other parts of the code aren't executed. This is by design. Only when the interrupt is finished does program execution pick up where it left off. Normally you wouldn't have such long delays in your interrupt routines.

Also notice I used the *delayMicroseconds* statement, rather than *delay*. Why? The *delay* statement uses some internal interrupts of the Arduino, so it's disabled while in an interrupt handler. If you want a delay, you need to use *delayMicroseconds*. The *for* loop is used to extend the delayed time in 1-millisecond (1000-microsecond) increments.

arduino-button_press.ino

```
const int led = 13;    // Built-in LED
int bumperA = 12;          // Digital pin D12
int bumperB = 0;           // Interrupt 0 (digital pin D2)

void setup()
{
  pinMode(led, OUTPUT);
  digitalWrite(led, LOW);
  attachInterrupt(bumperB, handle_interrupt, RISING);
}

void loop() {
  poll();
}

void poll() {
  digitalWrite(led, digitalRead(bumperA));
}

void handle_interrupt() {
  digitalWrite(led, HIGH);
  for (int i=0; i <= 1000; i++)
    delayMicroseconds(1000);
  digitalWrite(led, LOW);
  for (int i=0; i <= 1000; i++)
    delayMicroseconds(1000);
}
```

Mechanical Pressure Sensors

A switch is a go/no-go device that can detect only the presence of an object, not the amount of pressure on it. A pressure-sensitive detector senses the force exerted by the object onto the robot, or vice versa—how hard the robot has crashed into something.

There are a number of pressure-sensitive detectors you can use in your robots. Some you can make, some you can steal from old parts, and yet others are available as specialty sensors, available from numerous online sources.

FORCE-SENSITIVE RESISTORS

A force-sensitive resistor (FSR) is a strip or pad that produces a variable resistance depending on the amount of pressure, or force, exerted on it. FSRs consist of a main pad that forms the sensing element. The pad comes in various shapes and sizes. A couple of example FSRs are shown in Figure 30-9.

For use as a robot fingertip, you can opt for a small, 5mm to 10mm pad (about 1/4″ to 1/2″). For use as a robot bumper to detect collisions, you may want to go with a somewhat larger pad. But take care: The larger the pad the more expensive it is, and the greater chance it can be damaged from wayward impact. So treat these puppies with respect!

When selecting an FSR pay close attention to its minimum and maximum pressure ranges. For most small robots you should opt for a fairly low minimum pressure; for instance, no higher

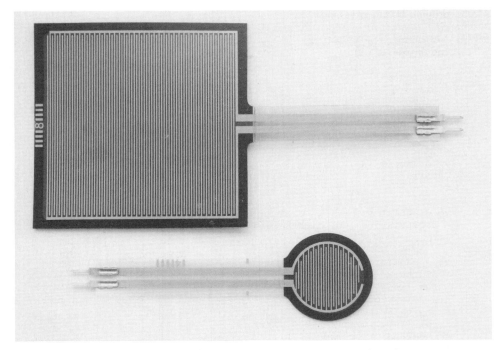

Figure 30-9 Two examples of force-sensitive resistor sensors. These provide a changing resistance depending on the amount of pressure on the center pad.

than 100 grams—that's slightly less weight than the uncooked hamburger patty in a Quarter Pounder at McDonald's.

Getting a Voltage Output from an FSR

An FSR appears to a circuit like an ordinary resistor, so it can be used within a voltage divider to output a voltage that can be registered by your microcontroller. Figure 30-10 shows a divider circuit you can use. Select the fixed resistor R1 to best match the free/pressed resistance values of the FSR.

- On many FSRs, the free (unpressed) resistors will be 100 kΩ or higher.
- The harder the force area is pressed, the lower the output resistance, to perhaps 1 to 10 kΩ.

With this general observation in mind, select a fixed resistor that is somewhere around the middle of the usable resistance range of the FSR when pressed. Use a volt-ohm meter to take readings of the sensor output, and write down a low and high range. Don't be alarmed if the readings vary somewhat as you apply pressure. FSRs are not known for their supreme accuracy.

So, for example, if the midrange value is about 30 kΩ for the typical pressures you expect, select the nearest standard resistor value, or 33 kΩ.

Connecting an FSR to Your Circuit

You've already read that FSRs can be a bit delicate. This applies to not only its pressure pad, but also its connecting leads as well. The typical FSR comes with connecting pins on its leads.

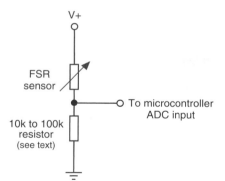

Figure 30-10 Use a voltage divider to convert the resistance value from an FSR to a voltage signal usable with your microcontroller or other circuitry.

These leads should always be used to plug into short (shorter the better) extension wires, and never directly soldered to a circuit board. The heat from the soldering iron can permanently ruin the device.

The lead portion of the FSR can likewise be damaged by impact, folding, crimping, and heat. Select the FSR for your application. Unless you need it for your application, get an FSR with just the right lead length to get the job done.

 You can enlarge the contact area of the pad by using rubber balls cut in half. A 2″ round ball has a cross section of 2″, yet at the opposite pole it has a contact area of mere millimeters—perfect for a smallish FSR. Let your mind wander and you'll think of plenty of clever homemade designs for enlarging the contact area of a force-sensitive resistor.

FLEX RESISTORS

A variation on the FSR is the flex resistor, similar to a resistive pad but greatly elongated (see Figure 30-11). This makes it more sensitive to the effects of bending. It's ideal as a bumper detector on the front of your robot.

Example: Mount the flex resistor on a very thin piece of plastic (such as 1mm polystyrene from the hobby store), and mount this strip on the front of the bot so that it creates a flexible bow. Push in anywhere along the strip, and the value of the flex resistor changes.

As flex resistors are just variable resistors, use them in voltage divider circuits, like the one previously discussed for FSRs.

You can make your own ersatz pressure-sensitive resistors using a small piece of antistatic conductive foam. I've written up a handy how-to for the RBB Support Site. Check it out!

ON/OFF OUTPUT FOR RESISTANCE SENSORS

Both FSRs and flex sensors provide as their output a changing resistance. When used in a simple voltage divider circuit, these components produce a variable voltage that can be routed to an analog-to-digital converter pin on your microcontroller.

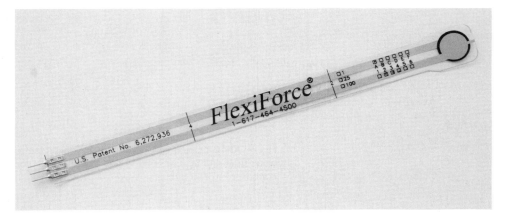

Figure 30-11   This resistive sensor detects when the plastic strip is flexed. Its resistance changes as the strip is twisted and deformed. Use the same kind of interface circuit as in Figure 30-10 to connect this to your microcontroller.

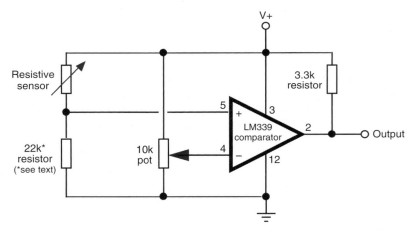

Figure 30-12   An LM339 voltage comparator IC can be used to construct an on/off switch using a pressure pad. Adjust the potentiometer for the threshold between on and off. This circuit works with any kind of resistive sensor.

But maybe you don't want to hook things up that way. You'd rather the sensor had an on/off or HIGH/LOW output. Measuring the exact pressure or bend isn't important; what you're interested is whether the sensor is detecting a meaningful amount of force or bend.

This calls for a 1-bit analog-to-digital circuit, and it can be done using the venerable LM339 IC. This chip contains four independent voltage comparators in one very inexpensive package. Figure 30-12 shows a wiring diagram.

As with the voltage divider circuit in Figure 30-10, experiment with the value of the dividing resistor connected to the sensor. For the FSR I used, I found good overall response with a 22 kΩ resistor, but you should feel free to try other values. What you want is a good voltage swing (sensor pressed/bend and not) without many false readings.

In operation, the conductive foam and series resistor produce a voltage, which is fed into the noninverting input of the comparator. Dial the 10 kΩ potentiometer to vary the reference voltage for the comparator. While applying pressure to the sensor, set the potentiometer so that the output of the LM339 just goes between LOW and HIGH. Release the pressure, and the output should return to LOW again.

From time to time you may need to tweak the potentiometer, in order to set a new reference point. The output characteristics of FSRs and bend sensors can vary somewhat with age.

LOAD CELL PRESSURE SENSORS

If you have a digital weight scale in your bathroom, odds are it uses what's known as a load cell to register the Earth's gravitational pull on you. A load cell is a sensor that detects strain or stress, such as that when a weight is placed on it. They're quite common in industry, not to mention digital bathroom scales. Because of how they function load cells are often referred to as strain gauges.

Not all load cells are intended to measure the weight of a human—or trucks, big machines, and other heavy objects. Some are intended to measure featherweights as low as a fraction of an ounce. Most FSRs don't register such light touches. As or more important, the typical load cell is far more robust than an FSR, not to mention more accurate. Of course it's also a little more expensive.

The typical load cell comprises just the sensor; you need an amplifier to get a useful reading out of it. The amp boosts the very low (almost minuscule) voltage output of the cell to something more useful by a microcontroller or other circuit.

While you could fashion your own load cell amplifier, ready-made amps are plentiful and fairly cheap. When selecting an amp be sure it matches the load cell's electrical specifications (supply voltage, output signals) and connector type. Once connected to an amplifier, the load cell provides a voltage to the microcontroller that you can read directly.

To get a useful reading from the cell, mount it so that weight or pressure can be exerted between the two ends of the sensor. For "bar style" sensors this is often done by physically mounting one end of the load cell to your robot, and allowing the other end to act as a pressure pad or lever, as shown in Figure 30-13. The output of the cell will detect any force against the free end.

If you just have to build your own load cell amplifier, do a Web search for a Wheatstone bridge amp. The amplifier is so named because it's intended to be used with the type of electrical connection—known as a Wheatstone bridge—used by most load cells.

Making Touch Sensors from a Microphone

Sound can transmit through solid objects. If the object—say, a metal beam—is connected to a small microphone, touching the beam can directly transmit the noise of the touch into the microphone. This is known, among other terms, as microphonic coupling.

If you've ever played in a garage band, you've probably needed to avoid this very thing with your microphones. A mic firmly attached to a stand has a tendency to pick up the sound of hands touching the stand, or feet hitting against its base.

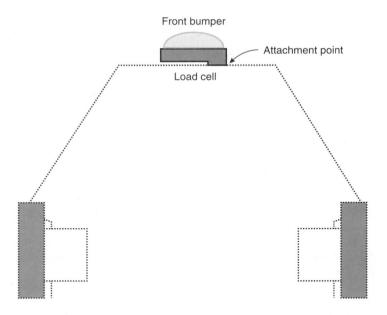

Figure 30-13 A load cell can measure the pressure exerted on a mechanical bar. Attach one end of the bar to the front of a robot to register the strain caused by direct contact against the bumper.

So the effect is generally an unwanted one, except where our robots are concerned. You can create a microphonic touch sensor using ready-made modules. Construct is easy: firmly attach the microphone or sound module to the side or front of the robot.

Touching the robot, especially rough handling, creates a kind of rustling that is acoustically transferred to the microphone. For this to work well,

- The sound module should be fairly sensitive. It should easily pick up normal conversation from within 5 or 6 feet.
- The microphone is coupled to metal. Plastic and wood may not have enough acoustic conductivity.
- The microphone/transducer coupling must be rigid. Consider gluing the microphone to the robot structure using regular Super Glue. Avoid the use of hot melt glue, epoxy, or other adhesives that may create a buffer area between the mic and robot.

For connection diagrams to various sound input modules refer to Chapter 36, "Visual Feedback from Your Robot."

Other Types of "Touch" Sensors

The human body has many kinds of "touch receptors" embedded within the skin. Some receptors are sensitive to physical pressure, while others are sensitive to heat. You may wish to endow your robot with some additional touchlike sensors. You can find projects for many of these ideas in the remaining chapters of this book.

- *Heat sensors* can detect changes in the heat of objects within grasp.
- *Air pressure sensors* can be used to detect physical contact. The sensor is connected to a flexible tube or bladder (like a balloon); pressure on the tube causes air to push into or out of the sensor, thereby triggering it.
- *Accelerometers* measure shock and vibration. If your robot hits something, it will cause a shock or vibration that'll be picked up by an accelerometer. Should it stop moving (the object is solid enough), the accelerometer will detect that as well.
- *Touch fabric* is fabric and threads that are electrically conductive. Originally designed for use in making garments, carpets, and other textiles resistant to static electricity buildup, you can use it in any robotic application where you need a flexible sensing area.

On the Web: Piezoelectric Sensor Detectors

The little sound element in your smoke detector is composed of a piezoelectric element: apply a signal to the element, and it produces a sound.

Piezo speakers also work in reverse: tap on one, and it produces a voltage. Connected to your microcontroller using a suitable interface, piezo elements make for handy contact sensors for your robot. While the most familiar piezo element is the metal and ceramic disc used in sound buzzers, it's also available as a flexible film or wire. Bending the film or wire produces a voltage, which can be detected by your microcontroller.

Sounds useful, and it is. I've written about this topic in past editions of this book, and this time around I've moved the projects to the RBB Support Site. There you'll find useful interface circuits, plus ideas for making a wide "touch bar" for the front of your robot, plus attaching a commonly available piezo film "tab" to a strip of plastic to make a large bend sensor, as shown in Figure 30-14.

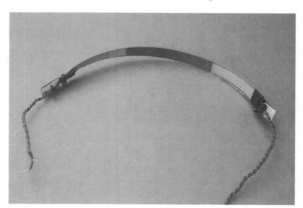

Figure 30-14   You can create a bumper sensor for your robot using two low-cost piezo film elements and a strip of flexible plastic cut from a school report cover. Piezo film registers the *change* in the bend anywhere along the length of the plastic.

 And there's more: See the RBB Support Site for details on how to implement capacitive touch sensors—useful when you need to enlarge the "touch zone" of human-interface buttons. You'll also find an unusual method of using lasers and fiber optics as "feature touch" feelers, and more.

CHAPTER 31

Proximity and Distance

You've spent hundreds of hours designing and building your latest robot creation. It's filled with complex little doodads and precision instrumentation. You bring it into your living room, fire it up, and step back. Promptly, the beautiful new robot smashes into the fireplace and scatters itself over the living room rug. You remembered things like motor speed controls, electronic eyes and ears, even a synthetic voice, but you forgot to provide your robot with the ability to look before it leaps.

Proximity and distance sensing is all about collision avoidance. These systems take many forms, and the most basic are easy to build and use. In this chapter you'll learn about passive and active detection systems you can use in your robots.

 Source code for all software examples may be found at the RBB Support site. To save page space, lengthier programs are not printed here.

Design Overview

For a robot to be self-sufficient in the human world, it must be able to determine its environment. It does this by sensing objects, obstacles, and terrain around it. This can include you, the cat, an old sock, the wall, the little hump on the ground between the carpet and the kitchen floor, and a million other things.

Robots sense objects using either contact or noncontact means. With contact sensing, the robot detects a collision after it's happened. With noncontact sensing, the robot is able to avoid a collision before it happens. Collision detection and collision avoidance are two similar but separate aspects of robot design.

- In Chapter 30, "Touch," you learned about techniques in *collision detection,* which concerns what happens when the robot has already gone too far and contact has been made with whatever foreign object was unlucky enough to be in the machine's path.
- With *collision avoidance,* the subject of this chapter, the robot uses noncontact techniques to determine the proximity and/or distance of objects around it. It then avoids any objects it detects.

Collision avoidance can be further broken down into two subtypes: near-object detection and far-object detection.

Robot builders commonly use object detection methods to navigate a robot from one spot to the next. Many of these techniques are introduced here because they are relevant to object detection, but for navigation they are developed more fully in Chapter 32, "Navigation."

NEAR-OBJECT DETECTION

Near-object detection does just what its name implies: it senses objects that are close by, from perhaps just a breath away to as much as 8 or 10 feet. These are objects that a robot can consider to be in its immediate environment, objects it may have to deal with, and soon. These objects may be people, animals, furniture, or other robots. By detecting them, your robot can take appropriate action, which is defined by the program you give it.

There are two ways to effect near-object detection: proximity and distance (see Figure 31-1):

- *Proximity* sensors care only that some object is within a zone of relevance. That is, if an object is near enough in the physical scene the robot is looking at, the sensor detects it and triggers the appropriate circuit in the robot. Objects beyond the range of a sensor are effectively ignored because they cannot be detected.
- *Distance measurement* sensors determine the distance between the sensor and whatever object is within range. Distance measurement techniques vary; almost all have notable minimum and maximum ranges. Few yield accurate data if an object is smack-dab next to the robot. Likewise, objects just outside range can yield inaccurate results. Large objects far away may appear closer than they really are; very close small objects may appear abnormally larger.

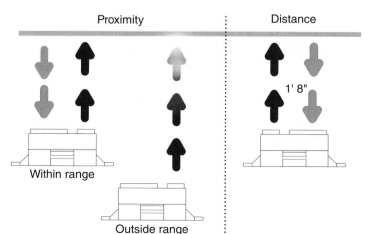

Figure 31-1 Proximity versus distance detection. Proximity provides a go/no-go result, while distance gives the actual closeness of an object.

Both detection schemes use similar technologies. The most common proximity detection schemes use infrared light or ultrasonic sound. If enough light (or sound) is reflected off the object, and received back to the robot, then an object is within proximity.

FAR-OBJECT DETECTION

Far-object detection focuses on objects that are outside the robot's primary area of interest, but still within a detection range. A wall 50 feet away is not of critical importance to a robot, but a wall 1 foot away is *very* important.

The difference between near- and far-object detection is relative. As the designer, builder, and master of your robot, you get to decide the threshold between near and far objects. Perhaps your robot is small and travels fairly slowly. In that case, far objects are those 4 to 5 feet away; anything closer is considered "near." With such a robot, you can employ ordinary sonar distance systems for far-object detection, including area mapping.

SENSOR DEPTH AND BREADTH

Sensors have depth and breadth limitations (see Figure 31-2):

1. **Depth** is the maximum distance an object can be from the robot and still be detected by the sensor. Except for some ultrasonic sensors, most proximity and distance-measuring detectors for amateur robotics are limited to less than about 6 feet.
2. **Breadth** is the height and width of the sensor detection area at any given distance. Some sensors have a very wide "beam" width, covering a large area at a time. Others are much more narrow and focused. There's a place for both.

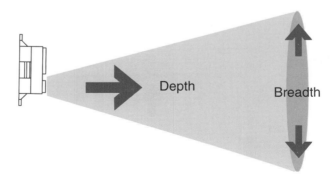

Figure 31-2 Robotic sensors have both depth and breadth. Depth defines the maximum distance of detection from the robot, while breath defines the field of view of the sensor.

Sensors that use light can use lenses to focus the beam into a smaller spot. In this way the sensor detects only what's directly in front of it. Ultrasonic sensors use various means to control their beam pattern. Some are very narrow, while others are wide. Makers of ultrasonic sensors for use in robotics provide a picture of the beam spread, so you can match the detector to your planned use.

Simple Infrared Light Proximity Sensor

Avoiding a collision is better than detecting it once it has happened. Short of building some elaborate radar distance measurement system, the ways for providing proximity detection to avoid collisions fall into two categories: light and sound. Let's start with a simple sensor using light.

Light may always travel in a straight line, but it bounces off nearly everything. You can use this to your advantage to build an infrared collision detection system. You can mount several infrared "bumper" sensors around the periphery of your robot. They can be linked together to tell the robot that "something is out there," or they can provide specific details about the outside environment to a computer or control circuit.

A basic (and very simplistic) infrared detector is shown in Figure 31-3, along with a suitable interface circuit. It uses an infrared LED and infrared phototransistor. The output of the transistor can be connected to any number of control circuits, including a microcontroller or voltage comparator.

- Experiment with the 47 kΩ resistor (above the phototransistor). Lower resistance values decrease sensitivity; higher values increase sensitivity.
- Experiment with the 220 Ω resistor to alter the amount of light emitted by the LED. A higher value acts to reduce the light brightness. Avoid reducing the value under 180 to 220 Ω.

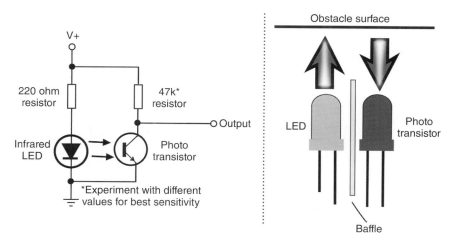

Figure 31-3 This simple proximity detector works by sensing infrared light reflected off an object. The output is a voltage proportional to the brightness of light falling on the phototransistor.

ADJUSTING SENSITIVITY

Sensitivity of the detector can be adjusted by changing the value of the resistor above the phototransistor; reduce the value to increase sensitivity. An increase in sensitivity means that the robot will be able to detect objects farther way. A decrease in sensitivity means that the robot must be fairly close to the object before it is detected.

 Objects reflect light in different ways. You'll probably want to adjust the sensitivity so the robot behaves itself best in a room with white walls. But that sensitivity may not be as great when the robot comes to a dark brown couch or the coal gray suit of your boss.

The infrared phototransistor should be baffled—blocked—from both ambient room light as well as direct light from the LED. Try placing a short piece of black heat-shrink tubing (unshrunk) over either or both the infrared LED or the phototransistor. The positioning of the LED and phototransistor is very important, and you must take care to ensure that the two are properly aligned.

You may wish to mount the LED-phototransistor pair in a small block of wood. Drill holes for the LED and phototransistor. Or, if you prefer, you can buy the detector pair already made up and installed on a circuit board. Two types are shown in Figure 31-4:

1. **Optics with interfacing electronics** (Figure 31-4A). Connect directly to a microcontroller.
2. **Optics without interfacing electronics** (Figure 31-4B). You must add components, such as the resistors shown in Figure 31-3, to complete the circuit.

Modulated Infrared Proximity Detector

The basic infrared emitter/detector system has a distinct disadvantage—namely, the phototransistor is susceptible to ambient lighting. This kind of sensor tends to work best in a darkened room, where there is little chance of light from a lamp or the sun striking the phototransistor.

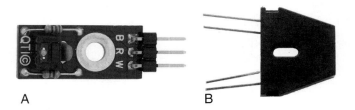

A   B

Figure 31-4   Two types of premade IR optics: (A) Integrated IR emitter and detector with interfacing electronics (*Photo courtesy Parallax Inc.*); (B) IR emitter and detector only; add interfacing electronics of your choice.

Another method is to use a simple technique that's grown much in popularity over the past years: the modulated *infrared proximity detector,* or *IRPD.* These also use infrared emitters and detectors. But to avoid the problems of ambient light spoilage, the system uses a beam of rapidly pulsating, or *modulated,* light. The detector looks only for light that is modulated at the proper frequency, that is flashing on and off at the correct speed. Everything else, it ignores.

Sounds complicated, but in practice it's fairly easy to do, thanks to the wide availability of remote control receiving modules, which form the heart of the IRPD. These are the same modules used in TVs, DVD players, and other devices to receive commands from an infrared remote control.

THE BASIC IRPD CIRCUIT

Figure 31-5 shows a typical IRPD circuit. It has two halves, working in concert:

- A timing circuit, in this case a low-power LM555 IC, pulses an infrared LED to about 38 kHz (38,000 cycles per second).
- The infrared receiving module is tuned to the same 38 kHz frequency. If there's no 38 kHz light beam around, the output of the module stays HIGH. The moment it detects pulses of 38 kHz light, the output of module goes LOW.

Many IRPDs use two LEDs, which are pulsed alternately. By checking which LED is pulsing at the time a signal is received through the module, the robot can detect if the object is to the left, to the right, or straight ahead. We'll look at this technique in a bit.

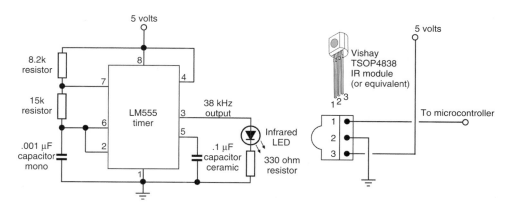

Figure 31-5 Modulated light overcomes many of the problems of using simpler forms of infrared detectors. The sensor is less likely to be influenced by ambient light. This circuit comprises a 555 timer IC providing a steady stream of pulses, that are then picked up by an infrared receiver/demodulator that is turned to the same pulse rate.

The frequency doesn't have to be *exactly* 38 kHz, though the closer it is to this value, the more sensitive the detector will be. There is some latitude in the accepted frequency range of the infrared receiver. With the components shown, and assuming exact values, the frequency of the timer is 37.8 kHz, which is certainly close enough.

In my prototype the measured frequency was actually 39.2 kHz, and the circuit still worked. The variance in measured frequency is due to normal component tolerance. I used resistors with 5 percent tolerance, and the .001 μF monolithic capacitor was rated at 10 percent.

If you wish to experiment with the frequency setting, replace the 15 kΩ resistor with a combination 12 kΩ fixed resistor and a 5 kΩ potentiometer. Wire the pot so that one of its legs is connected to the 8.2 kΩ resistor, and the wiper is connected to pin 6 of the LM555 IC. From one extreme of the pot to the other, the frequency range should span from about 34.2 to 44.8 kHz.

Detection range can be from inches to well over a foot, depending on the exact frequency of the timer and the power output of the infrared LED. You probably don't want a supersensitive detector. Alter the value of the 470 Ω dropping resistor connected to the infrared LED. Higher values weaken the strength of the light, which reduces detection range; lower values increases range. Don't go below about 200 Ω, or you risk smoking the LED by passing it too much current.

For best results:

1. **Point the LED and emitter away** from your solderless breadboard or circuit board. Otherwise you'll get false readings due to reflected light.
2. **Increase directionality (and decrease false readings)** by inserting a 3/4″ length of black heat-shrink tubing over the emitter LED. The longer the tube, the more discriminating the sensor will be.
3. **Add a baffle between** the IR emitter LED and the receiver module. Try a small piece of that black conductive foam they use to hold static-sensitive ICs and other components.
4. **Try different infrared LEDs** to experiment with range. Small IR emitters with 2 to 10-mW output are useful for "close in" detection of just a few inches. Check the datasheet for the LED to determine its power output.
5. **Keep lead lengths trimmed** (especially the capacitors) or else you could get erratic results. Once you get the circuit tested and working, transfer it from your solderless breadboard to a soldered board.
6. **"Detune" the frequency** to alter the effective range of the detector. Try a frequency that's off by 1 to 4 kHz from 38 kHz to make your detector a little less sensitive.

CONNECTING TO A MICROCONTROLLER

Interfacing the IRPD to a microcontroller is straightforward. You can generate the pulses for the infrared detector using a separate circuit, such as an LM555 timer as shown in Figure 31-5, do it in your microcontroller, or use a separate microcontroller.

Using an LM555 timer IC or a separate microcontroller to generate the modulation allows your main controller more freedom to do other, more important things.

LM555-Based Timer

Refer again to Figure 31-5. The LM555 timer is left free-running, generating a steady stream of (approximately) 38 kHz pulses. If the detector module is close enough to receive the pulses, its output goes LOW, which is registered in software running in the microcontroller. Figure 31-6 shows the simple hookup diagram for connecting the IR module to an Arduino.

The sketch *arduino-irpd.pde* for the Arduino demonstrates reading the IR sensor and lighting the built-in LED whenever the receiver detects a stream of 38 kHz pulses. The LED stays lit for a quarter of a second and turns back off if the pulse stream is no longer detected.

arduino-irpd-receive.ino

```
#define receiver 3;          // Connect receiver to pin 3
void setup() {
  pinMode(LED_BUILTIN, OUTPUT);
  pinMode(receiver, INPUT);
}

void loop(){
  // read the receiver, LED on if detection
  if(digitalRead(receiver) == LOW) {
    digitalWrite(LED_BUILTIN, HIGH);
    delay(250);
  }
  digitalWrite(LED_BUILTIN, LOW);
}
```

Auxiliary Microcontroller

Most infrared modules prefer that that modulation not be continuous. What it likes best is that the pulses turn on and off once every millisecond or so. This improves operating range. The on/off sequence works with the detector's internal autogain circuitry; it's how the detector adjusts to the ambient light of the room.

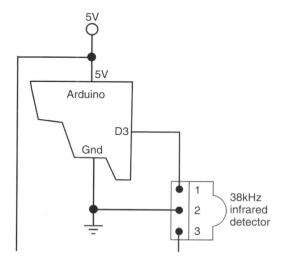

Figure 31-6   How to hook up the IR sensor to an Arduino. When the IR detects the 38 kHz pulses from the LM555 timer the Arduino's built-in LED turns on for a quarter of a second.

With the low cost of many microcontrollers, you can implement a full timing circuit that mimics the LM555 and also toggles the modulation on and off for about 30 milliseconds each. Many microcontrollers can be used for this, but out of consistency with many of the other examples in this chapter I'll use the venerable Arduino Uno to demonstrate it.

 Important! Because the sketch uses hardware specific to the Uno you should not expect to be able to use it on other Arduino boards without at least some modification.

Figure 31-7 shows the basic hookup. The Arduino is connected to an infrared emitting LED by way of a current-limiting resistor—the higher the value of this resistor the less strong the IR light will be, so consider this when deciding on the overall range of the detection system.

Program code running in the Arduino controls the modulation frequency and the on/off cycling by using the TIMER2 hardware timer in the Arduino. The Uno has three such hardware timers, named TIMER1, TIMER2, and TIMER3. A timer can be used for only one task at a time. TIMER2 is not as frequently used by the standard Arduino command statements, but you'll run into compatibility if anything else in your sketch taps it.

The sample program is shown in *arduino-irpd-basic.ino*. In operation it sets up TIMER2 to emit a steady stream of 38 kHz pulses on pin 11 of the Uno. As IR detectors prefer to not be continually blasted by constant pulses, a second pin (pin 12) is used to control the high-speed flashing of the LED.

To see the results of the detection open the Serial Monitor window. Place your hand or other object to block the detector. Note that 0=detection; 1=no detection. This is because the IR detector is active-low (its output goes LOW when it receives a signal).

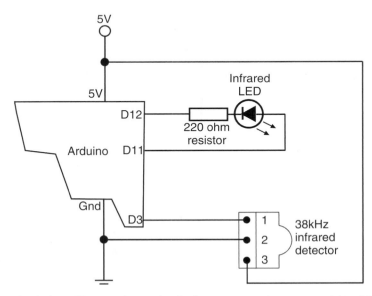

Figure 31-7 An Arduino Uno may be used to both generate and receive modulated light pulses. The LED and sensor should point in the same direction so that the detector only sees light when it's bounced off an object.

arduino-irpd-basic.ino

```
#define CLOCK_RATE 38000L
#define pwm 11        // IR PWM pin
#define ctrl 12       // Control pin
#define IR 3          // IR signal pin

void setup()   {

  // Use Serial Monitor to display results
  Serial.begin(9600);

  // TIMER 2oggle output for each overflow
  TCCR2A = _BV(WGM21) | _BV(COM2A0);
  TCCR2B = _BV(CS20);

  // 38kHz PWM
  OCR2A = (F_CPU/( CLOCK_RATE * 2L) -1);
  pinMode(pwm, OUTPUT);         // PWM output

  pinMode(ctrl, OUTPUT);        // LED control
  digitalWrite(ctrl, LOW);
  pinMode (IR, INPUT);          // Input for IR detector
}

void loop() {
  digitalWrite(ctrl, HIGH);
  delay(30);
  Serial.println(digitalRead(IR));
  digitalWrite(ctrl, LOW);
  delay(150);
}
```

ENHANCED IRPD CIRCUIT

As noted, by using a pair of infrared emitters, each producing a short burst of modulated light, your robot can determine whether an object is to the right, to the left, or straight ahead. While you can produce such a sequence using a couple of LM555 timers (or its LM556 dual-timer cousin), it's easier and perhaps even more cost effective to use a microcontroller. Once again, a good job for the Arduino.

The wiring diagram for the enhanced IRPD is shown in Figure 31-8. Two infrared emitters are used with one detector positioned between; as above LEDs and detectors all face in the same direction. The distance between the emitters and the detector is variable; a few inches is typical, but you can try 6″ to 8″, and angle the front of the emitters slightly inward to help pinpoint the infrared light so that the emitter adequately sees it. Feel free to play!

By placing the detector slightly ahead of the emitters, you can reduce problems of "crosstalk," where light from the emitters goes straight into the infrared detector. You can also try adding a heavy black paper or plastic baffle along the sides of the emitter, or outfitting the front of it with a small length of black heat-shrink tubing to keep stray light from the emitters.

Program code for the enhanced IRPD is found in *arduino-irpd-enhanced.ino*. Two infrared LED emitters are connected to the Arduino so that the 38 kHz pulses can be routed to one or the other. The LEDs are turned on by bringing pins 10 or 12 high. When an LED is *LOW* it is off; when *HIGH* it is sending 38 kHz pulses.

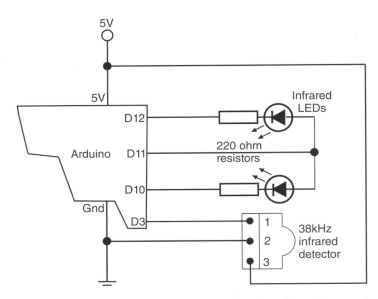

Figure 31-8 This enhanced version of the light modulator for the Arduino alternatively pulses two infrared LEDs; the receiver/demodulator is placed between the LEDs.

To see the action of the circuit upload the sketch and open the Serial Monitor window.

arduino-irpd-enhanced.ino

Check out the full in *arduino-irpd-enhanced.ino* sketch on the RBB Support Site.

Also see the RBB Support Site for alternative microcontroller setups, plus using commercially made IRPD modules. These self-contained modules provide a LOW/HIGH signal depending on whether an object is within detection range, and can be directly connected to any digital input pin. The modules are handy when your microcontroller is already busy doing other robotic things.

Infrared Distance Measurement

Not only can infrared light be used to detect if something is nearby, it can measure the distance between your bot and some object. Infrared distance measurement detectors use the displacement of reflected light across a linear sensor (see Figure 31-9).

Here's how the technique works: A beam of infrared light from the sensor illuminates some object. The beam reflects off the object and bounces back into the sensor. The reflected beam is focused onto what's known as a *position-sensitive device,* or *PSD.* The PSD has a surface whose resistance changes depending on where light strikes it. As the distance between sensor and object changes, so does the linear position of the light falling on the PSD. Circuitry in the sensor monitors the resistance of the PSD element and calculates the distance based on this resistance.

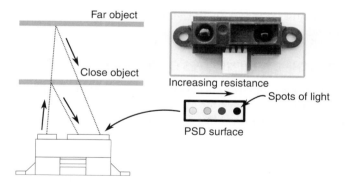

Figure 31-9   The inner workings of the Sharp infrared distance and proximity sensors. Light from an emitter bounces off an object and reenters the sensor at some angle, as shown. The reflected beam strikes against a position-sensitive detector. This sensor detects when light falling on it is off center.

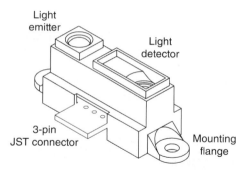

Figure 31-10   The important parts of a Sharp IR sensor. Note the three-pin JST connector. This is smaller than the typical .100″ connector you may be used to, and it requires a special cable.

Nearly all of the infrared distance-measuring modules you'll find are made by Sharp. Figure 31-10 is an outline view of the important parts on a typical Sharp IR sensor. All share better-than-average immunity to ambient light levels, so you can use them under a variety of lighting conditions (except very, very bright light outdoors).

The sensors use a modulated—as opposed to a continuous—infrared beam that helps reject false triggering. It also makes the system accurate even if the detected object absorbs or scatters infrared light, such as heavy curtains or dark-colored fabrics.

Most of the Sharp IR modules use a miniature JST-type three-prong connector. At a distance it looks like the typical 0.100″ connector, but it's smaller. When buying an infrared module, be sure to get the right connector with it.

The circuitry of the sensor can provide either a digital or an analog output. Both are routinely used in robotics, so we'll cover them both.

EXPLORING THE DIFFERENT TYPES OF SENSORS

Infrared distance sensors aren't made with robotics in mind—it's just that they're perfect for the job. Rather, they are intended for use in industrial control. Because of this, there are several models to choose from, each with a unique set of features.

- *Working distance* is the effective minimum and maximum distance of the sensor. "Effective" means the distance at which you get accurate detection. Distances are almost always noted in centimeters. The minimum working distance of the most popular Sharp sensor model is 10 cm, or roughly 4″. Maximum working distance is 80 cm (31.5″). Outside this range, the sensor may still pick up an object, but measurement accuracy may be affected.
- *Beam width* (or spread) relates to how much of an area the sensor "sees." Most IR modules tend to have a fairly narrow beam width, often just 5° to 10°. This means that the object needs to be almost straight on to the sensor to be detected. Some versions are designed to have a wider field, and some a narrow field of view. Pick the one you want based on your application.
- *Digital go/no-go output* sensors are the simplest to use. Their output is either LOW or HIGH, depending on whether an object is within a set proximity range. These are called *distance judgment* sensors, because they merely judge that the distance is within a certain range.
- *Analog measurement output* provides a varying voltage representing distance. The distance to voltage is not a linear (straight path) relationship, which makes very accurate readings a challenge. These are called distance measurement sensors, as they tell you the distance between the module and the object.

BASIC ELECTRICAL HOOKUP

Many (but not all) Sharp IR modules follow a standard connection scheme, as shown in Figure 31-11.

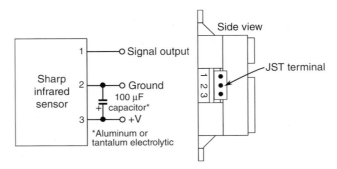

Figure 31-11 Typical connection diagram for the Sharp IR sensors. Not all follow this hookup, but most do. The output may be a digital (on/off) signal or an analog voltage. The capacitor across the power pins is optional, but recommended.

- Pin 1—Signal output. This terminal may provide a digital or an analog signal, as described in the previous section.
- Pin 2—Power supply ground.
- Pin 3—Power supply voltage, which on most units is 4.5 to 5.5 volts. These are suitable for use on 5 volt microcontrollers. On others, particularly the more compact variation provided in a DIP-like package, have a wider supply range of 2.7 to 6.2 volts, and can be used on micro-controllers that operate at 3.3 volts or 5 volts.

Several online sources, such as Lynxmotion, offer adapter cables that go from the specialty JST connector to a standard 0.100" connector. On many of these adapter cables *the pin order is altered,* so be mindful of this. Follow the color coding of the wires to keep track of what goes where.

Why do they flip the pins? They're following good practice by placing the +V connection in the center. That way, if the cable is plugged in backward, there is less chance of wrecking the sensor. It also standardizes on the same connection scheme used by radio control servo motors.

Always consult the Sharp datasheet for specific information about the sensor, especially its operating voltage. A sensor rated at 4.5 to 5.5 volts will not function properly when used with a 3.3 volt power supply.

USING A SHARP DIGITAL OUTPUT INFRARED SENSOR

Sharp units that provide a 1-bit LOW/HIGH output at a specific distance are termed *distance judgement* sensors. The LOW/HIGH signal depends on whether an object has been detected within a specific threshold range. Different models have different preset ranges. For example, the GP2Y0D810Z0F sensor has a preset distance of 2 to 10 cm. Anything closer or farther away is not detected.

The digital output sensors are the easiest to use. They operate just like a regular switch; the output of the sensor is either LOW or HIGH, depending on whether a nearby object is within range.

USING A SHARP ANALOG OUTPUT INFRARED SENSOR

Other Sharp models provide an analog output that varies following the distance between module and object. These are classified as *distance ranging* sensors, since you can measure the specific distance within its range. Starting from the minimum working distance, the voltage goes down as the distance to the object increases; the lowest voltage is present at the maximum working distance, and beyond.

The voltage span of many Sharp analog sensors is restricted to a smaller range of the unit's supply voltage, on the order of 0.4 to 3 volts. This range can vary from unit to unit, so be sure to read the datasheet for the sensor you are using to make sure. The ideal way to measure this voltage is with a microcontroller equipped with at least one ADC input. A sample program using the Arduino is shown in *sharp-analog.ino.*

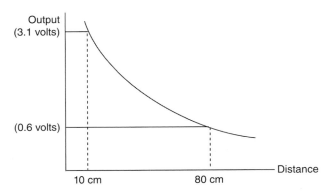

Figure 31-12   The analog voltage output of the Sharp IR sensors is not linear. The output curve depicted here is representative only; refer to the datasheet for the specific Sharp sensor you are using for specifics.

The voltage from Sharp analog sensors is not linear, as shown in Figure 31-12. Don't expect a 1:1 ratio between the value you get and the distance separating the sensor and the detected object. While you could write a math function that attempts to linearize the curved analog response of the sensor, many people use a short lookup table or *switch* statement that correlates voltage to approximate distance. Conduct tests with objects placed at set distances from the sensor (use a tape measure for accuracy), and use those as benchmarks.

The accuracy of the readings will depend greatly on the width of the target. You may wish to experiment by placing the sensor in front of a smooth, white wall. Vary the distance between wall and sensor, and note your results.

Personally, I think it's better to use an ultrasonic sensor (see next section) when accurate distance measurement is needed, and use IR sensors for ballpark approximations—instead of worrying about exact fractions of an inch, you're more interested in knowing whether an object is far, near, or really close.

Note that when an object is very close—less than the minimum detection range—the value returned by the sensor is meaningless. Depending on the exact sensor you use, it'll show the same value as no detection at all. So be sure to back up an IR sensor with, at the very least, a mechanical contact switch to detect physical collision.

sharp-analog.ino

```
int distance = 0;
int averaging = 0;

void setup() {
  Serial.begin(9600);      // Use Serial Monitor window
}

void loop() {
  // Get a sampling of 10 readings from sensor
  for (int i=0; i <= 5; i++) {
    distance = analogRead(A0);
    averaging = averaging + distance;
    delay(55);
  }
```

```
// Average out the 5 readings
distance = averaging / 5;
averaging = 0;
Serial.println(distance, DEC);  // Display result
delay(250);                     // Short delay
```

CONVERTING AN ANALOG OUTPUT TO DIGITAL

In times past some of the Sharp digital sensors provided an adjustment that allowed you to set the desired detection distance. These units were handy when you didn't want to bother with the analog signal in your microcontroller. Some microcontrollers, such as the Parallax BASIC Stamp, don't have analog inputs, for example.

You can construct your own adjustable analog-to-digital distance judgment sensor by combining any Sharp analog sensor with LM339 comparator IC, as shown in Figure 31-13. Don't forget the pull-up resistor on the output of the comparator.

Dialing the potentiometer alters the reference voltage to the comparator. Use this to set your own distance "trip point." Set this point empirically by placing an object in front of the sensor and dialing the 10 kΩ pot until the output of the LM339 changes.

SHARP IR MODULE GOOD CODING PRACTICE

Over the years, robot experimenters have discovered many useful tips and tidbits about how to best use the Sharp infrared distance measurement and distance judgment sensors. Here are several of the most important issues to keep in mind:

- The Sharp modules tend to draw a lot of current when operating—up to 50 mA, depending on model and the instantaneous measured distance. To reduce current consumption, you may wish to selectively turn the module off. Use a driver chip like the ULN2003, which can provide the current demanded by the module.

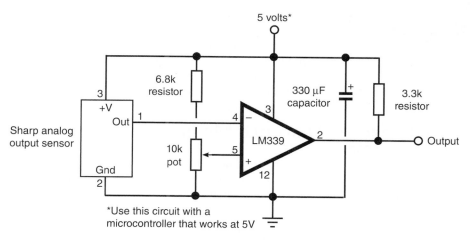

Figure 31-13 How to use an analog output sensor with a voltage comparator. The comparator triggers when the voltage from the sensor reaches a certain point, as set by the 10 kΩ potentiometer.

- If you have more than one IR module on your robot, you may wish to turn them all on at once, take a reading from each, then turn them all off. Again, use the ULN2003 (or something similar), and tie all the inputs of the chip to one I/O line on your microcontroller. That one line operates all the modules.
- Though not an absolute requirement, adding a 100 µF bypass capacitor between +V and ground of the sensor helps to reduce errant readings due to induced spikes in the power supply. You may also want to add a secondary 0.1 µF monolithic capacitor in the same way.
- After turning on a module, wait at least 100 ms (milliseconds) before taking a reading. This allows the device to settle; otherwise, the reading may be inaccurate.
- For greater accuracy, take several successive readings from the sensor, at no less than 50-ms intervals, and average them out.
- In addition to averaging, you may wish to disregard any readings within a group that are wildly different from the rest. For example, if you get readings of 100, 105, 345, 97, and 101, all within 55 ms of one another, you can be fairly sure the 345 result is spurious and should not be considered.
- Accuracy is somewhat diminished when using an infrared sensor on a fast-moving robot, or when attached to a rotating sensor turret. Movement affects the reading. During travel or motion, you can use the measurements for general proximity detection, but you may wish to slow down or stop the robot (or turret) to get a more accurate distance reading.
- The width of the infrared beam increases with distance. Use this to your advantage when placing multiple sensors on your robot. To maximize beam coverage of two sensors, point them so they cross. To minimize beam coverage (make the sensors more selective), point them away from one another.

On the Web: Using Passive Infrared Sensors

Passive infrared sensors detect the proximity of humans and animals. These systems, popular in both indoor and outdoor security systems, work by detecting the change in infrared thermal heat patterns in front of a sensor.

This sensor uses a pair of *pyroelectric* elements that react to changes in temperature. Instantaneous differences in the output of the two elements are detected as movement, especially movement by a heat-bearing object, such as a human.

Check out more about passive infrared detection on the RBB Support Site, where you'll find a useful how-to on using pyroelectric sensors to detect movement and objects.

Ultrasonic Distance Measurement

A police radar system works by sending out a high-frequency radio beam that is reflected off nearby objects—maybe your car as you are speeding down the road. The difference between the time the transmit pulse is sent and when the echo is received denotes distance.

Radar systems are complex and expensive, and most require certification by a government authority, such as the Federal Communications Commission for devices used in the United States. Fortunately, there's another approach you can use with robots: high-frequency sound.

Ultrasonic distance measurement—also called *ultrasonic ranging*—is now an old science. Polaroid used it for years as an automatic focusing aid on their instant cameras. To measure distance, a short burst of ultrasonic sound is sent out through a *transducer,* basically speaker that operates at ultrasonic frequencies. The sound bounces off an object, and the echo is received by another transducer (an ultrasonic microphone). The delay between sending and receiving the signal is used to formulate the distance.

Complete and ready-to-go ultrasonic distance-measuring sensors, like the one in Figure 31-14, are commonly available from a variety of sources. Depending on features, prices start at about $25 to $30. These ready-made modules are the ones we're learning about in this chapter.

FACTS AND FIGURES

First, some statistics. At sea level, sound travels at a speed of about 1130 feet per second (about 344 meters per second) or 13,560″ per second. This time varies depending on atmospheric conditions, including air pressure (which varies by altitude), temperature, and humidity.

The time it takes for the echo to be received is in microseconds if the object is within a few inches or even a few feet of the robot. The overall time between transmit pulse and echo is divided by two to compensate for the round-trip travel time between the robot and the object.

Given a travel time of 13,560″ per second for sound, it takes 73.7 μs (microseconds, or 0.0000737 seconds) for sound to travel 1″. If an object is 10″ away from the ultrasonic sensor, it takes 737 μs to travel there, plus an additional 737 μs to travel back, for a total time of 1474 μs. The formula is:

$$(1474/2)/73.7 = 10$$

First, divide the total transit time by 2, then divide by 73.7 (use 74 to avoid floating-point math), the time it takes sound to travel 1″ at sea level.

Figure 31-14 A typical ultrasonic distance sensor, using two transducers—one for sending the ultrasonic "ping," and the other to receive it.

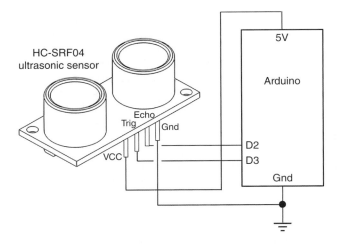

Figure 31-15   Basic connection diagram for the ultrasonic sensor to the Arduino microcontroller.

USING AN ULTRASONIC DISTANCE SENSOR

Ultrasonic distance sensors come in several varieties. The basic models require you to do all the calculation math yourself, though as the previous section showed, it's not that difficult.

The HC-SR04, available from numerous online resellers, serves as a basic ultrasonic sensor suitable for many robotics tasks. It is connected to a microcontroller or other circuit that contains its own timer.

To use, your circuit initiates a sound pulse, then measures the time until the receiver picks up the echo. Your control program, such as *hc-sr04.ino* for the Arduino, is required to perform the math to calculate time of flight. See Figure 31-15 for the very simple diagram for hookup to the Arduino.

hc-sr04.ino

(Rui Santos has written code for the control program at *https://randomnerdtutorials.com/complete-guide-for-ultrasonic-sensor-hc-sr04/*.)

To use, compile and upload the sketch, then open the Serial Monitor window to view the results.

ULTRASONIC SENSOR SPECS

Not all ultrasonic sensors offer the same characteristics. Read the specifications to better match the sensor to your task. Things to consider:

Working distance: Ultrasonic sensors have a minimum and maximum effective distance. For sensors like the SRF08, the minimum range is about 2″; the maximum about 20 feet. Some sensor ranges are more, some less. A very small minimum range is handy when using ultrasonic sensors in arms and grippers.

Sensitivity: The term "sensitivity" can refer not only to the resolving power of the sensor (see "Resolution," next), but also to the ability of the sensor to accurately detect objects at a distance. In addition to the design of the receiver electronics, sensitivity is affected by things you can't easily control: temperature, humidity, even wind, if you're using the sensor outdoors.

Resolution: The frequency of the sound dictates its ability to detect small or thin objects. The 40-kHz sound waves used in low-cost ultrasonic sensors are fairly coarse. That equates to roughly 8.5mm from peak to peak of each wave. This means the smallest object an ultrasonic sensor operating at 40 kHz can fully resolve is less than half an inch.

Beam spread and pattern: The typical ultrasonic sensor has a fairly wide cone-shaped echo pattern. The maximum distance of the sensor changes for objects that are off center (not aligned at 0°). On some, the far end of the cone is very wide; these are useful when you want to take in a large area. On others, the cone is very narrow. These so-called pencil-thin sound beams are useful when you want your robot to measure objects some distance away from itself.

Using a Laser Rangefinder

We already know about mounting lasers on the heads of sharks. What if you were to swap the shark for a robot? You might not be able to take over the world with it, but you can use the laser detect the presence and distance to nearby objects.

Laser rangefinding takes two general forms:

- *Trigonometric* sensors measure the relative position of a laser spot projected onto an object. The larger the distance the farther the object is from the sensor. The system is comprised of a laser and an image array. The laser light focused onto the array using a lens.
- *Time-of-flight or LiDAR (light detecting and ranging)* sensors use any of a number of techniques to measure the time it takes for a light signal to travel from the sensor, to an object, and back again. LiDAR stands for light detecting and ranging. While some high-end time-of-flight (TOF) sensors measure the actual speed of the light pulse, other units modulate the light with a very fast signal which is carried over the light beam. Distance is measured by detecting the changes in phase of this signal.

Regardless of the method, laser-based distance sensors require the light to be reflected from the object back to the sensor. Not all objects do this: very dark velvet fabric reflects very little light, making this type of material almost invisible to many laser-based sensors. For this reason, it's often a good idea to use multiple sensor types, each of which has its own detection pros and cons.

Laser sensors may sound expensive, and some are. But a growing cadre of low-cost distance detectors have been released that are intended for personal electronics such as mobile phones. Outfits like SparkFun and Parallax have developed affordable module-level sensors that use these inexpensive detectors. They're well suited for robotics.

One such module is the Parallax LaserPING, shown in Figure 31-16. The module has a maximum detection range of up to 6 feet, and can be used by both 3.3V and 5V microcontrollers. The output signal is comparable to an ultrasonic sensor; in fact, you can use similar programming code. In operation, you "trigger" the laser pulse, then measure how long it takes to get a response.

Using the LaserPING sensor involves setting the module's I/O pin momentarily HIGH. Following a brief delay, the module then returns a pulse whose width varies depending on the distance between the sensor and the detected object. The longer the pulse, the further away the object.

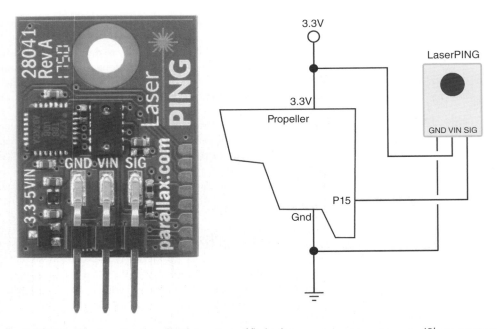

Figure 31-16 The Parallax LaserPING is a time-of-flight distance measurement sensor. (*Photo courtesy Parallax, Inc.*)

Though the LaserPING can be used with most any microcontroller, I'll demonstrate its use with the Parallax Propeller (Activity Board or Board of Education). Connect the LaserPING as shown in Figure 31-16, and load the code in *propeller-laserping.c* into the SimpleIDE programming environment. Open the Serial Window to view the distances reported.

propeller-laserping.c

```
#include "simpletools.h"
#include "ping.h"

int main() {
  while (1) {
    term_cmd(CLS);
    print("Distance");
    print("%d", ping_cm(15));
    pause(150);
  }
}
```

Widening the Field of View of Sensors

Most of the time you'll mount your proximity and distance sensors to the front of the robot looking ahead. This captures most obstacles in the direct path of the robot, but some objects

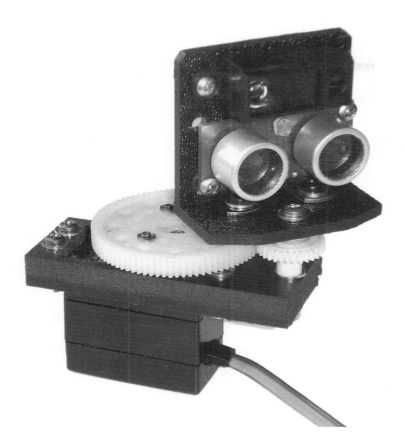

Figure 31-17   Use a servo motor (standard, mini, or micro will do) to scan proximity and distance sensors from side to side.

may crowd close to either side of the sensors field of vision, and might be missed as your robot travels its world.

Just as we turn our eyes and head to take in a larger view, you can broaden your bot's vision by mounting proximity and distance sensors on a rotating turret. To the front of your bot mount a standard servo motor, and then attach one or more sensors to the servo.

USING A SENSOR TURRET

Figure 31-17 shows both a Sharp IR and ultrasonic sensor attached to the same bracket mounted atop the servo. This particular turret uses a homebrew "gear-up" mechanism to spin the turret a full 360°, rather than the typical 180°. This is not strictly required; it's usually sufficient to scan the turret just 20° or 30° to either side of the robot.

USING MULTIPLE SENSORS

A disadvantage of using a sensor turret is that you may need to take three or more "snapshots" of the road ahead, and each snap takes a finite amount of time. When speed is important, consider adding multiple sensors to your robot, situated so that each sensor has a separate view.

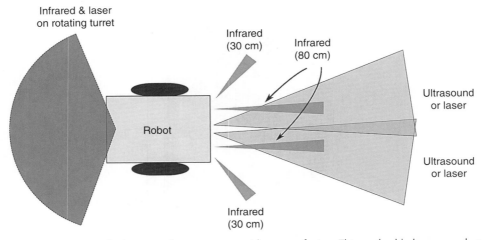

Figure 31-18 Use multiple types of sensors to provide sensor fusion. This method helps your robot overcome limitations inherent in each type of sensor.

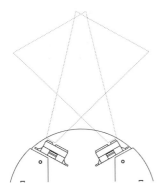

If your budget can handle it, a workable approach is to add multiple sensors around the periphery of your robot. Rotate each sensor to increase the angle of view. Attach two IR or ultrasonic sensors mounted at slight angles on the front of the robot.

Be sure you don't create a dead zone directly in front of the robot. Otherwise the robot may ignore objects directly in front of it. If the dead zone cannot be avoided, add a third sensor that points straight ahead.

MIXING AND MATCHING SENSORS

Sensor fusion is the technique of combining two or more sensors that use completely different detection techniques. Each sensor fills in the holes left by the others. IR, laser, and ultrasonic sensors each have limitations; light-based sensors rely on reflectance to detect things, and some objects—such as glass—may have poorly reflective surface. While a pane of glass might be invisible (or nearly so) to an IR or laser sensor, it is easily picked up by an ultrasonic sensor.

Figure 31-18 shows the general idea. Mount sensors to the front, back, and sides of the robot depending on the extent of vision you want to provide. Consider the special needs of your bot, but don't go overboard. Each sensor adds to the current consumption of your robot, not to mention its cost. Just because you *can* mount 20 ultrasonic sensors around the body of your robot doesn't mean you should!

Select the sensor characteristics based on its application. Sensors to the side of the robot generally don't need to have a lot of range. Pick sensors designed to detect closer objects. For forward (or aft) sensors, select these to provide a deep enough sensing range to allow your robot to have enough time to recognize an obstacle (3 feet away is enough), and work to avoid it.

CHAPTER 32

Navigation

Robots suddenly become useful once they can master their surroundings. Being able to wend their way through their surroundings is the first step toward that mastery.

The projects and discussion in this chapter focus on navigating your robot through space—not the outer-space kind, but the space between two chairs in your living room, between your bedroom and the hall bathroom, or outside your home by the pool.

The techniques used to provide such navigation are varied: line followers, wall followers, compass bearings, odometry, and more.

 Source code for all software examples may be found at the RBB Support Site. To save page space, the lengthier programs are not printed here.

Tracing a Predefined Path: Line Following

Perhaps the simplest navigation system for mobile robots involves following some predefined path that's marked on the ground. The path can be a black or white line painted on a hard-surfaced floor, a wire buried beneath a carpet, a physical track, or any of several other methods. This type of robot navigation is used in some factories. The reflective tape method is preferred because the track can easily be changed without ripping up the floor.

You can readily incorporate a tape-track navigation system into your robot. The line-following feature can be the robot's only means of semi-intelligent action, or it can be just one part of a more sophisticated machine. You could, for example, use the tape to help guide a robot back to its battery charger nest.

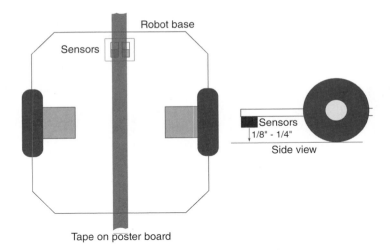

Figure 32-1   Placement of line-following sensors on the bottom of the robot. For best results, the sensors should be no more than 1/8″ to 1/4″ from the surface of the floor.

SETTING UP FOR FOLLOWING A LINE

With a line-following robot, you place a piece of white, black, or reflective tape on white poster board. Place the board on a table or the floor. One or more optical sensors are placed under the robot. These sensors incorporate an infrared LED and an infrared phototransistor. When the transistor turns on, it sees the light from the LED reflected off the tape.

You can build your own line-following sensors, or use a commercially available LED-phototransistor pair. Mount two detectors on the bottom of the robot, as shown in Figure 32-1, in which two detectors have been placed the same or a *slightly* farther apart than the width of the tape.

Figure 32-2 shows the basic sensor circuit and how the LED and phototransistor are wired. While you can use just two LED/phototransistor pairs for line following, you get better results when using three (or more) pairs that straddle the track:

- The pairs on either side tell the robot it's veered off too far left or right. To counter the mistracking, the robot steers in the opposite direction of the sensor pair that was activated. For example, if the right sensor pair sees the line, the robot corrects by steering to the left.
- The sensor pair in the center is used to indicate correct tracking. As long as this sensor sees the line, then the robot is headed in the correct direction.

The spacing of the sensors depends on the thickness of the line. A good starting point is 1.5 times the thickness of the line. That is, if the line is 0.25″ wide, then the sensors are spaced about 0.375″ from each another.

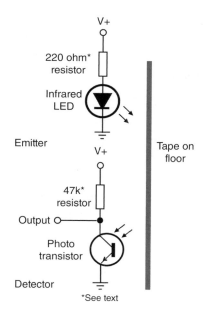

Figure 32-2   Circuit diagram for an LED emitter and phototransistor pair, used for line following (and other robotic tasks). The resistor above the infrared emitter should be selected to provide a high current to the LED, but without destroying it. See Chapter 21 on how to calculate the value of the resistor used for limiting current through an LED.

Some robot builders prefer using a photoresistor (CdS cell) rather than a phototransistor. Photoresistors are often desired because their sensor sees a larger area, and they are naturally slower to react to changes in light intensity.

However, since photoresistors are most sensitive to the middle of the visible light spectrum (see Chapter 36, "Visual Feedback from Your Robot," for more details), it's better to use green or yellow LEDs for illumination. What's more, the output of photoresistors is also not as consistent as that of phototransistors. This may require you to selectively adjust the sensitivity of the photoresistor by altering the value of its series resistor.

SELECTING PROPER RESISTOR VALUES

Specifications for infrared LEDs and phototransistors vary widely. You may need to play with the values for the two resistors shown in Figure 32-2 to determine what works best for the parts you have.

- The resistor above the infrared emitter LED controls the brightness of the LED.
- The resistor above the phototransistor controls the sensitivity of the phototransistor.

Both work hand-in-hand to provide the proper amount of light and detection.

Some LEDs don't emit as much light as others, so you may need to increase their current so that they glow more brightly. I've specified a 220 Ω resistor to start; *decrease* the value of the resistor (to *no less than* 180 Ω) to increase the output of the LED. Conversely, if the LED is too bright—it spills light everywhere and causes the phototransistor to falsely trigger—you'll need to *increase* the value of the resistor to perhaps 330 Ω to 1 kΩ.

Similarly, I've specified a 47 kΩ resistor for the phototransistor. You can experiment with values as low as 3.3 kΩ and as high as 470 kΩ and higher. The lower the resistance, the less sensitive the phototransistor is to light. If you don't seem to be getting much voltage swing when alternately hiding and exposing the phototransistor to a bright light source, try a higher value for this resistor.

USING READY-MADE EMITTER/DETECTOR PAIRS

You may opt for a ready-made solution whereby the IR LED and phototransistor are in a single module, perhaps mounted on a small circuit board that you can attach to your robot. Premade modules make it a snap to add line-following capabilities to any robot. Simple secure the sensors to the front center of the robot (see Figure 32-3), and space them approximately 3/8″ apart.

Adding more than four or five LED/phototransistor sets can get clumsy. In these instances it's usually better to use an array of emitter/detector pairs, such as the one in Figure 34-4 that provides eight light sensors on a single board. Two of the pairs can be snapped off, making for two or six pairs, if that better suits your design objective.

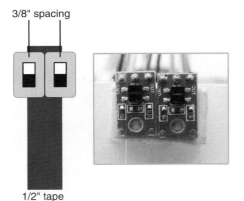

Figure 32-3   You might opt for a ready-made detector pair if you find your homebrew efforts lacking.

Figure 32-4   Infrared emitter/detector sensor array. Each pair is separated by 0.375″ (3/8″). Two of the pairs can be snapped off on their own. (*Photo courtesy Pololu.*)

AVOID OVERLY TIGHT TURNS

Your robot won't be able to make the turn if it's too tight. When drawing a line for your bot to follow, try more gradual turns rather than abrupt intersections. Otherwise, the robot might skip the line and go off course. The actual *turn radius* will depend entirely on the robot—its size, wheel base dimensions, and speed.

At the same time, you don't want to shy away from the challenge of making your robot handle increasingly tight turns. Line following is a learned art; start out simple, and progress from there. With the right design and programming, your bot may be able to literally turn on a dime. It takes experience to get to that point in your robot-building prowess.

LINE-FOLLOWING SOFTWARE

Most any microcontroller is up to the task of navigating a robot over a line. On a three-sensor line follower, three input lines are required. Your program monitors the status of the three sensors and adjusts the motors accordingly.

Figure 32-5 shows a wiring diagram for connecting a trio of infrared light emitters and detectors to an Arduino. Refer to the section "Selecting Proper Resistor Values," previously in this chapter, for more information on experimenting with the values of the resistors used in the circuit. The program *arduino-line_follow.ino* controls the typical two-wheeled robot using these three sensors.

In operation, when the sensor pair to the right sees the line, the left motor is very briefly stopped (it can also be slowed down or reversed for different effect). This causes the robot to veer to the left. The reverse happens when the sensor pair to the left sees the line.

 The "color" of the line is important, and ideally your line-following software should be selectable between black-on-white and white-on-black. This entails reversing the logic on the input pins: LOW is treated as HIGH, and vice versa.

 arduino-line_follow.ino

To save space, the program code for this project is found on the RBB Support Site.

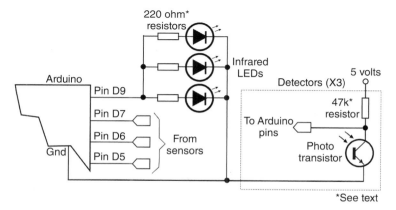

Figure 32-5 Arduino connection diagram for a three-pair emitter/detector line follower. The LEDs are driven through a digital output (so they can be turned on and off, or even pulsed). The output of each phototransistor is connected to a separate digital I/O pin of the Arduino.

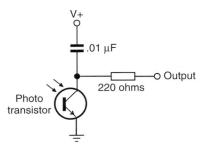

Figure 32-6 Adding a capacitor and resistor to the output of the transistor provides an RC network, a simple method for converting an analog output to digital.

ANALOG OR DIGITAL SENSORS

So far I've talked about using LED/phototransistor circuits using an analog output. This is a good approach when using an Arduino, BBC micro:bit, or other popular microcontrollers that have analog-to-digital conversion (ADC) built-in. Simply attach the output of the sensor to a pin that supports analog voltage. On the Arduino these are pins A0 through A5. You then use the *analogRead* statement to retrieve the instantaneous value of the sensor.

Alas, not all microcontrollers provide ADC inputs. On models like the Raspberry Pi and BASIC Stamp, for example, you can either use an ADC chip (adds complexity) or modify the photodetector circuit to provide a digital output.

Figure 32-6 shows the traditional approach using a resistor and capacitor. The technique is referred to as *RC*, for *Resistor Capacitor*. In use, your microcontroller must first briefly apply a voltage to the pin. This *charges* the capacitor. The microcontroller then checks how long it takes for the capacitor voltage to decay to 0 volts. The timing of the output signal represents the reflectance value.

Here's an example using an old skool BASIC Stamp with the Parallax QTI module, which has an RC circuit built-in.

```
' {$STAMP BS2}
' {$PBASIC 2.5}

SnsrIn CON 9          ' R(ed) wire of QTI sensor to Stamp pin 9
Duration VAR Word

Read_Sensor:
  HIGH SnsrIn
  PAUSE 1
  RCTIME SnsrIn, 1, Duration

  Display:
  DEBUG HOME
  DEBUG "-----", CR
  DEBUG DEC Duration, CLREOL
  PAUSE 250
  GOTO Read_Sensor
```

When you run this program with the Stamp connected to your PC the Debug Terminal will automatically open and display the results of the sensor.

A number of online stores offer IR modules with the RC circuit already added. This includes Pololu (single and 8-pair versions available), Parallax, and RobotShop. The Parallax QTI sensor is popular; though originally devised for the BASIC Stamp it can be used with any microcontroller. Connection is easy using a standard 3-wire servo cable. (The documentation for the device uses the White-Red-Black color coding to denote its pin functions.)

OTHER USES FOR LINE-FOLLOWING SENSORS

You can adapt line-following sensors for other robotic applications. For example, use the sensors to detect when the robot is about to go off the edge of a table or stairs. As long as the sensor(s) see light, the robot is over even ground. But when the reflected light is absent, it could indicate a steep falloff to the ground below.

Obviously, this technique works best for robots on which the line sensors are ahead of the drive wheels. It's less helpful if the wheels are in the front with the sensors; the robot won't know it's falling over a cliff until it's already heading down!

Wall Following

Robots that can follow walls are similar to those that can trace a line. Like the line, the wall is used to provide the robot with navigation orientation. One benefit of wall-following robots is that you can use them without having to paint any lines or lay down tape. Depending on the robot's design, the machine can even maneuver around small obstacles.

VARIATIONS OF WALL FOLLOWING

Wall following can be accomplished with any of four methods including various types of contact and noncontact sensors, as shown in Figure 32-7.

In all cases, upon encountering a wall, the robot goes into a controlled program phase to follow the wall in order to get to its destination. This might involve minor "zig-sagging" in order to keep a constant distance between robot and wall (Figure 32-8). Using a distance measurement sensor (infrared, ultrasonic) the robot can determine how far it is from the wall, and make course corrections as needed.

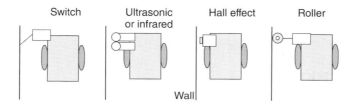

Figure 32-7   Several methods for wall following, including noncontact (ultrasonic, infrared, Hall effect) or contact (roller, whisker switch).

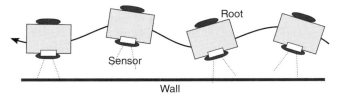

Figure 32-8   The actual course taken by the robot when following the wall will resemble a wave. This pattern is created when the robot self-corrects its course by adjusting the distance between itself and the wall

Odometry: Calculating Your Robot's Distance of Travel

Odometry is the technique of counting the revolution of a robot's wheels to determine how far it's gone. Odometry is perhaps the most common method to determine where a robot is at any given time (it also can provide other information, such as speed). It's cheap and easy to implement, and it is fairly accurate over short distances.

Odometers are often referred to as *encoders* or *shaft encoders* because they are mounted on the shafts of motors and wheels to count revolutions. The terms "odometer," "encoder," and "shaft encoder" are often used interchangeably.

Encoders allow you to measure not only the distance of travel of a robot, but its velocity. By counting the number of transitions provided by the encoder, the robot's control circuits can keep track of the revolutions of the drive wheels.

OPTICAL ENCODERS

Perhaps the most common method of adding odometry to robots is to attach a small disc (also called a *codewheel*) to the hub of a drive wheel. The disc works as part of an optical system, either reflective or transmissive. With either method, a pulse is generated each time the photodetector senses the light.

- With a *reflective* disc, infrared light strikes the disc and is reflected back to a photodetector. Both the infrared LED and the photodetector are mounted on the same side of the disc.
- With a *transmissive* disc, infrared light is alternatively blocked and passed by slots or small cutouts. The light is detected by a sensor mounted on the opposite side of the LED.

More about optical encoders in a bit.

MAGNETIC ENCODERS

You can construct a magnetic encoder using a Hall effect switch (a semiconductor sensitive to magnetic fields) and one or more magnets. A pulse is generated each time a magnet passes by the Hall effect switch. A variation on the theme uses a metal gear and a special Hall effect sensor that is sensitive to the variations in the magnetic influence produced by the gear.

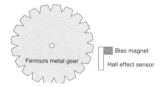

A bias magnet is placed behind the Hall effect sensor. A pulse is generated each time a tooth of the gear passes in front of the sensor. The technique provides more pulses on each revolution of the wheel or motor shaft, and without having to use separate magnets on the rim of the wheel or wheel shaft.

MECHANICAL ENCODERS

A low-cost form of encoder uses mechanical contacts rather than light or magnets. While technically referred to as encoders, these are used more as digital potentiometers; many have a "detent," tactile feedback when turning the dial, and may wear out after a few thousand rotations.

A CLOSER LOOK AT ODOMETRY

Your robot uses odometry to judge distance of travel (and speed, if needed) by counting electrical pulses.

Let's say the codewheel disc has 50 slots around its circumference. That represents a minimum sensing angle of 7.2°. As the wheel rotates, it provides a signal pulse to the counting circuit every 7.2°. By counting those pulses and calculating how many pulses there are each second, and knowing the diameter of the wheels on your robot, you can determine distance and speed.

Now suppose the robot is outfitted with a 7″ wheel (circumference = 21.98″; you calculate circumference by multiplying the diameter of the wheel by *pi*, or approximately 3.14).

Given a pulse every 7.2° of the wheel's rotation, it produces a resolution that is approximately 0.44 linear inch per pulse. This figure was calculated by taking the circumference of the wheel and dividing it by the number of slots in the codewheel disc. If your robot counts 100 pulses, it knows it's moved 0.44″ × 100, or 44″.

If the robot uses the traditional two-wheel-drive approach, optical encoders are attached to both wheels. This is necessary because the drive wheels of a robot are bound to turn at slightly different speeds over time. By integrating the results of both optical encoders, it's possible to determine where the robot really is, as opposed to where it should be. As well, if one wheel rolls over a cord or other small lump, its rotation will be hindered; this can cause the robot to veer off course.

ANATOMY OF A REFLECTIVE ENCODER

Perhaps the easiest type of odometer to build is the reflective encoder, like that in Figure 32-9. You can use the inside surface of a wheel to paste on a circular pattern of white/black stripes. This forms the encoder disc (codewheel) itself. You then mount an infrared emitter and detector close to the codewheel. As the wheel turns, the detector senses the light/dark pattern on the codewheel, providing a signal for use with your control electronics.

See the RBB Support Site for a variety of free codewheel designs that you can download and print.

When printing a pattern for a reflective codewheel, the paper should be thick (24# or more) gloss or semigloss. Photo-quality paper is a good choice to start. Be sure your printer has a relatively new ink or toner cartridge. Set the print quality to Extra Super Super High Fine (or whatever the highest setting is).

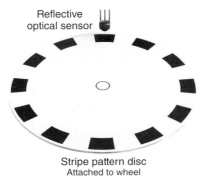

Reflective
optical sensor

Stripe pattern disc
Attached to wheel
or motor shaft

Figure 32-9   Single-stripe reflectance disc for wheel odometry. An emitter/detector pair is positioned over the stripe and senses the white/black pattern as the disc rotates.

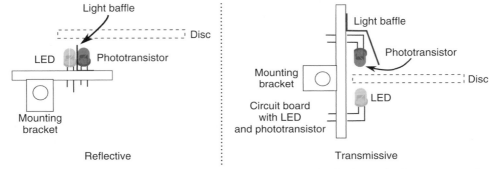

Light baffle

Disc

LED   Phototransistor

Mounting
bracket

Reflective

Light baffle

Phototransistor

Mounting
bracket

Disc

LED

Circuit board
with LED
and phototransistor

Transmissive

Figure 32-10   Mounting for both reflective and transmissive codewheels. For both, you may wish to use light baffles or tubes to prevent stray light from striking the detector.

As an alternative, you can take your pattern to a professional copy center and have them make a print using a wax-based or other dense-medium toner/ink.

ANATOMY OF A TRANSMISSIVE ENCODER

A transmissive encoder is a disc that has numerous holes or slots near its outside edge. Rather than reflect light off a printed pattern, transmissive encoders pass light through the disc. An infrared emitter is placed on one side of the disc, so that its light shines through the holes or slots. An infrared-sensitive detector is positioned directly opposite (see Figure 32-10) so that when the disc turns, the holes pass the light intermittently. The result, as seen by the detector, is a series of flashing lights.

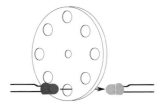

Transmissive encoders are harder to build because you need to cut or drill the holes/slots into a solid material. That material must be opaque to infrared light; many types of dark or black plastics actually let through infrared, making them unsuitable. A thin sheet of metal is the best overall choice, followed by an ABS or PVC plastic. Black-tinted polycarbonate or acrylic likely won't do.

You may be able to find already machined parts that closely fit the bill, such as the encoder wheels in a discarded mouse (the computer kind, not the live rodent kind). The mouse contains two encoders, one for each wheel of the robot.

ENCODER RESOLUTION

With both reflective and transmissive odometers, the number of stripes/holes/slots in the disc determines the resolution of the encoders. This in turn affects the accuracy of the distance measurement. The more stripes, holes, or slots, the better the accuracy.

For reflective encoders, increasing the number of stripes is straightforward. Depending on the type and quality of your printer, and the size of the pattern, you can easily make discs with 100+ stripes.

MEASURING DISTANCE

Odometry measurements can be made using a microcontroller that is outfitted with a *pulse accumulator* or *counter* input. These kinds of inputs independently count the number of pulses received since the last time they were reset. To take an odometry reading, you clear the accumulator or counter, then start the motors. Your software then counts the pulses; multiply the number of pulses by the known distance of travel for each pulse.

If the number of pulses from both encoders is the same, you can assume the robot traveled in a straight line, and you have only to multiply the number of pulses by the distance per pulse.

MOUNTING THE CODEWHEEL AND OPTICS

With the codewheels made, you need to mount them to the wheel (or motor shaft) and secure the infrared emitter/detector optics.

Mounting the Codewheel Discs

- For reflective encoders, cut out the disc and paste it to the inside of the wheel. If the wheel isn't smooth enough, find or make a thin plastic disc that can act as a backstop. Unless the paper for the disc is absolutely opaque, best use a white-colored backstop.
- For transmissive encoders, mount the disc 1/4″ or more away from the inside of the wheel. Use a spacer at the hub of the wheel to maintain separation.

Mounting the Sensor Optics

Using brackets, attach the infrared emitter and detector so they fit snugly near the codewheel. For reflective encoders, you'll want to place the sensor within about 1/8″ to no more than 1/4″ away from the disc. For transmissive encoders, be sure to orient the emitter and detector so that they face each other directly. You can bend the lead of the emitter and/or detector a bit to line it up with the holes.

For either encoder type, you'll want to mask the phototransistor detector with a light baffle (thick piece of black felt or something similar) so it doesn't pick up stray light or reflected light from the emitter. See Figure 32-10 for details. If the phototransistor you're using doesn't have a built-in infrared filter (and it should), you can increase the effectiveness of the sensor by adding an infrared filter over it.

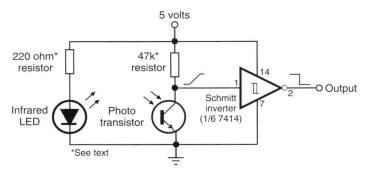

Figure 32-11 Electrical connection of infrared LED emitter and phototransistor used for wheel odometry, including a Schmitt trigger interface.

ELECTRICAL INTERFACE AND CONDITIONING

Figure 32-11 shows the hookup diagram for the emitter and detector of an optical encoder. The same circuit works for both reflective and transmissive, and from now on, for simplicity, I'll assume you're using the reflective type.

The outputs of the phototransistors might need to be conditioned before they can be connected to a microcontroller. Otherwise, you could get all kinds of false signals and inaccurate readings. The circuit uses a *Schmitt trigger,* a kind of buffer circuit that smooths out the wave shape of the light pulses. The output of the buffer is limited to only on and off. This helps prevent spurious triggers and provides a cleaner output.

The output of the Schmitt trigger is applied to the control circuitry of the robot. You can use either an inverting or a noninverting Schmitt trigger, as the circuit is interested only in on and off transitions.

The exact values of the resistors to use depend on the specifications of the infrared LED and phototransistor you use in your encoder. Try the values shown, and adjust as needed.

- The resistor above the LED controls the brightness of the LED: use a lower value to make the LED brighter; a higher value if the LED is too bright, and the phototransistor always sees light. Be sure not to choose too low of a value for the resistor (anything below about 180 Ω) or the LED could be damaged from too much current.
- The resistor above the phototransistor determines the sensitivity of the phototransistor. A lower value makes the transistor less sensitive; a higher value, the inverse.

 Note that the Schmitt trigger inverts the signal: as the output of the phototransistor *increases,* the Schmitt trigger changes from HIGH to LOW. If you'd rather it work the other way, flip the order of the transistor and its resistor (the resistor goes on the bottom). Or route the output of the trigger to another trigger—that inverts the inversion.

TESTING YOUR ENCODERS

To test your encoder, connect the phototransistor to a multimeter. Slowly rotate the wheel or motor shaft. If the system is working properly, you should see a definite voltage swing as the detector goes past the light and dark stripes. If you don't see much difference, double-check your work, and be sure the detector isn't being spoiled by room light (turn the work lamp on your desk away from the wheel).

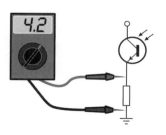

- If the voltage stays the same and is *too high,* the emitter may be over-driven. Try a higher value for the resistor above the emitter LED. This reduces the output of the emitter.
- If the voltage stays the same and is *too low,* the emitter may be under-driven. Try a lower value for the resistor above the emitter LED—but don't go below about 180 Ω, or the emitter may be damaged.
- If the voltage changes only slightly, then the stripe pattern may not be dark enough. The black stripes might be reflecting too much infra-red light. Reprint the discs, being sure to use fresh toner or ink car-tridges. If available, set the printing options to a darker level.

 Not sure if the infrared emitters are emitting infrared? A quick check is to point a digital camera or camcorder at them. You should see a white glow from the emitter. Though invisible to the human eye, digital imaging sensors are sensitive to the light from infrared emitters.

PROGRAM FOR COUNTING PULSES

Your wheel encoders are now ready to rumble and need only be connected to a microcontroller or other circuit. The program listing in *arduino-single_encoder.ino* shows how to implement a basic pulse-counting odometer using an Arduino microcontroller.

 **arduino-single_encoder.ino**
To save space, the program code for this project is found on the RBB Support Site.

ON THE WEB: ALL ABOUT QUADRATURE ENCODERS

By using two sensors you can enhance the optical encoder to provide higher resolution without needing to increase the number of stripes. This tech-nique, known as quadrature encoding, also allows you to determine which direction the codewheel is turning.

This is not quite as useful in robotic drive motors (you already know which way the wheels are turning), but it can be handy for such things as *compliant* robot arms where you want to manually adjust the position of joints. Your robot can keep track of the motion of its arm by knowing the direction and distance the joint is turning.

You can read more about quadrature encoders on the RBB Support Site. You'll find optical and electrical hookups, artwork for codewheels, and program sketches for the Arduino.

USING READY-MADE ENCODERS

You're not limited to using just homebrew encoders. Commercially made optical, magnetic, and mechanical encoders are available, some of them are affordable, and a few of them are readily adaptable to robotics.

Wheel encoders tend to be designed for specific wheels or motors, or at specific types of wheels or motors. Research to find what's available for your wheels and motors. The optics attach to the side of the motor; the codewheel slips around the output shaft.

High-quality (but still relatively low-cost) commercial wheel encoders include kits from US Digital (*www.usdigital.com*) that work with motor shafts from 1.5mm to 4mm (059″ to 0.157″). US Digital also sells transmissive optical discs of various diameters, reflective encoder modules with all the optics built-in and properly spaced, and many other solutions. These are useful for any intermediate or advanced robotics project.

An alternative to adding an encoder is to find motors where one is already included inside! This is often the domain of very expensive industrial-grade motors, but there's a growing number of low-cost motors for hobby and education that sport their own built-in encoders.

One example is the Parallax Feedback 360° servo. For its encoder it uses a magnet and Hall effect sensor to determine the angular position of the motor shaft. Distances can be determined by keeping track of the times the shaft has turned a full circle. Accuracy of fractions of an inch is possible.

UNDERSTANDING ERRORS IN ODOMETRY

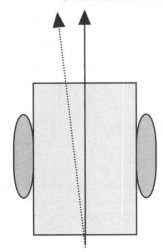

Wheel encoders of any design aren't perfect. Over a 30- to 50-foot range it's not uncommon for the average odometer to misrepresent the position of the robot by as much as half a foot or more!

Why the discrepancy? First and foremost: Wheels slip. As a wheel turns, it is bound to slip, especially if the surface is hard and smooth, or when there's an obstacle on the floor. Second, certain robot drive designs are more prone to error than others. Robots with tracks are steered using slip, odometry is basically useless on most any tracked vehicle design.

And there are less subtle reasons for odometry error. If you're even a hundredth of an inch off when measuring the diameter of the wheel, the error will be compounded over long distances. You're best off relying on odometry for short distances only. Assume that if your robot has made a turn, the actual distance traveled may be a bit different from what's being reported.

WHEN YOU DON'T NEED ODOMETRY

Keep in mind that some robotics applications simply don't need odometry, saving you the trouble of incorporating it into your designs. A good example is line-following robots. These are navigated using a predefined line, so their route is known.

Other examples include wall-following bots, though in some cases these can be augmented by simple distance measuring. And for obvious reasons, robots that walk on legs, swim over water, or fly in the air aren't suitable for the type of odometry covered here.

Knowing Acceleration, Rotation, and Direction

Knowing where something is and where it's heading is the essence of navigation. There are a number of methods for determining where we are in space, and each one has its particular uses and limitations. For robots used indoors and in small areas, sensors for detecting changes in speed, rate of rotation, and even the Earth's magnetic poles all aid in helping it to know where it's going.

The collection of electronics for keeping tabs on these senses are the compass, accelerometer, and gyroscope. Your robot may not need all three, or even any of them. But knowing they're available might help unstick a sticky problem you've run up against.

COMPASS

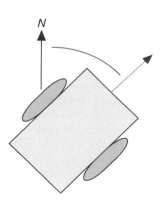

A compass tells you where your robot is headed compared to a known landmark, in this case Mother Earth's magnetic North Pole. You can use this magnetic beacon to orient your robot in the direction you want to go. Use compass readings to move in an absolute direction, or a new direction relative to where you are now.

- Moving in an absolute direction modifies the path of your robot based on world coordinates. For example, you might want to simply head due North: turn the robot until its compass reads N, then straight on 'til morning. (Well, it works for Peter Pan.)
- Moving in a relative modifies the path of your robot based on its own coordinates. If your bot is currently pointed at 30° on the compass, and you want to make a 90° right turn, you need only add 30+90, and then turn the robot until its compass reads 120.

ACCELEROMETER

An accelerometer measures changes in velocity or acceleration. The most basic accelerometer has a single axis of detection, usually "down." This axis measures linear acceleration in a single direction.

If the sensor is oriented so that it faces the center of the Earth, and it's not moving, it measures 1g, which represents the Earth's gravitation pull—about 10 meter per second squared. If the accelerometer moves up or down, then the g-forces on it change. Falling downward reduces the g-force; while rising upward increases it.

Accelerometers can combine more axes of detection. Each axis is a degree of freedom (DOF). Given a sensor with three DOF, the accelerometer provides measurements in the X, Y, and Z axes. Assuming the sensor is lying flat on the earth:

- The X and Y axes denote the front/back and right/left movement.
- The Z axis denotes movement up and down, but also the constant acceleration from the Earth's gravity.

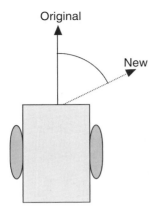

Original

New

GYROSCOPE

A gyroscope measures orientation and rate of angular velocity. Imagine yourself on a spinning top. Other than getting monstrously sick, you'd be spinning at the same speed of the top. Measuring your angular velocity tells you how fast you're going in circles each second. If you're spinning 60 times a minute, that equates to 360° each second—(60/60) × number of degrees in a full circle.

Gyros are their most useful, at least for robots, when dealing with much lower rotations per second. Suppose your robot makes a turn. A gyroscope can measure the rate of turn within a given timeframe, thus allowing you to compute the new direction of the robot. If the gyro measures a 70° turn, for example, you can use this information to plot the new heading of your wandering bot.

Navigation by Compass Bearing

Besides the stars, the magnetic compass has served as humankind's principal navigation aid over long distances. You know how it works: a needle points to the magnetic North Pole of the Earth. Once you know which way is north, you can more easily reorient yourself in your travels.

Robots can use compasses as well, and a number of electronic and electromechanical compasses are available for use in hobby robots. Electronic compasses with digital or analog outputs are now fairly common. In fact, the micro:bit comes with a compass already installed on it. The compass registers the direction of the micro:bit is facing as a number from 0 to 360, with magnetic North as 0.

It's important to bear in mind that like any magnetic compass, the compass built into the micro:bit is actually a magnetometer. What it tells you is the direction of the nearest strong magnetic field. Outside any other sources the Earth's magnetic field will likely be the strongest, so this is what the micro:bit's compass latches onto. Place the microcontroller next to a big magnet from a loudspeaker, and its magnetic field will overwhelm the compass. Your readings will not point to the speaker.

The compass.py program is written in Python for the micro:bit, and demonstrates continually reading the controller's compass and displaying its result as a clock hand in the LED grid. Note that the program begins by making a call to the *microbit-compass.calibrate* routine, which pauses execution so you can calibrate the compass. Calibrating involves waving the micro:bit in a figure-8 motion for several seconds.

microbit-compass.py

```
from microbit import *

compass.calibrate()       # calibrate first
while True:
    sleep(100)
    needle = ((15 - compass.heading()) // 30) % 12
    display.show(Image.ALL_CLOCKS[needle])
```

Compasses require careful placement on the robot to be useful. Because they sense magnetic fields, the compass must not be located near any large metal mass or near the strong magnetic fields of DC motors.

Using Tilt and Gravity Sensors

As every schoolchild learns, the human body has five principal senses: sight, hearing, touch, smell, and taste. But there's also a "sixth sense" of balance—seeing dead people notwithstanding.

Balance involves our body knowing where it is in space. Our sense of balance combines information about both the body's angle and its motion. At least part of the sense of balance is derived from a sensation of gravity—the constant pull on our bodies from the Earth's mass.

Similarly, robots can be made to "feel" gravity. The same forces of gravity that help us to stay upright might provide a two-legged robot with the same ability. Or a rolling robot—on wheels or tracks—might avoid tipping over and damaging something by determining if its angle is too steep.

ENTER THE ACCELEROMETER

One of the most accurate, yet surprisingly low-cost, methods for tilt measurement involves an accelerometer. Once the province only of high-tech aviation and automotive testing labs, accelerometers are quickly becoming common staples in consumer electronics.

New techniques for manufacturing accelerometers have made them more sensitive and accurate yet less expensive. A device that might have cost upward of $500 a few years ago sells in quantity to manufacturers for under $5 today. Fortunately, the same devices used in cars and other products are available to hobby robot builders, though the cost is a little higher because we're not buying 10,000 at a time!

The basic accelerometer is a device that measures *change* in speed. Many types of accelerometers are also sensitive to the constant pull of the Earth's gravity. It is this latter capability that is of interest to us, since it means you can use the accelerometer to measure the tilt, or "attitude," of your robot at any given time. This tilt is represented by a change in the gravitational forces acting on the sensor.

SINGLE-, DUAL-, AND THREE-AXIS SENSING

The basic accelerometer is *single axis,* meaning it can detect a change in acceleration (or gravity) in one axis only. While this is moderately restrictive, you can still use such a device to create a capable and accurate tilt-and-motion sensor for your robot.

A dual-axis accelerometer detects changes in acceleration and gravity in both the X and Y planes. If the sensor is mounted vertically—so that the Y axis points straight up and down—the Y axis detects up-and-down changes, and the X axis will detect side-to-side motion. Conversely, if the sensor is mounted horizontally, the Y axis detects motion forward and backward, and the X axis detects motion from side to side.

A three-axis accelerometer detects changes in acceleration and gravity in all three planes, basically in 3D. These types of accelerometers are frequently used, along with gyro sensors, in certain kinds of self-balancing robots.

EXPERIMENTING WITH AN ACCELEROMETER

Of accelerometers for hobby use, the 3-DOF variety is probably the most common, with the ADXL family of components from Analog Devices among the most popular. The ADXL parts come as surface-mount components and require a few external components. Most folks get an ADXL-based accelerometer already mounted on a convenient *breakout* or experimenter's board, like the one in Figure 32-12.

The ADXL accelerators include two- and three-axis versions, as well as components sensitive to different levels of gravity (or g) forces. For example, the ADXL335 is a three-axis model, sensitive to ±3g. Generally, an accelerometer with a 2g to 6g range will provide a good balance between precision and performance limits. Most robotics applications don't need to read gravity forces in excess of 5g.

 The higher the g's, the higher the force, impact, and acceleration you can measure, but the less sensitive the component to small changes. A model rocket could use an 18g or 50g accelerometer, but most robots not destined for space can make do with a lot less.

One other variation in accelerometers is the output signal. The three most common are analog, serial, and PWM.

- *Analog output* is a varying voltage, usually something a bit less than the full operating voltage of the accelerometer. Since accelerometers register both positive g-forces and negative g-forces, the output voltage at 0g is halfway between 0 and the operating voltage of the device.
- *Serial output* uses I2C, SPI, or other serial communications to relay the g-force readings to the microcontroller or computer. While a bit more difficult to use than analog output, the serial data is less prone to noise and glitches, and on some accelerometers you can send it commands on the serial link to alter its behavior.
- *PWM output* is a series of pulses whose duration changes depending on the g-forces on the accelerometer. To read the duration of the pulses, you need a microcontroller with an input capture (or similar) pin. The PWM approach is not as common these days.

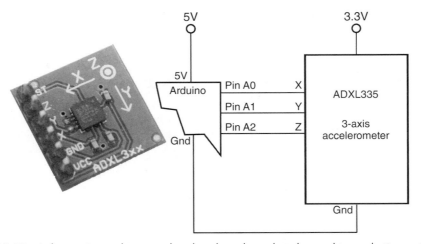

Figure 32-12   A three-axis accelerometer breakout board, ready to be used in a robotics project. This particular model provides an analog voltage proportional to the g-forces on the device.

Figure 32-12 shows a typical connection of an ADXL-based analog output breakout board to an Arduino microcontroller. Because the sensor is a 3.3V device, it must be powered from the Arduino's 3.3V supply pin, or some other regulated 3.3V source.

The various ADXL3xx breakout boards have X, Y, and Z outputs, though not all devices provide an analog output. The program *arduino-accel2.ino* demonstrates how incredibly easy it is to set up and read all three axis values of a 3-DOF analog output accelerometer. The values appear in the Serial Monitor window.

The ADXL335 outputs an analog signal that is proportional to the measured acceleration. Because the chip is powered by 3.3V the total voltage swing from the accelerometer will be a value from 0 to 3.3 volts.

Other members of the ADXL3xx family may use I2C and SPI to communicate with its host microcontroller. If you're using a 5V microcontroller, such as the Arduino Uno, you will need level shifting to bridge the two devices. See Chapter 29, "Interfacing Hardware with Your Microcontroller," for more information on level shifting, as well as "Experimenting with an IMU," later in this chapter.

arduino-accel2.ino

```
#define xpin A0          // X-axis to analog pin 0
#define ypin = A1        // Y-axis to analog pin 1
#define xpin = A2        // Z-axis to analog pin 2

void setup() {
  Serial.begin(9600);    // Set up for Serial Monitor
}

void loop() {
  Serial.print(analogRead(xpin));   // Show X-axis
  Serial.print("\t");

  Serial.print(analogRead(ypin));   // Show Y-axis
  Serial.print("\t");

  Serial.print(analogRead(zpin));   // Show Z-axis
  Serial.println("");
  delay(100);                       // Short delay
}
```

The values shown in the Serial Monitor window represent voltage from 0 (0 volts) to 1023 (5 volts). At 0g (the axis arrow sideways; i.e., pointing neither up nor down), the output of the axis is 512. This is midway between 0 and 5 volts. At +1g (the axis arrow pointing up), the output rises to approximately 650. And at −1g (axis arrow pointing down), the output lowers to 345.

These values were taken using an ADXL355, which has a +/− 3g response. The exact values you get will be different, depending on the specific model of accelerometer you use. You'll also encounter moderate variations in output levels between the outputs—due to the tolerance of the components. This is to be expected, and you can adjust for it in your software.

Given these sample values, one real-world example is knowing when your robot has tipped over on its side, or is about to. Let's assume the *X* axis of the accelerometer represents side-to-side tilt (you can mount your accelerometer in a number of ways, so you can choose what axis does what).

With the accelerometer flat at its bottom, the *X* axis will read 0g, or about 512. Any value above or below that reading indicates at least some tilt in the *X* axis. You might determine excessive tilt if the value is more than 75 points above or below 512.

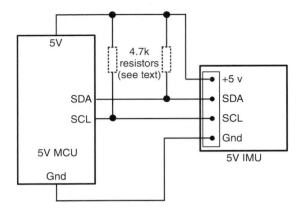

Figure 32-13 Basic wiring diagram for connecting a 5V microcontroller to a 5V IMU that uses I2C.

EXPERIMENTING WITH AN IMU

With the popularity of remote control quadcopters and R/C helicopters the functions of accelerometer, gyroscope, and compass are now provided in an all-in-one breakout board. These boards go by the common name of inertial measurement unit, or IMU.

Capabilities between boards vary: some contain only two of the three classes of sensors, and not all sensors may provide the full 3-DOF output. You can determine what the IMU offers by looking at its specifications: a 9-DOF unit can be assumed to be fully featured with accelerometer, gyro, and compass, with three-axis output for each sensor.

While an IMU board provides several sensors in one package, and is often less expensive than the constituent breakout boards purchased separately, they do represent a level of added complexity for hookup and programming.

The typical IMU connects to the microcontroller by way of a serial interface, usually I2C or SPI. To obtain a reading your microcontroller must first query the IMU for the value you want, then wait for that value to be returned via the serial interface. Each brand of IMU uses a different communications protocol; it's not as easy as simply connecting an output to a digital or analog pin on your microcontroller and taking a direct reading.

As coding complexity can vary, when selecting an IMU look for one that already has a programming library available for your microcontroller. These are most common for the Arduino.

Figure 32-13 shows typical I2C connection between a 5V microcontroller, such as the Arduino Uno, and a 5V IMU board. Both boards can use the same 5V power source, and the serial communication lines can be directly connected.

You'll probably need to use level shifting if the microcontroller and IMU do not share the same operating voltage, for example, a 5V Arduino Uno and 3.3V IMU. Figure 32-14 shows the addition of a level shifter module placed between the two devices. The same idea works in reverse as well: a 3.3V microcontroller and 5V IMU.

Note the pull-up resistors in both diagrams. These are commonly needed for devices that use I2C serial connections. The exact value of the pull-up resistor depends on the device and its operating voltage. The general rule of thumb is to start at 4.7 kΩ and if communication is sporadic try increasing lower values.

A few I2C devices have the pull-up resistors already embedded, though this is not common. And some level-shifting modules designed specifically for I2C include the pull ups. Check the specifications that come with your IMU and level-shifting breakouts.

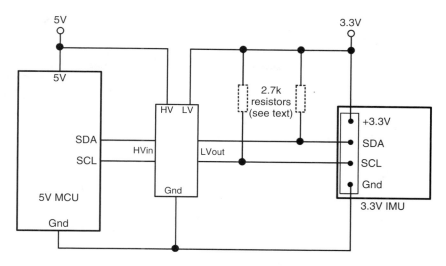

Figure 32-14   Wiring diagram for connecting a 5V microcontroller to a 3.3 I2C-based IMU, using a bidirectional level shifter. You can use the same idea in reverse for a 3.3V controller and 5V IMU.

Level shifting may not be required if the lower voltage device is 5V tolerant. In this case it's often sufficient to simply use a 1k resistor in series between the controller and external device. Best to check the datasheet for your microcontroller or IMU to be sure, as driving a 3.3 volt non-tolerant circuit at 5V may damage it.

More Navigational Systems for Robots

For under $100 your robot can know exactly where on earth it's located. Or precisely what room it's in within your house (or even what wall or furniture it's near). These techniques are more fully developed in a series of bonus articles you can find on the RBB Support Site.

- *Radio frequency identification,* or RFID, uses small transponders that send signals to non-battery-powered tags you have mounted on nearby walls or objects. Each tag provides a unique code that identifies itself and, therefore, can be used as a kind of beacon system.
- *Gyroscopes* sense changes in rotational velocity. When coupled with other navigation techniques such as odometry and acceleration your robot can keep accurate track of where it is. Gyros (the instrument, not the food) are also used to construct self-balancing bots.
- *Beacons* provide known landmarks for your robot to orient itself in a room or other space. One type of beacon emits a pulse of infrared light as a kind of lighthouse; your robot scans the room for this light and when found may proceed toward it. This technique is often used to enable your robot to automatically return to a recharging station.
- *Global positioning satellite* (GPS) is by now a well-known technique using satellites in space to provide location, altitude, even speed information. You can apply the same technology on your robot.

CHAPTER 33

Environment

For a robot to be truly useful it needs senses; the more senses, the better. It's easy to endow even the most basic robot with a sense of touch—all it takes is a couple of small switches, or maybe an infrared or ultrasonic detector here and there.

But to experience its full environment your robot needs other senses, too—sight, smell, and hearing especially. They're a bit tougher to implement, but not impossible. You just need the right electronics. In this chapter you'll learn about various environmental sensors for detecting light, sound, temperature, smoke, and gas.

 Source code for all software examples may be found at the RBB Support Site. To save page space, the lengthier programs are not printed here.

Listening for Sound

Sound detection allows your robot creation to respond to your commands, whether they take the form of a series of tones, an ultrasonic whistle, a hand clap, or even your voice. It can also listen for the telltale sounds of intruders or search out the sounds in the room to look for and follow its master.

MICROPHONE

Obviously, your robot needs a microphone (or *mic*) to pick up the sounds around it. The most sensitive type of microphone is the electret condenser, which is used in most higher-quality hi-fi mikes. The trouble with electret condenser elements, unlike crystal element mikes, is that they need electricity to operate. Supplying electricity to the microphone element really isn't a problem, however, because the voltage level is low—under 4 or 5 volts.

Most all electret condenser microphone elements come with a built-in field-effect transistor (FET) amplifier stage. As a result, the sound is amplified before it is passed on to the main amplifier.

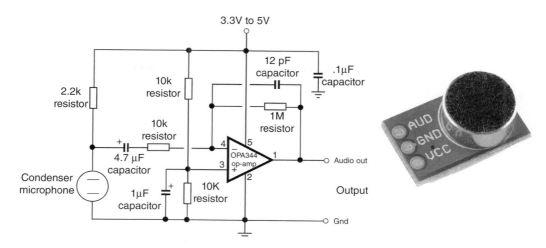

Figure 33-1   Microphone preamplifier circuit and module. You can build the amp yourself or get it ready-made. (*Circuit and photo courtesy SparkFun Electronics.*)

Electret condenser elements are available from a number of online sources for under $4. You should buy the best one you can. A cheap microphone isn't sensitive enough.

The placement of the microphone is important. You should mount the mike element at a location on the robot where vibration from motors is minimal. Otherwise, the robot will do nothing but listen to itself. Depending on the application, such as listening for intruders, you might never be able to place the microphone far enough away from sound sources or make your robot quiet enough. You'll have to program the machine to stop, then listen.

MICROPHONE AMPLIFIER

Use the circuit in Figure 33-1 as an amplifier for the microphone. The circuit is designed around an op-amp; the op-amp used is an OPA344, which you may not be familiar with. Unlike the ubiquitous LM741 op-amp, the OPA344 is intended to run using a *single-ended* power supply—that is, you don't need to provide both + and − power. It's also known as a *rail-to-rail* amp, meaning its output varies from 0 volts to the full supply voltage (or pretty close to it).

CONNECTING TO YOUR MICROCONTROLLER

With the sound detector hardware finished, you can connect the electronics to a microcontroller. Figure 33-2 shows an interface that will conform the output signal from the amplifier to a level that is easier to use with a microcontroller analog-to-digital input. The capacitor is included to remove the steady DC voltage present at the amplifier output. The diode makes the output signal positive-going only.

When connecting to an Arduino, you may have better luck using an external power supply, rather than powering the microcontroller via the USB port. If this isn't practical, try powering the microphone module from the 3.3V pin of the Arduino pin, or else using a separate regulated 5-volt supply for the amp.

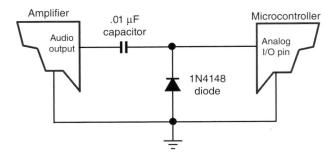

Figure 33-2 How to connect the microphone amplifier to a microcontroller. Be sure the ground connections are shared. See text for power source recommendations.

Upload the *arduino.sounddetect.ino* sketch to your Arduino, then open the Serial Monitor window to see the actual values registered through the amplifier. Assuming the module is powered by 5 volts, theoretically the values will vary from 0 to 1023, representing the full 0- to 5-volt voltage swing of the amplifier output.

Note the *threshold* value set at the top of the sketch. Sounds over this level cause the Arduino's built-in LED to light up. Even with no sound in the room, the amp will deliver some voltage to its output—and more than you might think because of thermal noise, which is the "hiss" you hear when you turn up an amplifier. Depending on your wiring, you should expect an ambient (quiet room) level between 0 and 25, especially if you experiment near your PC with its cooling fan. Therefore, set the *threshold* to some value above the ambient level, so that louder sounds trigger the LED.

arduino.sounddetect.ino

```
#define ledPin 13               // Built-in LED
const int soundSensor = A0;     // Audio output to pin A0
const int threshold = 105;      // Threshold for sound
int sensorReading = 0;

void setup() {
  pinMode(ledPin, OUTPUT);      // Make LED an output
  Serial.begin(9600);          // Use Serial Monitor
}

void loop() {
  sensorReading = analogRead(soundSensor);   // Read sound
  // If sound level is over threshold flash LED
  if (sensorReading >= threshold) {
    digitalWrite(ledPin, HIGH);
    delay(300);
    digitalWrite(ledPin, LOW);
  }
  // Display sound level
  Serial.println(sensorReading, DEC);
  delay(50);
}
```

Sound metering is a science in its own right, and the preceding example is a simplistic way to monitor audio levels. One method that you might like to research includes adding a peak detector circuit, useful for catching those fleeting high-output sounds that might otherwise get missed. The peak detector stores the highest volume detected in that period. After checking the volume level, the circuit is cleared.

Peak detector circuits can be constructed using op-amps and a small assortment of standard parts (resistors, capacitors, diodes, and maybe a transistor). Research on the Web to see what you come up with!

RESPONDING TO LOUD NOISES

Remember the "clap on, clap off" light switch control of a few years back? You know, clap once, and the lamp beside your easy chair turns on; clap again and the lamp goes out. You can do the same thing with your robot by using a noise activation sensor. These compact modules incorporate a microphone, amplifier, and all necessary interface electronics.

A small trimmer potentiometer lets you set the sound volume that triggers the sensor. A good example is the Parallax Sound Impact Sensor, which is both low cost and easy to connect to the Arduino or other microcontroller. Adjust the trimmer pot to set the triggering level.

 Though sound is analog, this type of noise activation sensor provides a digital off/on output signal. The signal is normally off (LOW) when the sound is below the threshold level you've set with the trimmer pot. Sound above the threshold causes the output to turn on (go HIGH).

Figure 33-3 shows how to wire the sound impact sensor to digital pin D2 on the Arduino. This pin is specifically selected because it is one of two Arduino I/O pins that support external interrupts: a change on this pin can trigger a function in the sketch. By using an interrupt the Arduino doesn't need to constantly poll the I/O pin to determine the state of the sensor.

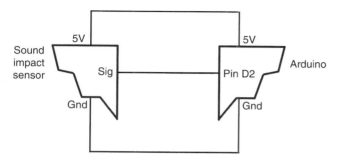

Figure 33-3   Simple wiring connection between the Sound Impact Sensor and the Arduino.

- *Polling* requires the sketch to constantly monitor the status of the input pin. If a sound event is very short, and occurs between polls, the Arduino may miss it. The more tasks handled by the Arduino in its *loop()* function, the more chance sound events will go unnoticed.
- *Interrupts* are monitored in the background. The interrupt is triggered only if and when a change occurs to the input pin. As sound events can be fleeting, using an interrupt to process them is the best method to ensure that each one is captured.

Refer to *arduino-soundimpact.ino* for a demonstration of using the Parallax sound impact sensor with an Arduino interrupt. There are two important pieces that make the sketch work: First is the *attachInterrupt* statement, found in the *setup()* function. This statement sets up the interrupt, specifying the interrupt number (interrupt 0 is on pin D2, interrupt 1 is on pin D3); the name of the function that is executed when an interrupt occurs; and the type of signal change that causes the interrupt to trigger.

The second important piece is the interrupt function itself—called an *interrupt handler* or *interrupt service routine* (ISR) in programming parlance. In the sketch the function is named *sounder*.

arduino-soundimpact.ino

```
#define soundPin 2
#define ledPin 13
volatile int state = LOW;

void setup() {
  pinMode(ledPin, OUTPUT);
  attachInterrupt(0, sounder, RISING);
}

void loop() {
  digitalWrite(ledPin, state);
}

void sounder() {
  state = !state;
  delayMicroseconds(1000);
}
```

The *arduino-soundimpact.ino* sketch demonstrates how each event that triggers the sound impact sensor toggles the Arduino's LED on or off. Set the trimmer pot on the sensor to adjust its sensitivity. Turning it clockwise reduces sensitivity, and turning it counterclockwise increases sensitivity. Use the built-in LED on the sensor to help you adjust the trigger level.

Here's a tip: Collisions with objects cause noise. You can use a sound impact sensor to detect if your robot has crashed into something. Increase sensitivity to hard contact by mechanically fastening the sensor to the base of the robot. Adjust the trimmer pot on the module to capture the sound of running into an object. Avoid setting the control so low that it'll trigger on the robot's own motors.

EXTENDING SOUND TRIGGERING TO CONTROL MOTORS

Now take a look at *arduino-soundimpact_servo.ino*. It takes the basic concept of reacting to events from the sound impact sensor to operate the servo motors of a robot. This sketch adds two *Servo* objects for the left and right servo motors, and a small assortment of functions for controlling the direction of the servos. The robot is started and stopped at each triggered sound event. The Arduino's LED (hard-wired to pin D13) shows the running status of the motors.

arduino-soundimpact_servo.ino

```
#include <Servo.h>
Servo servoLeft;           // Define left servo
Servo servoRight;          // Define right servo

#define soundPin 2
#define ledPin 13
volatile int state = LOW;

void setup() {
  pinMode(ledPin, OUTPUT);
  attachInterrupt(0, sounder, RISING);
  servoLeft.attach(10);      // Left servo on pin D9
  servoRight.attach(11);   // Right servo on pin D10
}

void loop() {
  if (state) {
    servoLeft.write(180);    // Go forward
    servoRight.write(0);
  } else {
    servoLeft.write(90);     // Stop
    servoRight.write(90);
  }
}

void sounder() {
  state = !state;
  digitalWrite(ledPin, state);
  delayMicroseconds(1000);
}
```

ADDING VOICE RECOGNITION

The ultimate in robot sound control is voice command: say a word, and your robot obeys. Voice recognition is common in desktop computers and even some smartphones, and with a separate module you can add the technology to your own robot.

Of the several commercial products available, the EasyVR coprocessor board from VeeaR is among the least expensive. And while operation is straightforward, programming and "teaching" the device requires some study and patience, so don't consider it a quick 1-hour project. VeeaR provides an 87-page manual to explain it all!

The EasyVR is available from a number of online sellers and comes in several form factors, both as an Arduino shield and as a stand-alone board (Figure 33-4) you can adapt to any microcontroller.

While the board has numerous input/output pins for various optional functionalities, basic hookup to listen for and respond to sound requires two connections per function: power, serial transmit and receive, and microphone.

For best results, plastic the microphone away from your robot's motors so that they don't interfere with the speech recognition. Still, any kind of speech recognition can be problematic if your robot is especially noisy. You may need to limit giving voice commands only when the robot is not moving, and is reasonably close to you.

Figure 33-4 The EasyVR voice command recognition board can be used with the Arduino and many other microcontrollers. (*Photo courtesy VeeaR by RoboTech srl.*)

EasyVR comes with a set of 26 speaker independent commands (in multiple languages) that you can use right out of the box. The commands are separated into hierarchical wordsets. To get the system's attention you speak the word "robot," followed by an action and then a direction (and an optional numeric parameter that can be used for things like setting speed).

For example, to move your bot to the right, you'd speak *robot, turn, right*. Or to stop the robot, you'd announce *robot, stop*. The index values of the words you speak are transmitted from the EasyVR to your microcontroller via a standard serial connection.

Google offers a speech recognition module billed as a "natural language recognizer." The Google AIY Voice Kit is available in kit form and connects to an included Raspberry Pi Zero W board. See Chapter 27, "Using the Raspberry Pi," for additional details.

Simple Light Sensors for Robotic Eyes

Most people think about "robot vision" as some full video-like snapshot, complete with auxiliary text explaining what's going on, a la *Terminator*. It's not always like that. A number of very simple electronic devices can be used as very effective eyes for your robot. These include:

- *Photoresistors,* which are also known as light-dependent resistors and photocells.
- *Phototransistors,* which are like regular transistors, except they are activated when light strikes them.
- *Photodiodes, which* are photo-sensitive diodes that begin to conduct current when exposed to light.

Photoresistors, photodiodes, and phototransistors are connected to other electronics in the same general way: you place a resistor between the device and either +V or ground. This forms

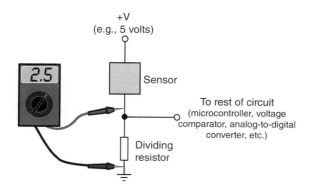

Figure 33-5   The output of photoresistors, phototransistors, and photodiodes can be converted to a varying voltage by using a divider resistor. You may measure the voltage with a multimeter.

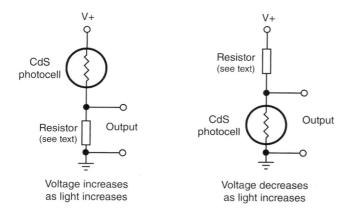

Figure 33-6   A CdS photocell, the most common type of photoresistor, may be connected so that the output voltage increases or decreases with more light. Experiment with the value of the fixed resistor for best sensitivity.

a voltage divider. The point between the device and the resistor is the output, as shown in Figure 33-5. With this arrangement, the device outputs a varying voltage.

PHOTORESISTORS

Photoresistors are typically made with cadmium sulfide, so they are often referred to as *CdS cells* or *photocells*. A photoresistor functions as a *light-dependent resistor* (also known as an LDR): the resistance of the cell varies depending on the intensity of the light striking it.

When no light strikes the cell, the device exhibits very high resistance, typically in the high 100 kilohms, or even megohms. Light reduces the resistance, usually significantly—a few hundreds or thousands of ohms.

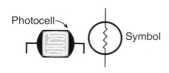

Photoresistors are easy to interface to other electronics. Figure 33-6 shows two ways to connect photoresistors to a circuit. The resistor is selected to match the light and dark resistance of the particular photocell you're using. You'll need to experiment a bit here to find the ideal resistor value. Start with 100 kΩ and work your way down. A 100 kΩ potentiometer works well for testing. If you use a pot, add 1 kΩ to a 5 kΩ resistor in series with it to prevent a near short circuit when the photocell is exposed to bright light.

Note that photocells are fairly slow reacting, and are unable to discern when light flashes more than 20 or 30 times per second. This trait actually comes in handy because it means photoresistor cells basically ignore the on/off flashes of AC-operated lamps. The cell still sees the light, but isn't fast enough to track it.

PHOTOTRANSISTORS

All semiconductors are sensitive to light. If you were to take the top off of a regular transistor, it would act as a phototransistor. Only in a real phototransistor, the light sensitivity of the device is much enhanced. A glass or plastic cover protects the delicate semiconductor material inside.

Many phototransistors come in a package that looks a lot like an LED. And like an LED, one side of the plastic case is flattened. Unless otherwise indicated in the datasheet for the phototransistor, the flat end denotes the collector (C) lead. The other lead is the emitter (E).

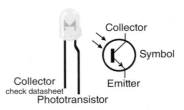

Unlike photoresistors, phototransistors are very quick acting and able to sense tens of thousands of flashes of light per second. Because of this, they can be used for optical data communications. It also means that when used under AC-operated or fluorescent lamps, the output of the sensor can rapidly vary as it registers the fluctuation in light intensity.

The output of a phototransistor is not "linear"; that is, there is a disproportionate change in the output of a phototransistor as more and more light strikes it. A phototransistor can become easily "swamped" with too much light. Even as more light shines on the device, the phototransistor is not able to detect any more change.

See Figure 33-7 for ways to connect a phototransistor to the rest of a control circuit. Like the photoresistor, experiment with the value of the fixed resistor to determine the optimum light sensitivity. Values of 4.7 to 250 kΩ are typical, but it depends on the phototransistor and the amount of light that you want to cause a trigger. Start with a lower-value resistor and work your way up. Higher resistances make the circuit more sensitive.

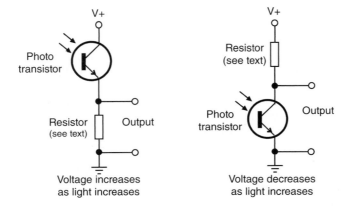

Figure 33-7 Methods of connecting a phototransistor to the rest of the circuit. The output is a varying voltage. Experiment with the value of the fixed resistor for best sensitivity.

You may instead wish to add a small, 50 kΩ potentiometer in series with a 1 to 5 kΩ fixed resistor. Dial the potentiometer for the sensitivity you want. Connect your multimeter to test the *voltage swing* you get under light and no-light situations. I like to select a resistor so that the sensor outputs about half of the supply voltage (i.e., 2.5 volts) under ambient, or normal, light levels.

PHOTODIODES

These work much like phototransistors but are simpler devices. Like phototransistors, they are made with glass or plastic to cover the semiconductor material inside. And like phototransistors, photodiodes are very fast acting. They can also become "swamped" when exposed to extra bright light—after a certain point, the device passes all the current it can, even as the light intensity increases.

Photodiodes are essentially LEDs in reverse; in fact, you can even use some LEDs as photodiodes, though as sensors they aren't very sensitive.

If you'd like to experiment with the LED-as-photodiode trick, try to find one that's in a "water-clear" casing. The light should enter straight into the top of the LED; off to the sides doesn't work. You might need a simple focusing lens to improve the light-gathering capability. The LED is most sensitive to light at the same wavelength as the color it produces. To register green light, use a green LED, for example.

One common characteristic of most photodiodes is that their output is rather low, even when fully exposed to bright light. This means that, to be effective, the output of the photodiode must usually be connected to a small operational amplifier or transistor amplifier.

SPECTRAL RESPONSE OF SIMPLE SENSORS

Light-sensitive devices differ in their spectral response, which is the span of the visible and near-infrared light region of the electromagnetic spectrum that they are most sensitive to. As depicted in Figure 33-8, CdS photoresistors exhibit a spectral response close to that of the human eye, with the greatest degree of sensitivity in the green or yellow-green region.

Most phototransistors and photodiodes have peak spectral responses in the infrared and near-infrared regions. In addition, some phototransistors and photodiodes incorporate optical filtration to decrease their sensitivity to a particular part of the light spectrum, usually infrared.

A few special-purpose photodiodes are engineered with enhanced sensitivity to shorter-wavelength light. This allows them to be used with ultraviolet emitters that cause paint and ink pigments to fluoresce in the visible light spectrum. These are used, for example, in currency verification systems. For robotics you might use such a sensor to follow a line on the ground painted with a fluorescent dye.

BUILDING A ONE-CELL EYE

A single light-sensitive photoresistor is all your robot needs to sense the presence of light. As noted, the photoresistor is a variable resistor that works much like a potentiometer but has no control shaft. You vary its resistance by increasing or decreasing the light shining on it.

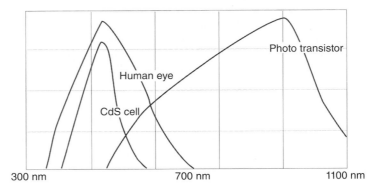

Figure 33-8   Comparison of the light spectrum sensitivity of the human eye, the CdS photoresistor, and the most common form of phototransistor. Visible light is between about 400 nanometers (nm) and 750 nm.

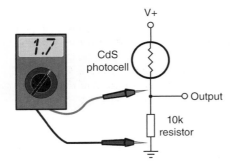

Figure 33-9   How to test a CdS photoresistor using a multimeter. Dial the meter to read DC voltage, then take the measurement as shown.

Connect the photocell as shown in the "voltage increases as light increases" version of Figure 33-6. As explained in the previous section, a resistor is placed in series with the photocell and the "output tap" is between the cell and resistor. This converts the output of the photocell from resistance to voltage. Voltages are a lot easier to measure in a practical circuit.

Typical resistor values are 1 to 10 kΩ, but this is open to experimentation. You can vary the sensitivity of the cell by substituting a higher or lower value. Test the cell output by connecting a multimeter to the ground and output terminals, as shown in Figure 33-9. For experimenting, you can connect a 2 kΩ resistor in series with a 50 kΩ potentiometer—these replace the single resistor that's shown. Try testing the cell at various settings of the pot.

So far, you have a nice light-to-voltage sensor, and when you think about it, there are numerous ways to interface this ultrasimple circuit to a robot. One way is to connect the output of the sensor to the input of a comparator. The LM339 quad comparator IC is a good choice. The output of the comparator changes state when the voltage at its input goes beyond or below a reference voltage or "trip point."

In the circuit shown in Figure 33-10, the comparator is hooked up so the noninverting input (marked +) serves as a voltage reference. Adjust the potentiometer to set the trip point higher or lower than what you want to trigger at. To begin, set it midway, then adjust the trip point higher or lower as you experiment with different light levels. The output of the photoresistor circuit is

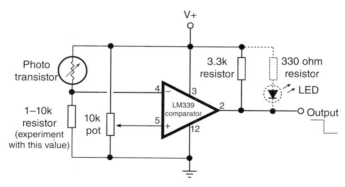

Figure 33-10 Using an LM339 voltage comparator to provide a basic on/off output from a photo-resistor (this also works with a phototransistor). Adjust the 10 kΩ potentiometer so that the comparator triggers at just the desired amount of light.

connected to the inverting input (marked −) of the comparator. When the voltage at this pin goes above or below the point with the potentiometer, the output of the comparator changes state.

With the wiring shown, the output goes from HIGH to LOW as the light increases. One application is to adjust the potentiometer so that under normal room light the output of the LM339 just goes HIGH. Shine a light directly into the sensor, and the output switches LOW.

If you want the opposite effect—the output goes from LOW to HIGH under increasing light—simply switch the connections to pins 4 and 5 of the comparator. This makes the logic go in reverse.

The circuit also shows an optional LED and resistor, useful as a visual indicator, for testing purposes only. The LED will wink on when the output of the comparator goes HIGH and wink off when it goes LOW. Remove the LED and resistor if you connect the comparator to another digital circuit, such as a microcontroller.

CREATING A LIGHT-RECEPTIVE ROBOT

One practical application of this circuit is to detect light levels that are higher than the ambient light in the room. Doing so enables your robot to ignore the background light level and respond only to the higher-intensity light, like a flashlight. You don't even need a microcontroller or other brain for your robot. It can be done using only simple components and a relay. Figure 33-11 shows one example of an old-skool bot that reacts to light falling on a phototransistor.

To begin, set the 10 kΩ potentiometer so that in a darkened room the circuit just switches HIGH (this deenergizes the relay). Use a flashlight with fresh batteries to focus a beam directly onto the phototransistor. Watch the output of the comparator change state from HIGH to LOW. When LOW, the relay is energized through the 2N3906 PNP transistor.

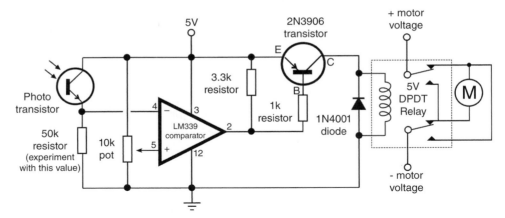

Figure 33-11 All-in-one control circuit for operating a motor based on the amount of light falling on a phototransistor. When the relay triggers, the motor reverses. Adjust the light sensitivity of the circuit by changing the value of the fixed resistor below the phototransistor.

You can use this so-simple-it's-dumb circuit to activate your robot so it will move toward you when you turn the flashlight on it. Or the inverse: It runs away from you. Just reverse the connections to the motor.

- When your robot advances toward the light, it's said to be *photophilic*—it "likes" light.
- When your robot shies away from the light, it's said to be *photophobic*—it "hates" light.
- When combined with other actions, these *behaviors* allow your robot to exhibit what appears to be artificial intelligence.

 See Chapter 37, "Make Light-Seeking Robots," for a hands-on project involving a light-smart bot. Also look for more light-dependent robot examples on the RBB Support Site.

DEALING WITH LIGHT SPOILAGE

The bane of any light-sensitive detector is *spoilage* from stray light sources. Examples include:

- *Infrared light coming from outdoors, a desk lamp,* or other source, and not from the infrared LEDs you have so carefully placed on your robot. You can help mitigate this by using tubes and baffles to block unwanted light. A simple light tube can be constructed using a small piece of black heat shrink tubing (not shrunk) cut to length and placed around the sensor.
- *Ambient (natural room) light striking the sensor from the sides,* or even from behind, rather than straight down its gullet. CdS photocells are sensitive to light coming from behind them (through their backing). To avoid this spoilage, always place your photoresistors against a black or opaque backstop that will block light.

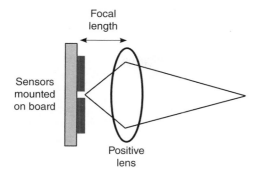

Figure 33-12 Use a lens to focus the light into the sensor. This makes the sensor much more sensitive to the light, and improves performance.

- *Light spilling from the side of the LED* and directly into the sensor. Use the heat shrink tubing trick, or add a heavy felt baffle between the two. It also helps to position the LED slightly forward of the sensor.
- *Visible-light LEDs and LCD display panels* that are located too close to the sensor. Their light can accidentally influence the sensor. Be sure to always locate your light sensors away from any potential light sources.

USING LENSES AND FILTERS WITH LIGHT-SENSITIVE SENSORS

Simple lenses and filters can be used to greatly enhance the sensitivity, directionality, and effectiveness of both single- and multicell vision systems. By placing a positive (magnifying) lens over a small cluster of light cells, you can concentrate room light to make the cells more sensitive to the movement of humans and other animate objects.

Figure 33-12 shows the basic idea behind focusing light onto a sensor. Position the lens so that it focuses into the sensor; you can determine this by pointing the lens at any point light source (like a bare bulb) and measuring the distance between lens and sensor when the focused spot is the sharpest.

And you can also use optical filters to enhance the operation of light cells. An IR pass filter allows infrared light to pass through, but blocks other wavelengths of light. This helps prevent the sensor from reacting to other light sources in the room. Colored gels will allow their own color of light to pass through, (roughly) rejecting the other colors. You might use this technique to create a light-following robot that is particularly responsive to blue light, but not red or green.

There's more to lenses and filters than just this. See the RBB Support Site for an extended discussion of selecting and using lenses and filters with robotic optical systems.

Vision Systems: An Introduction

The one sense that offers the most potential to the science of robotics is also the most elusive: vision. Robotic sight is something of a paradox. The sensor for providing a video image is actually quite mundane, you can now purchase a color video camera for under $25, complete with lens.

But there's a second part to the video equation: what to do with the image data once it's been acquired. There is a burgeoning science of *machine vision* that seeks to provide answers.

The single- and multicell vision systems described previously are useful for detecting the absence or presence of light, but they cannot make out anything except very crude shapes. This greatly limits the environment into which such a robot can be placed. By detecting the shape of an object, a robot might be able to make intelligent assumptions about its surroundings and perhaps navigate those surroundings, recognize its "master," and more.

The machine vision described here is different from video beamed from a robot back to its human operator. The latter technique is used in teleoperated robots, where you get to see what the robot sees. Video for teleoperation is covered in Chapter 34, "Operating Your Bot via Remote Control."

CAMERAS AND EQUIPMENT

A video system for robot vision need not be overly sophisticated. The resolution of the image can be as low as about 300 by 200 pixels. A color camera is not mandatory, though in some kinds of vision analysis techniques, the use of color is how the system tracks objects and movement.

Video systems that provide a digital output are generally easier to work with than those that provide only an analog video output. You can connect digital video systems directly to a PC, such as through a serial, parallel, or USB port. Analog video systems require the PC to have a video capture card, fast analog-to-digital converter, or similar device attached to it.

MICROCONTROLLER-BASED VISION

Analyzing video takes a certain amount of processor horsepower. Most low-end microcontrollers lack the speed and memory to do effective video processing, though this depends heavily on the techniques that are used. If you're using an Arduino, BBC Micro:bit, PICAXE, or similar microcontroller, you can opt for an off-board computer vision system.

A popular off-board solution is the open source Pixy (CMUcam5), shown in Figure 33-13 and available commercially from a number of specialty robotics stores; see the Sources lists on the RBB Support Site for more information on these. This device incorporates a color imager, lens, and image analysis circuitry. Pixy tracks objects by color and motion. It can connect to a computer or microcontroller via a serial data link.

More robust microcontrollers and single-board computers, notably the Raspberry Pi, have the processing aptitude to handle vision analysis. A number of compact color cameras are available for the Pi; the Camera Module 2 plugs directly into a connector slot on the board (see Figure 33-14). A number of open source vision libraries available, with varying degrees of complexity, leverage the open source OpenCV and SimpleCV vision libraries.

Google provides an enhanced vision development system for the Raspberry Pi Zero W board that uses an accessory board that extends the Pi's own functionality. AIY Vision Kit includes a *VisionBonnet* daughter board that connects to the Pi Zero; the Pi camera connects to the Vision-Bonnet, which in turn connects to the Pi. Refer to Chapter 27, "Using the Raspberry Pi," for more details.

PC-BASED VISION

If your robot is based on a laptop or desktop personal computer with a small-form factor motherboard, you're in luck! There's an almost unlimited array of inexpensive digital video cameras that

Figure 33-13 The open source Pixy (CMUcam5) is a self-contained vision analysis system that includes its own camera and processor. (*Photo courtesy Charmed Labs.*)

Figure 33-14 The Camera Module v2 for the Raspberry Pi sports a Sony eight megapixel color camera capable of up to 1080p resolution. It plugs directly into the Pi board via a ribbon cable.

you can attach to a PC. The proliferation of Webcams for use with personal computers has brought down the cost of these devices to under $50, and often less for bare-bones models. You can use a variety of operating systems—Windows, Linux, or Macintosh—though be forewarned that not every Webcam has software drivers for every operating system. Be sure to check before buying.

Connecting the camera to the PC is the easy part; using its output is far more difficult. You need software to interpret the video scene that the camera is capturing. An example is RoboRealm, at *www.roborealm.com,* a set of low-cost vision analysis tools that allows your *tobor* to recognize shapes and objects, and even track them in real time.

If you're a user of the .NET programming platform under Windows and are fairly familiar with C# or VB programming, be sure to investigate the DirectShow.Net Sourceforge open-source project at *directshownet.sourceforge.net.* DirectShow.Net is a managed .NET wrapper that allows you to tap into the incredibly powerful DirectShow architecture of Windows, without the need to use C++.

There are also a number of open source vision libraries you can use, including OpenCV, Simple CV, and AForge.

Smoke Detection

Don't rely on your robot sentry as the only means to detect smoke and dangerous gases. The methods described here are for *experimental use only.* For reasons I'll get into, a robot may not be the best bet for detecting smoke or gas in a room; for true safety, leave these tasks to approved fire, gas, and smoke detector appliances designed for the job.

"Where there's smoke, there's fire." For less than $15, you can add smoke detection to your robot's long list of capabilities and, with a little bit of programming, have it wander through the house checking each room for trouble.

You can build your own smoke detector using individually purchased components, but some items, such as the smoke detector cell, are hard to find. It's easier to use a commercially available smoke detector and modify it for use with your robot. In fact, the process is so simple that you can add one to each of your robots. Tear the smoke detector apart and strip it down to the base circuit board.

Many smoke detectors employ an ionization chamber that uses a mildly radioactive substance, Americium 241. This human-made element has a half-life of over 400 years, but the ionization chamber can become ineffective for smoke detection after less than 10 years, due to dust and other contaminants.

HACKING A SMOKE ALARM

You can either buy a new smoke detector module for your robot or scavenge one from a commercial smoke alarm unit. The latter tends to be considerably cheaper—you can buy quality smoke alarms for as little as $7 to $10.

In this section, I'll discuss hacking a commercial smoke alarm, specifically a First Alert model SA300, so it can be directly connected to a robot's computer port or microcontroller. Of course, smoke alarms are not all designed the same, but on many, the basic construction is similar to that described here. You should have relatively little trouble hacking most any smoke detector you happen to use.

However, you should limit your hacking attempts to those smoke alarms that use traditional 9-volt batteries. Certain smoke alarm models, particularly older ones, require you to use AC power or specialized batteries. These are not suitable for use on your battery-powered bot.

Checking for Proper Operation

Start by checking the alarm for proper operation. If it doesn't have one already, insert a fresh battery into the battery compartment. *Put plugs in your ears* (or cover up the audio transducer hole on the alarm). Press the "Test" button on the alarm; if it is properly functioning, the alarm should emit a loud, piercing tone.

If everything checks out okay, remove the battery and disassemble the alarm. Less expensive models will not have screws but will likely use a "snap-on" construction. Use a small flat-headed screwdriver to unsnap the snaps.

Getting Inside the Smoke Alarm

Inside the smoke detector is a circuit board that consists of the drive electronics and the smoke detector chamber.

Either mounted on the board or located elsewhere will be the piezo disc used to make the loud tone. Remove the circuit board, being careful you don't damage it. Examine the board for obvious "hack points," and note the wiring to the piezo disc. More than likely, there will be either two or three wires going to the disc:

- *Two wires to the piezo disc:* The wires will provide ground and +V power. This design is typical when you are using all-in-one piezo disc buzzers, in which the disc itself contains the electronics to produce the signal for audible tones, or the disc is used as a simple speaker.
- *Three wires to the piezo disc:* The wires will provide ground, +V power, and a signal that causes the disc to oscillate with an audible tone.

Find the wire that serves as ground. On the SA300 it's easy because the battery terminals are labeled + and – right on the board. Find the ground (negative or –) terminal, and connect the COM black lead from the multimeter to it. Connect the red test lead to one of the wires or connections to the piezo disc.

Replace the battery in the battery compartment, and depress the "Test" button on the alarm. Watch for a change in voltage. For a two-wire disc, you should see the voltage change as the tone is produced. For a three-wire disc, try each wire to determine which produces the higher voltage; that is the one you should use. If you are using an oscilloscope, find the wire that produces the cleanest on/off pulse.

Once you have determined the functions of the wires to the piezo disc, remove the disc and save it for some other project. Retest the alarm's circuit board to make sure you can still read the voltage changes with your multimeter. Then clip off the wires to the battery compartment, noting their polarity.

On the prototype smoke detector, the piezo disc has three leads; the one that produced the largest voltage change was in the upper right corner. So I soldered a long "tap-off" wire to it from the underside of the printed circuit board. Figure 33-15 shows the SA300 unit, before it was disassembled and gutted, with the piezo disc removed, and tap-off wire soldered to the PCB.

INTERFACING THE ALARM TO A MICROCONTROLLER

See Figure 33-16 for interfacing the output of a battery-operated smoke detector to a microcontroller or other circuit that expects 5-volt input. To be on the safe side, the circuit uses a 5.1-volt zener diode and current-limiting resistor to prevent a possible overvoltage condition. Most battery-operated smoke alarms operate at up to 9 volts (a few up to 12 volts), and you don't want more than 5 volts going to your microcontroller.

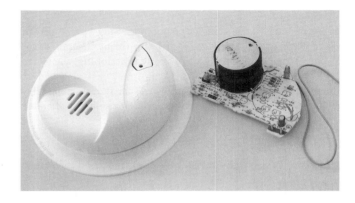

Figure 33-15 The guts of a typical battery-operated smoke detector.

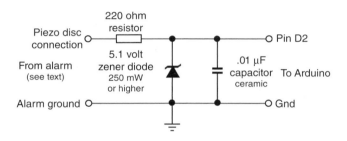

Figure 33-16 A 5.1-volt zener diode clamps the output of the smoke alarm to about 5 volts, helping to protect the input of the microcontroller.

 Remember: There's always a chance of damaging your microcontroller when connecting it to alien devices. Exercise care and *proceed at your own risk!*

 The circuit in Figure 33-16 assumes an output voltage to the piezo sounder of the smoke detector of 7 to 12 volts. You may wish to add an opto-isolator (see Chapter 29, "Interfacing Hardware with Your Microcontroller"), which adds protection between the detector and the robot's electronics.

By way of example, let's assume that the microcontroller periodically *polls* the input pin that is connected to the smoke alarm circuit board. The program, shown in *arduino-smoke_detector. ino* (for the Arduino), checks the pin several times each second. When the pin goes HIGH, the smoke alarm has been triggered and the Arduino's LED lights up.

arduino-smoke_detector.ino

```
#define alarmPin 2
#define ledPin 13

void setup() {
  pinMode(ledPin, OUTPUT);
  pinMode(alarmPin, INPUT);
}
```

```
void loop(){
  if (digitalRead(alarmPin) == HIGH) {
    digitalWrite(ledPin, HIGH);
  }
  else {
    digitalWrite(ledPin, LOW);
  }
}
```

TESTING THE ALARM

Once the smoke alarm circuit board is connected to the microcontroller or computer port, test it and your software by triggering the "Test" button on the smoke alarm.

RUNNING ON 5 VOLTS

I found that the First Alert SA300 unit operated at 5 volts, rather than the 9 volts from its regular battery. Or I should say that when operating at 5 volts it triggered and sounded its alarm when you are pressing the Test button. I have no idea if the unit's smoke detection abilities are diminished at the reduced voltage, but I suspect this may be the case. Anyway, running the board at 5 volts saves you from having to power it from a separate battery, so it's worth looking into.

On the other hand, at the reduced voltage there wasn't enough rise in signal to trigger the Arduino; the voltage at pin D2 wasn't enough to register as a HIGH. An option in this case is to connect a voltage comparator to the hacked output of the smoke detector and adjust its reference voltage to trigger at the reduced signal level (in my tests it was about 2 volts). See Chapter 29, "Interfacing Hardware with Your Microcontroller," for additional information on using voltage comparator circuits.

LIMITATIONS OF ROBOTS DETECTING SMOKE

You should be aware of certain limitations inherent in robot fire detectors. In the early stages of a fire, smoke tends to cling to the ceilings. That's why manufacturers recommend that you place smoke detectors on the ceiling rather than on the wall. Only when the fire gets going and smoke builds up does it start to fill up the rest of the room.

Your robot is probably a rather short creature, and it might not detect smoke that confines itself only to the ceiling. This is not to say that the smoke detector mounted on even a 1-foot-high robot won't detect the smoke from a small fire; just don't count on it. Always use a regular smoke alarm for actual protection, and treat the robot's smoke alarm as an educational toy.

Detecting Dangerous Gas

Smoke alarms detect the smoke from fires but not noxious fumes. Some fires emit very little smoke but plenty of toxic fumes, and these are left undetected by the traditional smoke alarm. Moreover, potentially deadly fumes can be produced in the absence of a fire. For example, a malfunctioning gas heater can generate poisonous carbon monoxide gas.

Just as there are alarms for detecting smoke, so there are alarms for detecting noxious gases, including carbon monoxide. Such gas alarms tend to be a little more expensive than smoke alarms, but they can be hacked in much the same way as a smoke alarm.

Combination units that include both a smoke and a gas alarm are also available. You should determine if the all-in-one design will be useful for you. In some combination smoke-gas alarm units, there is no simple way to determine which (smoke or gas) has been detected.

BUILDING A NOXIOUS GAS DETECTOR CIRCUIT

There may not be many smoke detector modules available for robotics experimentation, but the same is not true of gas detectors. Parallax, Seeedstudio, and many others offer a wide variety of single-board gas detection modules that you can readily incorporate into your robot designs. No hacking required.

Figure 33-17 shows an example gas sensor and module that is easily connected to a microcontroller. A sample program for the Arduino is provided in *arduino-gas.ino*. Connect the output of the sensor to analog pin A0.

Note that the circuit and demo programs work with many of the gas and air quality sensors made by the same company (but not all; some need a much more sophisticated connection and heat/cool cycle). For example, the MQ-3 alcohol sensor could be used to create a Breathalyzer-bot. Check the datasheet that comes with the sensor you're using for details.

Adjust the 20 kΩ pot for best sensitivity. Since you probably don't have a calibrated gas source to work from, you'll have to simply guess at the best setting. Start with the pot at its midpoint, and try adjusting it in small steps higher or lower.

 Follow the manufacturer's recommended procedure for testing and using this, or any, noxious gas sensor, or serious injury or death could result. Test only outdoors or in well-ventilated areas, and away from flame, sparks, or heat sources. *Use for experimental purposes only!*

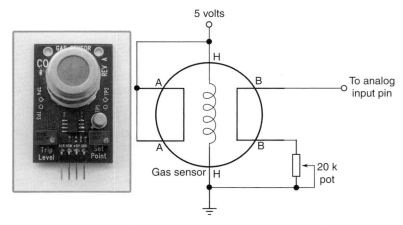

Figure 33-17   Gas sensors detect propane, LPG, and other toxic and flammable gases.

arduino-gas.ino

```
int val = 0;
void setup() {
   Serial.begin(9600);
}

void loop() {
   val = analogRead(0);
   Serial.println(val, DEC);
   delay(250);
}
```

WARM-UP, ADJUSTMENT, AND USE

Most gas sensors work by warming the air inside a chamber. They do this with a built-in heater element (this is why the sensor gets a little warm during operation). You should always allow the sensor to come to temperature before trying to take a reading. The datasheet for the sensor says to preheat the thing for 24 or more hours, but for general testing and use as a robotic gas sniffer, preheating for a minute or so will do the trick.

Run the program, open the Serial Monitor window, and let the sensor preheat. After preheating, dial the 20 k pot so the reading is right in the middle. On my prototype, turning the pot to the extremes resulted in readings from 0 to 330. So I dialed in 160 as the approximate midpoint.

The heater runs full-time and consumes about 750 milliwatts of power at 5 volts. That equates to 150 milliamps. Don't rely on a little 9-volt battery to operate your Arduino and gas sensor. You'll need a beefier battery pack, perhaps a set of six or eight D cells. Remember that when the Arduino is powered from the USB port you are limited to 500 milliamps. Don't use the gas sensor with other current-hogging components, such as servo motors.

HEAVY GAS, LIGHT GAS

As with smoke detection, where the gas detector is located has great bearing on whether it detects anything. The reason: Gases either rise or fall in air, depending on their specific gravity. If a gas is lighter than air, it rises—same thing as when you let go of a helium-filled balloon.

- Gases that are lighter than air, such as methane and natural gas, will rise to the ceiling and may not even be noticed (depending on the concentration) by a gas detector mounted on a robot that's only 6″ tall.
- Gases that are heavier than air, or about the same specific gravity as air, will sink to the ground. Sensors for these gases, which include benzene and propane, are ideal candidates for robot experimenting.

Gas	Specific Gravity	What It Does in Air
Acetylene	0.9	Slowly rises
Alcohol vapor	1.6	Sinks
Benzene	2.7	Sinks quickly
Butane	2.0	Sinks quickly
Carbon dioxide*	1.5	Sinks
Carbon monoxide	0.9	Slowly rises
Natural gas	0.6	Rises
Methane	0.5	Rises
Propane	1.5	Sinks

*Carbon dioxide is not considered a noxious gas unless its concentration is very high.

USING A WIDE PROFILE AIR QUALITY SENSOR

Sometimes you just want to know if the air is choked up with one noxious material or another. This is the job of an air quality sensor, such as the CCS811 Air Quality Sensor from AMS. It's available on a variety of easy-connect breakout boards from SparkFun, Adafruit, and others.

An air quality sensor like the CCS811 samples the molecules in the air to detect unhealthy levels of CO_2, volatile organic compounds (VOCs), and metal oxide. When mounted on a breakout board these devices use an I2C interface to communicate with a host microcontroller, which can query the sensor for levels of air quality.

Heat Sensing

In a fire, smoke and flames are most often encountered before heat, which isn't felt until the fire is going strong. But what about before the fire gets started in the first place, such as when a kerosene heater is inadvertently left on or an iron has been tipped over and is melting the nylon clothes underneath?

Realistically, heat sensors provide the least protection against a fire. But heat sensors are easy to build, and, besides, when the robot isn't sniffing out fires it can be wandering through the house giving it an energy check or reporting on the outside temperature or . . . you get the idea.

Figure 33-18 shows a basic but workable circuit centered around the popular LM34 temperature sensor. This device is relatively easy to find and costs just a few dollars. The output of the device, when wired as shown, is a linear voltage. The voltage increases 10 millivolts (mV) for every rise in temperature of 1°F. The *arduino-temperature.ino* sketch provides an example for the Arduino microcontroller for reading the sensor.

 The LM34 provides output relative to degrees Fahrenheit. Use the LM35 if you want to measure in centigrade.

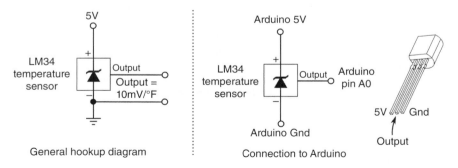

Figure 33-18   Circuit diagram for an LM34 Fahrenheit temperature sensor. For best results, mount the sensor on a small piece of aluminum or other metal, using a thermally conductive adhesive.

arduino-temperature.ino

```
int lm34 = A0;
int tempF = 0;

void setup() {
  Serial.begin(9600);
}

void loop() {
  tempF = (500.0 * analogRead(lm34)) / 1024;
  Serial.print("Current temperature: ");
  Serial.println(tempF, DEC);
  delay(500);
}
```

PART 8

Interacting with Your Robot

CHAPTER 34

Operating Your Bot via Remote Control

The most basic robot designs—just a step up from motorized toys—use a wired control box for operation. You flip switches to move the robot around the room or activate the motors in the robotic arm and hand.

This chapter details several popular ways to achieve links between you and your robot. You can use the remote controller to activate all of the robot's functions, or with a suitable onboard computer working as an electronic recorder, you can use the controller as a teaching pendant. You manually program the robot through a series of steps and routines, then play it back under the direction of the computer. Some remote control systems even let you connect your personal computer to your robot. You type on the keyboard, or use a joystick for control, and the invisible link does the rest.

 Source code for all software examples may be found at the RBB Support Site. See the Appendix for more details.

Commanding a Robot with Infrared Remote Control

You can use a TV remote control to operate a mobile robot. A computer or microcontroller is used to decipher the signal patterns received from the remote via an infrared receiver. Because infrared receiver units are common finds in electronics and surplus stores, adapting a remote control for robotics use is actually fairly straightforward.

It's mostly a matter of connecting the pieces together. With your infrared remote control you'll be able to command your robot in just about any way you wish—to start, stop, turn, whatever.

Figure 34-1   Example universal infrared remote control. To use a universal remote with the PICAXE microcontroller, you must select the code for a Sony TV.

SYSTEM OVERVIEW

Here are the major components of the robot infrared remote control system:

- *Infrared remote.* Most any modern *universal* infrared remote control, like the one in Figure 34-1, will work. These are priced starting at a few bucks at most any discount store, and I've even seen them at dollar stores. Important: You want a universal remote that supports Sony TVs and other Sony brand gear—98 percent of all universal remotes do, but check just to be sure.
- *Infrared receiver/demodulator.* This all-in-one module contains an infrared light detector, along with various electronics to clean up, amplify, and demodulate the signal from the remote control. The remote sends a pattern of on/off flashes of light. These flashes are modulated at about 38 to 40 kHz in order to reduce interference from other light sources. The receiver strips out the modulation and provides just the on/off flashing patterns.
- *Computer or microcontroller.* You need some hardware to decode the light patterns, and a computer or microcontroller, running appropriate software, makes the job straightforward. For this project we'll use the PICAXE microcontroller, because it's inexpensive and has a very handy built-in command for directly reading the codes sent by Sony remote controls.

INTERFACING THE RECEIVER/DEMODULATOR

The first order of business is to interface the receiver/demodulator to the PICAXE. Most any receiver module for 38 kHz infrared operation should work well. I've specified the Vishay TSOP4838 because (at least as of this writing) it's widely available and fairly inexpensive.

In developing this project I came across some very old infrared receiver/demodulators in my parts bin that only supported 5-volt operation—the modern ones work over a wider voltage range. If you happen to have one of these ancient modules, you can still use it in your project, but be sure to operate your circuit at 5 volts. It won't function, or will behave erratically, at anything under about 4.6 volts.

Figure 34-2 shows the simple interface for the infrared receiver/demodulator to a PICAXE 08M2 microcontroller. You can use any larger PICAXE microcontroller, as long as it supports the *infrain2* command. If you do use another version of the PICAXE chip, remember to change the pins for power, ground, and input, accordingly. Figure 34-2 also shows the pin assignment of the TSOP4838 receiver/demodulator.

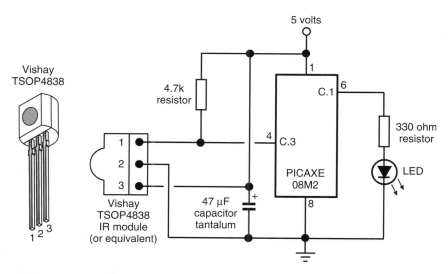

Figure 34-2 Connection diagram between infrared receiver/demodulator and the PICAXE 08M2 microcontroller. An LED provides visual feedback. Note: This schematic does not include required minimum parts if not attached to a download cable.

Various versions of the PICAXE support a number of built-in commands for decoding signals from an infrared remote control. There's also the *infrain* and *irin* commands, supported by various versions of the PICAXE chip. Refer to the PICAXE documentation for the details on these commands.

PROGRAMMING THE PICAXE

Sonyremote.bas is a simple demonstration program for use with the PICAXE, which appears below. Refer to Chapter 28, "More Microcontrollers for Robots," for more information on how to use the chip and access its programming editor. For now, I'll assume you're familiar with all these things and cut right to the chase.

To use the program, be sure to set your universal remote control to output Sony TV infrared codes. The remote control will come with an instruction booklet on how to select the proper code setting. Look under the TV listing for Sony and note which code number(s) to use. You may need to try several of them before you will have positive results.

picaxe-sonyremote.bas

```
main:
    symbol infra = b13
    irin, C.0, infra      'wait for new IR signal
    select case infra
      case 1              ' Button 2
        high C.1          'switch on output 1
      case 4              ' Button 5
        low C.1           'switch off output 1
    endselect
    goto main
```

DETERMINING CONTROL VALUES

In *picaxe-sonyremote.bas* the values 1 and 4 represent the 2 and 5 buttons. But how did I know that? I used a simple debug program that displayed the values in the PICAXE Debug window:

```
main:
    symbol infra = b13
    irin, C.0, infra
    debug 'open Debug window when run
    goto main
```

Like all programs that use the Debug window, your PICAXE must be connected to your computer via its serial or USB download cable, so that the controller can send back values. The b13 variable represents the data collected by the *irin* command. As you press buttons on the remote, the first column for b13 in the Debug window shows the numeric equivalent of the *iron* value. For example, pressing the On/Off button displays a 21.

I got the following results for my test remote programmed to output Sony TV codes. Your results may vary, depending on the Sony TV code you select for your remote.

Function	Value
Power	21
Volume up	18
Volume down	19
Channel up	16
Channel down	17
Mute	20
1	0
2	1
3	2
4	3
5	4
6	5
7	6
8	7
9	8
0	9
Enter	11
Antenna	42
Rew	27
Play	26
FF	28
Rec	29*
Pause	25
Stop	24

*Press twice.

CONTROLLING ROBOT MOTORS

 Let's assume you want to drive the traditional two-motor robot, using DC motors. You could use the PICAXE 08M2, which has just enough I/O pins for the job. But that doesn't leave any pins for other tasks. Unless you plan on using the 08M2 just as an infrared signal decoder, you'll want to select a PICAXE chip with more pins.

 Figure 34-3 shows a wiring diagram for using the PICAXE 18M2 to steer a robot based on buttons pressed on a universal remote. The 18M2 has more than enough I/O pins for what we want to do, allowing you room for other hardware expansion. The code is *picaxe-ir-bot.bas*.

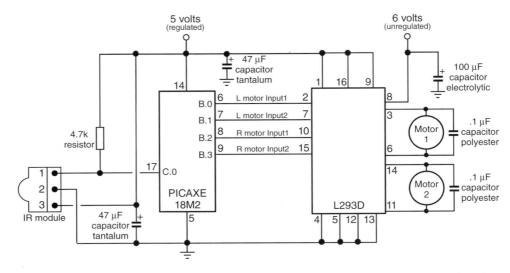

Figure 34-3 Using a PICAXE 18M2 microcontroller to receive signals from a compatible infrared receiver/demodulator and an H-bridge motor driver.

By itself, the PICAXE lacks the drive current to power the motors, so a motor driver circuit is used between the chip and the two motors. See Chapter 14, "Using DC Motors," for more information on motor driver circuits. Figure 34-3 shows how to create an H-bridge using the popular L293D motor H-bridge IC, which handles smallish motors up to 600 mA current draw.

picaxe-ir-bot.bas

To save space, the program code for this project is found on the RBB Support Site. See the Appendix for more details.

OPERATING THE ROBOT WITH THE REMOTE

Now that you have the remote control system working and you're done testing, it's time to play! With *ir-bot.bas* downloaded, disconnect the PICAXE from its programming cable, set your robot on the ground, and apply all power. In the beginning, the robot should not move. Point the remote control at the infrared receiver/demodulator, and press the following buttons to command the robot:

Key	Action	Key	Action
1	Left motor forward	3	Right motor forward
4	Left motor stop	6	Right motor stop
7	Left motor reverse	9	Right motor reverse
2	Left + right motor forward	8	Left + right motor reverse

You may press 5 to stop both motors.

SPECIAL CONSIDERATIONS

Keep the following in mind when experimenting with this project:

- Your robot spins when it should go forward. Reverse the leads to one of the motors to change its polarity. For example, switch the wires to pins 11 and 14 on the L293D.
- Press only one button on the remote at a time. If you want to create combinational events, use another button to code those events into your program. For instance, you might use Volume Up/Down or Channel Up/Down to move the robot straight forward and backward.
- Infrared links work best when used indoors, out of direct sunlight, and away from bright light sources.
- The better remote controls put out more light power, allowing you to operate your robot from greater distances. Depending on the model of your remote, you may be limited to about 6 to 8 feet away in controlling the bot.
- If your robot doesn't appear to react when you press buttons, or if it behaves erratically, double-check your wiring, or try another Sony product code for the remote. Or try a different remote.

USING IR REMOTE CONTROL WITH AN ARDUINO

The Arduino is also quite adept at being remotely controlled via a universal remote. You need just the same type of infrared detector module used for the PICAXE, plus a freebie third-party software library that decodes the signals sent from the remote. It's remarkably easy. For this project I'll use the ubiquitous Arduino Uno.

Begin by bending, cutting, and then attaching the IR detector as shown in Figure 34-4. The detector slides right into the I/O pins of the Arduino, so you don't need any additional components.

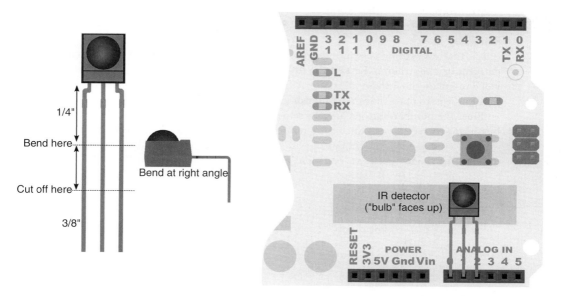

Figure 34-4 Trim and bend the leads of the IR detector so it can sit face up when inserted into the Arduino's I/O pins.

Install IR Remote Library

Next on the agenda is to install the IR library. There are a few available, but I like the *IRremote* library written by Ken Shirriff the best. His code can actually be used to both send and receive code remote signals, though for this project we're only using the receive components.

See the RBB Support Site (see Appendix) for links to the IRremote GitHub repository; the library is also included on the RBB Support with all full Arduino sketches that use the IR library. Once you've downloaded the library zip you need to add it to your Arduino IDE:

1. Download the IRremote.zip file to your computer. Don't unzip it.
2. In the Arduino IDE, choose Sketch→Include Library→Add .ZIP Library.
3. Navigate to the folder that contains the IRremote.zip file you placed in Step 1.
4. Select the IRremote.zip file, click OK to add it to your IDE.
5. Exit the IDE and restart it.

Program Remote Control for Sony TV Code

Different brands of electronic equipment use different infrared remote control codes to operate them. The *IRremote* library works with a wide variety of brands, but I've written the sketches for the project to use a universal remote control set to operate a Sony TV.

You don't need to own a Sony TV to make the required codes. Use a universal or programmable remote control that is made to operate all kinds of different brands of televisions, DVD players, and other gear. You don't need anything fancy; for this project I bought one for $1.99 at the local dollar store.

You need to look up SONY TV in the instruction manual for the remote, and enter the matching code. There's usually more than one code to try. In the IR testing sketch that follows, try each code until you find one that works.

Run IR Test Sketch

Check that everything is working properly by first double-checking the placement of the IR detector on the Arduino. Restart the Arduino IDE and load the *arduino_test_ir.ino* sketch, provided in full on the RBB Support Site. After compiling and uploading the sketch, click the Serial Monitor icon to open the Serial Monitor window.

Press the number buttons on your remote to check proper function. You'll know everything is working correctly when the button numbers you press appear in the Serial Monitor.

arduino_test_ir.ino (partial code only)

```
#include <IRremote.h>        // Include IRremote library
                             // Put in /libraries folder

#define showCode false       // Test mode
const int RECV_PIN = A0;      // Receiver input pin on A0
IRrecv irrecv(RECV_PIN);      // Define IR recever object
decode_results results;

void setup() {
  pinMode(A1, OUTPUT);       // IR power, ground pins
  pinMode(A2, OUTPUT);
  digitalWrite(A1, LOW);     // IR ground
```

```
    digitalWrite(A2, HIGH);    // IR power
    irrecv.enableIRIn();       // Start the receiver
    Serial.begin(9600);
}

void loop() {
    if (irrecv.decode(&results)) {  // If valid value was receiver
        if(showCode) {              // If showCode=trtue
            Serial.print("0x");
            Serial.println(results.value, HEX);
        } else {
            switch (results.value) {    // Match button against Sony codes
                case 0x10:
                    Serial.println("1");
                    break;
                case 0x810:
                    Serial.println("2");
                    break;
    . . .
            }
        }
        irrecv.resume();                // Receive the next value
        delay(10);                      // 10ms delay
    }
```

The *arduino_test_ir* sketch works by first importing the IRremote library using the #include directive. It then sets up analog pins A0, A1, and A2 for the IR detector. Note that power and ground for the detector are derived from the I/O pins; this is acceptable because the detector draws very little current.

Each time through the loop() the sketch checks if a button code has been received by the IR detector. If there's a valid code there, the code is checked against a series of possible matches. For example, the value 0x10 represents the 1 button; 0x810 is the 2 button, and so on (the 0x denotes the value is expressed in hexadecimal format).

The *arduino_test_ir* sketch includes an option to display the decoded button value. Set

```
#define showCode true
```

to use the testing mode if you cannot find a suitable matchup for any Sony TV code that your remote may produce. Press each number button on your remote and note the values you get. You can then use those values to modify the switch statement to match with other code values.

When you're done testing check out the playpen robot in Chapter 39, "Make Line-Following Robots," for Arduino code that integrates the IR remote control, two servo drive motors, and various sensors.

Controlling a Robot with Zigbee Radio

Remember those mad scientists from the old B-movies who were bent on taking over the world? Their primary choice of weapon: a robot they operated by remote control. I'm not so sure your desktop bot projects will help you achieve world domination, but building a remote control bot will certainly help you learn about *telerobotics*—the *tele* means "at a distance."

There are numerous methods to control a robot. You've already seen how to do it with an infrared signal beamed over the air. Another way is to use radio waves. Instead of voice or music over this radio, the link sends and receives packets of digital data. You push buttons on your transmitter and the receiver on your robot catches them. It then acts upon your command.

There are numerous types of digital data radios you can use with your robot. One of the most handy and easy to use is Zigbee. It's one of the most widely used *ad hoc* wireless networks used for home applications. I'll show you how to create a remote controller using two Zigbee radios and two Arduinos.

ABOUT ZIGBEE

Zigbee is a wireless data standard. Two Zigbee radios form a link for sending and receiving serial data over the air. The standard Zigbee is based on the IEEE 802.15.4 specification, intended for low-speed, low-power wireless data. Zigbee is only one of the technologies that uses the 802.15.4 specification—MiFi and WirelessHART are a couple others—but it's the one that's been most embraced by the robotics community.

Most Zigbee radios operate at the 2.4 GHz band, though this can differ somewhat by country. This band is shared by a number of other wireless systems, including home Wi-Fi, cordless phones, and security video systems. Each Zigbee radio has a Data In pin (transmit) and a Data Out pin (receive). These two pins are referred to as DIN and DOUT, respectively.

Though the typical Zigbee radio module has almost two dozen connection pins, other than a power and ground, these are the only wires needed to establish an over-the-air serial link. A self-contained module that conforms to the Zigbee standard, and complete with integrated antenna and mounted on a carrier board, is shown in Figure 34-5.

Figure 34-5   An XBee IEEE 802.15.4 (Series 1) radio, on a Parallax carrier board. The board provides regulated 3.3V to the XBee.

There are numerous variations among Zigbee radios. If you're interested in learning about them check out the Wikipedia page on *Zigbee* for more information. Radios of different sets aren't compatible with one another, so be sure to get two modules that match.

This project uses a pair of low-cost XBee Series 1 modules. XBee is a popular brand of Zigbee RF radios that support the 802.15.4 protocol. The form factor—the physical layout—of the XBee radios is widely supported with various adapters that allow you to plug them directly into solderless breadboards. As the transmission range need not be extensive, the low-power 1 milliwatt (1 mW) modules are more than adequate.

XBee modules use connector pins that are set 2mm apart. In order to plug the XBee module into a solderless breadboard you must first connect it to a carrier or adapter. These carriers come in various styles, with a variety of features.

Series 1 modules are simple to use. If you don't mind the default 9600 baud communications speed, they provide out-of-the-box factory settings for quick and easy setup. While you can always modify the factory settings (communications speed, channel, ID number, and so on), none of it is required if you're only needing to establish a basic wireless serial link between two nodes.

PREPARING THE REMOTE

Strictly speaking you don't even need a microcontroller to interface to an XBee radio, but I'll use two Arduino controllers to demonstrate the flexibility of the project.

- A transmitter XBee sends wireless commands to the robot and serves as the remote control. The remote can use any of several sensors to provide actuation. To keep things simple I'll show how to interface a 5-position switch to control to the Arduino to operate forward/back and right/left motion of your robot.
- A receiver XBee is connected to a second Arduino that's mounted on the robot. The received commands are used to control the robot's motors.

The remote consists of an Arduino Uno development board, prototyping shield with mini solderless breadboard, battery, Series 1 XBee radio, 3.3V/5V XBee adapter board, and a Parallax 5-position switch.

The 5-position switch is literally five switches in one, arranged to provide up/down/left/right navigation. (The fifth switch is a center push button.) Each of the five switches has its own pull-up resistor, and is interfaced to the Arduino with only a simple wire connection.

Figure 34-6 shows a drilling and cutting layout for the hand-held remote. The Arduino is powered by a 9-volt battery, which you can hold in place with a battery clip or piece of double-sided foam tape.

Cut and drill the base plate as shown; exact dimensions aren't critical. Mount the Arduino to the plate using a set of four 1/2″ long nylon standoffs and 4-40 machine screws. Use 4-40 × 1/2″ flat head screws on the underside of the plate, and 4-40 × 3/8″ screws to mount the Arduino board to the standoffs. If using metal screws add a plastic washer to prevent any possible shorts.

Attach a standard protoshield over the Arduino. If the shield does not already have a mini breadboard on it, attach one using double-sided foam tape (most breadboards have the tape already applied).

All three of the control sensors are linked, by way of the Arduino, to a Series 1 XBee radio. The wiring diagram is shown in Figure 34-7. The radio is mounted on a 22-pin adapter board carrier

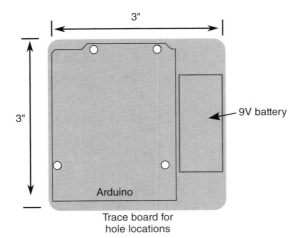

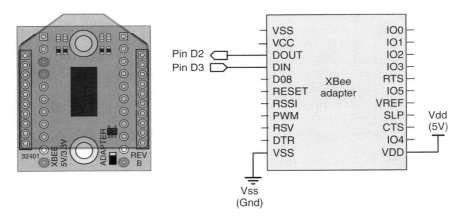

Figure 34-6 Layout guide for the remote baseplate. Use the Arduino marking holes for drilling.

Figure 34-7 Connection diagram for the XBee radio (mounted on a 3.3/5V carrier board from Parallax) to remote Arduino.

available from Parallax. Remember that XBees are designed for 3.3 volt operation. I'm using a carrier made for either 3.3V or 5V operation. The carrier has its own built-in 3.3V regulator, so you can use the Arduino's 5V supply.

 Don't use the Arduino's 3.3V pin to power the XBee. The 3.3V regulator on the Arduino Uno (and similar boards) is limited to providing only 50 mA of current. That's about what the XBee *alone* uses when transmitting.

Though the XBee works on 3.3 volts, the adapter board carrier has level shifters to convert between 3.3V and 5V. This means that as long as you use this or a similar carrier, you can connect the XBee to a 5V device like the Arduino Uno, without the need for external level-shifting electronics.

The wiring diagram for the 5-position switch is shown in Figure 34-8. The switch comes on an 8-pin breakout board that's a bit oversized but will still fit onto the breadboard with the XBee module. Programming code is provided in *arduino-tele_transmit.ino*, available in full on the RBB Support Site. The sketch sends out a single byte character, indicating the direction of the switch: press *u* for up, *d* for down, and so forth. The completed remote control is shown in Figure 34-9.

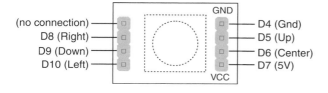

arduino-tele_transmit.ino `(partial code only)`

```
#include <SoftwareSerial.h>
const int xb_rx = 2;
const int xb_tx = 3;
SoftwareSerial Xbee(xb_rx, xb_tx);
...
void loop() {
    if(digitalRead(ltBttn) == LOW)
      Xbee.print("l");
    ...
  }
}
```

(no connection) ── GND
D8 (Right) ──   ── D4 (Gnd)
D9 (Down) ──   ── D5 (Up)
D10 (Left) ──   ── D6 (Center)
                 ── D7 (5V)
VCC

Figure 34-8 Wiring diagram for the Parallax 5-position switch to the remote Arduino.

Figure 34-9 The completed remote control, with Arduino, protoshield, and other parts.

The code works by first setting up a software serial (serial communications on any pins), specifying pins D2 and D3 for the link to the XBee. When a button is pressed, the Arduino sends a single character to the XBee for transmission. For example, pressing the button toward the Left sends a l (lower case ell).

ADDING THE RECEIVER

To turn your Arduino bot into the telerobot you need only to add an XBee module to receive the signals from the remote control unit. You can use an XBee with the same kinds of carrier board as the transmitter. Electrical connections are the same as in Figure 34-7.

As noted, when using two Series 1 XBee radios, you can keep the default factory settings and the two radios in your link will automatically know how to communicate with one another. Transfer speed is 9600 baud, which is plenty fast enough for the kind of simple data used for a telerobot.

Programming code is provided in *arduino-tele_receive.ino,* available in full on the RBB Support Site. The sketch sets up the XBee for reception, and controls two servo motors to move the robot.

arduino-tele_receive.ino (partial code only)

```
. . .
void loop() {
  readXbee();
  delay(50);
}

void readXbee() {
  if(Xbee.available()) {
    char val = Xbee.read();
    controlMotor(val);
  }
}

void controlMotor(char val) {
  switch (val) {
    case ('c'):
      stopRobot();
      break;
    case ('u'):
      forward();
      break;
. . .
  }
}
// Motion routines for motors follow
```

As with the *tele_transmit* code, the receiver uses the software serial library included with the Arduino IDE to provide communications with the XBee. The *loop()* function repeatedly calls the *readXbee* routine, which checks if anything has been received on the serial port.

```
readXbee();
delay(50);
```

The *readXBee* function continuously reads the XBee receiver, and if there's data, processes it using the *controlMotor* routine, which operates the servos to steer the bot. Each value received from the remote has a corresponding motor routine. For example

```
case ('u'):
   forward();
```

runs both motors forward when receiving the *u* (Up) signal.

You don't have to use a 5-position switch. You can substitute most any other type of switch if you'd like, or adapt the project to use a joystick, accelerometer, or a tilt compensated compass (includes both a compass and an accelerometer).

An analog joystick adds the ability to control both the direction and speed of the robot. Of course you'll need to modify how the XBees' send and receive data to accommodate the additional speed information. One method might be to use more letters to represent both direction and speed: *a* thru *e* for five forward speeds, for example, and *f* thru *j* for five reverse speeds. You could also send multiple bytes: the first byte is the direction, and the second byte is the speed.

Bluetooth Remote Control

Bluetooth is another international wireless standard for data transmission. For microcontrollers and robots, we're most interested in a subset of the technology known as Bluetooth Low Energy, or BLE. This variation has a lower current requirement than "classic" Bluetooth, but enjoys a similar range.

You can get Bluetooth (BLE) radios as add-on module for connection to most any microcontroller. They work in a similar fashion as Zigbee, where you connect a two-wire serial interface (transmit and receive) between the module and the controller.

A common low-cost Bluetooth BLE transceiver module is the HC 05. You can use it to talk to another microcontroller that has the same or similar module, or with devices (like smartphones) that support Bluetooth BLE.

Some controllers come with a Bluetooth radio already in them. This includes the BBC micro:bit, described in Chapter 26. Its "radio" function is surprisingly easy to use, and allows you to connect with other micro:bit modules in the same room.

The micro:bit's *radio* function is different from the Bluetooth connectivity that the controller uses for programming via a tablet or smartphone.

You can send messages to any and all micro:bit controllers in the vicinity, or just those that match a particular "channel." The channel number, from 0 to 100, is set in the program.

The following short MicroPython example skips channel selection so it defaults to channel 7. It broadcasts a simple message to any other micro:bit in the room that is also using the default channel selection. The message is a single asterisk character, which when received appears briefly on the LED grid. Activate transmission by pressing the A button.

microbit-radio.py

```
import radio
from microbit import display, button_a, sleep
myCode = '*'
radio.on()
while True:
    if button_a.was_pressed():
        radio.send(myCode)
        incoming = radio.receive()
        if incoming == myCode:
            display(myCode)
            sleep(2000)
            display.clear
```

- *import radio* brings in the radio functionality.
- *radio.on* activates the built-in radio; because the radio consumes extra power you should turn it off (*radio.off*) when not using it.
- *radio.send* transmits a text message—it can be more than a single character, but no more than 251 bytes.
- *radio.receive* receives the message.

This is a fairly simplistic example. See the micro:bit documentation (both MicroPython and MakeCode Blocks Editor) for additional ideas.

> Go to the RBB Support Site (see Appendix) for more useful information articles on wireless control, including tips on extending radio range and using low-cost RF modems.

Broadcasting Video

Added to a telebot, a video transmission lets you see what the robot sees, even when the robot is out of sight. If you've gotten as far as operating your robot via any kind of remote control, adding video to the mix is fairly simple. There are two general approaches:

Feed the video signal back through the data radio. This works if you're using 802.11g (or faster) Wi-Fi, as you need the higher data rates to keep up with the video. This is a doable approach if the video is already in digital format; if you're using an analog camera, you'll need to digitize it before passing it back through the communications link.

Use an analog video transmitter. This approach works with any kind of remote control, because it doesn't use the data link for the transmission of video. You use your regular wired, infrared, or radio link, but the video is sent back using its own transmitter.

Analog wireless cameras are pretty inexpensive, and many are already designed for use with low-voltage supplies. Figure 34-10 shows a camera and transmitter combo. Mount the camera on your robot, attach a 9V battery, and your robot is transmitting pictures.

The matching receiver connects to any monitor or TV with an analog video input. Check the video format; this particular camera/receiver pair uses the NTSC video standard, so it'll connect

Figure 34-10 A video camera/transmitter and receiver pair. The receiver plugs into a TV or monitor.

to any NTSC-compatible monitor. A small 7″ LCD monitor for in-car video is ideal, as these have a power jack on them for low-voltage (usually 12 volts or less) operation. Rig up a battery for both the receiver and monitor, and you can create a fully portable video reception unit.

AVOIDING INTERFERENCE

The standard frequency band used by wireless analog video cameras is 2.4 GHz, which so happens is the same band used by your XBee radios. There is a chance that the video camera will interfere with the XBee, and vice versa. However, this problem is mitigated by using a multi-channel wireless camera, where you can select a different RF frequency within the 2.4 GHz band.

You can tell if your camera and XBee are lousing up one other by monitoring the video image when you place the XBee remote next to the camera. If there's a problem you'll see wavy lines and other interference patterns.

If your wireless camera lacks a channel-changing feature, or selecting none of its channels remove the interference, you can try altering the XBee's channel; there are 16 to choose from, and each channel occupies its own 5 MHz space. The channel on both the transmitter and receiver XBee must be tuned to the same channel.

Refer to the documentation on the XBee radios on how to set channels. It can be done directly in the Arduino sketch (using something called *Command Mode*), though most people prefer to use the separately available X-TCU program, which provides a handy graphical interface. You need an XBee carrier board with USB, in order to communicate between your PC and the radio. The X-TCU program is available for download from *digi.com,* makers of the XBee.

USING WI-FI (DIGITAL) VIDEO

Cameras and receivers that transmit using analog video can suffer from reduced picture quality from interference from other electronics that use the same radio band. An alternative is transmitting the video digitally, often using Wi-Fi. A benefit is that you can then receive the pictures on any device that connects to Wi-Fi, such as a tablet or smartphone.

Connect any compatible small analog video camera to the transmitter (some transmitters contain their own camera). When using a Wi-Fi setup note the power requirements of the camera and transmitter; some designed for automotive use may need 12V power. A model that operates at 5.8 GHz will help reduce the possibility of interfering with an XBee or other data communications system.

If you're going the Wi-Fi video route, look for a rig that includes an app for your model of tablet or smartphone. You can use it to receive the video directly on your mobile device. For the ultimate in telerobotics, place your smartphone inside a pair of first person video (FPV) goggles.

CHAPTER 35

Producing Sound

In the movies robots are seldom mute. They may utter pithy sayings—"Don't you call me a mindless philosopher, you overweight glob of grease!" Or squeak out blips and beeps in some "advanced" language only other robots can understand.

Voice and sound make a robot more "human-like," or at least more entertaining. What is a personal robot if not to entertain?

What's good for robots in TV shows and movies is good enough for us, so this chapter presents a number of useful projects for giving your mechanical creations the ability to make noise. The projects include using recorded sound, generating warning sirens, playing music, and even speaking. And if you're interested in having your robot respond to sound, be sure to check out Chapter 33, "Environment."

 Source code for all software examples may be found at the RBB Support Site. To save page space, the lengthier programs are not printed here.

Preprogrammed Sound Modules

At the bottom of the sound food chain is the preprogrammed, or "canned," sound module, typical in such products as greeting cards and musical ornaments. Most are programmed with a song, though a few—like the electronic whoopee cushion—are meant to emit a sound effect. Humor notwithstanding.

Most sound modules are completely self-contained, including a small speaker (piezo or dynamic magnet), a button battery, and some means to set it off, usually a small push-button switch. Several are shown in Figure 35-1. You can salvage the sound module from a greeting card or other product and reuse it in your robot. Craft stores are a good source of new sound modules that can be added to homemade ornaments.

Figure 35-1 An assortment of sound modules with prerecorded sounds and songs on them. Most modules can be hacked to operate from electronic control.

Controlling a sound module from a microcontroller involves triggering the module by electrical signal, rather than by using its mechanical push button. Depending on the design of the module, you may be able to trigger it simply by replacing its push-button switch with the circuit shown in Figure 35-2A. An opto-isolator and accompanying parts bypass the switch. You trigger the module by momentarily bringing the microcontroller line that is connected to the opto-isolator to HIGH.

If the sound module is powered by more or less 5 volts, you can try hardwiring the trigger switch to on, and then the module with operating juice from the microcontroller line itself (Figure 35-2B). If your microcontroller doesn't provide enough current to drive the module, try adding one of the buffers in a ULN2003 chip. The driver also acts as a buffer to help protect the microcontroller.

And finally, if the module operates at only 1.5 volts (a single button battery), you may need to use something like Figure 35-2C. Yank the mechanical switch from the module, and replace it with a small relay.

Commercial Electronic Sound Effects Kits

Among premade electronic kits, ones for sound effects are always popular. Several companies manufacture and sell sound effects kits that you can use as self-contained modules in your robot projects. For example, the light-sensitive Theremin Kit from Chaney Electronics produces distinctly outworldly sound effects by altering the amount of light falling on two sensors. The company also sells a 10-note sound kit and several others.

Most sound effects kits are designed to be self-contained. That means they come with an amplifier, if they need one, and a speaker. Controlling them on a robot requires interfacing the selector buttons to the outputs of your microcontroller, in much the same way as for the sound modules in the previous section. Most kits come with a schematic, and you can readily determine where to hook things up.

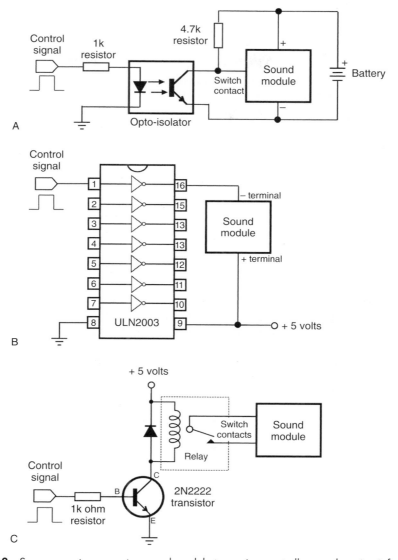

Figure 35-2 Some ways to connect a sound module to a microcontroller or other circuit for activation. Techniques include opto-isolator, power driver (turn battery supply on and off), and relay to replace the mechanical switch.

Using a driver IC, as previously described, to power and depower the board is among the best methods to turn the sound on and off, as it reduces overall current draw from the robot's batteries when the sound effect isn't needed. But there are other methods, such as using a CD4051 analog switch. See the "Stupid Sound Tricks for Your Robot" AppNote on the RBB Support Site for details on controlling which sound-producing element is routed to the amplifier and speaker.

Making Sirens and Other Warning Sounds

If you use your robot as a security device or to detect intruders, fire, water, or whatever, then you probably want the machine to warn you of immediate or impending danger. The warbling siren shown in Figure 35-3 will do the trick, assuming it's connected to a strong enough amplifier (some amp circuits are provided later in the chapter).

The circuit is constructed using two 555 timer chips; alternatively, you can combine the functions into the 556 dual-timer chip, but I prefer the separate chips because they provide a bit more room on the breadboard to experiment. The "warble speed" and pitch can be altered by changing the capacitors connected between pin 2 and ground of each chip.

● The timer on the left toggles HIGH and LOW about once a second. It makes the "warble."
● The timer on the right produces the high-pitched siren sound and alternates between two frequencies. The two frequencies are controlled by the left timer, as it toggles HIGH and LOW.

When using an 8 or 16 Ω dynamic speaker, the sound output is pretty loud—enough for family members to come into your workshop and complain about it. But if you need more oomph, connect the output of the second 555 to a high-powered amplifier. You can build an amplifier using an LM386 IC, as described later in the chapter, or get a ready-made amplified speaker or an audio amplifier kit with a wattage of 8, 16, or more.

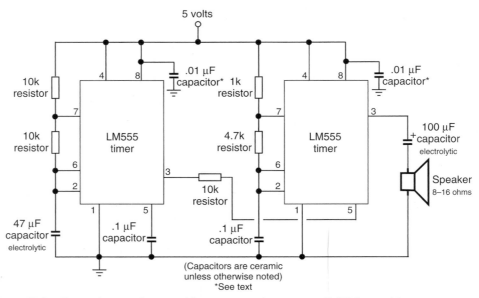

Figure 35-3   Circuit diagram for a warbling siren, made using two LM555 timer ICs.

The circuit needs several small-value ceramic (monolithic or disc) capacitors to prevent the inevitable power supply glitches that are generated by the versatile but electrically "noisy" LM555 chips. I've specified 0.1 μF capacitor as a bypass capacitor connected to pin 5 of the "warbler." More critical are the 0.01 μF capacitors as power supply decoupling, connected between the power pins (pins 8 and 1) of both chips. Try to get these decoupling capacitors as physically close to the IC as you can.

If you breadboard this circuit, keep the lead lengths to a minimum, especially for the capacitors. After you have tested the circuit, you will want to transfer it to a soldered breadboard and clip all the leads as short as possible.

Using a Microcontroller to Produce Sound and Music

Any microcontroller with a pulse width modulation (PWM) feature can be used to produce sound, and even music. See Figure 35-4 for an overview of the general idea. When the PWM frequency is within the range of hearing—about 20 Hz to 20 kHz—we hear it as a tone. The sound is heard by passing the I/O line through a speaker or amplifier. It's not great sound—your robot won't win a Grammy for it—but it's cheap and easy to do.

By varying the tones, you make sound. You can produce music with one PWM output (monophonic), or you can combine the outputs of two or more PWM I/O lines to create polyphonic sounds, like a music synthesizer. It's easy to produce warning sirens, warblers, bio-sounds (a la R2-D2), and other effects.

SOUND WITH A PIEZO SPEAKER ELEMENT

Most any microcontroller can produce sound using a basic piezoelectric speaker element. This is the kind without internal electronics—they're not buzzers, but simply transducers that use a piezoelectric disc to produce sound from an AC signal source. That AC signal can be the PWM waveform of a microcontroller.

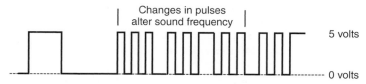

Figure 35-4 By altering the width of pulses from a microcontroller your robot can produce various types of sound effects.

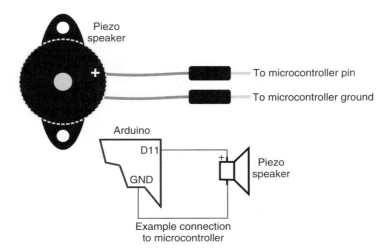

Figure 35-5   Basic connection from a microcontroller, such as an Arduino, to a piezo speaker element.

Connect the piezo element to a pin of the microcontroller as shown in Figure 35-5. You want an element in a nice housing with wires or pins coming out of it. These produce louder sound than a bare disc. If the element has a polarity marking on it, the + wire goes to the microcontroller pin, the other wire goes to Gnd.

If your piezo sound maker produces noise simply by attaching it to a battery then it's really a buzzer and not a speaker. Buzzers have their own tone generation circuit in them, and produce just that tone. The frequency of the buzz can't be changed.

Complicating matters is that some sellers call basic piezo elements "buzzers." So before buying anything read the description carefully to make sure you're getting the right thing.

Most microcontrollers have a *tone* (or similarly named) code statement that produces a PWM signal on a pin. I'll show some examples for a couple of popular microcontrollers here.

Making Tones on an Arduino

When you download the Arduino IDE you get several tone-making examples that come with it. I recommend you try these, but to get you started, I've boiled down the code to the core concepts to make it easier to follow and adapt. Check *arduino-siren.ino* for a simple siren demo. It works by alternating two frequencies, 440 Hertz (Hz, or cycles per second) and 880 Hz, once every half second.

The essential part of the code is the *tone* statement, which produces a tone of a specific frequency. (An alternative variation of the statement lets you optionally specify a duration. I've omitted this and used a standard *delay* statement instead.)

```
tone(pin_number, frequency);
```

Figure 35-6   Use a piezo element in a housing for better sound. For a louder sound firmly attach the element to the body of your robot.

arduino-siren.ino

```
void setup() {
    pinMode(11, OUTPUT);        // Speaker connected to pin 11
}

void loop() {
    tone(11, 440);             // Concert pitch 'A'
    delay(500);                // Wait 1/2 second
    tone(11, 880);             // Octave higher
    delay(500);                // Wait 1/2 second
}
```

In case you're interested in such things, 440 Hz is the generally accepted pitch for concert A (the A above middle C on a piano). And 880 Hz—also an A tone—is exactly one octave higher.

Provided on the RBB Support Site is an enhanced tone-making sketch, *arduino-maketones.ino*. It's adapted from the *Tone* example at *arduino.cc/en/Reference/Tone,* but stripped down to make it simpler to follow. The sketch cycles through two sets of short musical notes. The *frequency* of each note is specified as one of four defined constants; the *duration* is a numeric value representing a fraction of a second: 1/1 is one second, 1/2 is a half second (and therefore a half note), and so on.

arduino-maketones.ino

To save space Arduino-tones is provided in its entirety on the RBB Support Site.

Mounting the Piezo Element

The siren sound from a piezo element isn't terribly loud, but it's enough to be used as a signal. Attach the element on the top of your robot so it's not muffled. If you mount the Arduino using 1/2″ or longer standoffs you can slip the element just under it, and save a little space. Firmly attaching the element to the body of your robot helps increase sound output. See Figure 35-6.

Making Tones on a BBC Micro:bit

The BBC Micro:bit uses a similar coding statement called *playTone*. By default, the PWM signal is passed through pin P0, but this can be changed to P1 or P2 by using the *analogSetPitchPin* statement. I'll show just the basic coding here.

```
playTone(frequency; duration)
```

Frequency is the frequency of the tone, in Hz; *duration* is how long the tone plays, in milliseconds. In the following example the Micro:bit toggles between two tones to make a siren effect.

microbit-siren.py
```
while True:
  playTone(440; 500)
  playTone(480; 500)
```

MicroPython on the BBC Micro:bit also comes with a powerful *music* library, complete with built-in melodies. For example, to play the birthday song you'd use:

```
import music
music.play(music.BIRTHDAY)
```

Refer to the MicroPython *music* entry for more details.

USING A DYNAMIC SPEAKER INSTEAD OF A PIEZO ELEMENT

Piezo elements aren't known for their musical fidelity. Most are engineered to produce fairly high frequencies, making the lower tones quiet or even indistinguishable. But for their limitations they offer an important advantage when using a microcontroller. Piezo elements are preferred (for getting started anyway) because they have a high input impedance, and so they won't draw a lot of current from the Arduino's I/O connections.

The obvious alternative is a dynamic speaker, the kind in your TV or stereo system. These use a magnet and paper or plastic cone to make sound. Dynamic speakers produce a wider range of frequencies. But their downside is that most dynamic speakers have a low impedance, typically 4 to 32 ohms. The lower impedance can cause the speaker to draw excessive current from the microcontroller, and that could possibly damage it.

You can always limit current from microcontroller by placing a resistor inline with speaker, but this tends to excessively diminish the sound output. You might also select a higher impedance speaker—say 200 Ω or higher, but these are not common.

Better methods involve using a transistor as a speaker driver, and an amplifier. Both techniques are discussed next.

Using Audio Amplifiers

More likely you will want to connect the output of the microcontroller to an amplifier to actually increase the volume. You can either make the amplifier out of discrete parts, built a ready-made amp board or kit, or use a portable amplifier that includes its own speaker.

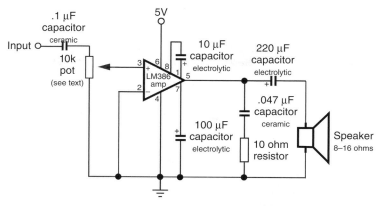

Figure 35-7 LM386 audio amplifier, wired for a gain of about 200.

MAKING YOUR OWN AMP

Figure 35-7 shows a rather straightforward audio amplifier that connects directly to an 8- or 16-ohm speaker. For best results, make sure the 10 kΩ potentiometer is the *logarithmic* (or *audio*) *taper* kind, not the typical linear taper. When using a linear taper pot, the sound volume will be affected at only one end of the dial. With a logarithmic potentiometer, the volume change will be more evenly spread across the dial.

USING A PREMADE AMP BOARD

If there's one thing I hate it's making audio amplifiers. Though chips like the ML386 are inexpensive and ubiquitous, they're a bit behind the technology curve, being solid 60s technology. Better ICs are offered, but they're not always as easy to use or widely available. They also tend to be bulky, thanks to the big capacitors.

Enter the premade amplifier board. These (usually) come assembled, and may use surface mount components, so they are more compact than homebrew amps. To use, connect power, sound input, and speaker. Check out the offerings from Adafruit, SparkFun, and Pololu for current models.

If you prefer to put something together, Velleman, among others, offers a handy kit of parts for small monophonic amplifiers. See Figure 35-8 for an example amplifier board wired to a cheap computer speaker purchased from the thrift store. You get better sound when the speaker is in an enclosure.

USING A SELF-CONTAINED AMP AND SPEAKER

Amplifiers, speakers, and all the associated wiring for sound can add weight and bulk to a robot. When this is a problem opt for a self-contained amplifier and speaker, already in a convenient housing. Many of these amp/speaker combos run on their own battery power, sometimes rechargeable.

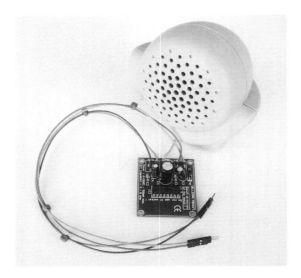

Figure 35-8 Velleman 3-watt audio amplifier kit, shown constructed, attached to a surplus computer speaker. Velcro or double-sided foam tape can be used to secure the speaker to your robot.

Figure 35-9 A commercially made amplifier and speaker, ready for immediate use on your robot. The amplifier contains its own rechargeable battery.

Figure 35-9 shows a capsule or "pill" amplified speaker tucked into a recess of a small robot. The speaker has its own internal rechargeable battery, and plugs into an audio output jack using a standard 1/8″ (3.5mm) phone jack. Some boards—Parallax Propeller Activity, Raspberry Pi, others—have this jack already, and you can easily add one by soldering. Secure the amp/speaker to your robot using Velcro or double-sided tape.

Sound and Music Playback with a Microcontroller

My first robot used the mechanism from an old cassette recorder for its sound and music playback. It was, shall we say, clunky. Thanks to several ready-made add-on boards for the Arduino and other popular microcontrollers, you can incorporate melodies, effects, and other sounds

Figure 35-10   All-in-one WAV/OGG playback board from Adafruit, which can be connected to a switch or a microcontroller for fully automated sound clip selection. Copy music and sound effects files from your PC to the card using USB.

where you have full control over what's played—from short, quarter-second gunshots to 10-minute-long lullabies.

These add-ons use prerecorded sounds that you create on your computer. The sound files, which can be in WAV, MP3, WMA, OGG, or a proprietary format, are then transferred to a solid-state memory card, typically SD or μSD (micro SD, the tiny version of standard SD). The memory card is inserted into the playback board, which is controlled via simple commands from a microcontroller.

There are a number of breakout boards and shields for adding sound to the Arduino or other microcontroller. Most sound boards use solid-state memory cards to store sound, music, and effects files. The card is formatted in FAT16 (some support FAT32), meaning you can prepare the card on your computer and copy sound files to it.

Sound boards come in two general types, coprocessor and manual processing.

COPROCESSOR

Sound boards with a coprocessor are the easiest to use. They contain their own microcontroller to actually process the sound data. The controller reads the data from Flash memory, and processes it through a specialized sound chip one byte at a time. Because the sound board does all the heavy lifting, your microcontroller is free for its other robot tasks.

Figure 35-10 depicts a compact board (this one made by Adafruit) that stores sound files to an internal Flash memory (no need for a memory card). It supports WAV and OGG format; sound files play by activating one of the inputs. You can use a switch or a simple connection from your microcontroller to start playing a sound clip.

The programming libraries to support reading the content of a memory card use up a lot of RAM in the microcontroller. For this reason, it's always a good idea to use these projects with an MCU equipped with at least 2KB of RAM, and 32KB or more of Flash memory. When using the Arduino, you want the version with the ATmega328 (or higher) controller chip.

How files are played depends on the features of the sound board.

- *Triggering* allows you to play a file just by activating a pin on the board. For this to work the sound files are given specific names that associate them with the trigger pins on the board— *file001.ogg* is triggered by activating pin 0, *file002.ogg* is triggered by activating pin 2, and so on.
- Serial control allows you to connect your microcontroller to the board and activate a sound clip using a defined command language. It's usually easier than it sounds. Using your microcontroller you send a sequence of bytes that specify the sound file to play.

MANUAL PROCESSING

Sound boards have specialized chips on them that do the actual data-to-sound conversion. With manual processing your microcontroller must do all the data transfer between the Flash memory and the sound chip. This is usually accomplished using a serial interface, most often SPI.

If you get this type of sound board be sure it comes with programming examples for your microcontroller. The examples must show how to precisely meter out the data when the sound chip needs it. If the data are sent to the sound chip too fast or too slow, the audio will have annoying clicks and skips in it.

Depending on the data rate of the sound clip, and the speed of your microcontroller, there may not be very much time left over for your robot to do other things. Programming complexity aside, for this reason alone you're usually better off getting a sound board with a coprocessor on it.

PREPARING SOUND CLIPS

Let's suppose you already have the music or sound effect you want to use with your robot. You need to not only prepare the clip so it's in the correct format for your sound board, you must transfer that clip into the memory of the board.

Prepare the clip using Audacity or other audio editing program. You can spend a lot of money on audio software, but you don't need anything fancy for robot sound clips. Figure 35-11 is a

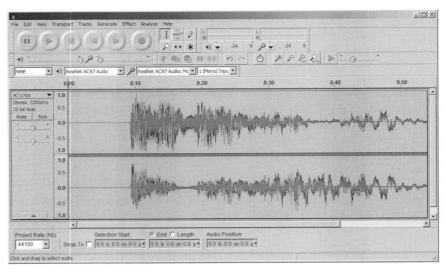

Figure 35-11 Use the freeware Audacity software to prepare your sound clips for use in a sound board.

sample screen from the freeware Audacity software. The program allows you to trim off parts of a sound file you don't want, change bit rates and sampling, convert stereo to mono, and save the resulting creation in any of a number of popular formats.

 Audacity is complex software, and the sound board of your choice may require a specific file format. If you're new to this topic, check out the *AppNotes* on sound production on the RBB Support Site.

Typically, sound files are stored as 8+3 DOS-style filenames on the memory card. Sound files can be any size, up to the capacity of the memory card, but as you can only place files into the root directory of the card, you are limited to 512 files total. That's generally more than enough for any robotics application.

Once the sound clips have been prepared, you can transfer them to the sound board.

- If the board uses a Flash card, mount the card as a drive in your PC. Use your operating system to copy the files to the card.
- If the board has only built-in Flash memory, it'll likely have a USB connection for transferring files. Plug the board into your PC and it should register as a thumb drive. Copy the files to the drive.

WAV, MP3, OGG, OR. . .

Sound can be stored in many different formats. The three most commonly used are WAV, MP3, and OGG. Here's a quick rundown:

1. **WAV** is a "raw" data format that is comprised just of the sound data. There is no compression of the sound data, so WAV files tend to be fairly big.
2. **MP3** is a compressed file format well known to most anyone who's ever worn a pair of earbuds. With MP3 sound data are compressed so it takes up much less space.
3. **OGG** is another file format for compressed audio. It's similar to MP3, but was created in order to avoid the patents associated with creating and playing MP3 files. Though most patents related to MP3 have now expired, OGG is still a useful alternative for storing and playing compressed audio. Some of the lower cost sound boards do not process MP3, but do process OGG.

Which to choose? If storage space is not a major concern, and you're using a sound board with its own coprocessor, opt for WAV format. It can take a finite amount of time for a compressed file to start playing. It may only be a 200 millisecond delay before the sound clip plays, but for robotics applications, you may want rapid-fire playback, especially if the sound effect is continually repeated.

Creating Music and Sound with MIDI

PWM is a fast and convenient way of creating sound from your microcontroller, but it leaves much to be desired. Quality is greatly improved by using WAV, OGG, or MP3 files, but then you're limited to only what you've pre-recorded. What to do?

A MIDI-based synthesizer produces high-quality sound and even music, and can be created on-the-fly. You can now get a MIDI synthesizer on a small board or Arduino shield. Connect it to an amplifier and speaker, and your robot can create a symphony of music and sound effects.

WHAT IS MIDI

MIDI stands for *Musical Instrument Digital Interface*. It's a standard method for controlling electronic instruments—it does other jobs, too, but music is what we're interested in here. The MIDI specification covers the data transmission itself, the electrical connection, even the hardware used to link everything together.

For a self-contained robot we're mainly interested in the data talking part. MIDI speaks by sending short *messages* over an asynchronous serial connection.

- The data sender is referred to as a *controller*. A common MIDI controller is an electronic keyboard, but there are many other kinds—for the Tunebot the controller is an Arduino.
- The data receiver is referred to by various names, such as *sound module, synthesizer,* and *sound bank*. Its job is to listen for commands sent by the controller, and turn them into musical notes. The sound module is connected to an amplifier and speaker so you can hear the music.

Most messages are only two or three bytes long. Each message starts with an 8-bit *command* (or *status*) byte, followed by one or more 7-bit *data* (or *parameter*) bytes. The combination of command and data byte(s) of a single message is an *event*.

Figure 35-12 shows a simplified example of a 3-byte MIDI message. By starting every message with an 8-bit byte, and then only using 7-bit data bytes, the sound module can more easily keep sync between itself and the controller.

Somewhat of a convention in some MIDI examples you'll see on the Web, the Command byte and first data byte are often expressed in hexadecimal format—this is mostly programmer's preference. The second data byte may be in hex or decimal. From an operation standpoint it doesn't matter what number base you use.

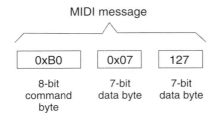

Figure 35-12 MIDI messages are composed of just a few bytes, an 8-bit *command* byte (it always has its eighth bit set to 1), and one or more 7-bit *data* bytes.

Figure 35-13   The Music(al) Instrument Shield from SparkFun is a low-cost MIDI sound module that plugs directly on top of the Arduino. Similar boards are available from other online sellers.

CONNECTING A MIDI SOUND MODULE TO THE ARDUINO

Thanks to the "maker movement" there are now a number of affordable MIDI sound modules available for the Arduino and other microcontrollers. Figure 35-13 shows a MIDI sound module shield from SparkFun, designed for the Arduino Uno (and compatible).

Despite all the pins on these boards, most modules require only a couple of connections for MIDI operation:

- *MIDI input.* This pin accepts asynchronous serial at a specific baud rate of 31250 bits per second. Use the Arduino *SoftwareSerial* object library to transmit serial data directly to the board.
- *Reset.* This pin provides a method of resetting the MIDI chip on the sound board to a known state. Do this after the board has been powered up, but before sending any MIDI data. Resetting is accomplished by momentarily bringing the pin LOW, then HIGH again.

No matter what MIDI sound board you use, be sure it has a MIDI serial input pin. It may be labeled something like *MIDI Interactive* or *MIDI Realtime*. This allows the board to use arbitrary data for MIDI, as opposed to only the more formal data contained in a complete MIDI file.

On some boards you must set a jumper or input pin in order to activate MIDI Realtime. Check the documentation that comes with the board you're using.

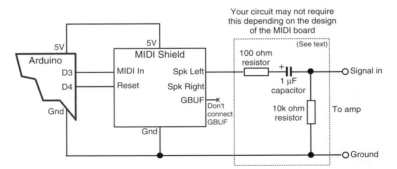

Figure 35-14 Connection diagram between an Arduino and a MIDI sound board such as the SparkFun Musical Instrument Shield. When the board is used in MIDI Realtime mode you need only provide connections for MIDI In and Reset.

Figure 35-14 shows the basic wiring points between the Arduino and MIDI board. If you're using a shield the connections are made for you, and no additional wiring is needed. Should the shield use different pins than those shown you'll need to alter the pin designations in the program sketch.

Make note of the additional components inside the dotted box. They provide a filter to help smooth out the audio. Your circuit may not require this depending on the design of the MIDI board you use, and the type of amplifier you connect to. You can leave off these components if the board you use already has them built-in, or if you're using an external amplifier. The amp likely already has these types of components inside it.

The GBUF output pin is used when connecting headphones directly to the board. It isn't intended for connecting to amplifiers. Especially never connect GBUF to Gnd. The GBUF pin is not circuit ground, but is set to about 1.2 volts in order to drive stereo headphones.

MAKING TUNES

The *arduino-basicmidi.ino* sketch demonstrates basic functionality of making MIDI sounds from the Arduino. The sketch begins with including the *SoftwareSerial* object library (it comes with the Arduino IDE). The library creates an object named *MIDI*, which is set to transmit data on pin D3.

Don't worry if some of the code looks a bit alien to you. For reasons of limited space, I'll only talk about the main components of the sketch here. Read more details about it on the RBB Support Site.

arduino-basicmidi.ino

```
#include <SoftwareSerial.h>
SoftwareSerial MIDI(255, 3); // Rx, Tx (Rx not used, Tx pin=3)

void setup() {
  MIDI.begin(31250);               // Set up serial comm to MIDI
  resetMidi(4);                    // Reset MIDI device (pin 8)
  ctrlMIDI(0, 0x07, 127);          // Set channel volume (0-127) to full
  ctrlMIDI(0, 0x00, 0x79);         // Select 'melodic' bank
  sendMIDI(0xC0|0, 0, 0);          // Select instrument in bank
}

void loop() {
  noteOn(0, 60, 127);              // Turn on middle-C note, velocity=127
  delay(1000);                     // Wait one second
  noteOff(0, 60,  127);            // Turn off note
  delay(1000);
}

// Turn note on (press key)
void noteOn(byte channel, byte note, byte attackVelo) {
  sendMIDI( (0x90 | channel), note, attackVelo);
}

// Turn note off (release key)
void noteOff(byte channel, byte note, byte releaseVelo) {
  sendMIDI( (0x80 | channel), note, releaseVelo);
}

// Set controller value with channel
void ctrlMIDI(byte channel, byte data1, byte data2) {
  sendMIDI( (0xB0 | channel), data1, data2);
}

// Send bytes to MIDI device
void sendMIDI(byte cmd, byte data1, byte data2) {
  MIDI.write(cmd);
  MIDI.write(data1);
  // Pass only if valid 2nd byte
  if( (cmd & 0xF0) <= 0xB0 || (cmd & 0xF0) == 0xE0 )
    MIDI.write(data2);
}

// Reset MIDI device
void resetMidi(int resetPin) {
  pinMode(resetPin, OUTPUT);
  digitalWrite(resetPin, LOW);
  delay(100);
  digitalWrite(resetPin, HIGH);
  delay(100);
}
```

The *setup()* function starts the MIDI serial link at 31250 baud, then resets the board by toggling the Reset pin LOW, then HIGH.

The remainder of the *setup* function sends a series of messages to the MIDI board. These messages are transmitted with the assistance of some helper functions, *ctrlMIDI* and *sendMIDI,* found later in the sketch.

- *ctrlMIDI* is for passing control change messages. The function takes three arguments: the channel number, the specific control to change, and the value to change it to. Example: The line *ctrlMIDI(0, 0x07, 127)* alters the volume on channel 0 to *127.*
- *sendMIDI* is for passing generic messages. This function also takes three arguments: the command byte, and up to two data bytes. The statement line *sendMIDI(0xC0|0, 0, 0)* sets the instrument on channel _0 to 0.

The *loop()* function cycles through turning on and off a note at 1-second intervals. Two function calls are used here: *noteOn* and *noteOff.* They're simply wrappers to the *sendMIDI* function, and make the coding simpler. Both take the same arguments, but they do different things:

1. *noteOn* turns on a note of a specific pitch, and at a specific velocity (loudness). The pitch of the note is a numeric value indicating a key on a keyboard. For example, Middle C is 60. Key values count both the white and black keys, as shown in Figure 35-15.
2. *noteOff* turns the note off. You want a *noteOff* for every *noteOn,* or otherwise the notes will just congeal into one another. (This continues until the MIDI chip runs out of internal resource space.)

There are many other options available to you, which you can use to produce anything from unusual sound effects to melodic music. For example, you can play multiple notes and instruments together, allowing your robot to produce complex chords. You might use these chords as a way for your robot to communicate with you. Or you might code your robot to produce different notes as it navigates a room, or speeds through a line-following course.

There's lots more to robot MIDI than what's here. See the RBB Support Site for additional resources for using a MIDI synthesizer to augment your robot's sound and music repertoire.

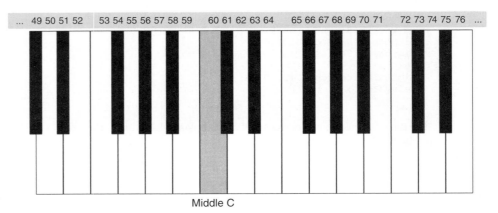

Middle C

Figure 35-15 Note pitch is defined using numbers that represent the white and black keys on a piano. For reference a pitch of 60 is Middle C.

Speech Synthesis: Getting Your Robot to Talk

Not long ago, integrated circuits for the reproduction of human-sounding speech were fairly common. But with the proliferation of digitized recorded voice, unlimited vocabulary speech synthesizers have become something of an exception. Using only software and a sound card, it is possible to reproduce a male or female voice. In fact, both Windows and Macintosh OS X come with free speech-making tools for their operating systems.

These days solutions for adding affordable speech to a robot or other microcontroller-based project is slim pickin's. These days there are only a few options available, and the price is a little steep compared to other types of sound-making boards. Fortunately, the speech quality of these products has much improved over the barely intelligible speech of grandpa's robots—so much that you can actually understand what your robot is saying to you!

USING THE EMIC 2 TEXT-TO-SPEECH MODULE

Among the most well-regarded voice solutions is the Emic 2 Text-to-Speech Module from Parallax. The module accepts a string of text from an Arduino, Raspberry Pi, or other microcontroller, and speaks the result. Predefined voice styles include male, female, and child.

Communicating with the Emic 2 involves setting up a serial connection between it and your microcontroller, then sending regular text. The Emic 2 has its own text-to-speech translator which provides correct pronunciation of most words, but if things don't sound just right, you can alter the spelling until you get it just right.

Check out *arduino-emic2-demo.ino* on the RBB Support Site for a sample demo of setting up the board and then talking. The actual speech is provided with the lines:

```
emicSerial.print('S');
emicSerial.print("My name is Mister Robot.");
emicSerial.print('\n');
```

To change your robot's speech just modify the middle text string.

See the RBB Support Site for additional resources for speech synthesis, including other modules similar to Emic 2. You'll also find useful links on the science and technology of synthesized speech.

OTHER OPTIONS FOR SPEECH SYNTHESIS

Of course, there are other options for robot speech synthesis. Hardware speech synthesizers for the Arduino and other microcontrollers tend to come and go, and your best bet is to search the Web to explore the current offerings.

As of this writing there are several commercially available single-chip solutions available from a smattering of online sources. These include the Speakjet and Soundgin voice coprocessors. The Soundgin is an interesting device in that it also incorporates three separate waveform synthesizers for creating fun robotic sound effects. On the downside, availability of the chip comes and goes.

Speech doesn't strictly require hardware; several intrepid experimenters have developed software-only speech for Arduino, Raspberry Pi, and other microcontrollers. In order to keep reasonably current I maintain a selected list of sources for speech and other sound synthesis solutions on the RBB Support Site.

CHAPTER 36

Visual Feedback from Your Robot

Robots and children—sometimes they just like talking back. Using any of several feedback techniques, your robot can communicate back to you, so you know that your programming is working the way it should. Several easy-to-implement feedback techniques are covered in this chapter.

There's another aspect of robot-to-human interaction, bridging that psychological gap between machine and person. You can draw in human spectators by using movement, sound, lights, and color. Previous chapters dealt with providing movement and sound; in these pages you'll learn some simple techniques to add pizzazz and flair (15 pieces or otherwise) to your robot. The more *personal* your bot is, the more others will take an interest in it.

 Source code for all software examples may be found at the RBB Support Site. To save page space, the lengthier programs are not printed here.

Using LEDs and LED Displays for Feedback

One light-emitting diode is all it takes for your robot to communicate with you. The language may not be elegant, and the conversations are strikingly short, but it gets the job done. When you don't need a talkative bot, you can use a single LED, or, for more words in the language, use multiple LEDs or 7-segment LED display panels.

FEEDBACK WITH ONE LED

The basic LED feedback circuit is shown in Figure 36-1: it's one of the I/O pins of your robot's microcontroller connected to a current-limiting resistor and an LED. Code running in your controller, like the supersimple Arduino sketch shown in *arduino-led.ino,* turns the LED off and on. Vary the *delay* value to make the LED flash on and off at a faster or slower rate.

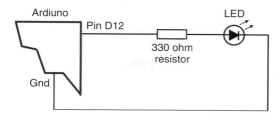

Figure 36-1 Basic connection for illuminating an LED via a microcontroller pin. Bringing the pin HIGH turns the LED on.

arduino-led.ino

```
void setup() {
   pinMode(12, OUTPUT);         // Make D12 an output
}

void loop() {
   digitalWrite(12, HIGH);      // Turn LED on
   delay(500);                  // Wait 1/2 second (500 ms)
   digitalWrite(12, LOW);       // Turn LED off
   delay(500);
}
```

Remember that you're not limited to lighting an LED merely to show status—either good or bad. You can flash the LED, using various patterns, to communicate more variations. Use Morse code to relay a variety of conditions or to "speak" phrases.

If you don't know Morse code, you can invent your own system. Here are just a few ideas. Note that in each case, there's an *Off* pause between the three-flash sequence.

Sequence	Meaning
Short-Short-Short	Status A-OK
Long-Long-Long	Unknown trouble
Long-Short-Long	Battery low
Short-Long-Short	Cannot read sensors
Short-Long-Long	Goal reached
. . . and so forth	

Put the LED where it's easy to see, and select a component large and bright enough to make it visible from across the room. I like to use large, 5mm or even 10mm, bright LEDs mounted on the top of the robot that can be seen at any angle.

FEEDBACK WITH MULTIPLE LEDS

Use multiple light-emitting diodes when you want to quickly convey operating or sensor status. For example, you might light an LED each time one of the bump switches or proximity detectors senses an object.

When only a small number of LEDs are needed, you can wire them directly to the I/O pins of your microcontroller. You simply duplicate the circuit and code in the preceding section, "Feedback with One LED," and use a different output pin for each light-emitting diode.

If you want to use more than four or five LEDs, then you probably don't want to dedicate an I/O pin for each one. One approach: With a simple serial-in, parallel-out (SIPO) shift register, you can turn three pins into eight.

Refer to Figure 36-2 for a schematic using a 74595 serial-in, serial-out (SIPO) integrated circuit. This chip is widely available and inexpensive. You can select any member of the '595 family, such as the 74HC595 or 74HCT595, whatever is available to you.

Sample program code for the Arduino is shown in *arduino-multi_led.ino*. The shift register works by first setting the Latch line to LOW and keeping it there for the time being. Then 8 bits of data are sent, bit by bit, to the Data pin of the chip. For each bit, the Clock pin on the '595 is toggled to tell the chip that new data have been sent.

The Arduino makes sending serial data easy because it packages up the Data and Clock activities in one simple statement, *shiftOut*. To use this statement you specify the number of the Data pin and Clock pin, the order of the data to be sent, and the value—from 0 to 255—that you wish to use.

```
shiftOut(dataPin, clockPin, MSBFIRST, numberValue);
```

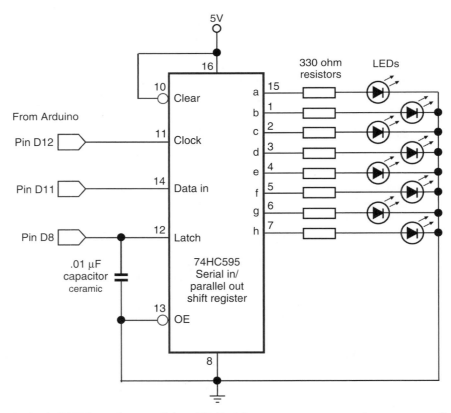

Figure 36-2 A 74595 serial-in, parallel-out (SIPO) shift register converts just a few microcontroller I/O lines to eight outputs.

MSBFIRST tells the Arduino that you wish to send the data starting with the *most significant bit,* which is the most common. For example, to send the value 127, the Arduino first converts it to binary form, which is 10000000. It then sends the bits left to right, starting with the 1.

The other variation, *LSBFIRST,* sends the *least significant bit* first, or right to left. Some circuits you interface to may require this order.

After all the shifted-out data have been sent, the program returns the Latch pin to HIGH. This sets the output pins of the 74595 chip, illuminating the LEDs as desired.

arduino-multi_led.ino

```
#define latchPin 8      // Connected to Latch pin
#define clockPin 12     // Connected to Clock pin
#define dataPin 11      // Connected to Data pin

void setup() {
  // Set all pins to output
  pinMode(latchPin, OUTPUT);
  pinMode(clockPin, OUTPUT);
  pinMode(dataPin, OUTPUT);
}

void loop() {
  // Count 0 to 255
  for (int val = 0; val <= 255; val++) {

    // Disable update during count
    digitalWrite(latchPin, LOW);

    // Shift out bits
    shiftOut(dataPin, clockPin, MSBFIRST, val);

    // Activate LEDs
    digitalWrite(latchPin, HIGH);

    // Short delay to next update
    delay(100);
  }
}
```

Another way to add multiple LEDs on fewer I/O pins is to use addressable LEDs. With these you simply connect them like Christmas tree lights, and use programming code to control a specific LED. See the section "Using Addressable LEDs" for more details.

USING BBC MICRO:BIT BUILT-IN LEDS

The BBC Micro:bit makes it extremely easy to send messages via flashing lights. The microcontroller sports a 5×5 matrix of small red LEDs, controlling using simple programming statements. You can flash individual lights, or create custom shapes by illuminating just certain LEDs in the matrix. The Micro:bit also comes with dozens predefined graphic icons, and routines for displaying text as scrolling messages.

Accessing Individual LEDs

The Micro:bit's LEDs are arranged in five columns and five rows, for a total of 25 light-emitting diodes. You can control any specific LED by using its X and Y coordinates. The coordinates are in a grid, where the column and row numbers start at 0.

0, 0	1, 0	2, 0	3, 0	4, 0
0, 1	1, 1	2, 1	3, 1	4, 1
0, 2	1, 2	2, 2	3, 2	4, 2
0, 3	1, 3	2, 3	3, 3	4, 3
0, 4	1, 4	2, 4	3, 4	4, 4

When using MicroPython on the Micro:bit you can set and unset individual LEDs using the statement

```
microbit.display.set_pixel(x, y, value)
```

- *x* and *y* are the X and Y coordinates of the pixel, respectively.
- *value* is a brightness level, with 0 as off, and 9 as fully on. Use a value of 1 to 8 to set an intermediate brightness.

You can reset all pixels in the grid to off with a single statement:

```
from microbit import *
microbit.display.clear()
```

Drawing Predefined Icon Images

Use any of over 60 predefined image icons built into the Micro:bit to convey a message. These include a happy and sad face, various arrows pointing in different directions, common shapes like circles and diamonds, and more. The MicroPython documentation lists them all; if you're using blocks to program your Micro:bit all the icon images are shown in a pull-down list.

Display the icon shape of your choice with the statement

```
microbit.display.show(image)
```

where image is the constant name of the icon you want to use. These all take the form *Image. ICON_NAME*, such as

```
from microbit import *
microbit.display.show(Image.YES)
microbit.display.show(Image.NO)
microbit.display.show(Image.SMILE)
```

See microbit-icons.py for a hands-on example of displaying a specific icon based on whether Button A or B is pressed.

microbit-icons.py

```
from microbit import *
while True:
  if button_a.is_pressed():
    display.show(Image.SAD)
```

```
elif button_b.is_pressed():
  display.show(Image.HAPPY)
else:
  display.show(Image.CONFUSED)
```

Drawing Your Own Images

When there isn't a premade graphic for the message you want to convey, you can make your own.
Do it by defining your own Image grid:

```
myimage = Image(
  "90909:"
  "09090:"
  "00900:"
  "09090:"
  "99999")
display.show(myimage)
```

The image data is composed of 25 values, here separated out into separate rows to make it easier to visualize. A value of 0 means that LED is off; 9 means it's on. Figure 36-3 shows the result on the LED grid. (Remember that you can also use intermediate values between 0 and 9 to control LED brightness.)

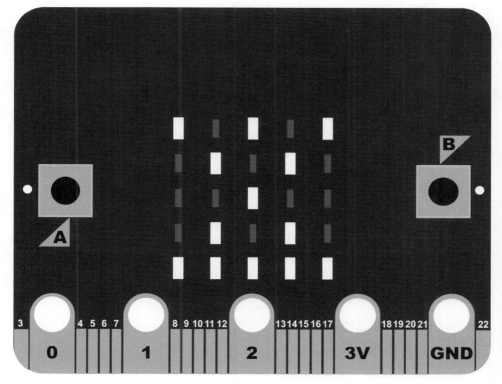

Figure 36-3   You can define your own graphical images simply by defining a string of values (0 to 9) in row order.

You don't need to separate the grid into rows. Either of these work just as well:

```
myimage = Image("90909:09090:00900:09090:99999")
myimage = Image("90909/n09090/n00900/n09090/n99999/n")
```

Displaying Scrolling Text

Even though the Micro:bit's display is just a grid of 5×5 LEDs, that's enough to form a single letter at a time. You can display whole messages by having these letters scroll. There's a built-in function for this that makes it easy. Here's an example:

```
import microbit
microbit.display.scroll('I Am A Robot!', wait=False, loop=True)
```

- The text within the single quotes is the message you want to show.
- The *wait* parameter tells the Micro:bit if you want to control the speed of the message. You can set this to *False,* which means no scrolling delay. If you do want to slow down the message, set *wait* to *True,* and add the parameter *delay=###,* where ### is the amount of the delay in milliseconds. Example: *delay=250.*
- The *loop* parameter specifies if you want the scrolling animation to keep repeating. Set it to True when you do.
- Not shown in this example is an optional *monospace* parameter, which when set to True makes all the characters of equal width: *monospace=True.*

Except for the text to show, all the other parameters are optional if you just want to use the default values: *delay=150, wait=True, loop=False, clear=False, monospace=False.*

Single letters can use the display.show statement instead:

microbit-display_show.py
```
from microbit import *
while True:
  if button_a.is_pressed():
    microbit.display.show("A")
  elif button_b.is_pressed():
    microbit.display.show("B")
  else:
    microbit.display.show("-")
```

Repurposing the LED I/O Pins

The Micro:bit uses several of the I/O pins available on the edge connector to address the rows and columns of the LED grid. This means the pins are out of commission for other tasks if you're using the LEDs. As several of these pins are analog inputs, you may want to reserve the use of the LEDs to only when you're not using an analog sensor.

Pin	Function	Type
3	Column 1	Analog
4	Column 2	Analog
6	Row 2	Digital
7	Row 1	Digital
9	Row 3	Digital
10	Column 3	Analog

Use the *display.off()* statement to turn off the LED grid, and reclaim the I/O pins for other tasks, and the *display.on()* statement when you want to activate the LEDs.

```
microbit.display.off()
microbit.display.on()
```

FEEDBACK WITH 7-SEGMENT LED DISPLAYS

Your robot can also talk to you using a 7-segment numerical LED display. You can light up the segments to produce numerals, in which case your robot can output up to 10 "codes" to indicate its status. For instance, 0 might be okay, 1 might be battery low, and so on.

You can also illuminate the segments to make nonnumeric shapes. You can light up each segment individually or in combination. Figure 36-4 shows some variations, including an E for Error, H for help, and numerous symbols that can mean special things.

Displaying Numerals

The easiest way to show numerals on a 7-segment display is to use a display driver IC, such as the CD4511 or 7447. These chips have four inputs and eight outputs—seven outputs for the numeric segments and an eighth for the decimal point, which we won't be using in this example. To display a number, set its binary-coded decimal (BCD) value at the A to D input lines.

Figure 36-4 Unusual nonnumeric shapes can be created by activating selected segments of a 7-segment LED display. Use this system to display numbers (of course) or codes. Reminds me of the wrist display in the movie Predator.

	4511 Inputs					
BCD	A	B	C	D	4511 Outputs	Numeral
0000	0	0	0	0	1111110	0
0001	1	0	0	0	0110000	1
0010	0	1	0	0	1101101	2
0011	1	1	0	0	1111001	3
0100	0	0	1	0	0110011	4
0101	1	0	1	0	1011011	5
0110	0	1	1	0	0011111	6
0111	1	1	1	0	1110000	7
1000	0	0	0	1	1111111	8
1001	1	0	0	1	1110011	9

Refer to Figure 36-5 for a diagram for hooking up the CD4511 to a common-cathode 7-segment LED display. See *arduino-segment.ino* for a simple Arduino sketch that sends different values to the CD4511, lighting up different segments. (You can use the same general concept with individual LEDs.)

The circuit shows a common-cathode 7-segment LED module; that is, all the LED segments in the display share the same cathode connection. Be sure yours is also a common-cathode display and not a common-anode display.

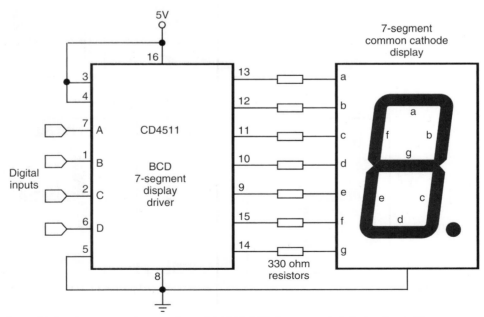

Figure 36-5   Connection diagram for a CD4511 BCD (binary-coded decimal) to a 7-segment driver. Numbers appear in the display based on the binary-coded values on the A to D input pins.

arduino-segment.ino

```
#define outA 8      // Connected to A pin on 4511
#define outB 9      // Connected to B pin
#define outC 10     // Connected to C pin
#define outD 11     // Connected to D pin

void setup() {
  // Set all pins to output
  pinMode(outA, OUTPUT);
  pinMode(outB, OUTPUT);
  pinMode(outC, OUTPUT);
  pinMode(outD, OUTPUT);
}

void loop() {
  // Display 3
  digitalWrite(outA, HIGH);
  digitalWrite(outB, HIGH);
  digitalWrite(outC, LOW);
  digitalWrite(outD, LOW);
  delay (1000);

  // Display 9
  digitalWrite(outA, HIGH);
  digitalWrite(outB, LOW);
  digitalWrite(outC, LOW);
  digitalWrite(outD, HIGH);
  delay (1000);
}
```

Displaying Arbitrary Shapes

Seven-segment displays are really just multiple LEDs that share a common cathode (or anode) connection. You can light up the segments separately, as in the section "Feedback with Multiple LEDs," earlier in the chapter.

Refer to Figure 36-6 for how you can connect a 74595 SIPO chip to a 7-segment display. Use the program *arduino-segment_shapes.ino* to send out 8 bits to light the seven segments, plus decimal point. With this setup, you're still able to produce all 10 digits, of course, plus Predator-type alien language symbols that only you (and your robot) understand.

arduino-segment_shapes.ino

```
#define latchPin 8
#define clockPin 12
#define dataPin 11

void setup() {
  pinMode(latchPin, OUTPUT);
  pinMode(clockPin, OUTPUT);
  pinMode(dataPin, OUTPUT);
}
```

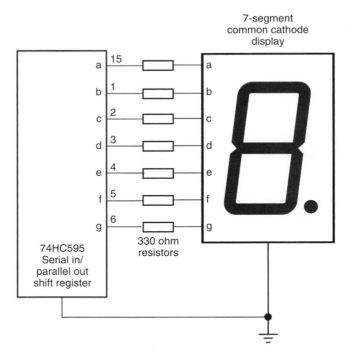

Figure 36-6 The 74595 SIPO (serial-in, parallel-out) shift register driving a 7-segment LED display. Using the shift register allows any combination of segments to light.

```
void loop() {
  // Write a regular 7
  digitalWrite(latchPin, LOW);
  shiftOut(dataPin, clockPin, LSBFIRST, B11100000);
  digitalWrite(latchPin, HIGH);
  delay(1000);

  // Write some funky character
  digitalWrite(latchPin, LOW);
  shiftOut(dataPin, clockPin, LSBFIRST, B01110011);
  digitalWrite(latchPin, HIGH);
  delay(1000);
}
```

You might find it easier to issue numbers in binary format, least significant bit first, as shown in the example. The first bit is segment A, the second is segment B, and so forth. Here's a handy binary bits guide for producing standard numerals via the 74595 shift register. Even if you don't use the decimal point, be sure to include its bit at the end, or the display won't look right.

Binary Bits	Numeral
11111100	0
01100000	1
11011010	2
11110010	3
01100110	4
10110110	5
00111110	6
11100000	7
11111110	8
11100110	9

ALL ABOUT ADDRESSABLE LEDS

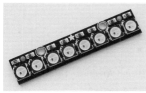

You've already seen how to use a serial-in, parallel-out shift register to control multiple LEDs at once. Another, and perhaps better, method is to use *addressable LEDs*. These are modules with an integrated light-emitting diode and control electronics. Addressable LEDs connect in daisy-chain fashion—think beads on a string. They are "addressable" in that you can specify which LED in the string you want to control.

Another advantage to addressable LEDs: they're intrinsically multicolor. Inside the module are three LED elements for red, green, and blue. By varying the intensity of any and all of the RGB elements you can produce any color you want, including white.

Most addressable LEDs are based on the WS2812 from World Semi. Unlike a traditional LED, which has solderable through-hole leads, the WS2812 come in a plastic surface mount "flat pack" that's harder for hobbyists to use. So these are most commonly provided as module or breakout boards, which make it simple to add the WS2812 LED in projects. Everything is included on the module—just wire it up and go.

An example WS2812 module, this one from Parallax, is shown in Figure 36-7. Addressable (or "smart") LED modules are available in singles or already connected in series to form a multi-LED display.

Using Addressable LEDs

Addressable LEDs contain their own simple microcontroller that receives input via a single-wire serial connection. You need power and ground as well, of course. The control signal contains the data, in 24-bit format, to control which LED in a string is to be activated, along with the relative intensities of the individual red, green, and blue diodes in the LED. By altering the value of each diode in 256 steps, you can obtain over 16 million of colors, including white (all colors on).

When viewed at close distance, you can see the individual R, G, and B colors of the diodes. But from further back, especially when using a diffuser lens or when the light is projected onto a surface, the rainbow colors are more defined.

How this addressability works inside the WS2812B requires careful timing of the control signals. Fortunately, you don't need to understand the internal signal format and timing

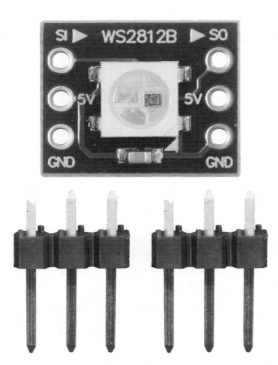

Figure 36-7 A WS2812B RGB LED mounted on a breakout board. This unit, available from Parallax, also includes two sets of solderable 3-pin headers. (*Photo courtesy Parallax Inc.*)

requirements, as special interfacing libraries are available for the Arduino, Parallax Propeller, Raspberry Pi, and many other popular microcontrollers. These libraries do all the heavy lifting for you. You need only specify the LED you wish to control, and its color and brightness.

Most libraries set the color using the familiar RGB triplet, with values ranging from 0 (off) to 255 (fully on). For example, 0,0,0 is all colors off; 255,255,255 is all colors on at full brightness. A triplet of 128,128,128 sets all colors on at half brightness—that's white, but not a bright white. To make a color, mix the R, G, and B values: 128,0,128 sets the red and blue LEDs to half brightness, creating a pleasing purplish hue.

Addressing is as equally easy. In the Arduino library provided by Adafruit, for example, you merely indicate which LED you wish to change by referencing its number in the strand, starting at 0. Suppose you have two LEDs. The first LED is referenced as 0, the second as 1, and so forth.

The number of WS2812B LEDs you can cascade from a single microcontroller pin depends on the specific library you are using, as well as the speed of the microcontroller and its internal memory capacity, among other things. A string of a hundred or more lights is readily doable, but in reality this stretches the practical number of LEDs you'd want on a robot. Remember the little thing called battery power.

At full brightness, the WS2812B consumes about 50 milliamps (mA). A string of just 10 LEDs, at maximum brightness, is half an amp (500 mA). That can quickly reduce battery reserves. Obviously use sparingly.

Hooking Up

An addressable LED mounted on a breakout board module is ridiculously easy to connect to your microcontroller of choice. There are only three pins: signal, power, and ground. Due to the nature of the device, the LED should be powered from a regulated 5V source.

The LED can be controlled via a 5V I/O pin. For best results, if your microcontroller is a 3.3V device, you may wish to use a 3.3V to 5V level shifter; see Chapter 29, "Interfacing Hardware with Your Microcontroller" for more details.

Addressable LED modules have both an "input" and "output" pin. This allows you to daisy-chain the LEDs from one to another to make a string. Power and ground are routinely tied together, but you'll want to take note of the "In" and "Out" pins on your module, and connect them properly, as shown in Figure 36-8. Figure 36-9 shows four addressable LEDs strung together on a standard breadboard.

Basic Arduino Control Examples

Refer to the *arduino-addressable-1led.ino* sketch for an example of how to control a single address-able LED. Of course, "simple" is relative: all the grunt work is done by the Adafruit NeoPixel library, which is provided as open source software for free download. See the RBB Support Site for additional information.

To use this sketch with your Arduino, you must first fetch the Adafruit_Neopixel library from its GitHub repository, unzip it, and place it in your Arduino libraries directory. This directory, named libraries, is located under your sketchbook folder—on Windows it's typically (My) Docu-ments\Arduino\libraries. If your sketchbook folder doesn't contain a libraries directory, you'll need to first create it.

Important! After the Adafruit_Neopixel library has been copied or moved to the sketchbook libraries directory, you must exit and restart the Arduino IDE.

arduino-addressable-1led.ino

To save space, the full sketch is available from the RBB Support Site.

The example demonstrates controlling a single LED connected to Arduino pin D6 (digital pin 6). Each primary color—red, green, blue—fades in from off to fully on, and the cycle repeats itself. Changing the LED color is done with two statements:

```
leds.setPixelColor(0, leds.Color(64, 0, 15));
leds.show();
```

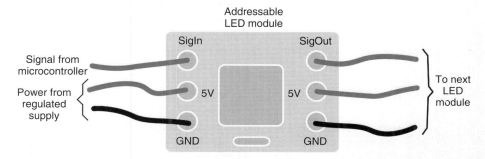

Figure 36-8   The WS2812B LED can be daisy chained to make light strings of dozens, even hundreds, of lights. Modules provide "in" and "out" pins for connecting them together.

Figure 36-9 Addressable LEDs connected in daisy-chain fashion. (*Photo courtesy Adafruit Industries.*)

In this example, the *setPixelColor* method sets LED 0 (zero, the first one in the string) to the RGB colors 64, 0, 15 (R=64, G=0, B=15).

Note also the *leds.show* method. This is required to actually set the new color value. If you omit it, the color change will be ignored.

Going further, *arduino-addressable-multiled.ino* shows the same basic code, this time control-ling a small cascade of two LEDs. The concept is the same: you first identify the LED to control—either 0 or 1—and change its RGB component values as desired.

arduino-addressable-1led.ino

To save space, the full sketch is available from the RBB Support Site.

In both sketches you'll notice some defined values at the top. You'll need to modify these if you change the Arduino pin connections, or the number of LEDs being controlled:

```
#define PIN          6
#define NUMLEDS      1
```

- *PIN* indicates the Arduino I/O pin connected to the first LED. Here it's pin 6, which was chosen arbitrarily. You can use any unused Arduino pin, though I'd advise avoiding pins 0, 1, and 13, as these have additional hardware functions you'll want to preserve (specifically, pins 0 and 1 are for the serial communications with your computer, and pin 13 controls the Arduino's built-in LED).

- *NUMLEDS* specifies the number of LEDs in the strand. Don't forget to change this if you add or subtract LEDs. If you have 10 LEDs, but only indicate 9, the last LED will not turn on.

HARDWARE DEPENDENCE AND ADDRESSABLE LEDS

For all their coolness, addressable LEDs require some hefty overhead that can tax your microcontroller. The LEDs need a fairly exacting signal, and for performance and reliability may use the controller's hardware timers. All fine and good, but those timers may be pressed into service by other hardware you've attached.

This is certainly true of the Arduino, where many code libraries for addressable LEDs use a hardware timer that's also shared with the servo library. Mixing code for servos and addressable LEDs will likely result in unexpected behavior.

Fortunately there are workarounds, including the obvious: using DC motors instead of servos. But if you're sold on servos for your bot, you can try adding a third-party servo library that avoids conflicts with the LED library. For example, Adafruit provides an alternative "TiCoServo" servo library for use with their NeoPixel LED library.

Additional options include using a microcontroller that's not bogged down by shared hardware (a Parallax Propeller can run different code in its eight separate cores), or by using a serial servo module—command the module via serial communications, and the module takes care of managing all your servos.

Using LCD Panels

Liquid-crystal display (LCD) panels let your robot talk to you in complete words, even sentences. Even the smallest of LCD panels can show up to eight characters—enough for a couple of words. If needed, your robot can display a verbose error code that indicates its state.

LCDs are particularly handy in testing and debugging when your robot is untethered from its programming computer. You can keep the LCD updated with the current program flow, indicating such things as when a sensor is activated and what program subroutine the robot's microcontroller is currently running.

TEXT- OR GRAPHICS-BASED

LCD panels are broadly available in two general forms: text and graphics. The differences are obvious, but let's discuss the two in brief:

- *Text-based* LCDs produce only text and other characters (such as dollar signs or symbols) stored inside the display module. The capacity of display is defined as the number of characters per line and the number of lines. For example, a 16×1 LCD has one line that can display up to 16 characters. A 32×2 LCD has two lines, and each line can display up to 32 characters.
- *Graphics-based* LCDs (GLCDs) are more like computer monitors, where text and other images are produced using a horizontal and vertical array of dots. The graphics LCD, or GLCD, is defined by the number of dots wide and high; 256×128 means the display panel has 256 pixels horizontally and 128 dots vertically.

Many color GLCDs use a technology known as thin-film transistor, or TFT, which provides for sharper images and brighter colors. Monochrome GLCD may use traditional LCD technologies, TFT, or organic LEDs. See "OLEDs" below for more information on this last type.

Regardless of the underlying display science, GLCDs always need to be paired with controller electronics to handle the hefty demands of writing text and drawing images. As there are many types of controllers for GLCDs each model will have its own communications protocol. That makes them tougher to program. Always look for a GLCD that has a decent programming library available for it.

Some GLCDs come with (or have as an option) a touch screen overlay, allowing you to press directly on the display to provide feedback. Most touch screens are resistive, meaning you can (and probably should) point on them using a rubber stylus.

Many text and graphics panels share standardized interface controllers, making it much easier to work with them. The controller is the electronics built into the LCD that provides the communications gateway.

For text displays, the most common controller is the Hitachi HD44780. For GLCD displays, the de facto standard controller is the Samsung KS0108. Other manufacturers make controllers that are compatible with these standards.

COLOR, MONOCHROME, BACKLIGHTING

Both text and graphics LCD products are available in either monochrome or color versions. Color is more common with GLCD, and color adds to the cost and the complexity in programming. Unless you specifically need it, stick with monochrome displays. (Note that the actual display color can be yellow on green, white on black, or numerous other variations.)

Many of the better LCD panels have their own backlighting, which increases the contrast under many types of indoor and outdoor light conditions. Though backlighting is not strictly required, it's a nice feature to have.

LCD INTERFACE TYPES

Before you can use an LCD panel, you must interface it to a microcontroller. There are two ways, parallel and serial:

- *Parallel interfacing* involves connecting separate I/O pins from the microcontroller to the LCD. Text-based LCDs require seven I/O lines; GLCDs require about 16 I/O lines, which obviously makes these harder to interface. Your program communicates directly with the controller onboard the LCD panel.
- *Serial interfacing* involves connecting two or three I/O lines from the microcontroller to the LCD. Your program communicates with the LCD via serial commands. Additional electronics on the LCD convert these commands to the parallel interface used on the panel.

As text-based LCDs are by far the most common (and many controllers have built-in functions to support them), I'll concentrate just on these. The example that follows shows how to connect an Arduino microcontroller to an HD44780-based text LCD. To keep it simple, the example uses a 16 × 2 LCD, which is common and inexpensive.

There are few established standards for the command set used with serial LCDs, so if you're planning on using a serial LCD, refer to the instruction manual that came with it. It'll tell you how to connect the panel to your controller, how to set up communications (such as setting the baud rate), and how to send commands to the LCD.

Figure 36-10 shows the hookup diagram between the Arduino microcontroller and an HD44780-based 4-bit parallel-character LCD module. It also shows how to use a 10 kΩ potentiometer for adjusting the contrast of the display. Dial the pot for the clearest lettering against the display background.

 Remember: Your LCD display needs to be compatible with the Hitachi HD44780 driver. Most are, but you'll want to check to make sure. The pinout order shown in Figure 36-10 is the most common you'll encounter, but variations exist in oddball LCD panels. Your panel may have 14 pins (no LED backlight) or 15 or 16 pins (with LED backlight).

Sketch *arduino-lcd16x2.ino* provides a basic programming example that displays "Robot Builder's Bonanza. 5th Ed." on the two lines of the display.

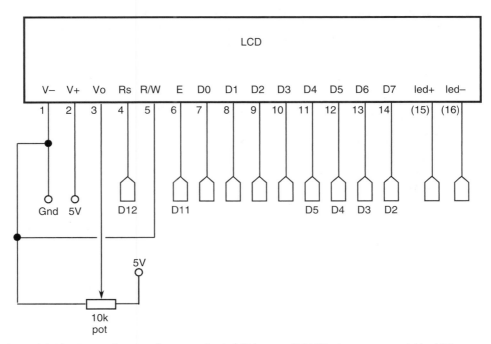

Figure 36-10   Pinout diagram for a standard 4-/8-bit parallel LCD character panel (the LCD must use the Hitachi HD44780 driver). The LCD is used in 4-bit mode to save I/O pins.

arduino-lcd16x2.ino

```
#include <LiquidCrystal.h>

// Initialize interface pins to LCD
LiquidCrystal lcd(12, 11, 5, 4, 3, 2);

void setup() {
  // Configure numbers/rows for LCD
  lcd.begin(16, 2);

  // Print the message

  lcd.setCursor(0, 0);
  lcd.print("Robot Builder's");
  lcd.setCursor(0, 1);
  lcd.print("Bonanza, 5th Ed.");
}

void loop() {
}
```

Once again, the 10 kΩ potentiometer adjusts the contrast of the display. For a quick setup, you can try simply tying pin 3 (*Vo*) of the LCD to Gnd. Or connect it to pin D9 of the Arduino, and add this line at the start of the setup() function:

```
analogWrite(9, 20);
```

This uses the PWM feature of the Arduino to set a very low voltage to pin 3 of the LCD. Play around with different values for the second parameter of the *analogWrite* statement, from 0 to 255. On many LCD panels, the higher the number, the less contrast there is.

OLEDS

Organic LEDs (OLEDs) are a type of digital display often used for graphical content. The display is composed of a matrix of emissive light elements. The screen can be in color or monochrome; in both cases the brightness of an OLED display is often several times that of a traditional LCD.

As with traditional LCD displays, there are numerous sources of OLED modules where the display and all interface electronics are on one compact board. Most use I2C or SPI for fast serial communications, and rely on a function library specially written for a given microcontroller. So when hunting for your OLED of choice, be sure to verify that the product comes with a supporting coding library. Otherwise you'll be twisting in the wind as you slog through the technical documentation on how to light a single pixel in the middle of the screen!

Fortunately, many of the online retailers such as SparkFun and Adafruit that offer OLEDs provide some kind of functional support library for at least the Arduino. Figure 36-11 shows an Adafruit monochrome OLED 1.3″ 128 × 64 display. The module is powered by 3.3V, but is 5V tolerant, meaning you can use it with a 5V microcontroller such as the Arduino Uno. It's programmed via either SPI or I2C, with code libraries for both Arduino and Raspberry Pi.

Figure 36-11 Adafruit's subminiature OLED display has a 1.3″ display, and can be used by either 3.3V or 5V microcontrollers.

Most OLED modules are designed to be powered by 3.3V. Many (if not most) aren't 5V tolerant, meaning that they may be damaged if connected to 5V logic pins. If your OLED module of choice doesn't support 5V you'll need to add level-shifting circuitry between it and your microcontroller. See Chapter 29, "Interfacing Hardware with Your Microcontroller" for more information on level shifting.

Programming for the OLED display relies heavily on using library functions. Since the functions are specific to the module you use I'll just show demo code to demonstrate the concepts.

Suppose you want to draw a white circle 24 pixels wide in the middle of the display. To do that you'd use

```
display.fillCircle(display.width() / 2, display.height() / 2, 12,
WHITE);
```

The OLED module is defined as an object, here named display. So everything you do with it begins with its object name, followed by the method you want to use. The text *display.drawCircle* draws a centered circle with a radius of 12 pixels.

Individual pixels can be placed anywhere on the screen by specifying its X and Y coordinates.

```
display.drawPixel(20, 30, WHITE);
```

This draws one lonesome pixel at X=20 and Y=30.

Robot-Human Interaction with Lighting Effects

Few onlookers will ever ask you, "What does it do," if your robot makes funny sounds or blinks lots of lights. Robot interaction includes making it interesting to humans. The more engaging the robot becomes, the more interaction it engenders. That increases the impact the robot has on its human watchers.

There are plenty of ways to attract attention to your robot, such as using musical tones and sound effects, and even making your bot stop every once in a while and do a little booty shake.

I'll let you figure out the booty shake part (hint: it has something to do with starting and stopping the drive motors), but for music and other auditory effects, be sure to read Chapter 35, "Producing Sound."

The premier method of adding light effects to your bot is with light-emitting diodes. LEDs have gone well beyond small, dim, red pinpoints of light. They're now available in all colors of the rainbow. Brightness has been drastically improved to the point where LEDs are used as flashlights.

Read more about LED basics in Chapter 22, "Common Electronic Components for Robotics." Included is how to select the current-limiting resistor used to prevent the LED from burning up.

MULTIPLE LEDS

Don't be content to use just one LED. Use multiple LEDs, of the same or a different color, mounted at various places on your robot. From a calculation standpoint, it's easiest to connect each LED to the robot power supply through its own current-limiting resistor. This also helps ensure an even brightness from each of the LEDs. As long as you stay under the maximum forward current specification of the LED, you can vary the value of the current-limiting resistor to change the brightness of each LED.

SUPERBRIGHT AND ULTRABRIGHT LEDS

The typical LED produces a fairly low amount of light—a few millicandles (a *candle* is a standard unit of light measurement; a millicandle is 1/1000 of a candle). Superbright and ultrabright LEDs produce 500 to 5000 millicandles; some go even higher. A few are so bright that they can cause eye damage if you stare into their beam.

Superbright and ultrabright LEDs are particularly striking on small robots. Turn down the lights and let your bot roam the floor. If you have a camera with an open-shutter (also called *open-bulb*) feature, you can take a long-exposure picture that shows the path of the robot around the room.

When selecting superbright and ultrabright LEDs, pay particular attention to beam pattern. The brightest LEDs have a narrow beam pattern—just 10° or 15°. Select a broader beam pattern if you want the LED to be visible at different viewing angles.

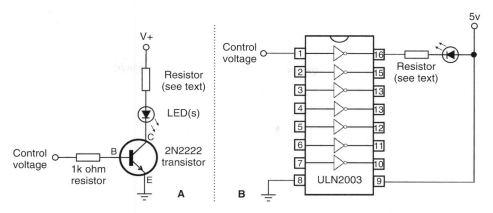

Figure 36-12 Superbright and similar high-output LEDs may require more drive current than a microcontroller I/O pin can supply. The transistor or ULN2003 driver chip boosts the current driving the LED. Select the resistor to maintain a safe current through the LED.

Many very bright LEDs require more current than can be provided by the output pins of some microcontrollers. You'll need to boost the current to the LED using a transistor (Figure 36-12A) or a current driver (Figure 36-12A). The driver shown is the ULN2003 Darlington transistor array, where each of its seven drivers can provide up to 500 mA (half an amp) of current, depending on the version of the chip. If you need eight drivers, you can use the ULN2803. It's functionally identical to the ULN2003, except for the extra driver. Of course, the chip has more pins.

Again, remember: LEDs need a current-limiting resistor, or else they'll quickly burn out. Refer to Chapter 21, "Robot Electronics—The Basics," for the formula. In order to calculate the value of the resistor, you need to know the forward voltage through the LED, as well as the maximum current that can be safely passed through the device. Refer to the datasheet that came with the LED you're using.

MULTICOLOR LEDS

You've already seen that addressable LEDs can display red, green, and blue colors in any combination. But maybe you don't want the complexity of an addressable module. You can still achieve multiple LED colors:

- A *bicolor* LED contains red and green LED elements (other color combinations are possible, too). You control which color is shown by reversing the voltage to the LED. You can also produce a mix color by quickly alternating the voltage polarity.
- A *tricolor* LED is functionally identical to the bicolor LED, except it has separate connections for the two color diodes.
- A *multicolor* LED contains red, green, and blue LED elements. You control which color to show by individually applying current to separate terminals on the LED.

Figure 36-13A shows how to connect a bicolor LED to two pins of a microcontroller. To turn the LED off, put both pins LOW. To turn on one color or the other, put one of the pins HIGH.

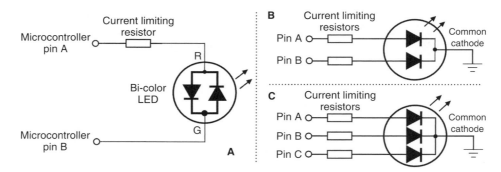

Figure 36-13 Connecting a multicolor LED to two or more I/O pins of a microcontroller. To display a color, make its pin HIGH. Quickly alternating between the two pins produces a combination hue of the two colors combined.

Pin A	Pin B	LED Output*
LOW	LOW	Off
LOW	HIGH	Green
HIGH	LOW	Red
Pulse†	Pulse†	Orange

*The actual color depends on how you connect the LED. And, of course, some bicolor LEDs display colors other than red and green.
†When pulsing, one pin is LOW while the other is HIGH. Pulse at a rate of at least 10 times per second. That way, your eyes will blend flashes into a single combination color.

As an example, the following code for the Arduino toggles between the two colors of a bicolor LED.

```
digitalWrite(led1, HIGH);
digitalWrite(led2, LOW);
delay(1000);
digitalWrite(led1, LOW);
digitalWrite(led2, HIGH);
delay(1000);
```

For the mixed-color effect, simply toggle between the two colors more quickly. Try reducing the delay from 1000 milliseconds (1 second) to 10 milliseconds (1/100th of a second).

Only one diode in a bicolor LED can be on at any time. But recall that in a tricolor LED, each diode has its own connection lead, so you can switch either on or off. You can also have both on at the same time, mixing the colors together. Figure 36-13B shows how to connect a tricolor (three-lead LED) to a microcontroller. Note that *each* diode in the LED gets its own current-limiting resistor. See *arduino-tricolor.ino* for a short demonstration program of toggling a tricolor LED from red to green to orange.

Pin A	Pin B	LED Output*
LOW	LOW	Off
LOW	HIGH	Green
HIGH	LOW	Red
HIGH	HIGH	Orange

*Once again, the actual colors depend on the LED you're using.

arduino-tricolor.ino

```
#define redPin 9          // Red diode of LED
#define greenPin 10       // Green diode of LED

void setup() {
  pinMode(redPin, OUTPUT);
  pinMode(greenPin, OUTPUT);
}

void loop() {
  digitalWrite(redPin, HIGH);
  delay(500);
  digitalWrite(redPin, LOW);
  delay(500);

  digitalWrite(greenPin, HIGH);
  delay(500);
  digitalWrite(greenPin, LOW);
  delay(500);

  digitalWrite(redPin, HIGH);
  digitalWrite(greenPin, HIGH);
  delay(500);
  digitalWrite(redPin, LOW);
  digitalWrite(greenPin, LOW);
  delay(500);
}
```

And finally, a multi- or RGB-color LED has red, green, and blue diodes. These three primary colors can be displayed independently or in different combinations to produce many other colors. You use these the same as with a tricolor LED, except that you need a third microcontroller pin to control the additional color. Figure 36-13C shows the connection scheme for an RGB multicolor LED.

Recall that you can mix colors by turning on more than one diode in the LED at a time. With a microcontroller with a PWM (pulse width modulation) output, you can vary the intensity of the light and create thousands of color combinations. The *arduino-rainbow-led.ino* sketch shows how to use the *analogWrite* statement, which produces a PWM signal on specific pins of the microcontroller. We'll be using digital pins D9, D10, and D11 for connecting to a multicolor LED (don't forget the current-limiting resistors!).

arduino-rainbow_led.ino
To save space, the program code for this project is found on the RBB Support Site.

> The hookup diagrams for the tri- and multicolor LEDs show common-cathode devices—that is, the cathode (negative) ends of all the diodes in the device are tied together. Multiple-color LEDs are also available in common-anode style, where the anode (positive) ends are linked together. The concepts behind using these are the same, though, of course, you must reverse the wiring. The LED is turned on when the cathode connection is brought LOW.

ON THE WEB: MORE STUPID LIGHT TRICKS

And there are more ways you can trick out your robot with light effects. Find these and other ideas on the RBB Online Support site:

- Using electroluminescent (EL) wire, fiber optics, and lasers.
- Ornamenting your robot with self-contained body lighting (glow-in-the-dark sticks, rave lights, magnetic LED earrings).
- Outfitting your robot with different colors of cold cathode fluorescent tubes.
- Using passive decoration, such as decals, fluorescent paint jobs, and transfer film.

Finally, Go Out and Do!

Few other moments in life compare to the instant when you solder that last piece of wire, file down that last piece of metal, tighten that last bolt, and switch on your robot. Something *you* created comes to life, obeying your commands and following your preprogrammed instructions.

I started this book with a promise of adventure—to provide you with a treasure map of plans, diagrams, schematics, and projects for making your own robots. I hope you've followed along and built a few of the mechanisms and circuits that I described. As you finish reading, you can make me a promise: improve on these ideas. Make them better. Use them in creative ways that no one has ever thought possible. Create that ultimate robot that everyone has dreamed about!

Online Robot Projects

CHAPTER 37

Make Light-Seeking Robots

With just a flashlight you can command your robot from afar—well, at least from the other side of the room. Two inexpensive sensors detect light, and based on which sensor receives more light, steers the robot toward the source.

We'll call it LightBot. Full details and parts list are provided on the RBB Support Site (see Appendix). This is an online open-source community project that you can contribute to.

Design Goals

Robots that respond to light are known as *phototropic*. What happens when the bot sees light depends on the wiring of the light sensors, and the robot's programming:

- *Photophilic* robots "love" light and move toward it
- *Photophobic* robots "hate" light and move away it

The same sensors are used for both kinds of responses. Typical light sensors for robots include the CdS photocell—also known as a light-sensitive resistor—and the phototransistor. Of the two, photocells have a spectral response closer to the human eye; they are most sensitive to the yellow-green region of the visible electromagnetic spectrum.

A flashlight with a bright, narrow beam makes for a guide light source. The beam must be wide enough so that its light can fall on both of the robot's sensors at the same time.

There's no rule that says a robot must behave one way or another when it encounters light. It's perfectly fine to have a robot that scurries away from light in some situations, and seeks it out in others. You'll use this behavior to couple the light seeking abilities of the robot with crude hearing: if the robot detects a loud sound, it will temporarily shy away from any bright light source.

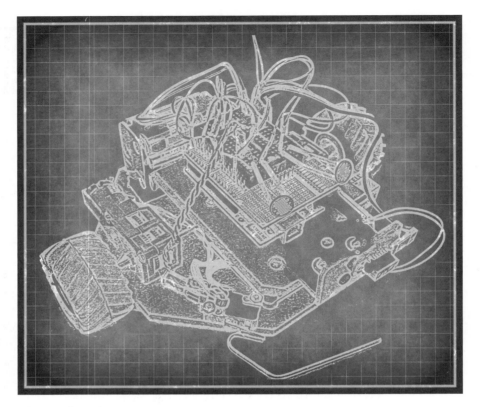

Figure 37-1   The LightBot.

The LightBot Platform

LightBot does not need fancy hardware. The basic robot is built on a small platform measuring less than 6″ in diameter. I built the prototype LightBot using 1/4″ expanded PVC plastic, but you can use thin plywood from the hobby store, foamboard, acrylic, polycarbonate, or most any other material you have around the house. Construction is not critical.

The LightBot design uses two DC gear motors built from kits. It's powered by a set of AAA batteries, and uses the following readily available sensors:

CdS photocells (2), for detecting light; see Figure 37-2.
Sound impact sensor, for listening for sudden and loud sounds, like a hand clap.

You are free to add additional sensors to your LightBot. Suggestions include a proximity sensor to prevent the robot from getting too close to any objects in front of it, and a bumper switch to stop motion if it runs into anything.

Figure 37-2 LightBot uses two commonly available CdS photocells as "eyes" for detecting light from a flashlight.

Microcontroller Support

LightBot exhibits a number of insect-like behaviors, using a microcontroller to manage its actions. The prototype LightBot supports several microcontroller platforms, including:

- Arduino or Arduino-compatible
- BBC micro:bit, with solderless breadboard adapter

Both microcontrollers require an outboard dual motor H-bridge, as noted in the parts list. For my prototype bots I used low-cost (under $7) H-bridge plug-and-play modules. Alternatively, when using the micro:bit you can opt to use the SparkFun moto:bit motor board, which includes a dual-motor H-bridge and ready connections for attaching to power and motor terminals.

CHAPTER 38

Make R/C Toys into Robots

Low-cost radio-controlled (R/C) toys make for terrific robots. The toy already has its own motors, wheels, battery compartment, and body. Add a microcontroller or other brain circuit and you have a functional robot that does your bidding.

Let's name this contraption R/CBot. Full details and parts list are provided on the RBB Support Site. This is an online open-source community project, and your contributions are encouraged!

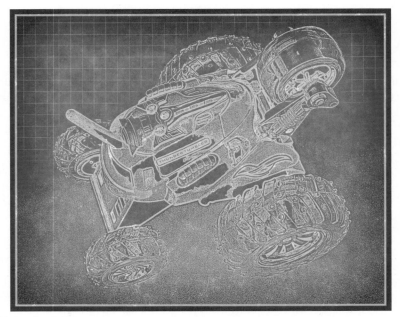

Figure 38-1 The R/CBot.

Design Goals

R/CBot is based on low-cost battery-operated toys, the kind intended for indoor playtime—no gasoline-powered engines here. These toys are generally available in two styles, based on how they steer:

- Toys with *car-type steering* have a drive motor in the back that powers two rear wheels; two more wheels in the front steer the vehicle right or left
- Toys with *differential steering* have two driven wheels on either side, with a third wheel in the front of back provides balance

While either style works as a robot, differentially steered robots are easier to maneuver in tight places. While R/C toys with differential steering aren't as common, look long enough and you'll find several under $20 to choose from. I used a "stunt car" toy that's available at many online and retail outlets. See the RBB Support Site for an updated list of sources.

R/CBot is controlled completely by its own internal microcontroller; the radio control aspect of the toy is disregarded. Consider using the transmitter and receiver electronics for another project.

The R/CBot Platform

Because R/CBot already comes with its own chassis, motors, and wheels, there's virtually no mechanical construction involved. Building the robot entails removing the plastic body, and once opened, chopping out the control circuit board; see Figure 38-2. This leaves you with:

Wires from the battery compartment. The toy uses four AA cells, kept in a battery compartment accessible from the bottom of the chassis. These cells serve as power for the electronics and motors.

Wires to the two motors. The revised R/CBot circuitry connect to these motors directly, giving you full control over their operation.

Figure 38-2 R/CBot doesn't need its old control board. It's surgically removed, and replaced with a microcontroller and H-bridge module.

Electronics fit in and on the cavity left over from removing the toy's control board. A small piece of foamboard mounted over the cavity provides additional mounting space for sensors. The prototype R/CBot uses a (low-cost) laser time-of-flight rangefinder as a proximity sensor. Of course you are free to add additional sensors of your choosing.

Because radio-controlled vehicles can seriously chew up operating time on a set of batteries, R/CBot is designed to use four rechargeable NiMH AA cells. You'll need a suitable recharger for NiMH batteries.

Microcontroller Support

I crafted the prototype R/CBot to use an Arduino Pro Mini 328, with a 3.3 volt operating voltage. This board may be powered directly from the 4.8 volt NiMH cells without the need for a boosting regulator.

A small dual H-bridge module connects the Arduino Pro Mini to the motors. As with the Arduino the H-bridge is selected to run from a 4.8 volt power supply. All the electronic parts (except for the laser sensor) mount on a mini-breadboard.

Other microcontroller options include the BBC micro:bit, Parallax Propeller FLiP, and Raspberry Pi Zero.

CHAPTER 39

Make Line-Following Robots

A robot that follows a line on the floor is a classic example of control mechanics. The robot is steered along its path using two or more sensors on its bottom. If the robot starts to veer off, its electronics apply just the right amount of corrective signal to keep it on track.

Meet LineBot, an easy-to-build automaton that follows a track of black electrical tape that you've stuck to a poster board. Full details and parts list are provided on the RBB Support Site (see Appendix). This is an online open-source community project that you can contribute to.

Figure 39-1 The LineBot.

Design Goals

Line-following bots have a simple mission in life: to follow a line on the ground without losing track of the line. It's not uncommon for LineBots to race against one another, pitting builders against one another to test who can make the fastest robot.

At the heart of any line-following robot is a pair (or more) of optical sensors. These sensors face straight down, and measure the boundary between a line and its background. Because 3/4″ wide electrical tape makes such a great line, LineBot uses a stripe of it to form a circular course on a piece of poster board or other white background.

Sensors for LineBot are cheap and easy-to-use. The sensors produce an analog voltage that depends on the reflectivity of the surface underneath the bot. The robot knows when it has encountered the edge of the line when the voltage value of the sensor abruptly changes.

Construction of the line-following sensors is simplicity itself: an infrared LED to emit some light, and a phototransistor to detect the light. You can make your own emitter/detector sensors, or purchase them in ready-made compact modules for a few bucks each. LineBot employs three emitter/detector sensors, placed next to each other underneath the front of the robot.

The LineBot Platform

LineBot needs to be fairly light and fast—the faster the bot will go the greater the odds you'll win in a line-following contest! The prototype uses the following parts:

- *Body made from foamboard.* Its chassis is composed of one main piece of the board, cut to size with a utility knife.
- *Fast DC gear motors.* The Internet is awash with inexpensive DC gear motors with a low reduction ratio—LineBot uses the commonly available 1:48 geared motors to keep the robot fast and nimble. Wide rubber tires keep a good grip as the robot maneuvers over its track.
- *Line sensors (x3).* As noted above—see Figure 39-2.

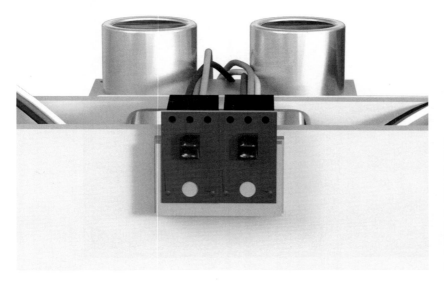

Figure 39-2 Three line sensors placed on the underside of the LineBot sense the black/white transition of the black electrical tape. The robot uses the relative brightness values from each sensor to keep on course.

- *3.7V LiPo battery.* Lithium-polymer rechargeable batteries pack a lot of wallop for their size and weight. You'll need a suitable recharger.
- *BBC micro:bit microcontroller.* The micro:bit is low-cost and easy to program; I used JavaScript for the prototype.
- *SparkFun moto:bit board.* The micro:bit plugs directly into this board, and provides voltage regulation, motor H-bridge, and handy 3-pin header connectors for the sensors.

All assembly is done with hot melt glue.

Microcontroller Support

While the prototype LineBot uses a BBC micro:bit, the same functionality can be readily provided by most any microcontroller. See the plans and programming code for tips on how to adapt the project to the Arduino and other controllers.

CHAPTER 40

Make Robot Arms

The first real robots were nailed to the floor and moved things from one place to another following a canned script of movements. Robots that are really automated arms are the backbone of modern factories. You can build your own robot arm even if you don't run your own factory.

Which brings us to BallBot, so named because this robot's *raison d'être* is moving balls or other small objects from one place to another, following an organization script you define. Full details and parts list are provided on the RBB Support Site, noted in the Appendix. This is an online open-source community project, and your contributions are encouraged!

Design Goals

BallBot is designed to move objects—balls, plastic discs, whatever—between prescribed points on a workplane. It does this by picking up an object, moving to a new location, and dropping the object here.

Robotic arms are composed of two main parts:

- *Articulation* is provided by the "bones" and joints of the arm, and defines the work envelope, the area where the arm can reach
- A *gripper* at the end of the arm acts as hands and fingers

Collectively, robotic arm/gripper mechanics are known as *manipulators,* and they come in many shapes, sizes, and varieties. (Read about the main kinds in Chapter 20, "Build Robotic Arms and Grippers," for more details.)

BallBot is a revolute robotic arm, using joints similar in function to a human arm. Each joint defines one degree of freedom (DOF); BallBot has a total of three degrees of freedom: A shoulder joint moves the arm in a circular path left and right; an elbow joint lifts the arm up and down; and two fingers in the gripper open and close to grasp an object.

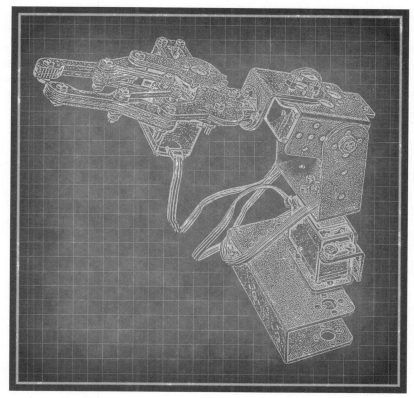

Figure 40-1 The BallBot.

The BallBot Platform

Robotic arms can be difficult and time consuming to build, so BallBot uses off-the-shelf parts when possible. No cutting, shaping, or milling is required, though you need to drill a couple of holes in a wooden base to mount the arm.

The arm uses these parts:

- *3x standard size servos.* The arm isn't designed to lift anything heavy, and its own mechanics don't add a lot of weight, so motors with a minimum of 50 to 55 oz-in torque are acceptable.
- *Stamped aluminum "servo erector set" parts* to make the rotating joints; see the RBB Support Site for an itemized list of parts.
- *Aluminum gripper,* either assembled or in kit form. The prototype BallBot uses a commonly available gripper that costs about $17 online (minus servo). There are many low-cost grippers available, and you are free to pick among your favorite as long as it's fairly lightweight.
- *Microcontroller, servo shield,* and *wall adapter* plug-in power.
- *Wood or plastic mounting base,* plus small rubber balls, shallow bowls.

Microcontroller Support

For the BallBot prototype I settled on a good-old Arduino Uno, sandwiched with an optional servo shield. I used the shield to make it easier to connect the servos to the Arduino. With the shield the servos are powered from their own source, and don't use current from the main Arduino board. To save your budget you may opt to skip the shield and wire the servos via a small breadboard.

RBB Support Site

This book comes with free online content at the **RBB Support Site.** Go to

www.robotoid.com

What you'll find:

Project Parts Finder—where to find parts for the projects in this book
Bonus projects—fun stuff that couldn't fit here!
New and updated links—curated list of Web sites and manufacturers
Cutting and drilling templates—for robot bases and other parts, in convenient printable PDF
AppNotes—Timely articles on a variety of robotics topics
Technical reference—electronics and mechanics
Book updates—get the latest
Robo-building Archive—chapters from previous editions of *Robot Builder's Bonanza* that are still relevant but were removed to make space for the new stuff!
. . . And more

Backup Support Site

In case the main Support Site is temporarily unavailable for whatever reason you'll find a backup of all content at:

github.com/LOCATION

INDEX